PROCEEDINGS

International Symposium on
Marine Positioning

PROCEEDINGS

International Symposium Marine Positioning

Symposium Directors:

Muneendra Kumar

Defense Mapping Agency, Washington, D.C., U.S.A.

and

George A. Maul

AOML/NOAA, Miami, Florida, U.S.A.

U.S. Geological Survey
Reston, VA 22092
October 14–17, 1986

D. REIDEL PUBLISHING COMPANY

A MEMBER OF THE KLUWER ACADEMIC PUBLISHERS GROUP

DORDRECHT / BOSTON / LANCASTER / TOKYO

Library of Congress Cataloging in Publication Data

International Symposium on Marine Positioning (1986: U.S. Geological Survey)
Proceedings.

1. Marine geodesy—Congresses. 2. Oceanographic instruments—
Congresses. I. Kumar, Muneendra. II. Maul, George A. III. Title.
QB275.I587   1986   681'.755   87–9592
ISBN-13:978-94-010-8226-6   e-ISBN-13:978-94-009-3885-4
DOI:10.1007/978-94-009-3885-4

Published by D. Reidel Publishing Company,
P.O. Box 17, 3300 AA Dordrecht, Holland.

Sold and distributed in the U.S.A. and Canada
by Kluwer Academic Publishers,
101 Philip Drive, Assinippi Park, Norwell, MA 02061, U.S.A.

In all other countries, sold and distributed
by Kluwer Academic Publishers Group,
P.O. Box 322, 3300 AH Dordrecht, Holland.

PREFACE

The International Symposium on Marine Positioning (INSMAP) was conceived by the Marine Geodesy Committee at OCEANS 84, Washington, DC. It became clear at that time, that timing is appropriate to focus attention on individual specific problem areas under the broad umbrella of Marine Geodesy. After scheduling INSMAP 86 by the Marine Technology Society, we were fortunate to generate strong support from our co-sponsors. All their assistance and support are gratefully acknowledged.

Our special thanks are expressed to the U.S. Geological Survey; Charting and Geodetic Services, NOS/NOAA; Office of Naval Research, and Naval Ocean Research and Development Activity for their support through financial grants (ONR No. N00014-86-G-0107, NOS/NOAA No. 40AANC601637, and USGS No. 14-08-0001-G1207) as partial funding to the INSMAP 86. We are also grateful to the U.S. Geological Survey for providing the auditorium and other logistic support in making the symposium a success.

A total of 165 persons attended INSMAP 86, of which 20 percent were from outside the United States. Nine technical sessions and five special workshops were held within a four-day format.

Invited speakers included Dr. Alan Berman, Dean, Rosensteil School of Marine and Atmospheric Sciences; RADM J. R. Seesholtz, Oceanographer of the U.S. Navy; RADM John D. Bossler, Director of Charting and Geodetic Services, NOS/NOAA; Mr. Chris von Alt, Woods Hole Oceanographic Institute; and RADM L. H. van Opstal, Hydrographer of the Royal Dutch Navy.

These proceedings contain the written versions of the talks, presentations, and the reports as submitted by the authors or workshop chairpersons. A few papers have also been included (with appropriate footnotes) which were submitted but could not be presented at the INSMAP 86 due to last minute exigencies.

We acknowledge with gratitude the support of all of the committee members, advisors, authors, chairpersons, and participants of INSMAP 86, and are very grateful to the National Geodetic Survey's editor, Mrs. Eleanor Andree, for her effort in organizing these proceedings.

George A. Maul
Muneendra Kumar
Symposium Co-Directors

ACKNOWLEDGMENTS

General Chairperson

RADM John D. Bossler

Symposium Directors

Dr. Muneendra Kumar
Dr. George A. Maul

Tech. Prog. Coordinator

Dr. George A. Maul

Registration Coordinators

Dr. Roop Malhotra
Mr. James McNamara

Budget Director

CAPT Scott E. Drummond

Admin. Coordinators
Dr. Bonnie McGrego
MAJ Mark Breach

Publicity Coordinator

CDR Max Ethridge

Audio/Visual Coordinator

Dr. Robert Rowland

Hospitality Coordinator

Mr. John G. Gergen

Publication Coordinator

Mr. Stanislaw Grzeda

Proceedings Editor

Ms. Eleanor Andree

Exhibits Coordinator

Mr. Morris Glenn

ADVISORS

USA

Mr. O. W. Williams
Dr. W. J. Klepczynski
Mr. R. J. Beaton
CAPT N. H. Keeler
Dr. Richard Swim
Dr. N. A. Renzetti
Dr. Ross Stein
Dr. George DuPont
Dr. Don Perkins
Mr. Carl Savit
Dr. Eugene Silva
Dr. James Collins
Dr. Javad Ashjaee

International

Prof. D. E. Wells
Prof. Gunter Seeber
Dr. J. C. Blankenburgh
Prof. A. Stolz
Prof. D. M. Fubara
Dr. J. Rais
Prof. F. Fajemirokun
Dr. Minoru Sasaki
Prof. Su-yen Chain
Dr. M. G. Arur
Prof. J. B. Andrade
Prof. H. G. Henneberg
Prof. E. J. Krakiwsky

TECHNICAL PROGRAM COMMITTEE
Dr. K.I. Daugherty, Chairperson
Dr. Muneendra Kumar, Co-Chairperson

Prof. I. I. Mueller	Prof. C. N. K. Mooers	Dr. Gary Hill	Dr. Carl Bowin
Prof. B. D. Tapley	CAPT John Fuechsel	Dr. George Born	Dr. William Ryan
Prof. Fred Spiess	CAPT P. J. Kies	Dr. Lanny Yeske	Dr. Tom Rossby
Prof. N. K. Saxena	Mr. Herb Johnson	Dr. Paul Perreault	Dr. Tom Curtin

SPONSORING ORGANIZATION

Marine Geodesy Committee,
Marine Technology Society

CO-SPONSORS

International Association of Geodesy

International Hydrographic Office

National Research Council

American Geophysical Union

Institute of Navigation

Joint Oceanographic Institute

Institute of Electrical and Electronics Engineers/
  Oceanic Engineering Society

American Congress on Surveying and Mapping

Radio Technical Commission for Maritime Service

American Society of Civil Engineers

The Hydrographic Society

U.S. Department of Transportation

U.S. Navy
  Office of Naval Research
  Naval Ocean Research and Development Activity
  Office of the Chief of Naval Operations

Defense Mapping Agency

U.S. Geological Survey

National Oceanic and Atmospheric Administration/
  National Ocean Service

U.S. Coast Guard

CONTENTS

SESSION 4. GPS IN MARINE POSITIONING II
Chairperson: G. H. Born

SESSION 5. MARINE MAPPING AND CHARTING I
Chairperson: Capt. P. J. Kies, USN

AFTERNOON BANQUET

SESSION 6: MARINE MAPPING AND CHARTING II
Chairperson: B. A. McGregor

SESSION 7: POSITIONING IN OCEANOGRAPHY
Chairperson: A. Gallegos

SESSION 8: CALIBRATION AND INTERCOMPARISON
Chairperson: H. G. Henneberg

SESSION 9: APPLICATIONS/REQUIREMENTS
Chairperson: E. Groten

EVENING BANQUET
Guest Speaker: R.Adm. J. D. Bossler, NOAA Corps

PLENARY SESSION
Chairperson: G. A. Maul

CLOSING SESSION
Chairperson: Capt. J. L. Hammer, III, USN

The Titanic Operations
C. J. von Alt, Woods Hole Oceanographic Institute

Closing Remarks
R.Adm. L. H. van Opstal, Royal Dutch Navy

WORKSHOPS

KEYNOTE SPEECH

Dr. Alan Berman
Dean, Rosensteil School of Marine
and Atmospheric Sciences
University of Miami
Miami, FL 33149

I am indeed honored to have been asked to give the keynote speech for the International Symposium on Marine Positioning. The subject of the symposium is one of
the few subjects that I still know enough about to say something meaningful.

My personal involvement in this field goes back almost 40 years to a day in the
summer of 1947 when I began working as an apprentice in the laboratory where I
was ultimately to do research for my doctorate. My first assignment was to
support the work of Professors Kusch and Rabi who under navy support were
building the world's first atomic clock. The device was an atomic beam Cs
magnetic resonance machine. Today it is in the Smithsonian, where it is part of
an exhibit on the history of time standards. While this device was large and
immobile, it did achieve a previously unheard of frequency stability of 1 part in
10^9.

In 1949, General Dwight Eisenhower made a tour of the laboratory and asked about
the device--"What good was it and why would the Navy support such work?" My boss
responded, "This is basic research, it has no possible application to anything."
It is hard to imagine a Nobel Laureate being so wrong.

I left the world of frequency standards and precision clocks in 1952 to become a
sea-going acoustician and oceanographer. I rapidly discovered that all of my
professional life was dominated by the problem of navigational precision. I
struggled with experiments that required precise knowledge of position. All I had
to work with was a loran A or the primitive loran C receivers that began to be
available in the late 50's and early 60's.

At that time, I was in charge of a large scale acoustic program that required
the location of 10,000 sensors to a relative accuracy of 1/2 meter ·and the
location of a 20,000 ton surface vessel to an accuracy of 1 meter for a period of
about 60 seconds.

I could solve the first requirement using explosive echo ranging and what in
1960 passed for a lot of calculation. The second problem was a lot harder. Even
though we deployed bottom transponders all over the ocean floor, we never quite
achieved the location accuracy we needed. I began to think that there had to be
a better way.

In 1963, I was involved in the Thresher search. That experience convinced me
that significantly better location accuracy was necessary. For those of you who
were not familiar with the Thresher problem, it can be summarized as follows:
the submarine was known to have sunk in a 10 mi x 10 mi area. The basic search
tool was a side scan sonar, with a detection range of about 100 yards. Using
rather erratic loran C signals, navigational accuracy was at best between 200 and
300 yards. There was simply no way of knowing if we had a 100% map of the
region. With current technology it would have been a routine search. Because of
navigational problems, It took four months to complete and was successful almost

1

M. Kumar and G. A. Maul (directors), Marine Positioning, 1–6.
© 1987 by the Marine Technology Society.

accidentally. As a result of this experience, the development of improved technology for marine positioning became a high priority item in my life.

When I became Director of Research at NRL in 1967, I was delighted to find,
that the primitive atomic time standards I had helped develop 18 to 20 years
earlier had become reliable, compact and five orders of magnitude more stable and
accurate than those I knew.

More importantly, an NRL team led by Roger Easton, and encouraged by NRL
veterans of the Thresher search, was experimenting with placing precise clocks in
satellite orbits.

Roger Easton's program was remarkably successful. Each Timation satellite was
more effective than the previous launch. A concept was brought forth to deploy a
large constellation of satellites each carrying precise clocks. This concept,
now known as GPS, would provide precise worldwide navigation.

One more bit of history, the decision to go with GPS took place in a DSARC
meeting held about 1972 or 1973, my memory of the date is a bit hazy, but I do
remember that when the navy was asked to comment on the system, a very senior
admiral got up and said the navy really did not need such accurate navigation. I
thought that my hopes for improved position accuracy were forever dead.
Fortunately, a rather crisp young one-star air force type arose and stated that
the Air Force absolutely needed the accuracy which could be provided by GPS. His
voice carried the day, and the air force became the GPS program manager.
Incidentally, the current navy position is that it absolutely needs all of the
navigational accuracy that it can get from GPS, and then some. Times,
institutions, and people change.

The important issues in marine positioning are clearly recognized by the papers
and sessions of this symposium. From my prospective, the seven most demanding
requirements for marine positioning are given in the first vuegraph which I
copied from a 1983 panel report of the National Academy of Science.

Note that the problem of marine positioning is described in terms of accuracy,
range and lifetime. None of these problems stresses these requirements
simultaneously.

For example, if radioactive wastes are disposed on the ocean floor, I, or since
I am not immortal, my remote descendants may wish to revisit the disposal site
any time within the next 10^5 years. Presumably a disposal site is defined in some
area, possibly 10 km x 10 km. Ideally, marine positioning technology should
allow anyone revisiting a site to identify an object of dimensions of about
1 meter and be able to distinguish it from another 1 meter sized object that is 10
to 100 meters away. The problem in this case is not navigational accuracy per
se, it is time. Can navigational accuracy be maintained for the next 10^5 years?

Some problems stress positioning technology more than others. The bore
hole re-entry problem is undoubtedly the most demanding of positioning accuracy. Some of you may be familiar with the deep ocean drill ship, the R/V Joides
Resolution. It is capable of drilling a hole in the ocean bottom at water
depths of 6000 meters. If a hurricane comes during the course of drilling
operations, the drill string must be removed and the hole is temporarily aban-
doned. After passage of the disturbance, drilling operations are resumed if
the bore hole can be re-entered. Since there is only about 1 mm of clearance

between the drill string and the bore hole wall, this problem is the ultimate in marine positioning. Accomplishing this type of precise repositioning involves the use of multiple technologies and location techniques. Electro-magnetic navigation systems are needed to bring the drill ship to within 10 to 100 meters of the location of the bore hole. Acoustic pingers and TV cameras allow the final positioning of the drill string with the required accuracy.

Of the three variables in marine positioning, time is perhaps the most difficult to guarantee. If positioning depends, as it generally does in underwater positioning problems, on acoustic beacons, the lifetime of a beacon becomes a limiting factor. Current technology limits the lifetime of autonomous beacons to about a year. Batteries are usually the limiting component. The use of RTGs might stretch this to 10 to 100 years. It is unlikely to go much longer. Thus, if I wish to do an experiment deep below the ocean surface that required me or my heirs to return to a precise location every 100 years, I would not know how to design the equipment.

Suppose I planned a 1000 to 10,000 year sequence of annual measurements to measure the distance between tectonic plates. Such measurements will require the best attainable GPS accuracy. Can we guarantee that the U.S. Government will continue to fund and fly GPS satellites for the next 1000 or 10,000 years? For that matter, can we guarantee that our government will exist that far into the future. I hope so, but none of us can guarantee it.

Despite the existence of a variety of excellent electromagnetic navigation systems, problems of underwater navigation still present some difficulties.

Seafloor-oriented navigation requirements are of many types; most of these do not strain the state-of-the-art. Seafloor exploration, search, and geologic studies can be carried out quite well with local systems having internal uncertainties of the order of 10 m and uncertainties relative to global coordinates of a kilometer. The latter requirement arises from a need to be able to return to a particular area to carry out subsequent operations, while the 10 m local accuracy assures that one could return within visual range to photograph, sample, or inspect a particular seafloor feature without undue position determination difficulty. Control of the position of the camera or sampling device is generally a more difficult requirement than position determination.

In some instances there are needs for much better accuracy (centimeter or even millimeter) in localized situations in which relative positioning is the important element. In such cases the allowable uncertainties translate into greater than one part in a thousand of the ranges over which the system must operate—for example, in the guidance of seafloor work devices or in placement of large objects precisely in register with some already-in-place indexing system or some other object (e.g., placing the end of a pipe against its termination or re-entering a drill hole). These cases can be considered as successive approximation problems. For example, the requirement for millimeter accuracy in order to join two pipes or connectors together can be met by a sequence of systems each having successively better resolution, but without requiring any single one to achieve better than a few parts in a thousand. As the distance of separation decreases, the magnitude of the allowable uncertainty decreases proportionally, thus never becoming particularly demanding as a position-determination activity.

At present, and for the foreseeable future, the need for high accuracy in establishing the coordinates of locations on the seafloor, either relatively

within localized areas or on a global scale, arises primarily in connection with
studies of the motions of the seafloor on various temporal and spatial scales.
Requirements for reference points in other contexts arise primarily in relation
to definition of national boundaries and in description of plots being exploited
for their mineral resources.

Most of today's seafloor-oriented navigation problems can be solved with
systems capable of precision of the order of 1 part in 10^4 of the range over
which the system must operate. Most requirements imply durations of no more
than a few years and coverage of areas about 10 km across, although some survey
or search activities may cover substantially larger extent (10^3 km).

Seafloor-related geodetic positioning requirements arise primarily in two
classes or situations--boundary marking and geodynamics. The former, whether
related to political interactions or resource exploitation, lead to accuracy
requirements in the 1- to 10-m range that can be met under the sea with currently
available near-bottom acoustic transponder systems. Ability to tie these acoustically determined positions to an appropriate global coordinate system will
vary with location. In nearshore areas electromagnetic systems can provide an
adequate tie. In midocean regions only the global positioning systems (GPS) can
be expected to achieve the desired accuracy.

Geodetic problems arise in a variety of geodynamic contexts. Large-scale
slumping of thick sediment columns and earthquake-induced crustal motions are
the most obvious. Aspects directly associated with plate tectonics enter as
well--in terms of interplate motion, intraplate deformation, and strain buildup
patterns at plate edges (ridge crests, transform faults, and subduction zones).
These all lead to much more stringent requirements, falling in the 1- to 20-cm
accuracy range. The problems also tend to fall into two separate categories,
partly based on technological considerations. Those involving areas having
lateral extent of 10 km and less can be attacked primarily with underwater
acoustic and optical approaches, while those concerned with longer ranges must
include ties from the seafloor to near-surface points that can provide access to
systems utilizing transmission of electromagnetic signals, whether from shore,
satellite, or other intermediate platforms.

The largest spatial scale sections of the earth's crust are the tectonic
plates--hundreds to thousands of kilometers across. These are in continuous
motion (measured on a geologic time scale) relative to one another, with speeds
ranging of a few centimeters per year. As they move apart, new crust is produced on the seafloor, with attendant hydrothermal effects. Where they slide
past one another or collide (producing trenches or mountain ranges) there are
major earthquakes and, behind the subduction zones, extensive volcanic activity.
Knowledge of how and why these motions take place on a time scale of years or
less can help us to understand a wide variety of geologic (and in some instances
biological) problems. This understanding may lead to other important impli-
cations--particularly for earthquake prediction and mineral-resource development.

There is also the question of whether tectonic plates are truly rigid. This
involves comparison of measurements of displacements at boundaries with those
made on a larger scale between interior points of adjacent plates. Some of the
plates are primarily oceanic and, while island locations can be visualized as
useful measuring sites, they can be suspect because they may individually be
subject to local motions owing to isostatic readjustments as they ride on the

cooling crust away from the source ridges. It would thus be desirable to be able to link deep-seafloor points in midocean with others in midcontinent.

Geophysical surveys utilize the measurement of gravity. This has been applied effectively on land and in shallow water to define certain geologic formations (e.g., salt domes) that have a high probability of being associated with oil reservoirs. In such surveys relative gravity measurements are made over any given area with uncertainties of 1 part in 10^7-10^8. On the ocean floor similar uncertainties are achievable if the gravimeter is placed on the bottom. While this is in principle feasible for any water depth, it has not been done at greater than about 200 m except in a few instances in which gravity meters have been operated in small submersibles such as Trieste and Alvin resting on the bottom. Generally, at-sea gravity measurements have been made in large, near-surface-operating submarines or surface ships. When this is done, it is necessary to correct for the fact that the gravity meter is not rigidly attached to the earth and thus does not feel the same centrifugal acceleration component as would an on-bottom measuring instrument. In order to obtain measurement accuracies comparable with those on land one would need to know the instrument's east-west velocity component relative to the seafloor to an accuracy of about 10^{-3} m/sec. This is the most demanding requirement for velocity information that I am aware of in the area of marine positioning. Lack of means for achieving this, limits present-day systems for at-sea gravity surveys to large-scale problems in which relative uncertainties of 10^{-5} or 10^{-6} are acceptable.

A variety of aspects of offshore oil and gas field exploitation provides examples from which navigational requirements emerge.

Seismic profiling is used to produce two-dimensional (2-D) sections. During a 2-D survey, the surface is sampled every 25 to 50 m along lines that form a grid extending over an area that may extend for many hundreds of kilometers; the grid consists of many 2-D seismic lines spaced a few hundred meters to a few kilometers apart. The method of data collection is such that redundant subsurface coverage is obtained from successive source-receiver positions. This allows the data to be stacked to enhance the signal-to-noise ratio. The quality of the stacked data is directly related to the positioning accuracy of both the sequence of shots and the streamer hydrophones. The exploration objectives and economics usually set the positioning goals of a 2-D survey. The shot interval can be maintained to an accuracy of a few meters. Because of sea currents and tides, the streamer (typically 2 to 3 km long) does not follow directly in line with the ship's track, feathering angles of 1 to 5 degrees are common. Tolerances of 5 to 10 degrees are usually set as the allowable limit for the feathering angle during the collection of 2-D data. The positioning accuracy of the seismic lines is specified to be within 50 to 500 m with reference to the local geodetic control.

The issues change dramatically during the oil or gas field development and production phases. Here the interest is in description of the reservoir with emphasis on the fine-scale details of the sediment structure. Resolution of the details of a subsurface structure requires that the line spacing be equal to the in-line sampling interval. This implies line separations of as little as 30 m. Significant degradation of coherent combination of signals will occur if they are out of phase by as much as an eighth of a wavelength, which at 200 hz implies an accuracy of just under 1 m. Any positional error will appear as a distortion in the process of combining the signals. Since the purpose is to achieve improved performance against noise and to produce good spatial resolu-

tion, it is desirable to maintain the distortion due to navigation errors at a
negligible level. The conclusion is that 10-cm accuracy for individual position
determinations should be the goal. The areal extent of such a 3-D survey is
typically 10 km x 10 km, and the time over which the precision tracking system
must operate continuously is of the order of a few weeks. It should be emphasized that the accuracy requirement is not related to the accuracy with which
the geographic coordinates of the subbottom structure must be known but arises
from the manner in which the various signals are to be combined with one
another.

During oil field development, many tasks on the seafloor require local high-
accuracy positioning. These tasks are generally accomplished by a combination
of television and high-frequency acoustic systems. Invariably, the total system
is designed specifically for the project at hand. The critical positioning
requirement in marine pipeline construction is to lay the pipe along a
previously surveyed corridor. This is achieved either by using the same
positioning system that was used by the original corridor survey or by using
transponders left in position by the corridor survey crew. Here again the
requirements are of the order of meters in most instances but become tighter as
one approaches specific connection points. Fortunately, these involve only
very short ranges, and thus special-purpose, very-high frequency systems can be
used.

My conclusion with regard to technology is that current capabilities,
including seafloor acoustic transponders and a variety of electromagnetic
systems, can meet the 1 part in 10^4 requirement if they are carefully used.
Today's technology is thus adequate to cope with most situations.

In general, attendees at this symposium are fortunate to live their professional lives at a time when the technology is available to allow them to solve
important and intriguing problems in research and applied engineering. I envy
you. I was born a generation too soon.

REQUIREMENTS FOR MARINE POSITIONING

PROBLEM	ACCURACY	RANGE (in km)	LIFETIME
3-D seismic survey	10 cm	10 km	1 mo
Missile-firing evaluation	10 m	10,000 km	1 mo
Bore hole re-entry	0.1 cm	1 km	10 years
Radioactive waste disposal	10 m	10 km	10^5 years
Geodynamic studies (spreading centers, transform faults, slump zones)	1-10 cm	10 km	10 years
Geodynamics (subduction inter-plate motion, interplate deformation)	1-10 cm	1000 km	10 years
Bottom survey and object recovery	1 m	100 km	10 years

SEA BEAM MAPPING SYSTEMS

Joyce Miller
Robert Tyce
Randy Edwards
NECOR Sea Beam Development Center
Graduate School of Oceanography
University of Rhode Island
Narragansett, RI 02882

ABSTRACT

The introduction to the oceanographic community of deep ocean, multibeam mapping systems such as Sea Beam has had a profound effect both on the way we view the sea floor and on the way in which sea floor surveys are conducted, particularly with regard to open ocean navigation. These systems provide more than a hundred-fold increase in survey resolution as well as practical potential for complete survey coverage with deep water resolution of 100-200m. Thus they have not only placed severe demands on shipboard navigation but have also by themselves permitted highly accurate terrain-following navigation in areas previously surveyed. Map production from these systems generally relies on topographic matching for final adjustment of ship navigation, even when extremely accurate navigation, such as GPS, is available. There are now a dozen Sea Beam installations worldwide (5 in the U.S.) with each system contributing on the order of 1/2 gigabyte of sea floor survey data each year. This paper will detail some of the history of multibeam sea floor mapping as well as some of the promises and problems of these systems. The systems installed and operated aboard the R/V's CONRAD and ATLANTIS II by the URI/NECOR Sea Beam Facility will be described. The CONRAD is a well-equipped geophysical research ship with multichannel and precision gravity capabilities. Its navigation equipment includes Loran-C, two-axis doppler speed log, SATNAV and GPS with atomic clock. The ATLANTIS II is the support ship for the submersible ALVIN, and is equipped for acoustic transponders and range/bearing navigation in addition to speed log, Loran-C, SATNAV and GPS.

1. INTRODUCTION

Sea Beam was developed from multibeam sonar mapping systems used by the Navy and introduced to the oceanographic community in 1975. In less than a decade, Sea Beam has profoundly changed not only the way we view sea floor surveys but also the way in which such surveys are conducted. The increased resolution, coverage and response provided by these systems, when integrated with modern navigation and data processing software and hardware, has led marine geologists to suggest that no research ship should leave port without Sea Beam or some comparable system. It is system integration which continues to present a challenge, mainly because the potential of such systems is so considerable and varied. Real-time terrain following, navigation adjustment and dynamic ship positioning are only a few of the possible applications. Perhaps the greatest potential benefit is the ability of these systems to provide continuous and accurate topographic coverage of large survey areas in real time aboard ship. The high resolution of the bathymetric data, however, places considerable demands on the navigation accuracy, both in real time and in post-processing.

M. Kumar and G. A. Maul (directors), Marine Positioning, 7–16.
© *1987 by the Marine Technology Society.*

2. SEA BEAM HISTORY

In 1964, General Instrument (GI) introduced a Narrow Beam Echo Sounder for detailed bathymetric work in deep water, which eleven years later would be transformed into the sophisticated multibeam mapping system called Sea Beam (Tyce, 1986).   Narrow Beam Echo Sounders were installed on a total of ten Navy and NOAA ships. This system uses a transmit array of twenty transducers, similar to side-looking sonar.   The transducers are mounted along the keel of the ship to project a beam 2-2/3 by 54 degrees. Instead of using these same transducers as receivers, as is common for side-looking sonar, a 40-transducer receiver array is mounted on the hull perpendicular to the transmit array. With the aid of electronic beam steering, 16 received beams (2-2/3 by 20 degrees) are formed perpendicular to the transmitted beam. A narrow 2-2/3 by 2-2/3 degree formed beam is electronically derived from the intersection of the transmit beam and the received beam. A vertical reference gyroscope is necessary to stabilize the transmitted beam and to select from the sixteen received soundings the one which is most nearly vertical. It is this single stabilized vertical beam which is then digitized and displayed as a bottom profile on a graphic recorder.

As the result of military experience with a multibeam bathymetric system, GI realized that more than just the Narrow Beam Echo Sounder's vertical beam could be used for mapping. Digital computer processing and control hardware were added to the original sonar hardware in 1975 to produce Sea Beam, which processes and combines all 16 beams into a 2-2/3 degree swath 42-2/3 degrees wide. This produces the equivalent of 16 narrow beam echo sounders stabilized in roll and pitch. These beams form a narrow line across the direction of travel, resulting in 16 times 16, or 256 independent beams for each distance traveled equal to the swath width. In an area no bigger than the area ensonified by a traditional single-beam echo sounder in one measurement, Sea Beam yields hundreds of independent points. The swath width is 80 per cent of the water depth or 3.2 km in 4 km deep water.

More important than the number of measurements, however, is the fact that the 16 beams are extremely well navigated relative to ship's position. Roll errors of less than 1/10 degree are detectable in the data. Thus, if only a single narrow beam echo sounder were mounted on a ship and the ship surveyed the same area as Sea Beam, the resulting soundings would be much less accurate than the Sea Beam data, since each beam point would depend on ship navigation accuracy over sixteen subsequent passes. As a result the swath topography supplied by Sea Beam, even before being combined into a survey map, can provide us with sea floor topographic information more precisely detailed than ever before (Tyce et al, 1986).

While multibeam sonars had been used previously by the NAVY, they were not introduced into the civilian oceanographic community until the beginning of this decade, when NOAA in the U.S. and CNEXO in France acquired Sea Beam sea floor mapping systems. The scientific potential of such detailed bathymetric data spurred the addition of two Sea Beam systems at NECOR (Northeast Consortium for Oceanographic Research), a second one at NOAA as well as one at Scripps Institution of Oceanography in the U.S. There are also more than half a dozen systems in other countries.

3. NECOR SHIP OPERATION

NECOR, formed through a cooperative agreement between Lamont-Doherty Geological Observatory (LDGO), University of Rhode Island (URI) and Woods Hole Oceanographic Institution (WHOI), operates two Sea Beam mapping systems. The first was

installed aboard the Lamont Research Vessel CONRAD during January of 1984; the
second was installed aboard the Woods Hole Research Vessel ATLANTIS II during
January of 1985. Together these systems have been utilized at sea for more than
800 days, and have produced more than two gigabytes of data to date in extremely
varied environments and applications.

4.1. Sea Beam Hardware

To display and interpret Sea Beam data in real time aboard ship requires, in
addition to the sonar, a considerable amount of hardware and software. This
includes navigation, computer processing and display systems. The capability to
do such real-time display has been a principle goal in the development of the
Scripps and NECOR Sea Beam facilities. Figure 1 is a block diagram outlining the
data collection and processing configuration utilized aboard the R/V CONRAD. The
heart of the system consists of two identical VAX 11/730 computers which serve as
their own spare parts. Generally one computer is dedicated to data acquisition
and real-time display and the other is devoted to post-processing of data from the
present or previous cruises. In the case of a failure, either computer can handle
both tasks at a reduced rate. It is anticipated that the VAX 11/730's will be
updated to MICROVAX II's in 1987.

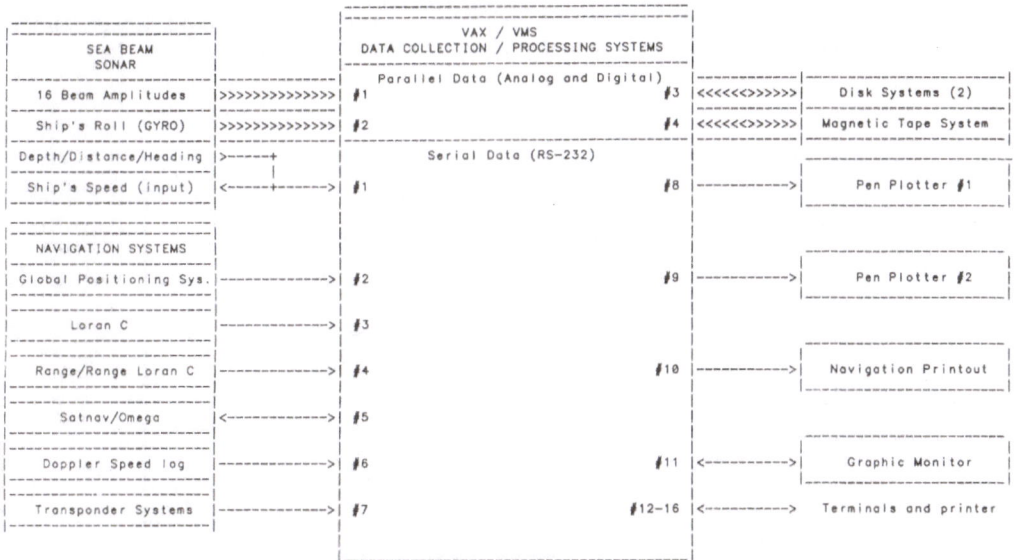

Figure 1. NECOR Sea Beam data collection configuration

4.2. Navigation

The greatest limitation to Sea Beam's performance at present is navigation. On
board both the R/V CONRAD and R/V ATLANTIS II, Sea Beam requirements for accurate
navigation have accelerated upgrades in navigation hardware. Both ships now carry
Transit Satellite (SATNAV), GPS with atomic clock and Loran-C. The ATLANTIS II
also has an Omega receiver integrated into its SATNAV. In addition there is
range-range Loran capability on the CONRAD. The CONRAD has a 2-axis Doppler speed
log, while the ATLANTIS II has a single-axis Doppler speed log. To maximize the

utility of navigation aboard CONRAD, a micro-computer system was recently added to simultaneously plot ship's position data from all navigation systems in real time, so that the best fixes can be selected. Radar transponders are occasionally used aboard the CONRAD. Acoustic transponder navigation is used on the ATLANTIS II in conjunction with ALVIN work.

The NECOR ships operate world-wide, usually in deep ocean areas.   This often precludes using limited range navigation systems such as Loran-C. We must be able to conduct a survey with a single fix navigation source, such as transit satellite, combined with gyro heading and speed. Conversely we are expected by the scientists to log and utilize in real time multiple navigation sources when available.   In the past two years we have conducted surveys in deep ocean areas as varied as the North and South Pacific and Atlantic, the South China Sea and the Indian Ocean, where limited range systems were unavailable. However, we also have conducted surveys in the Red Sea, the Bay of Naples, offshore Nova Scotia, and in the North Pacific (Shor and Chayes, 1986) using Loran-C.

4.3. Real-Time Sea Beam Operations

For real-time operations one navigation source, either SATNAV, GPS or Loran, is merged with Sea Beam bathymetry data and plotted as isobath contours within a few minutes of data acquisition (Figure 2). Real-time data often show offsets and/or gaps from discontinuous real-time navigation.   Ideally, the navigation source should be spatially and temporally continuous. At present none of the three

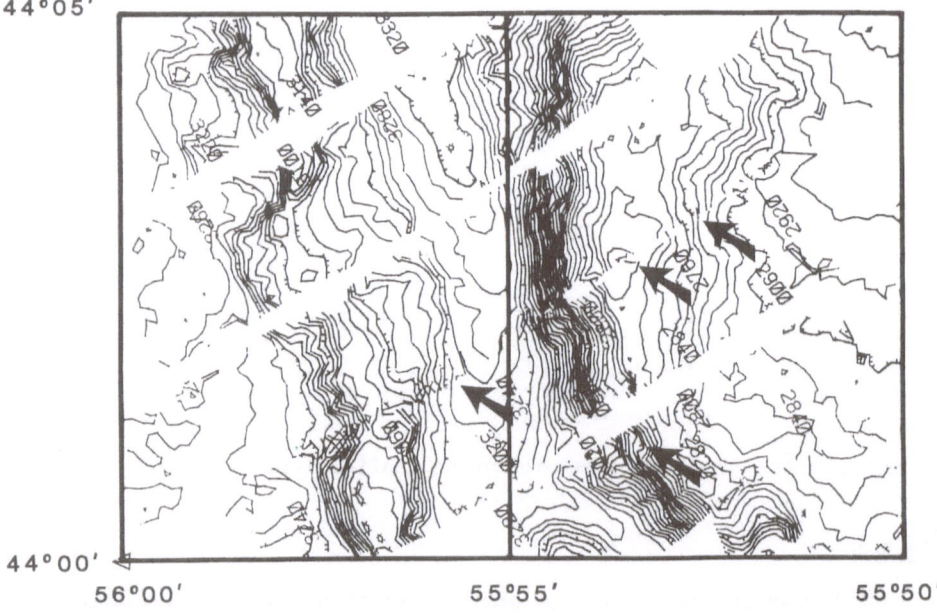

Figure 2. Sea Beam contour swaths collected using Range-Range
           Loran navigation. Arrows show contour mismatches
           between adjacent swaths.

available sources is satisfactorily continuous. For example, the survey data
shown in Figure 2 were collected off of Nova Scotia in August of this year during
a survey using range-range Loran as primary navigation . The contours were
originally plotted in real time using GPS, transit satellites or Loran and had
frequent jumps and data gaps. Smoothing of Loran data during post-processing on
board ship resulted in the continuous plot shown. Even with the best available
navigation, discrepancies between adjacent swaths are common.

The ATLANTIS II is the support ship for the diving submersible ALVIN, and thus
requires accurate navigation both to position the ALVIN on a dive site and to
navigate the ALVIN during a dive. Since the introduction of Sea Beam to the
Atlantis II, real-time bathymetric maps are commonly used to locate dive sites
relative to the sea floor topography. By surveying a potential dive site with Sea
Beam before sending the submersible down, a detailed bathymetric map is available
for the ALVIN dive. The bathymetric map is used as a base map, and the ship is
positioned dynamically to match the real-time contours (as an overlay) to the base
map. Figures 3 and 4 are maps used on a recent ALVIN survey of the Kane Fracture
Zone on the Mid-Atlantic Ridge. The survey contour map (Figure 3) was produced
from data collected on a 1984 cruise. This map represents approximately 3 days of
survey time from a 30-day survey. Figure 4 shows a dive site for a neovolcanic
feature referred to as the "Snake Pit", which included massive sulfide deposits;
the circled area on figure 3 indicates the dive site. The dotted line on Figure 4
is the ALVIN trackline for the dive. Maps similar to this one are provided to the
scientists before each dive and greatly facilitate submarine navigation.

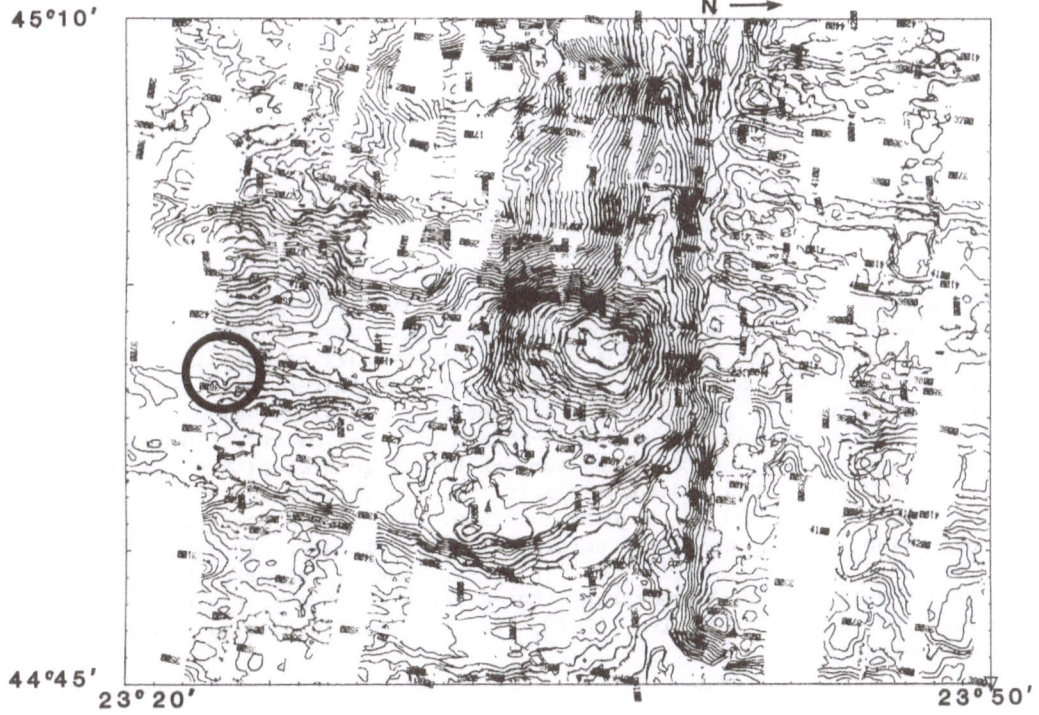

Figure 3.   Sea Beam contour map from Kane Fracture Zone survey.
Circle indicates ALVIN dive site shown in Figure 5.

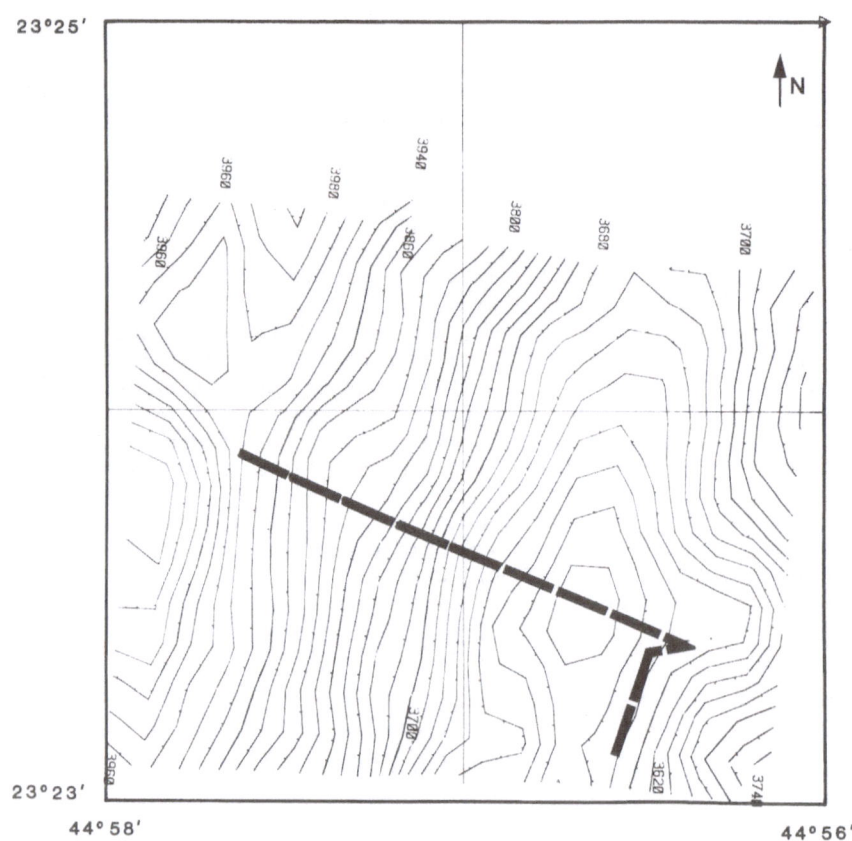

Figure 4. ALVIN dive site map from Sea Beam contour map.
Dashed line shows trackline followed by ALVIN during dive.

ALVIN dives are sometimes navigated using a transponder net, a series of ranging transducers which have been surveyed into position using the available navigation sources. This requires a at least eight hours of ship time to position a single transponder net. If only transit satellites are available, the navigation error will be on the order of one-half nautical mile, even though relative positioning may be on the order of a few meters. In addition, the transponder net can only position the submersible in an area with a radius equal to the water depth. This is acceptable if the dive site is known and if a series of dives will be conducted in the same location. Most ALVIN operations, however, are more wide ranging and cover several dive locations within a period of a few days. Prior to Sea Beam installation, if no transponder net were used, valuable dive time could be used searching for a targeted dive site. With a speed of approximately 1/2 knot for the ALVIN, it becomes imperative that ship be positioned accurately.

5. INTEGRATION AND POST-PROCESSING OF DATA

Processing of Sea Beam data is done both while at sea and at URI after each
cruise.   More than half of the processing time is spent correcting navigation to
produce the accurate maps needed for further applications.   A preliminary
navigation track is produced after plotting all available navigation and selecting
the "best" navigation source at any given time. Position fixes from GPS, Loran
and transit satellites are then combined with course and speed information using a
program which navigates between fixes (if they are more than five minutes apart)
to produce a dead-reckoned ship's position, set, and drift along track.   Set and
drift figures are examined for inconsistencies and suspect position fixes are
deleted.   Successive calculations are made until set and drift factors are
minimized and consistent. Developmental work is being done to use an interactive
graphics least-squares method for fitting a most likely ship's track to the
available data.

After navigation track errors are minimized, navigation is merged with the Sea
Beam data and plotted as contours.   Figure 5 displays a small portion of a
Galapagos Rift survey conducted in late 1985; the only fixes available at that
location and time were transit satellites.   For comparison, Figure 2 shows
contours produced using Loran-C navigation.   As indicated by arrows in the
figures, Sea Beam contours reveal navigation offsets in both well-navigated areas
(Loran and GPS) and areas where positions were dead-reckoned between fixes.

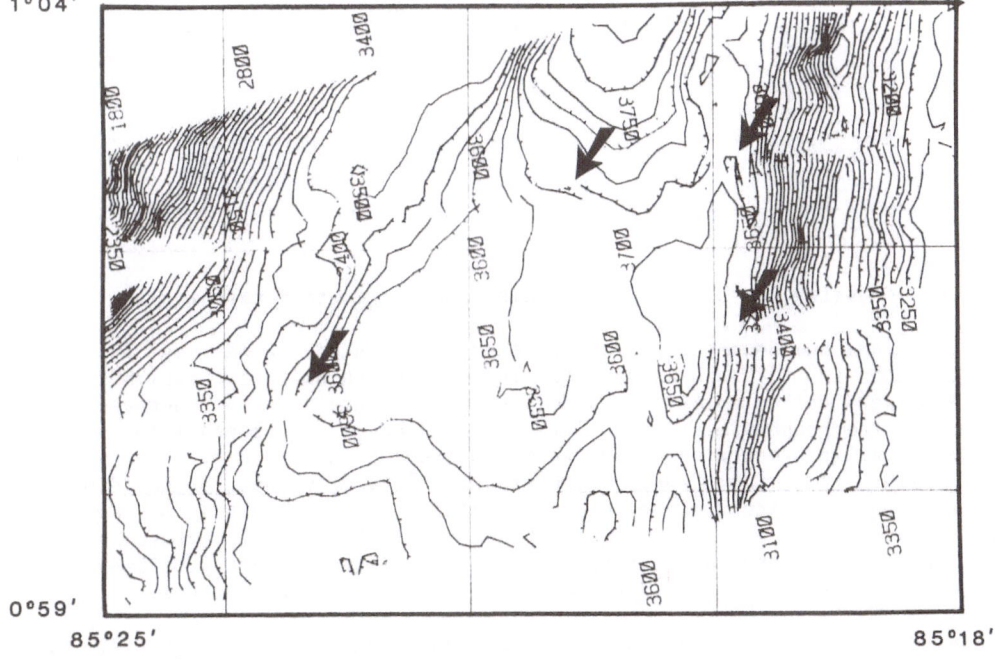

Figure 5.   Sea Beam contour swaths collected using transit satellite
                navigation. Arrows show mismatches between adjacent swaths.

To correct navigation offsets we first re-evaluate the navigation and try to improve its accuracy by ignoring position fixes of marginal quality (for instance, where a satellite fix was received from a satellite very low on the horizon). Still, it is often necessary to manually shift navigation tracks in order to align Sea Beam contours. The most precisely navigated lines are first selected and other crossing contours are shifted to match. Figure 6 shows two lines from the Kane survey area (Figure 3) which have a well matched crossing.

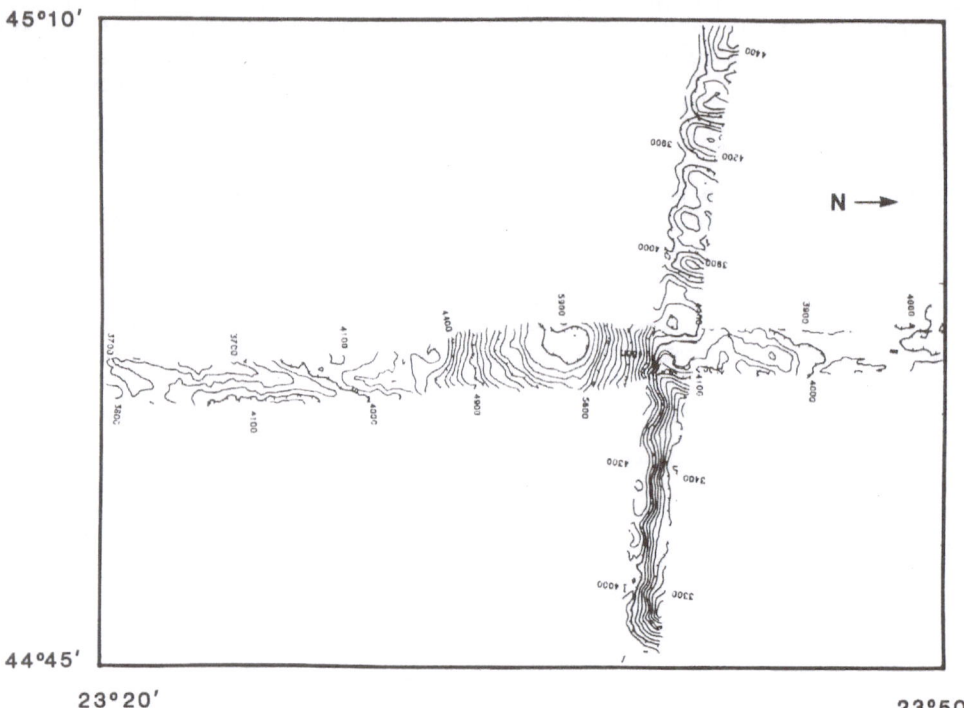

Figure 6. Crossing Sea Beam contour swaths showing contour matching.

When a typical Sea Beam survey is completed, the study area may be nearly completely covered by Sea Beam swaths. These data (an example of which is shown in Figure 3) were collected at irregularly spaced intervals. To increase their usefulness to the geophysicist, such data are gridded, i.e. placed into a regular x-y-z framework, to produce maps such as the one shown in Figure 7. A modified MISP algorithm from the Naval Oceanographic Office is used for gridding. To produce an accurate gridded map of an area, fairly complete coverage by Sea Beam swaths as well as corrected navigation are required. Offsets between adjacent swaths, bad sonar data and sharp turns can produce tears, bullseyes, or pseudofaults in a gridded map. These are eliminated whenever possible.

Gridded data can be displayed by any of numerous methods. An example of a surface mesh plot is shown in Figure 8. Further applications of the gridded data include three-dimensional contour plots, solid surface modelling, and three-dimensional displays which can be manipulated on a computer graphics screen.

N →

45°10'

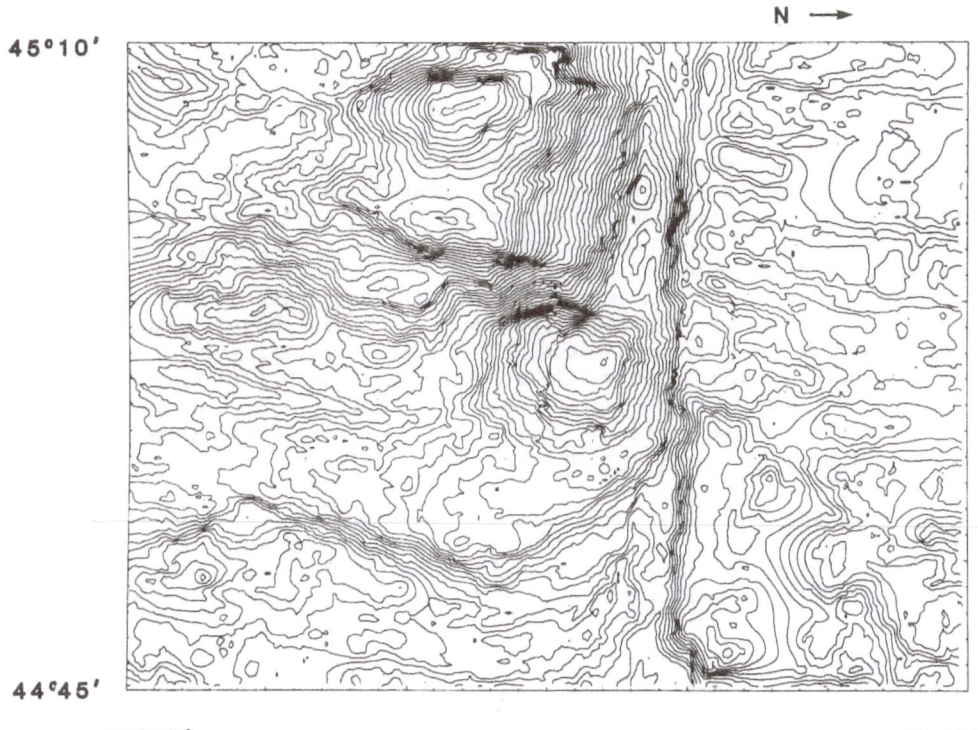

44°45'

23°20'   23°50'

Figure 7. Gridded Sea Beam contour map of the Kane Fracture Zone.

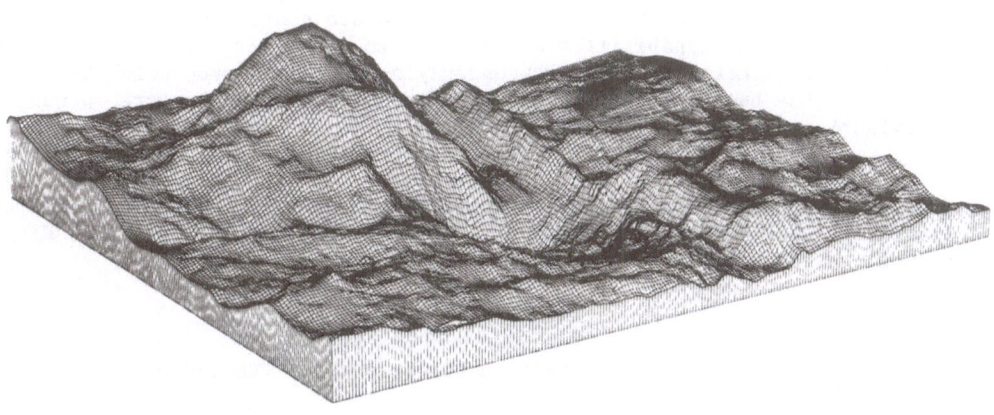

Figure 8. Three-dimensional surface mesh plot of the Kane Fracture
           Zone.

6. PROBLEMS AND IMPROVEMENTS

In order to utilize Sea Beam data to its full potential, there are numerous improvements which could be made in hardware and software. Shipboard navigation upgrades hold the most promise. Continuous global GPS coverage would improve the navigation of Sea Beam surveys immensely. Though the availability of GPS has increased remarkably in recent years, complete coverage will take several more years and we must still deal with the errors inherent in the present GPS. One of the critical factors we have found in processing the navigation is the accuracy of the ship's speed log. The speed logs on the two ships we operate are quite different in capability, which is reflected in the amount of post processing necessary to produce a final map. When navigating using transit satellites, position fixes often are not available for two to three hours between fixes. Ship's position is dead-reckoned between available fixes using speed and heading data. If the speed log has a systematic error as small as 0.1 knot, errors on the order of 500-600 meters can easily accumulate during long time spans between position fixes. Cumulative errors of this magnitude become quite significant when data are later processed to estimate the actual ship's track.

In early Sea Beam operation, clock errors of 30 seconds in navigation vs. Sea Beam data recording occurred. These errors are easily detected in the maps and represent a distance traveled of less than 200 m, essentially the size of a beam cell. As a result, we now update our computer clocks via satellite to minimize timing errors.

Another problem found on numerous surveys has been the inability to produce a single trackline which is usable for both Sea Beam and gravity applications. Gravity surveys, which are frequently run simultaneously with Sea Beam surveys, require that accelerations be continuous. Sea Beam swath matching and navigation through points causes discontinuous accelerations. Increasingly more sophisticated processing software is needed for smoothing Loran and GPS data, for determining quality of fixes, and for interpolating between fixes with continuous acceleration.

These are some of the hardware and software problems encountered when using a precision mapping system such as Sea Beam. We are continuing to improve our navigation and processing capabilities as well as to discover new methods for using Sea Beam bathymetric data. The underlying lesson in processing Sea Beam data up until now has been that Sea Beam bathymetry is consistently more accurate than any navigation source available to us.

The NECOR/URI Sea Beam Development Center is a research group jointly funded by the National Science Foundation (Grant OCE - 8418919) and the Office of Naval Research (Contract N00014-81-C00062).

REFERENCES

Shor, A., and Chayes, D., 1986. "Navigation for Surveys of Trans-Pacific Fiber-Optic Cables." INSMAP 1986.

Tyce, R., 1986. "Deep Seafloor Mapping Systems - A Review." Marine Technology Society, Special Journal on Sonar Technology, in press.

Tyce, R., Miller, J., Edwards, R. and Silver, A., 1986. "Deep Ocean Pathfinding - High Resolution Mapping and Navigation." OCEANS86. V. 1, IEEE.

STARFIX: A NEW HIGH-PRECISION SATELLITE POSITIONING SYSTEM

Arthur R. Dennis
Analytical Technology Laboratories, Inc.
301 Wells Fargo Drive, Suite 5
Houston, TX   77090

ABSTRACT

This paper describes a system which uses conventional geostationary C-band communication satellites as the basis for a precise positioning system. The system, called STARFIX*, became operational in early 1986 after three years of development effort. It has demonstrated an ability to provide 5 meter accuracies for marine users, which substantially exceeds the proposed accuracy to be provided by the Global Positioning System (GPS) standard positioning service of 100 meters. Present coverage of STARFIX includes the continental United States plus several hundred miles into the offshore areas. This paper describes STARFIX and discusses various sources of error that affect positioning accuracy.

INTRODUCTION

The STARFIX satellite-based positioning system makes use of conventional C-band geostationary communication satellites to provide users with 5 meter accuracy within the coverage area, which includes the continental United States plus offshore areas. STARFIX supports moving users having speeds of up to 50 knots, 24 hours a day, in all weather. It represents a joint development of Analytical Technology Laboratories and John E. Chance and Assoc., of Lafayette, Louisiana.

The STARFIX system is comprised of several major components:

- A constellation of at least three satellites.

- Uplink facilities for transmitting signals to each of the satellites.

- A nationwide tracking network for determining the positions of the satellites to send to each user.

- User equipment, including a (passive) receiver and antenna system.

In this paper we will discuss the geodetic aspects of STARFIX, including predicted and measured accuracies, the effect of various system errors on user fix accuracy, and so on. A more general description of the system and user equipment can be found in Reference [1].

BASIC SYSTEM CHARACTERISTICS

STARFIX operates on the differential pseudo-ranging principle. That is, the user makes a set of three or more pseudo-range measurements, one for each satellite. These pseudo-ranges are proportional to range but contain a common, unknown

---

*STARFIX is the registered trademark of the PANAV Joint Venture, Houston, TX

M. Kumar and G. A. Maul (directors), Marine Positioning, 17–29.

bias due to clock epoch offset. Simultaneously, a set of pseudo-range measurements is made at a known, reference location, typically the Master Site. These also contain a common offset bias. This is illustrated in Figure 1.

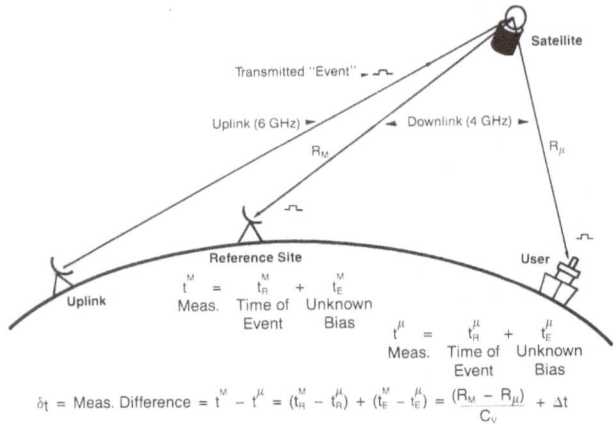

$$\delta_t = \text{Meas. Difference} = t^M - t^\mu = (t^M_R - t^\mu_R) + (t^M_E - t^\mu_E) = \frac{(R_M - R_\mu)}{C_v} + \Delta t$$

Figure 1
Basic Principles

The reference site pseudo ranges are compressed into a Pseudo Range Parameter Set (PRPS) which is transmitted out to all users at short (i.e. 30-40 second) intervals. Together with appropriate timing information, the PRPS allows the user to reconstruct the reference pseudo range data at precisely his measurement time. He then subtracts his own measurements to generate a set of differential measurements which contain a constant bias (user/System Time offset). Knowing the satellite positions (also transmitted periodically from the Master Site), the user's computer can solve for his own position as well as the common clock offset (in a manner identical to that used for Differential GPS operation).

The pseudo ranges are computed by measuring the code phase of spread spectrum signals that are transmitted to the satellites. The chipping rate is approximately 2.5 MHz and the signal itself resides in a 5 MHz bandwidth within the satellite transponder space.

SATELLITE CONSTELLATION AND UPLINK FACILITIES

The present STARFIX system utilizes three commercial geostationary communication satellites operating at (the Common Carrier) C-band. Thus the uplink frequencies are in the 6Ghz range while the downlinks are in the 4 GHz range. The satellites being used were manufactured by various companies, including Hughes and RCA. Their locations in the equatorial plane extend from about 75 degrees West Longitude to about 140 degrees West Longitude.

The uplinks consist of fully redundant transmitters and associated communication equipment. The uplink antennas are large parabolic dishes (ranging in size from 5 to 11 meters in diameter). The transmitted power is carefully set to permit downlink radiation levels of about 16 dbw (Effective Istotropic Radiated

Power, or EIRP). The transmission facilities are fully licensed as Common Carriers by the Federal Communications Commission. No operating license is required by the user since the receivers are completely passive.

As mentioned previously, the uplink signals consist of spread spectrum modulated carriers with a chipping rate of about 2.5 MHz. The chipper is in turn modulated with a digital data signal having a rate of 150 baud. This data signal contains the information necessary for the user to determine his position, including the PRPS, satellite coordinates, and special timing words that allow the user to synchronize his local oscillator to System Time. It also contains messages for users who subscribe to the STARFIX Network Message Service.

USER EQUIPMENT

The STARFIX operator, PANAV, manufactures a receiver that can track and demodulate the STARFIX signals and compute user position. The Model 5200A is shown in Figure 2.

Figure 2
Model 5200A Receiver

This receiver contains room for three independent receiving modules plus one in-line spare. It also contains a 16-bit CPU that performs all the required position computations using the pseudo ranging data from each channel plus the decoded digital data. The output is displayed on the built-in high-luminescence display and can be supplied to the user via an RS232 communication port. The 5200A weighs about 100 pounds and requires 17 inches of rack space. It can operate off of 24 volts dc or 110 volts ac with an external converter.

PANAV also supplies several types of antennas that can be used in marine applications. Each consists of a platform on which is mounted four (three operational and one in-line spare) horn-type antennas and associated Low Noise Amplifiers (LNA's). The platform is rotated in azimuth by a controller which uses an externally-supplied North-sensing device such as a magnetic compass (or gyrocompass) which will keep the antenna units pointed at the same location in the sky as the vessel moves.

Figure 3 shows the newly-developed Model ST-100 antenna system. It weighs about 100 pounds and requires only about 1 cubic meter of deck space.

Figure 3
Model ST-100 Antenna System

TRACKING AND ORBIT DETERMINATION

STARFIX is supported by a nationwide tracking network comprised of 10 sites: a Master Site and nine Remote Sites. The Master Site is located near Austin, Texas, while the Remote Sites are scattered throughout the country to provide as long baselines as possible.

Each Remote Site is in effect a user at a precisely surveyed location. A set of pseudo-ranges measured against a local atomic clock is transmitted in real time back to the Master Site at a high rate (two times per second) via leased telephone lines.

The Master Site contains a set of receivers that also generate pseudo-ranges. The incoming data from the Remote Sites is differenced from the Master Site data to yield up to nine sets of differential pseudo-ranges that are used in the tracking software to estimate the positions of each of the satellites. The software must also solve for the clock offsets between the Master Site clock and each of the Remote Site clocks.

Since geostationary communication satellites frequently undergo station-keeping maneuvers to keep them within a prescribed "box" in the sky, the tracking software cannot easily use classical equations of motion to determine ephemerides. Therefore a strictly geometric approach is used in which point-by-point solutions are obtained in earth-centered-earth-fixed coordinates. This in turn requires an overdetermined set of observations, which is easily provided by the nine Remote Sites.

The Master Site also generates the set of reference pseudo-ranges (the PRPS) that is transmitted to each user. This is accomplished by analyzing three

independent sets of receivers which produce three independent sets of pseudo-
ranges from separate antenna farms that are in close proximity to each other (to
eliminate orbital effects). The antennas are precisely surveyed onto the World
Geodetic System 1972 (WGS-72) datum, and these coordinates are used in each user
receiver as the reference locations for the position fix computations.

A round-trip time delay is also measured at the Master Site which allows each
user to synchronize his local clock to within about 20 milliseconds of System
Time. He can then reconstruct the exact value of differential pseudo-range
sufficiently well to maintain 5 meter position-fixing accuracy at speeds of up to
about 50 knots. Future plans are to increase this "coarse" time sync accuracy to
allow resolution of the 6.66 millisecond ambiguity associated with the 150 Hz code
length. This will allow time sync for the user clock to the full accuracy of the
clock offset solution, which is in the neighborhood of 3-5 nanoseconds.

All data (the PRPS, satellite coordinates, and time sync words) are modulated
onto the spread-spectrum carriers and transmitted to each of the satellites via
the uplinks. The rates for each type have been adjusted to minimize uplink
loading while providing satisfactory user fix accuracy and acceptable "time to
first fix".

ERROR SOURCES AND RELATED CONSIDERATIONS

As far as the STARFIX user is concerned, there are several sources of error that
he should be aware of if the full capability of the system is to be realized at
all times. These are:

- Position fixing errors resulting from errors in the estimate of user
  elevation;

- Instrumentation errors, including pseudo-range measurement and time-sync
  errors;

- Errors resulting from differential ionospheric propagation delays;

- Errors resulting from errors in the satellite position estimates emanating
  from the Master Site.

In the following paragraphs we will describe the sources of these errors, their
general magnitudes, and how they are and can be controlled.

The Effect of Unknown User Altitude

For a constellation of geostationary satellites, there is not enough spatial
volume to allow precise determination of a user's full three-dimensional position.
Therefore, an estimate of his height above the reference ellipsoid must be
provided to the position-fix computation. The effect of an error in the height
estimate on the user position fix computation (which is a Kalman filter that
solves for user latitude, longitude and clock offset) is shown in the accompanying
Table I where a 5 meter height uncertainty is assumed.

Increase In Latitude Est. Uncertainty
Due To 5 M Height Uncertainty

LATITUDE (DEG)	INCREASE IN LATITUDE UNCERTAINTY (M)
20	10.1
30	6.08
40	3.97
50	2.57
60	1.51

Table I

Note that the effect is strictly a function of user latitude, and decrease as his latitude decreases, roughly in proportion to the tangent of latitude. Also, the error appears only in the user latitude estimate; longitude is totally unaffected. For marine users, it is possible to predict mean sea level in most areas around the U.S. to the meter accuracy, so that errors in the latitude estimates can generally be controlled to the 2 to 3 meter level or better.

Instrumentation Errors

By instrumentation error, we mean errors in the measurement of the differential pseudo-range in the user's receiver. In order to achieve high accuracy, they must be consistent, i.e., can be calibrated. These errors are therefore generally hardware-related, and involve the performance of the receiver RF sections, the code phase tracking and measuring subsystems, and the time-recovery (time sync) accuracy. The best way for a user to evaluate this class of errors is simply to set a receiver in a fixed location and watch its position fixing performance over a long period of time, in various environments. This has been undertaken at PANAV for many months, and typical results show that the hardware is consistent to at least the 5 meter level (in terms of the lat/lon solution).

These sort of tests also evaluate the quality of the pseudo-range parameter set (PRPS) being created at the Master Site since any errors there will directly effect the user fix accuracy as well as noise content. For example, Figure 4 shows a typical static test result in the Houston area. Note that latitude contains the predominant noise component, which is predicted by the statistics of the fix solution, and result from the weak North-South geometry of the geostationary satellite constellation. What is important, however, is that the solutions are consistent over long periods, which indicates that the instrumentation errors are under control.

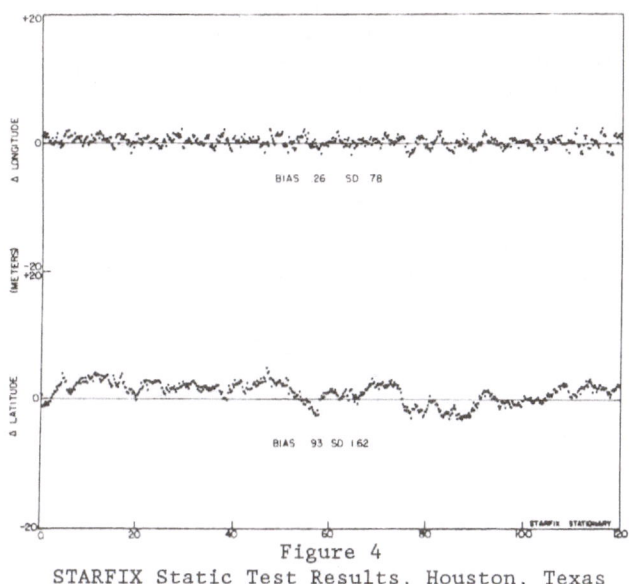

Figure 4
STARFIX Static Test Results, Houston, Texas

There will also be some small effects due to time sync errors in static tests, but these errors are better evaluated under moving conditions through comparisons with other systems. Some of these types of tests are described later on in this paper.

Ionospheric Errors

The 4 Ghz downlink signal undergoes delays due to interaction with the earth's ionosphere. Extensive analyses of these types of delays have been performed in conjunction with the Global Positioning System, where, taking into account the higher frequencies involved, we would expect to see some maximum group delay variations in the 5 to 10 nanosecond range (about 3 meters). If these were present to this level, we would be seeing diurnal variations in position fixing up to 20 meters. But we have yet to notice anything near this level, even over very long baseline distances from the reference site. This has certainly been a pleasant surprise since we had prepared ourselves to combat this problem with various approaches, including mathematical modeling and other compensation techniques.

The reason for this is not yet totally clear, although the differential nature of STARFIX plus the stationary satellite geometry may have a lot to do with cancellation of almost all ionospheric effects (since common-mode errors are eliminated). Until ionospheric-induced errors become clearly evident in some results somewhere, we are not going to be concerned with this source, although analysis is continuing.

Satellite Coordinate Errors

This error source merits a fairly detailed discussion so that STARFIX users can understand its nature and how they can control its effects.

Unlike a direct-ranging system such as GPS, satellite orbital errors in a differential-mode system such as STARFIX propagate into the users's fix equations in a differential manner, that is, they are zero at certain reference sites and then grow as the user moves from the reference. The mathematical relationships that define this kind of behavior are shown in the Appendix, and these results can be used to evaluate the effect of various levels of orbital error on the user's fix computation using statistical error analysis techniques (such as those described in Reference [2]). For example, Figure 5 shows user fix latitude error contours for 10 meter uncorrelated (one sigma) errors in all satellite coordinates for the reference site located at the STARFIX Network Master Site in Austin, Texas. Figure 6 shows the corresponding fix longitude errors (for these plots it was assumed that there were no user height-induced errors).

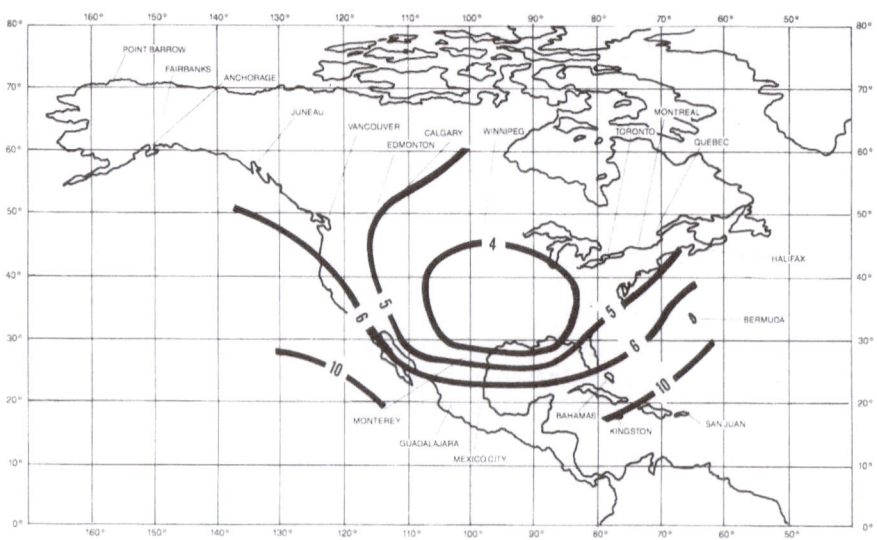

Figure 5
Latitude Positioning Error Contours (meters)
for 10 meter (one sigma) Satellite Coordinate Errors
(Austin Master Site Reference)

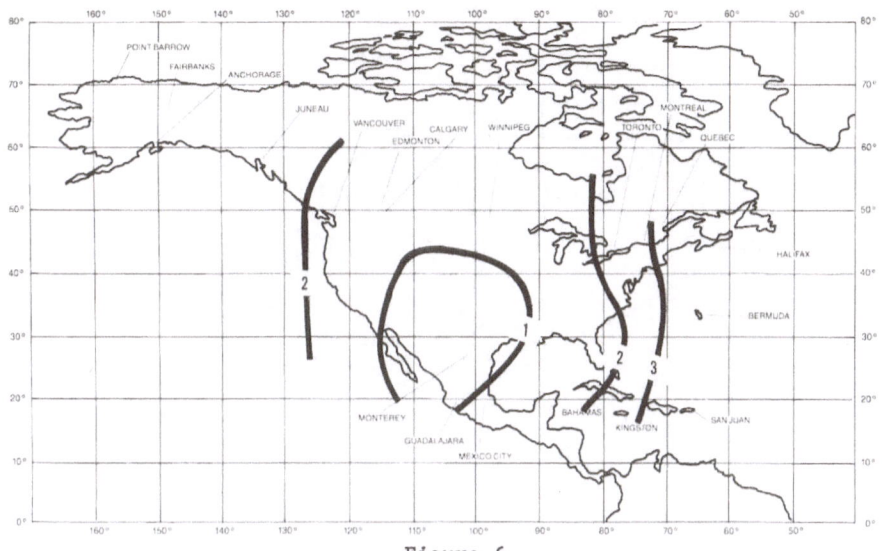

Figure 6
Longitude Positioning Error Contours (meters)
for 10 meter (one sigma) Satellite Coordinate Errors
(Austin Master Site Reference)

These figures clearly demonstrate the well-known error growth effects as the user moves away from the reference site. *However, it is very important to note that, as shown in the Appendix, any point can be made a reference site simply by performing a calibration at that point.* Therefore, error growth is from the calibration point, and not necessarily the point where the reference pseudo-ranges are computed (i.e. the Master Site). This is illustrated in Figure 7 which shows the same conditions that produced Figure 5 except that a calibration was assumed to have been performed at the southern tip of Louisiana in order to provide more accurate coverage for the Gulf of Mexico. This demonstrates the fact that the STARFIX user who's positioning requirements are restricted to a geographical area should calibrate in the neighborhood of that area.

One final point should be made for the benefit of STARFIX users regarding satellite orbital errors and their effect on fix accuracy. As described in a previous section, the STARFIX Tracking Network utilizes data from Remote Sites which have hardware configurations that are basically identical to that found in the user's receiver. Thus, by minimizing the errors of the fit of the satellite positions to these data, the Network in fact guarantees that user fix errors will be controlled, even though the absolute satellite positions may exhibit certain types of relatively large correlated errors. Error analyses of the type described above therefore tend to be on the pessimistic side since we generally assume uncorrelated satellite errors when conducting the simulations.

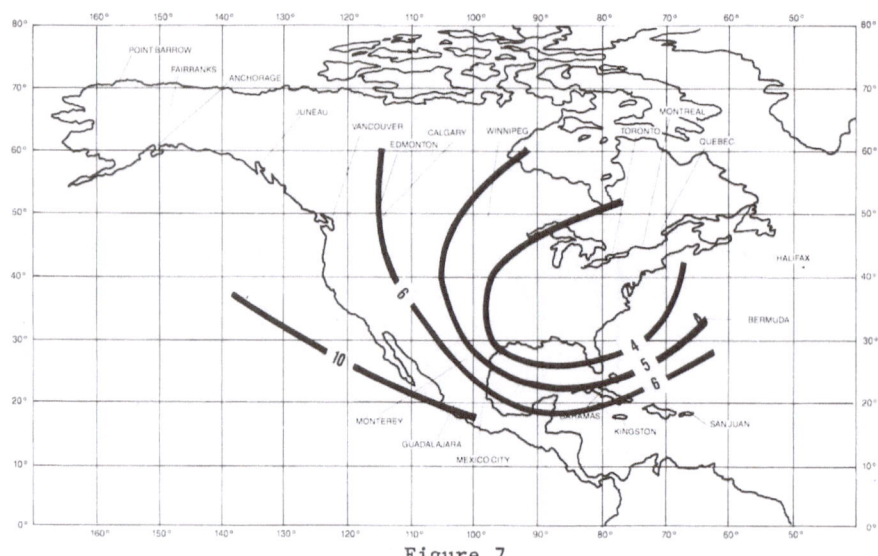

Figure 7
Latitude Positioning Error Contours (meters)
for 10 meter (one sigma) Satellite Coordinate Errors
(Southern Louisiana Reference Point)

OPERATIONAL RESULTS

STARFIX has been tested and used at sea since the beginning of 1986. Most of the significant tests have been performed in the Gulf of Mexico, where extensive comparisons can be made against existing high-precision shore-based radionavigation systems used throughout the offshore oil industry.

After STARFIX has been installed on the vessel at dockside, the antenna is precisely surveyed using standard techniques, and the surveyed coordinates are entered into the 5200A receiver to perform a calibration of the differential pseudo-ranges. Thus the dockside point becomes the point of reference as described above, and generally it has been desirable to try and return to the same point after the sea trial to verify that the calibration remained consistent.

On a number of sea trials comparisons were made with state-of-the-art microwave radionavigation systems that exhibit high accuracy over short ranges (i.e. line-of-sight to the transmitter). One such comparison made around 150 km offshore in the Gulf of Mexico is shown in Figure 8. The accuracy of the comparison system ("SYLEDIS") is generally accepted to be in the 5-7 meter range at the transmitter distances employed (about 25 km), so the results clearly indicate that STARFIX accuracy was in that range also.

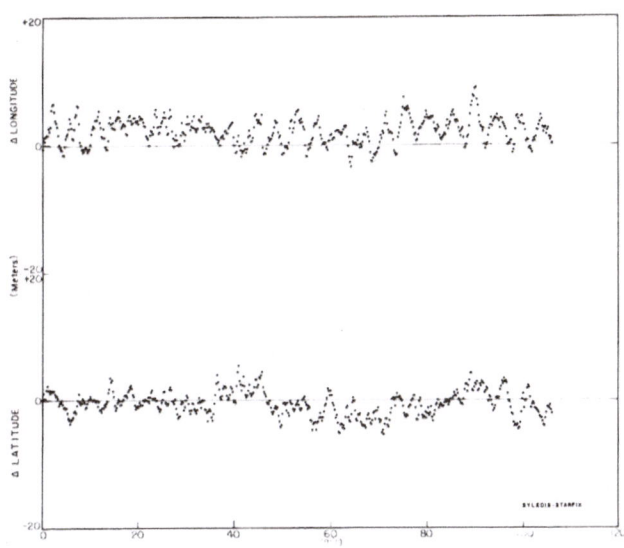

Figure 8
Gulf of Mexico Dynamic Test Comparisons

COMPARISONS WITH OTHER SATELLITE SYSTEMS

STARFIX is most closely related to the Global Positioning System (GPS) in the basic form of measurement, i.e., the use of pseudo-ranging. However, whereas GPS requires the use of stable atomic clocks in the satellites to generate System Time, with STARFIX this task is transferred to the Master Site, so there is no need to depend on the variations of the satellite transponders. In this way STARFIX is similar in operation to the differential form of GPS.

There are other comparisons that can be made with GPS: GPS is a global system, while, at least at the present time, STARFIX is not. If all 18 GPS satellites become operational, it will allow full three-dimensional fixing capability, while STARFIX alone does not. On the other hand, the Government plans to restrict GPS accuracy available to civilians, which is not the case with STARFIX. Also, the constant geometry of the STARFIX satellite constellation yields constant and predictable fixing results, with no "holes" due to changing geometry as is the case with GPS (this is particularly advantageous in differential operation). Furthermore, under the conditions mentioned in this paper, STARFIX can provide generally higher accuracies than GPS, even if one is able to use the restricted high-accuracy PPS. This is not only because of the differential nature of the system, but also because of the higher operating frequencies (which reduce ionospheric effects) and the great distances involved which tend to reduce the rate of error growth from a reference point. Finally, and perhaps most important, GPS is several years away from being operational, while STARFIX is available now.

The only other operational system is the Navy's TRANSIT. It provides positioning based upon the doppler principle using one satellite at a time, so that a fix can only be obtained every 1.5 hours on the average. Thus, other than the fact that TRANSIT is a global system, it does not compare with STARFIX in terms of accuracy and fix availability.

REFERENCES

[1] Dennis, A. R; "STARFIX". Proceedings of the 1986 Position Location and Navigation (PLANS) Symposium, Las Vegas, Nevada, November 4-6, 1986.

[2] Gelb, A., editor "Applied Optimal Estimation", MIT Press, 1974.

APPENDIX

In this Appendix we present the mathematical relationships that define the effects of satellite orbital errors on the STARFIX differential pseudo-range measurements.

Let:

(X_{s1}, X_{s2}, X_{s3}) be the coordinates of the satellite

(X_{m1}, X_{m2}, X_{m3}) be the coordinates of the Network
Master Site

(X_{o1}, X_{o2}, X_{o3}) be the coordinates of an observer

Then the error δZ in the observer's differential pseudo-range measurement due to satellite position errors is, to first order,

$$\delta Z = \sum_{n=1}^{3} W_n E_n \qquad (A1)$$

where

$$W_n = X_{sn}\left(\frac{1}{R_m} - \frac{1}{R_o}\right) + \frac{X_{on}}{R_o} - \frac{X_{mn}}{R_m}$$

and

$$R_m = [\Sigma(X_{mn} - X_{sn})^2]^{\frac{1}{2}}$$
$$R_o = [\Sigma(X_{on} - X_{sn})^2]^{\frac{1}{2}}$$

and E_n = error in the 'nth' satellite
coordinate.

Now, at a calibration point, the error is

$$\delta Z = \sum_{n=1}^{3} W_{cn} E_n \qquad (A2)$$

where

$$W_{cn} = X_{sn} \left(\frac{1}{R_m} - \frac{1}{R_c} \right) + \frac{X_{cn}}{R_c} - \frac{X_{nm}}{R_m}$$

(X_{c1}, X_{c2}, X_{c3}) are the calibration point
coordinates.

and

$$R_c = [\Sigma(X_{cn} - X_{sn})^2]^{\frac{1}{2}}$$

The calibration procedure involves subtracting (A2) from (A1) since we assume that
we can measure (A2) perfectly. Thus, the calibrated error is given by:

$$\delta Z_c = \sum_{n=1}^{3} (W_n - W_{cn})E_n = \sum_{n=1}^{3} \tilde{W}_n E_n$$

where

$$\tilde{W}_n = X_{sn} \left(\frac{1}{R_c} - \frac{1}{R_o} \right) + \frac{X_{on}}{R_o} - \frac{X_{cn}}{R_c}$$

so that the calibration point becomes the new point of reference in terms of error
propagation.

THE HYDRORANGE ACOUSTIC SYSTEM FOR SEISMIC STREAMER TRACKING
FOR INSMAP 86*

A. William Marchal
Offshore Navigation, Inc.
5728 Jefferson Highway
Harahan, LA 70123

EVOLUTION OF STREAMER LOCATION TECHNIQUES

Since the advent of 3-D seismic technology, with its requirement for increased positioning accuracy, much attention has been devoted to surface navigation quality control in an attempt to achieve absolute vessel positioning accuracy on the order of a few meters. What has been neglected, and deserves even greater attention, is the dynamic position over the ground of the seismic energy source (guns) and receivers (streamer mounted hydrophones). Knowledge of the vessel's position is undeniably of great importance, but knowledge of the hydrophone positions is critical when it comes to trace-stacking and the resultant interpretation of seismic data.

Early attempts at streamer location consisted merely of estimating the "feathering" of the cable by visually measuring the angle between the vessel's midline and the tail buoy, with the unjustified assumption that the cable remained straight in the water. Later, towpoint fixtures, or angle of departure devices, were used in conjunction with tail buoy tracking to provide a small increment of sephistication in streamer profiling. Some data smearing was removed by these means, but there is an inherent weakness in any attempt to correlate cable position with tail buoy tracking. Marine seismic contractors have made a considerable effort to decouple the tail buoy from the streamer itself by the use of tail-end stretch sections behind the last active cable section and 500 meters or more of nylon line beyond that. The intent is to prevent the tail buoy from dragging the cable about and inducing noise in the sensitive hydrophones. Also, the cable is subject to submarine phenomena and quite different current forcing functions than is the tail buoy. To thus decouple the tail buoy from the cable end, then assume that one is an indication of the other is a contradiction in reasoning. The proper argument is that tail buoy tracking provides more information that doing nothing at all.

As 3-D work became feasible and dynamic binning techniques were developed, knowledge of individual hydrophone positions become more important, and the use of in-cable magnetic compasses arose, providing an order of magnitude improvement in streamer profiling. Standard techniques of magnetic profiling combine known down-cable distances with compass azimuths to fix points along the streamer in a local or relative coordinate system. A least squares curve is then fitted to those points using a cubic spline or preferably a Chebychev polynomial, and that curve is taken to represent the streamer's profile in the water. Some current systems utilize a combination of magnetic and acoustic techniques to combine the best features of each system. The acoustic poritions of these systems are hindered in their ranging geometry by the short baselines used, thus limiting the down-cable distances at which their reported positions are accurate.

_____

*Paper was submitted for inclusion in proceedings but was not presented at symposium.

M. Kumar and G. A. Maul (directors), Marine Positioning, 31–47.
© 1987 by the Marine Technology Society.

SYSTEM DESCRIPTION

This paper introduces a system providing the next logical step in the evolution of streamer profiling accuracy by adding optimal geometry acoustics to the measurement process. The system, known as HYDRORANGE, requires in its full implementation a separate utility vessel provided with navigation sensors (Syledis, Microphase, Argo, SPOT, etc.) to follow alongside the towed streamer emitting acoustic pulses from a submerged acoustic transducer (see Figure 1). These pulses are received at the streamer by externally attached acoustic transponders, and are retransmitted back to]6-channel acoustic receivers aboard both vessels. A low power bidirectional VHF datalink is employed to transfer position and acoustic range data between vessels for purposes of station keeping and acoustic profiling.

The known positions of the two vessels are combined with the acoustic signal transit times and the known separation distance between acoustic transponders to produce extremely accurate position coordinates for each transponder, as illustrated in Figures 2 and 3. Ranging geometry is optimized by having the auxiliary vessel keep station half the cable length behind and half the cable length abeam of the master vessel, making the angle of intersection between the slave vessel and both cable ends a right angle. The transponder position coordinates form a set of points which are then fitted to a curve, and that curve is taken to be the streamer profile. The curve fitting algorithm utilizes a least squares Chebychev polynomial, with order equal to half the number of transponders or compasses. The orthogonal Chebychev was chosen for ease of computation. The identical algorithm is applied in the magnetic compass case, the only difference being that when magnetic compass data is utilized, distances and bearings are used to determine the points in a piecewise linear fashion rather than the direct range measurements of the acoustic case. Once the polynomial which describes the streamer shape is known, the coordinates of any point on the cable can be calculated. This is particularly useful with quarter-point steering, where navigation is centered on a downcable point rather than on the navigation antenna or stepback position. Also, since a derivative can be taken at any point along the curve, an accurate azimuth can be calculated, providing a compass correction factor and significantly reducing compass calibration time. Current compass calibration techniques are very consuming and are applicable only over a limited area of operations.

ERROR SOURCES

Before proceeding to a presentation of collected field data, a brief discussion of potential sources of error for both magnetic compass and acoustic systems is in order. Each error source contributes to inaccuracies in modeling the cable shape. Some errors can be readily detected and compensated for, others are more difficult and expensive to detect and correct.

1. Acoustic Error Sources

Foremost in importance in acoustic measurements is the velocity of an acoustic wavefront propagating through seawater. This propagation velocity is a function of temperature, salinity, and pressure, with temperature dominating. Figure 4 displays a typical acoustically derived streamer profile with varying values for propagation velocity used in the profiling model. Note that all the Figures of cable profilers are foreshortened 15:1 in the longitudinal direction due to scale differences and tic sizes in the longitudinal and vertical axes. Vertical tics

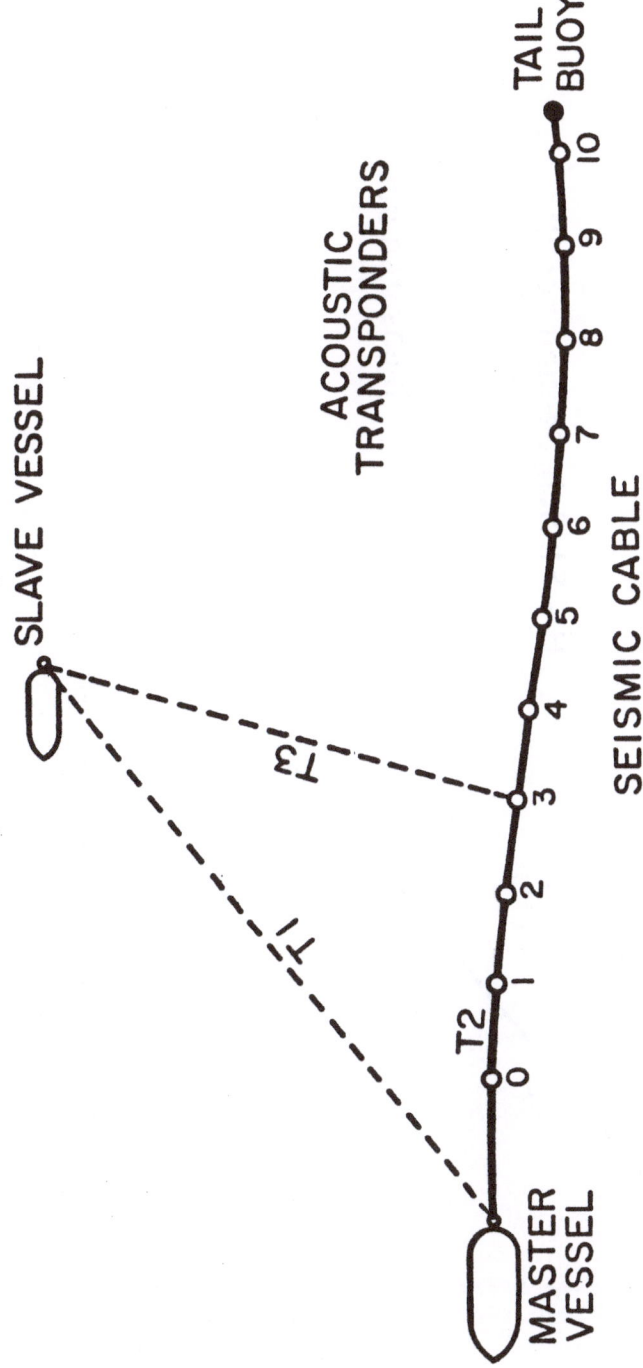

Fig. 1 PLAN VIEW OF OPERATIONAL SYSTEM

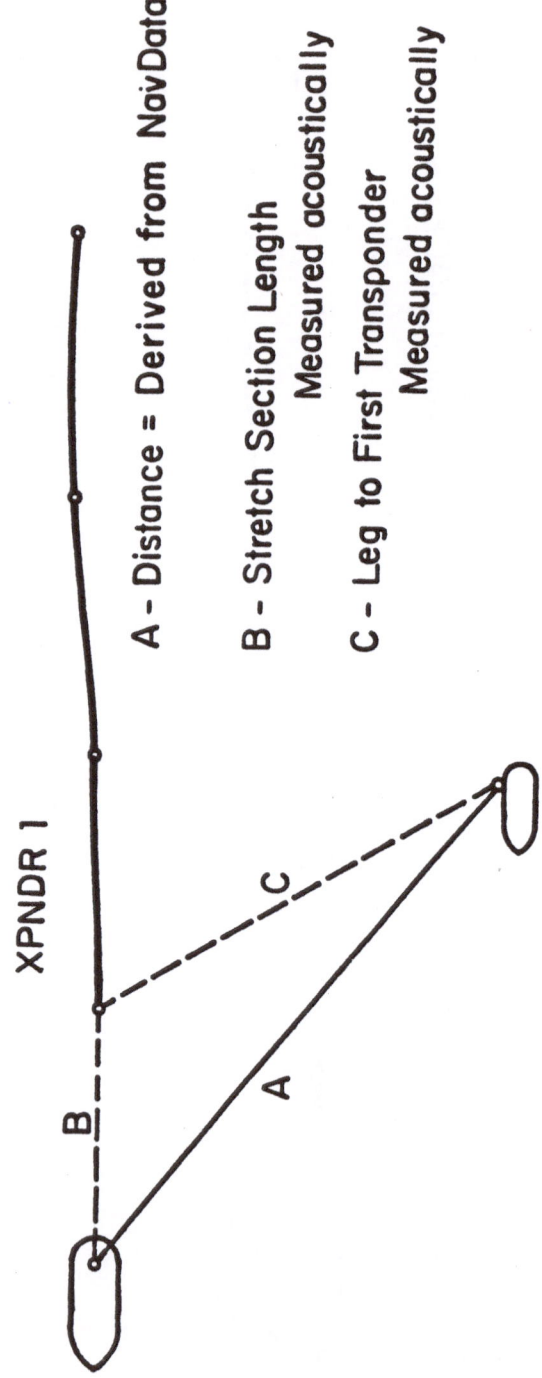

XPNDR 1

A – Distance = Derived from NavData

B – Stretch Section Length
    Measured acoustically

C – Leg to First Transponder
    Measured acoustically

Fig. 2   COORDINATES OF TRANSPONDER 1 COMPUTED

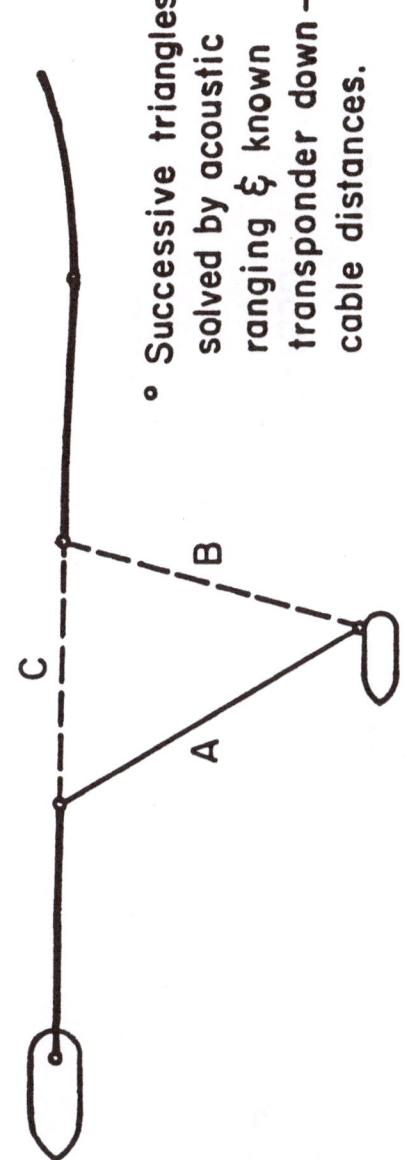

° Successive triangles solved by acoustic ranging ξ known transponder down - cable distances.

Fig. 3   COMPUTATION OF TRANSPONDER COORDINATES

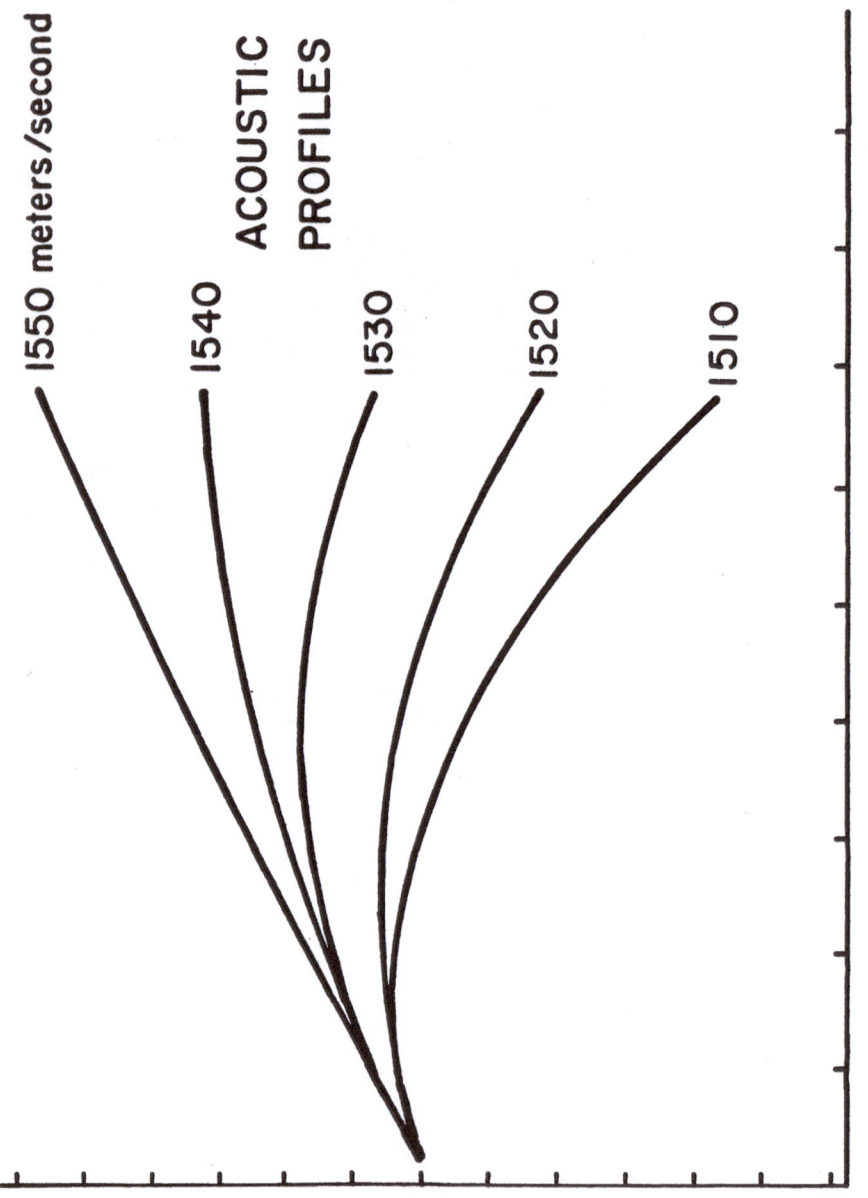

Fig. 4   PROPAGATION VELOCITY EFFECTS.

are 20 meters, longitudinal tics 500 meters. We see in Figure 4 that an error of
10 meters per second in acoustic velocity can result in a 50 meter lateral error at
the end of a 3000 meter streamer. Since the streamer profile is so sensitive to
acoustic propagation velocity, a means must be found to monitor changes in velocity
in real time. This is accomplished in the Hydrorange system by continuous acoustic
ranging between the two vessels. The acoustic transit times are scaled by the
known vessel separation distance, as determined by the navigation systems, to solve
for the velocity. A moving average of these measurements may then be applied to
the hydrophone position solution.

A related acoustic phenomenon is that of ray bending, or refraction, as the
acoustic wavefront traveles through water of different temperatures. This is most
pronounced when the acoustic transducer and transponders are separated in the vertical plan, thus subject to the variation in temperature experienced by th acoustic
wavefront travelling through the various thermoclines. In temperate zone waters,
velocity may vary by 2 meters per second per degree (f) change in temperature.
This ray bending results in a non-linear acoustic path, distorting the acoustic
transit time. This problem is minimized in the Hydrorange system by making every
attempt to keep all acoustic devices (transducers and cable-mounted transponders)
at a uniform depth so that the acoustic pulses will have an isothermal path over
which to travel. In addition, velocity profilers are available which measure the
actual acoustic propagation velocity of the medium in which they are submerged.
Correction factors may then be applied to acoustic transit times to assure accuracy.

2. Magnetic Compass Error Sources

If we turn our attention now to magnetic phenomena, we see several potential
sources of error, some fixed, some variable, each with its impact upon cable profiling. Chief among these is uncertainty in applying the appropriate magnetic
variation (declination) for an area of operations. Magnetic variation, or the
angle between a compass needle and true north, changes rapidly with geographic
position. In field seismic operations, the value for magnetic variation is usually
taken from the compass rose on a nautical chart. Figure 5 illustrates a single
cable profile contrasting acoustically and magnetically derived shapes at a single
point, the only variable being the magnetic variation applied to the magnetic profile solution. Note that a difference in magnetic variation of the magnitude of
0.1 degree can result in an apparent lateral displacement of about 7 meters at the
end of the 3000 meter cable. Note also that a variation error in indistinguishable
from any other fixed bias.

Another major source of error originates from the resolution of the compass'
digital readout. One widely used cable compass has a resolution of 0.35 degrees
due to its 10-bit readout. This translates to an average uncertainty of plus or
minus about 0.2 degree. Thus, with no other sources of error considered, compass
resolution alone can account for an uncertainty in cable position of about 15
meters at the end of a 3000 meter streamer.

A compass deviation bias also contributes to the error budget for magnetic
compasses. Figure 6 shows a compass deviation correction chart similar to those
seen near a ship's compass. It quantifies the directional errors experienced by a
compass at various points as it is swung throughout a circle. Cable mounted
compasses also experience some deviation error. The error curve is roughly sinu-
soidal in nature and decomposes into the function of Figure 6. The 'A" coefficient

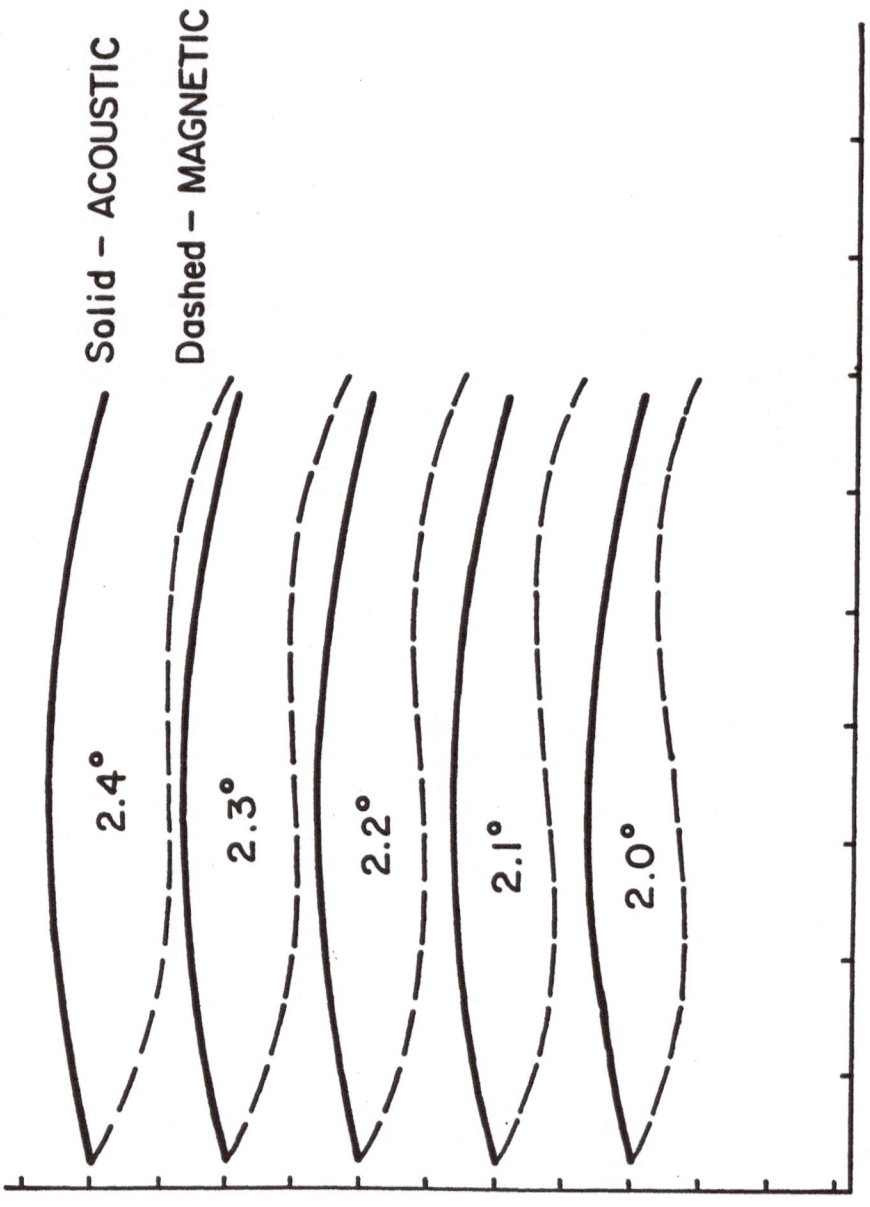

Fig. 5   PROFILE SENSITIVITY TO MAGNETIC VARIATION

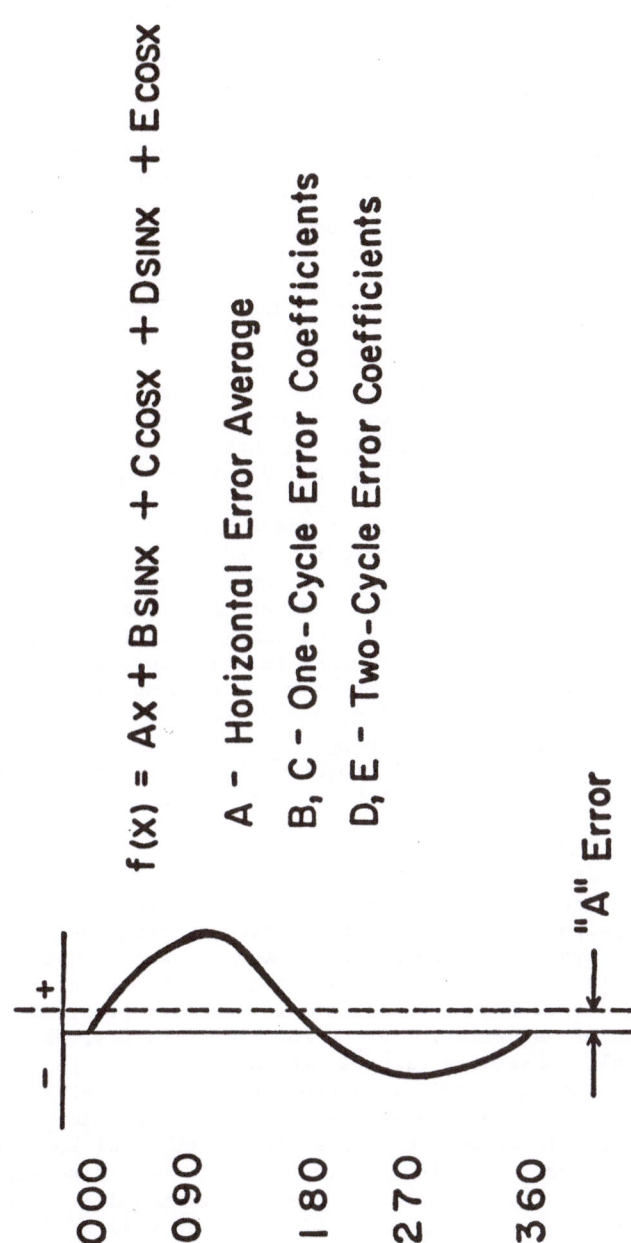

f(x) = Ax + Bsinx + Ccosx + Dsinx + Ecosx

A - Horizontal Error Average

B, C - One-Cycle Error Coefficients

D, E - Two-Cycle Error Coefficients

Fig. 6   COMPASS DEVIATION CORRECTION

represents the average horizontal error component and is often stamped on the case
of a magnetic compass, but seldom used. Its magnitude is typically less than one
degree, but may be as much as 2 degrees.

Several other minor sources of error are present in magnetic compasses, such as
mounting bias, or the misalignment of the compass within its case, and the case
within the cable section. There is also the question of whether the ferrous mass
of the seismic vessel itself may deflect a near compass to an appreciable degree
by which a far compass may not be affected. In addition, the stainless steel
stress member which runs through the center of a seismic streamer to absorb towing
tension may have impurities which are ferrous in nature, thus causing a deflection
in a compass mounted nearby.

As all error budget components, some of these error sources will add and some
will compensate, but even viewed in a root sum square sense, the combined sources
of error become quote significant when translated to uncertainties in cable profile
and hydrophone position, of cardinal importance in 3-D seismic surveys.

In an effort to control these errors, it is standard practice aboard seismic
vessels to go through a cable calibration procedure. Ordinarily this involves de-
ploying the compass sections behind the vessels with no intervening hydrophone
sections (to minimize cable length and justify straight cable assumption), then
steam back and forth in various patterns at various compass points to develop correction factors to apply to each individual compass in order to average out the
total error effect. These procedures have been known to take up to 2 or 3 days to
accomplish, and are thus costly, time consuming, and applicable only to a limited
area of operation. Acoustic streamer positioning, used alone or in concert with a
cable compass system, provides a substantial improvement in cable profiling accuracy and in the confidence with which hydrophone positions are reported.

PRESENTATION OF FIELD DATA

The Hydrorange system underwent extensive field trials in December of 1985 in
the Gulf of Mexico's Mississippi Canyon area. Navigation control was provided for
both vessels by the Syledis system, and comparative magnetic and acoustic data
were collected simultaneously. There was typically good agreement between the two
systems, except for a constant bias in the magnetically derived profiles caused by
the inappropriate selection of magnetic variation for the compasses, as illustrated
in Figure 7. In this and all subsequent Figures, the numerals next to the profiles
indicate the time of day of each data point. Where the acoustic system really
came into its own is in the detection of a strong, localized magnetic anomaly of
undertermined nature. A magnetic anomaly may arise from a physical body capable
of deflecting a magnetic compass. Drilling platforms, wellheads, pipelines, concentrated sub-surface ore deposits, and submarines are some examples. Magnetic
storms may affect the orientation of the geomagnetic field as well as its strength.
The anomaly in this case was detected in post mission analysis of the comparative
magnetic and acoustic data, but with the added feature of a real-time profiling
display could just as easily have been detected "on-line".

In the area of the magnetic anomaly, extreme differences between the magnetic
and acoustic profiles were observed. Figures 8 and 9 are an illustration of 9
consecutive cable profiles as determined by raw magnetic compass readings over a
15 minute period as the vessel entered the anomaly area. The apparent distortion

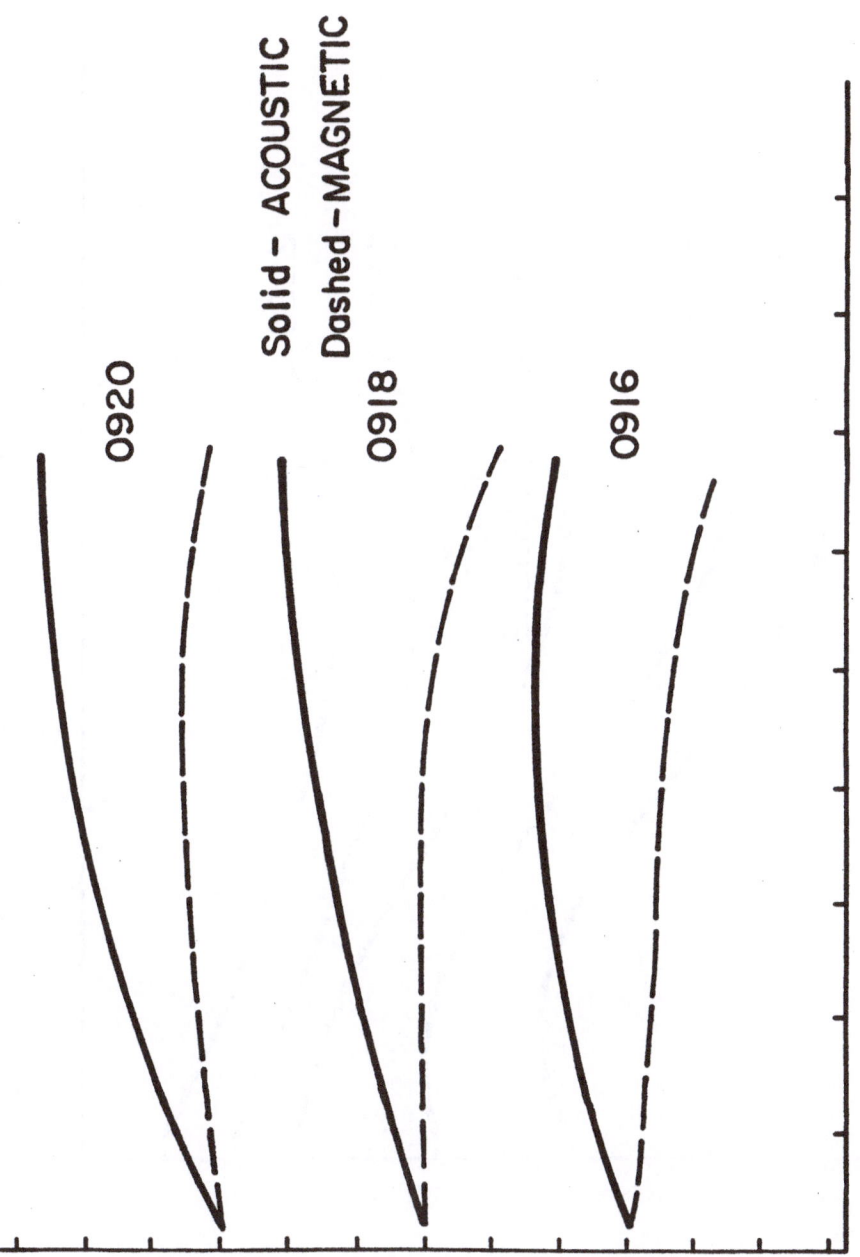

Fig. 7 TYPICAL PROFILES

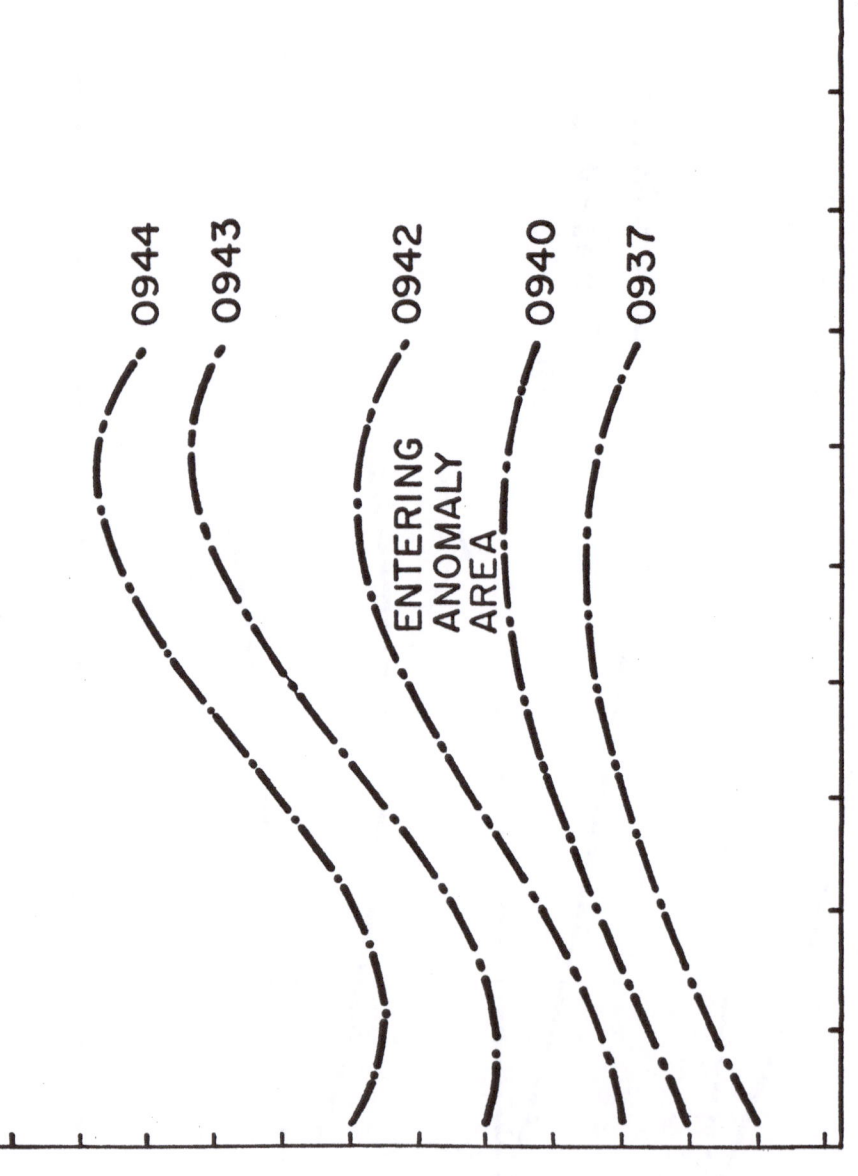

0944

0943

0942

ENTERING
ANOMALY
AREA

0940

0937

Fig. 8   MAGNETIC PROFILES IN ANOMALY AREA (1)

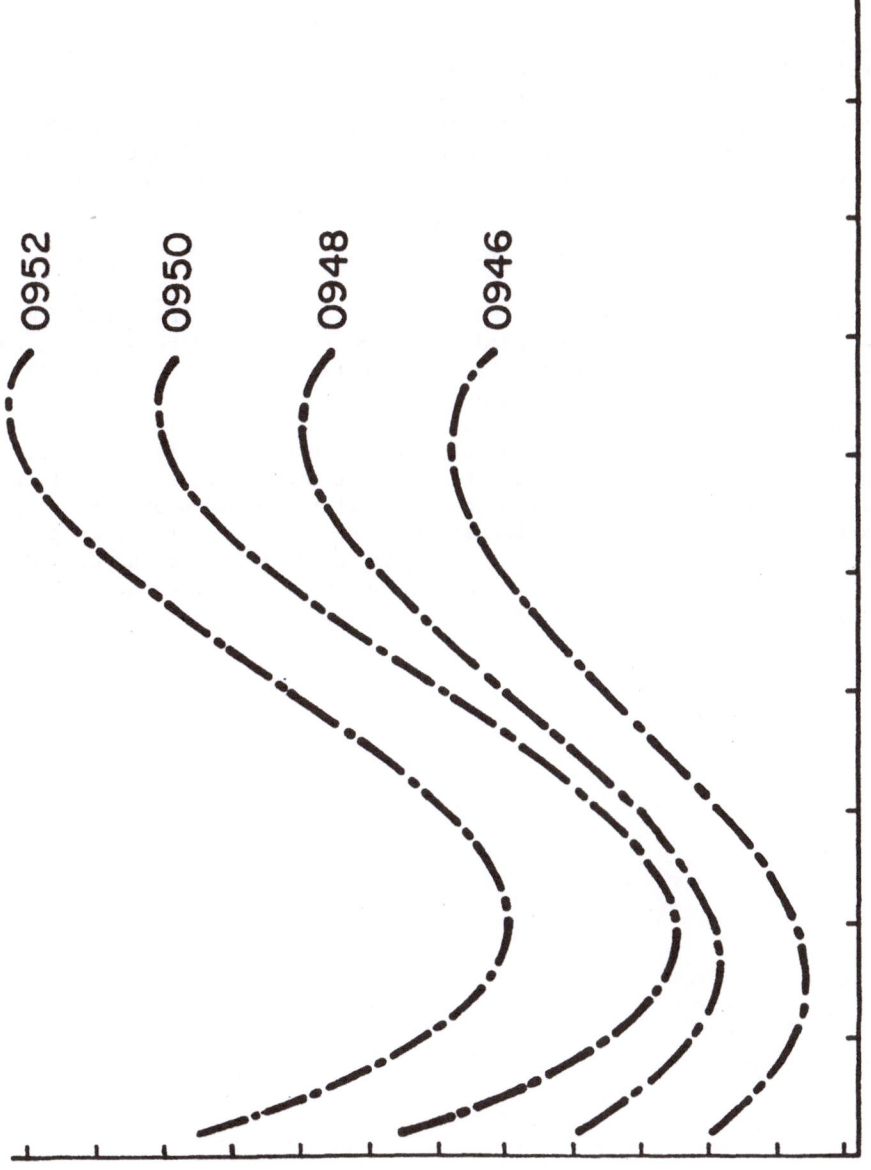

Fig. 9 MAGNETIC PROFILES IN ANOMALY AREA (2)

of the cable shape is quite severe, and a "hump" propagating down the streamer from
towpoint to tail is clearly visible. Contrast these profiles with Figures 10 and
11, which show data points from the same time period with the acoustic profiles
(solid lines) superimposed on the magnetic profiles (dashed lines). The difference
in magnetic and acoustic profiles is striking, and it is evident that considerable
"smearing" would result in data interpreted from trace stacking based solely on
hydrophone locations as determined by magnetic compasses. Given typical bin sizes
in dynamic binning, the use of compass derived hydrophone locations in this case
would lead to filling bins far removed from the ones actually in the streamer's
path. Obviously, real streamer synamics do not permit such drastic fluctuations
in cable shape as indicated by the magnetic data within one cable length. Figure
12 illustrates the excellent agreement between the magnetic and acoustic systems
as the vessel emerged from the influence of the magnetic anomaly.

CONCLUSIONS

    The conclusions to be drawn from these data are indisputable. That real time
acoustic range measurements to cable mounted transponders give a more accurate and
realistic indication of streamer profile than do magnetic compasses is generally
accepted in the geophysical industry. The additional expense involved in supporting a two-boat operation is ameliorated by the time saved in calibration
efforts and the resultant increase in resolution of the data collected, especially
when using dynamic binning techniques and quarter point steering. A combination of
magnetic and acoustic systems is currently the most effective means of determining
the actual position over the ground of streamer hydrophones. Acoustic transponders
may also be attached to the seismic sources (air gun rays) when long and/or wide
arrays are used to pin down the exact location of the arrays in reference to the
vessel and the streamer.

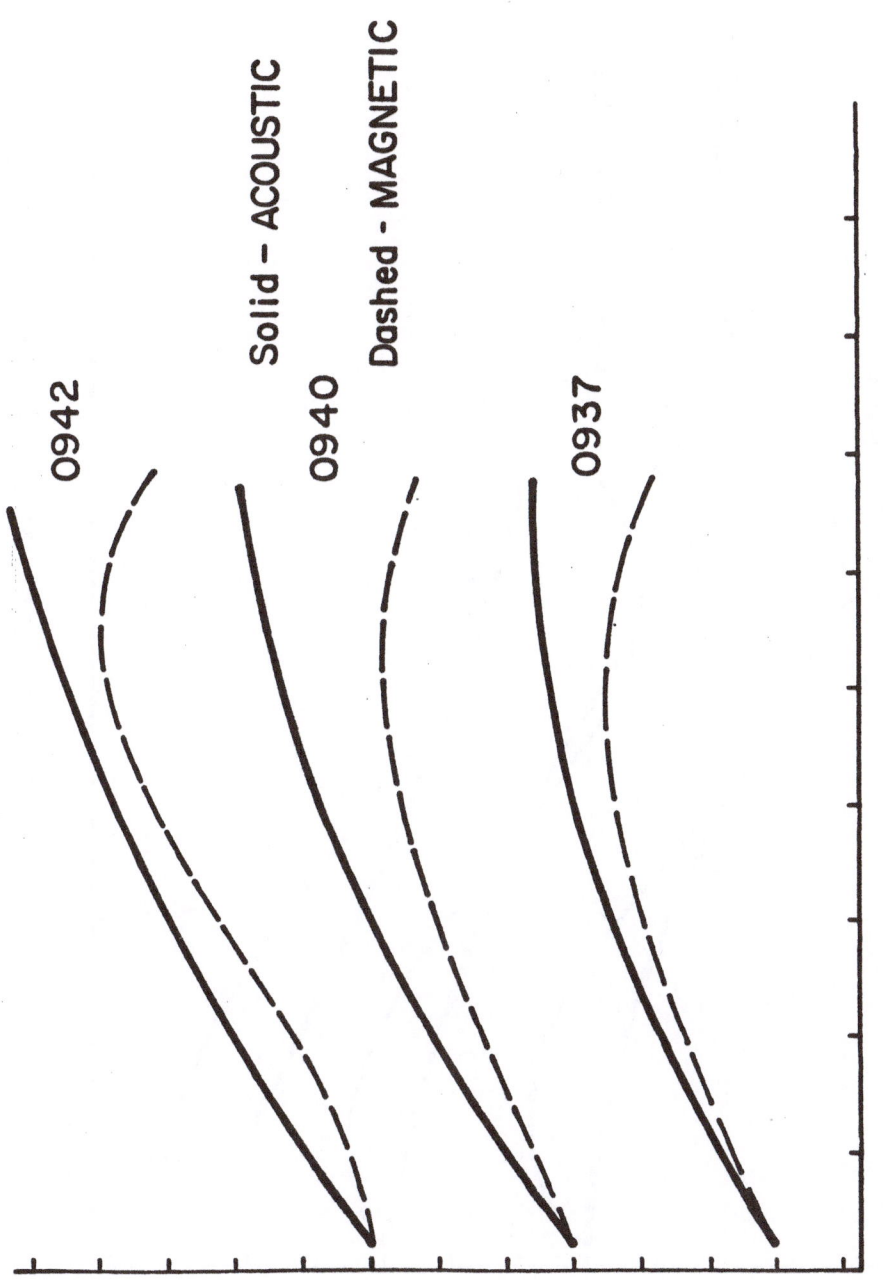

Fig. 10   ACOUSTIC PROFILES SUPERIMPOSED (1)

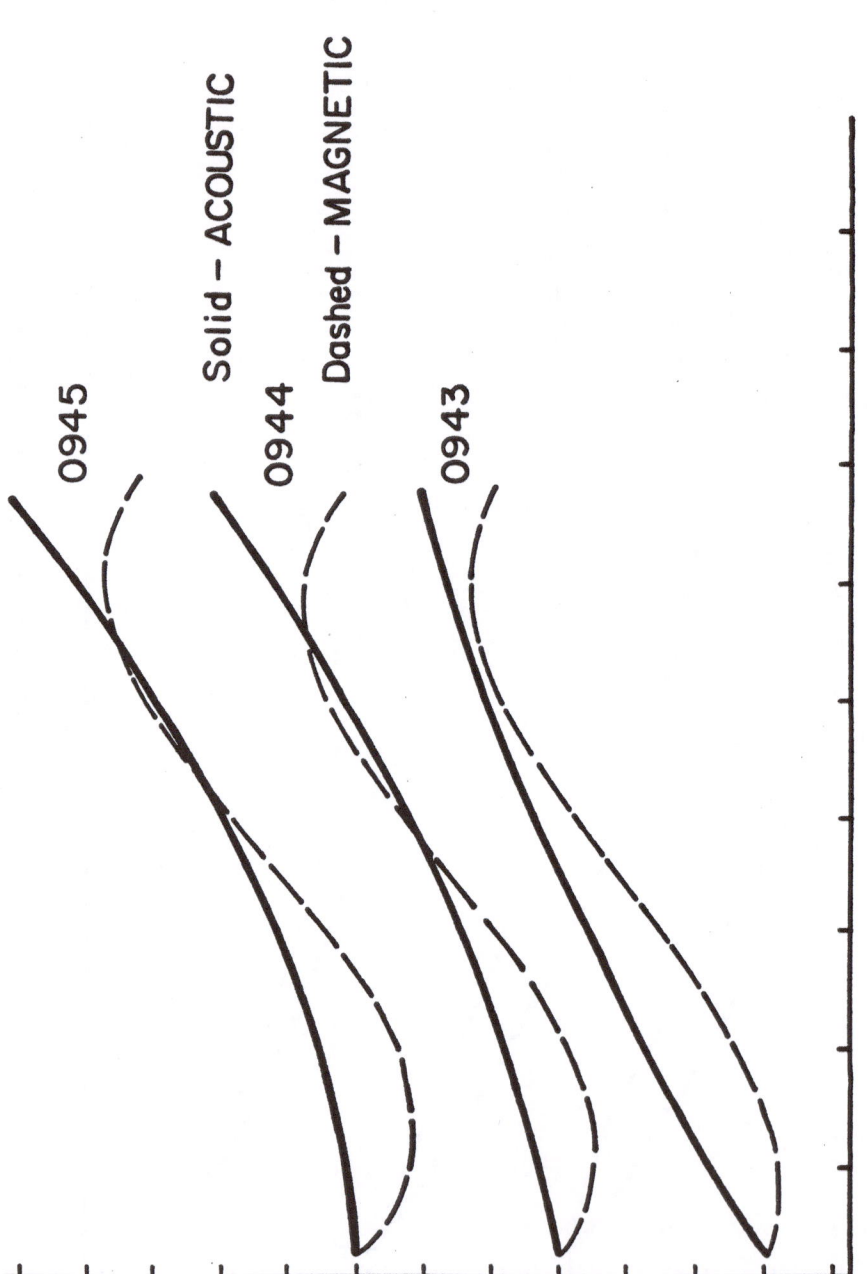

Fig. 11   ACOUSTIC PROFILES SUPERIMPOSED (2)

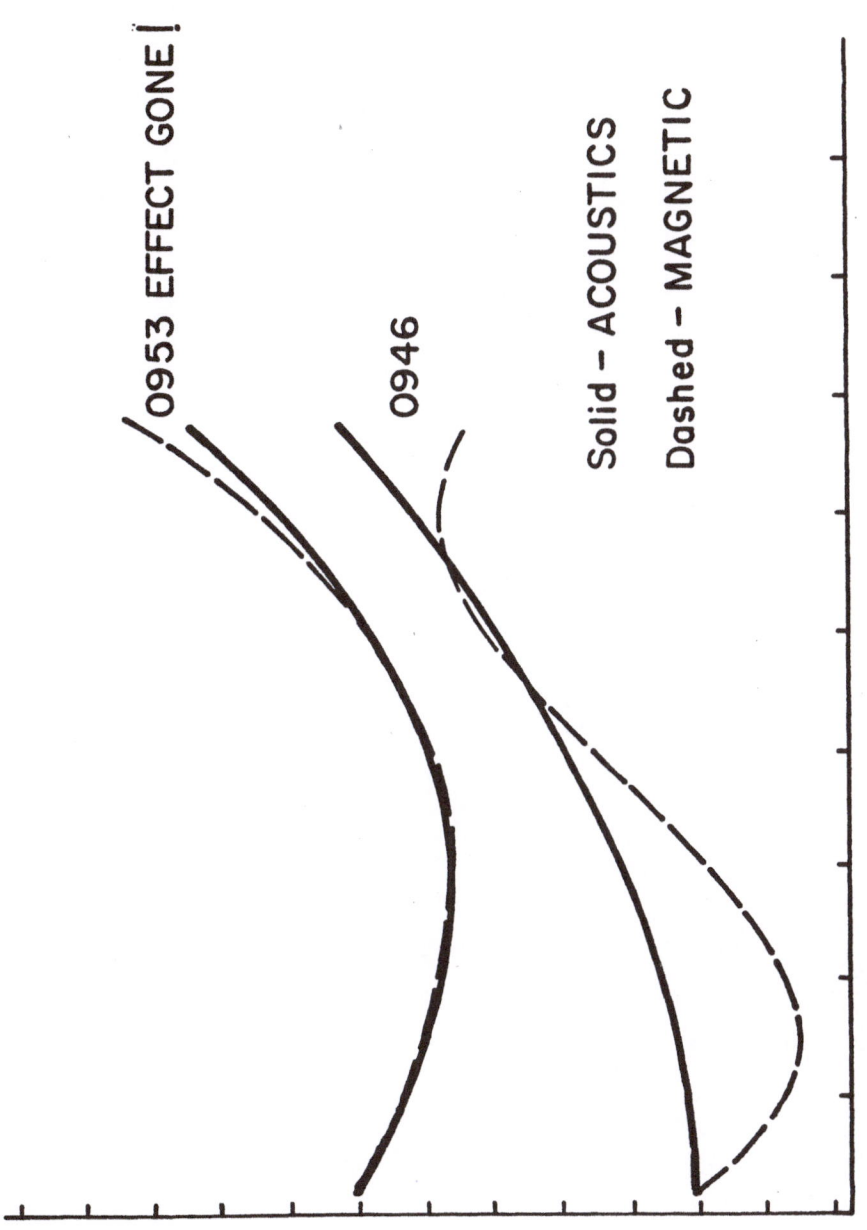

Fig. 12   STREAMER PROFILES LEAVING ANOMALY AREA

NEW TECHNOLOGY AND PROVEN TECHNIQUES COMBINE TO PROVIDE DYNAMIC PRECISE POSITIONING

Paul K. Dano
Del Norte Technology, Inc.
1100 Pamela Drive
Euless, Texas 76040

ABSTRACT

A new method of spread spectrum radiolocation offers real-time extraction and correction of system biases to eliminate fixed timing errors. An unlimited number of users may receive differential signals as well as system description data in such a manner as to facilitate complete "blind" entry into the system while attaining full operational capability. Utilizing a proprietary technique, the passive user obtains additional lines of position as well as calibration information while using the traditional number of shore reference stations. A single frequency can be used world-wide since "networks" are identified by code. Adjacent networks can be indicated to the receiver using the system description data thus facilitating network to network operation without operator intervention. Although the system accuracy is excellent for hydrographic survey, the automation of dynamic precise positioning is most advantageous in pilotage and navigation.

1. INTRODUCTION

The quest of the surveyor is to find the ideal measurement tool which will give the location of any point at his whim. We are all eagerly awaiting the implementation of GPS to provide this for us. However, from the volume of papers presented over the last several years, the question arises as to whether GPS is going to provide the answer for dynamic precision positioning. A rocking, moving boat is not the ideal vehicle for gathering sufficient repetitive data to take back to the office for reduction. What alternative or backup system does the dynamic surveyor have? If one was to propose an alternate system, what might it be? It appears logical that an alternate system should be capable of augmenting the GPS, not necessarily replacing it. Also, an alternate system should be capable of standing on its own, being implemented at will, being useable by anyone and be of reasonable cost.

The following discussion is a collection of techniques learned over the last forty years. Use of the latest technology in software and hardware is the glue which holds these concepts together to form a new system.

M. Kumar and G. A. Maul (directors), Marine Positioning, 49–58.

2. PROVEN TECHNIQUE

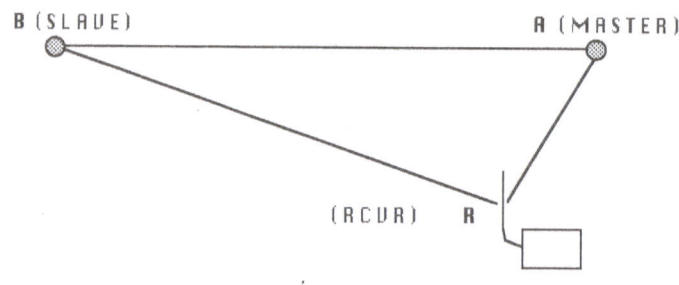

FIGURE 1

Figure 1 illustrates two shore stations and a mobile receiver of a passive system. Station A is the master, and station B is a slave. Master A emits a pulse which is radiated in all directions. The slave station B, upon reception of the master A pulse, repeats the pulse. The mobile receiver R would see two pulses arriving at different times. The time difference would depend upon the propagation time of the master station A pulse from the master position to the slave position (AB), the propagation time of the repeated pulse of slave station B from the slave position to the mobile receiver R position, the turn-around-time (TAD) of the slave station B and the time of propagation from the master station A position to the mobile receiver R position. The differential time relationship can be summarized by the following equation.

$$T_{AB} = [AB + BR + (TAD \text{ of } B)] - AR \quad (1)$$

This is the basic concept of most passive systems. Of course, there are many departures at this point to overcome the problems of this elementary concept such as superclocks and pseudoranging; however, let's continue with the basics. Since the remainder of this discussion continues to refer to baseline distances (such as A to B) and the propagation time intervals over these distances inter-changeably, please excuse the misnomer.

A standard method in which to check the calibration or measure the turn-around-delay of the slave station B is to have the mobile receiver R cross the baseline (AB) extension at the slave B end as shown in Figure 2.

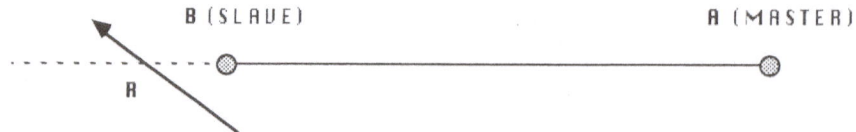

FIGURE 2

At the exact point of the baseline crossing R (special case), the basic hyperbolic equation (1) can be reduced as follows:

$$T_{AB} = [AB + BR + (TAD \ of \ B)] - (AB\text{-}BR) \quad (2)$$

$$T_{AB} = (TAD \ of \ B) \quad (3)$$

If it is not convenient to cross the baseline extension at the slave end, Figure 3 depicts the special case when crossing the master A end of the baseline.

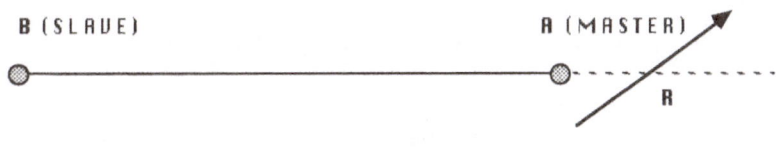

FIGURE 3

As before, the special case of basic hyperbolic equation (1) can be reduced as follows:

$$T_{AB} = [AB + (BA + AR) + (TAD \ of \ B)] - AR \quad (4)$$

$$T_{AB} = 2(AB) + (TAB \ of \ B) \quad (5)$$

3. NEW USE OF PROVEN TECHNIQUES

Most of the time, it is not convenient to cross any baseline and, in fact, it would be best if we could calibrate at any time and any place. If we return to Figure 1 and let B_{CAL} represent TAD of B, equation (1) can be written as follows:

$$T_{AB} = [AB + BR + B_{Cal}] - AR \quad (6)$$

If we leave the mobile receiver R in position as in Figure 1 and reverse the roles of the network stations as in Figure 4,

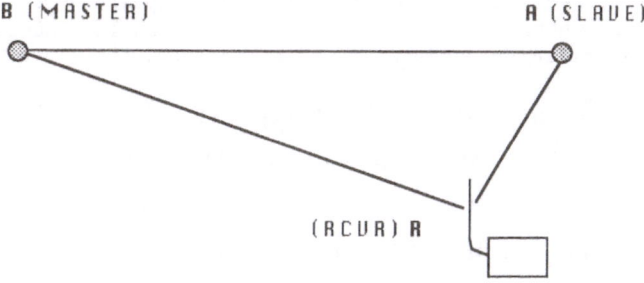

FIGURE 4

equation (1) can be written as follows:

$$T_{BA} = [BA + AR + A_{Cal}] - BR \quad (7)$$

By combining equations (6) and (7),

$$T_{AB} + T_{BA} = [AB + BR + B_{Cal}] - AR + [BA + AR + A_{Cal}] - BR \quad (8)$$

Which reduces to:

$$T_{AB} + T_{BA} = 2(AB) + A_{Cal} + B_{Cal} \quad (9)$$

By interchanging the roles of the network stations (A and B), we have in effect performed a similar action as if crossing the baseline (AB) extensions; however, now we have two unknowns, A_{CAL} and B_{CAL}. This reciprocity concept effectively equates the errors introduced by the system electronics to a constant eventhough the mobile receiver may be anywhere within the signal reception area.

4. THE NETWORK

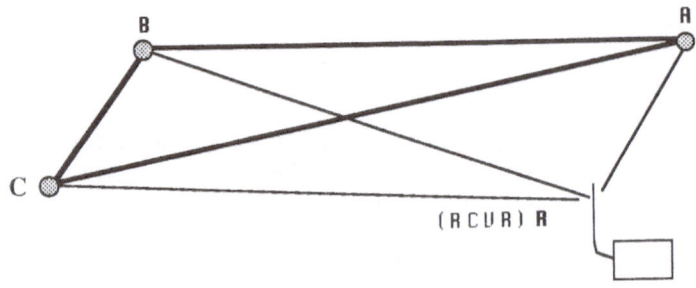

FIGURE 5

A three station network where each network station, in turn, becomes the master can be summarized with Figure 5. After a complete cycle in which each network station has assumed the master role, the mobile receiver will have accumulated six time difference measurements as follows:

$$T_{AB} = AB + BR + B_{Cal} - AR \quad (10)$$
$$T_{AC} = AC + CR + C_{Cal} - AR \quad (11)$$
$$T_{BA} = BA + AR + A_{Cal} - BR \quad (12)$$
$$T_{BC} = BC + CR + C_{Cal} - BR \quad (13)$$
$$T_{CA} = CA + AR + A_{Cal} - CR \quad (14)$$
$$T_{CB} = CB + BR + B_{Cal} - CR \quad (15)$$

Extending the reciprocity concept and combining equations:

$$T_{AB} + T_{BA} = 2\ (AB) + B_{Cal} + A_{Cal} \quad (16)$$

$$T_{BC} + T_{CB} = 2\ (BC) + C_{Cal} + B_{Cal} \quad (17)$$

$$T_{AC} + T_{CA} = 2\ (AC) + C_{Cal} + A_{Cal} \quad (18)$$

Since the station positions would be known, the baseline distances and, thus, the propagation intervals can be computed. The three equations are solved for the three unknowns (A_{CAL}, B_{CAL} and C_{CAL}).

The six mobile receiver measurements are composed of three sets of reciprocity pairs. Although the pairs must be considered redundant data and constitute only one line-of-position (LOP) per pair, it can be seen that three network stations provide three distinct LOPs.

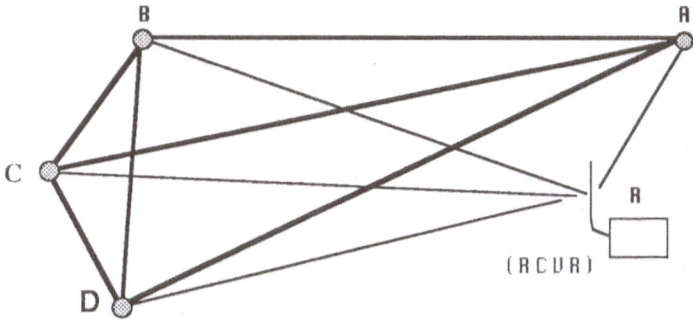

FIGURE 6

$$T_{AB} = AB + BR + B_{Cal} - AR \quad (19)$$
$$T_{AC} = AC + CR + C_{Cal} - AR \quad (20)$$
$$T_{AD} = AD + DR + D_{Cal} - AR \quad (21)$$
$$T_{BA} = BA + AR + A_{Cal} - BR \quad (22)$$
$$T_{BC} = BC + CR + C_{Cal} - BR \quad (23)$$
$$T_{BD} = BD + DR + D_{Cal} - BR \quad (24)$$

$$T_{CA} = CA + AR + A_{Cal} - CR \quad (25)$$
$$T_{CB} = CB + BR + B_{Cal} - CR \quad (26)$$
$$T_{CD} = CD + DR + D_{Cal} - CR \quad (27)$$
$$T_{DA} = DA + AR + A_{Cal} - DR \quad (28)$$
$$T_{DB} = DB + BR + B_{Cal} - DR \quad (29)$$
$$T_{DC} = DC + CR + C_{Cal} - DR \quad (30)$$

If the number of network stations is increased to four, as in Figure 6, the LOP advantage increases. The twelve readings obtained by the mobile receiver reduce to six distinct LOPs. Thus a four station passive system utilizing role switching not only has the advantage of self-calibration but will provide twice the number of distinct LOPs as a standard passive system of the same number of active stations and two additional LOPs over a four station active ranging system.

5. A NEED FOR NEW HARDWARE TECHNIQUES

The network stations must communicate with each other. The area would be implemented for reasonable geometric dilution of position (GDOP) for the work area. Passive hyperbolic systems present excellent GDOP conditions when the mobile receiver is within the network baselines; thus, placing the network stations around the work area is ideal. Such an arrangement also tends to ease the network station intercommunications.

The frequency of choice for such a local area system must be high enough not to be ground wave influenced and low enough not to be totally dependent upon line-of-sight (LOS). Due to world frequency allocations, this puts the frequency in the 400 megahertz band. At 400 MHz, two to three times the LOS communication distance is obtainable with reasonable equipments; thus, a separation of network stations of 50-90 kilometers is easily obtainable with moderate station heights.

Since the mobile receiver will be at different distances from the individual network stations, the signal levels will be different and, in some cases, very large differences will occur. The receiver must be capable of receiving these signals which are almost adjacent since, in this discussion, the slave is a repeat of the master signal. Standard automatic gain controls would not be effective in this case; therefore, a little technology is necessary. By using a limiting receiver with over 100 decibels of useful dynamic range, successive signals of large amplitude variation can be received without undue circuit propagation variation (this error must be kept to less than one meter equivalence).

The limiting receiver then necessitates that the pulse be of a frequency type modulation. By using a spread-spectrum technique, not only the frequency modulation is accomplished, but also a signal signature is transmitted which can be uniquely identified. By using Surface Acoustic Wave Devices (SAWDs), the generation as well as the reconstruction of the "Pulses" is accomplished.

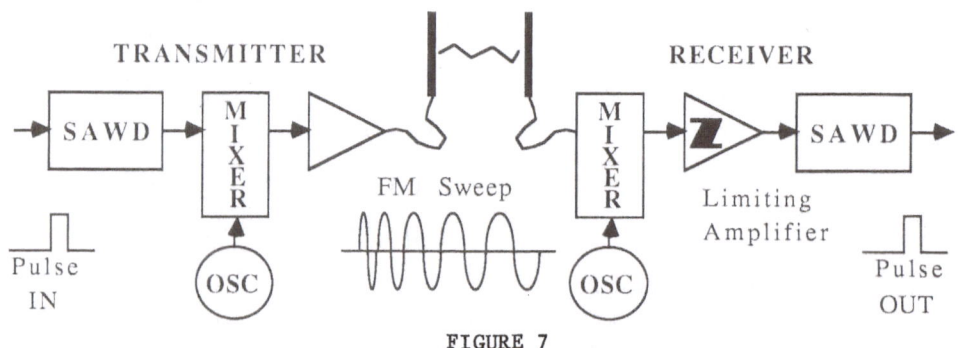

FIGURE 7

Figure 7 is a simplified block diagram of the signal process. A pulse is introduced into the SAWD which produces a frequency sweep output as a function of the geometric construction of the SAWD. The original input pulse of approximately 100 nanoseconds is expanded to sweep 10 megahertz in 20 microseconds, a gain of 200x in pulse power, without increasing peak power. The sweep signal is heterodyned up to the desired frequency band such as 427.5 megahertz center frequency of the sweep, then transmitted. The high gain limiting intermediate frequency (IF) amplifiers have a very wide dynamic range. The signal sweep is reintroduced into the SAWD which reconstructs the originating pulse.

The use of SAWDs to produce the spread spectrum technique was introduced to commercial ranging systems in early 1982. The technique has proven to produce very accurate operations under adverse signal conditions. Subsequent use of limiting receivers and digital correlation of the received pulses has shown a marked increase in maximum useable range while improving the signal-to-noise ratio.

6. WHOSE ON THIRD?

The mobile receiver must be able to recognize the difference between the network stations as well as adjacent network stations. Also, the network stations must be able to recognize which station to repeat. A very simple coding arrangement of pulse repetition intervals (PRIs) is assigned to each network with an individual code for each network station. Each network has an A code, B code and so on. When a network station determines it is to be master, it simply sends its code. This identifier alerts the other network stations and all mobile receivers that a sequence is about to start. The master, say A station, then transmits the next code in the network order; i.e., code B. The B network station would repeat.

Next the master would transmit code C and so on. Each second, a network station assumes the master role and identifies itself and goes through the remaining network codes. Thus, if there are three network stations, a complete cycle takes three seconds; four stations, four seconds; and so on. Adjacent networks, although on the same frequency, have different codes.

7. ALL THE GUSTO YOU CAN GET

Since the individual network stations communicate with each other for signal repeating and role timing, super accurate clocks are not needed either in the network stations or the mobile receivers. Note that the mobile receivers are measuring relatively short pulse-to-pulse intervals and are not pseudoranging; therefore, only inexpensive crystal clocks are needed, and one less unknown (time) is involved in the computations.

Each time a network station assumes the master role, it transmits an identifier which is used by the other network stations and all mobile receivers. Within the identifier is embedded data which can be the position of the particular network station. Of course, other information can be placed within the identifier. Each network station has the capability of having data loaded via a RS232 port. Data within the identifier can be a constant, such as the station position and adjacent network codes, or changed periodically, such as GPS differential information.

8. EVERY OUNCE COUNTS

Since the mobile receiver computes the baseline propagation interval using positions of the network stations and a constant for the velocity of propagation, it may be deemed more exacting to supply a correction factor for local variations.

The propagation correction factor can be obtained by the network stations. For example, referring to Figure 6, the receiver of network station A will see the signals of the other network stations. Station A, if considered as a mobile receiver, will see a three station network (B, C and D). Using the following two equations from the list of twelve possible measurements of a four station network, the correction factor can be computed.

$$T_{BC} = BC + CR + C_{Cal} - BR \quad (23)$$

$$T_{DC} = DC + CR + C_{Cal} - DR \quad (30)$$

Rewriting the above equations for A position instead of R and letting K be the propagation constant.

$$T_{BC} = C_{Cal} + K(BC + CA - BA) \quad (31)$$

$$T_{DC} = C_{Cal} + K(DC + CA - DA) \quad (32)$$

Subtracting equation 32 from equation 31 and solving for K.

$$K = (T_{BC} - T_{DC})/(BC + DA - BA - DC) \quad (33)$$

Since all the items on the right-hand side of the equation are either measured or known (derived from the network station positions), the local propagation constant can be solved. Other stations in the network can likewise solve for the constant. However, four of the six baselines are involved in the solution of equation 33. A correction factor could be included in the network station identification.

$$Correction \ Factor = K \ / \ K_{(Standard)} \quad (34)$$

9. HERE AND NOW

The all solid state network stations (see Figure 8) are composed of a microprocessor controlled transmitter and receiver housed in an all weather, waterproof case. These small 72 cm. long units are easily mounted. Their low 10 to 32 volt DC power requirements are easily met using wind generators, solar cells or power line adapters. Hand held terminal programming and control make the unit easily tested and setup as well as on-line controllable. The antennas, cables and connectors are readily available from most UHF suppliers.

The light weight mobile receiver consists of an antenna, low noise preamp antenna, low noise preamp, antenna cable and a receiver in a waterproof housing Again, the low power requirements of 10 to 32 volt DC are easily met aboard the smallest to the largest vehicles. The output of the receiver is simply utilized by a terminal display, computer or electronic map display via standard RS232 or RS422. This allow the receiver to be placed most anywhere out of the way, thus not taking up valuable space.

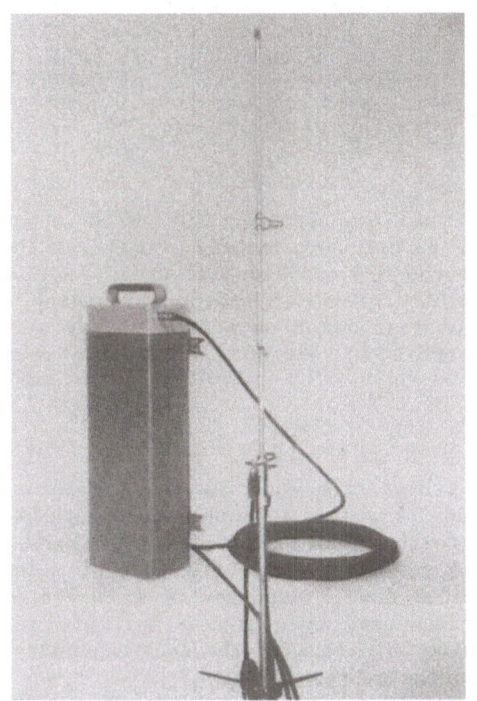
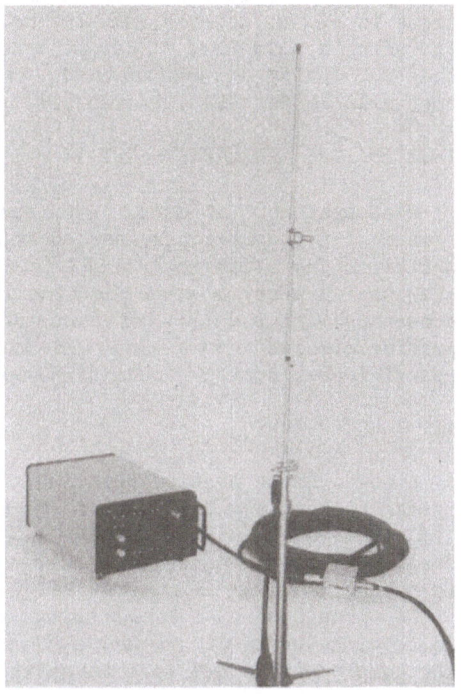

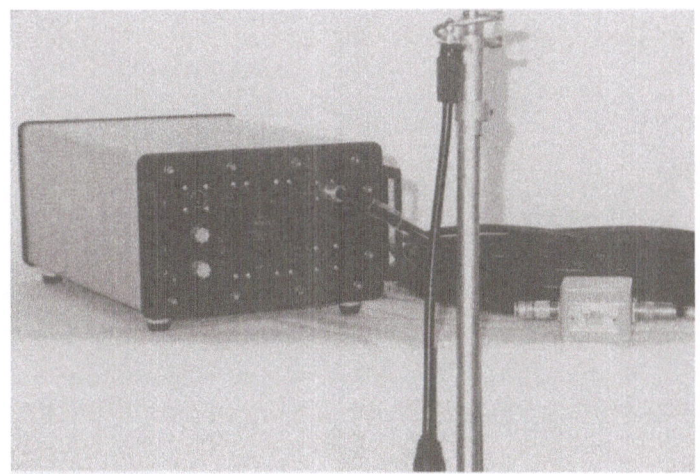

FIGURE 8

Although the receiver is programmable via its RS232 port, no operational information is lost in case of power loss, and automatic operation will continue upon power restoration. The receiver normally performs all positioning computations in WGS72; however, it can be programmed to use up to 9 different spheroids including a wild card in which spheroid parameters can be entered. UTM and Lat-Long outputting is available in controllable formats. Data outputting is also controllable for GPS differential operations.

10. IF IT'S AVAILABLE - GET IT

Most systems, to date, are stand alones and are incompatible with other systems. A complimentary system would improve an existing system. Providing the data link for differential GPS is a help, but the GPS positional data as well as the complimentary system positional data should augment each other providing a consensus with a figure of assurance. The combination of data from the dynamic precise passive system and GPS with differential correction should be a most powerful combination with a high assurance.

11. CONCLUSION

A new dynamic precise positioning system is available which uniquely combines proven techniques with newer technology and provides a complement path to GPS. The needs of the coming age of electronic mapping and electronic chart displays for more accurate sensing of position can be met with the concept. The fact that this low cost passive precision system can be implemented in selected areas throughout the world using only one frequency, even though multiple and adjacent networks may be necessary, is a definite advantage to the user community. The complement to GPS is a terrestrial positioning system (TPS).

LASERS AND NEARSHORE POSITIONING

Anthony Sprent
School of Surveying
University of Tasmania
GPO Box 252C
Hobart
Tasmania 7001

ABSTRACT

The paper addresses the problems of near shore positioning, in particular those associated with accuracy requirements and the availability of existing instrumentation. A description is given of a prototype instrument - Anglescan, developed by the author which uses a scanning laser beam system interfaced to a micro-computer to position a vessel in near shore conditions. Automatic angular measurements by the scanning laser system to fixed shore stations are used to compute both the XY coordinates of the vessel and its heading. Accuracies of better than 0.1m in X,Y coordinates with an update rate of one per second were obtained in tests. Possible applications are examined together with a discussion on recent developments of the Anglescan system.

1. INTRODUCTION

As the result of preliminary investigations into methods of automatic measurement of directions without the need for an observer to make actual pointings to targets, an instrument called the Laser Anglescan was developed. This instrument was designed to be mounted on board a vessel and provide a highly accurate determination of both position and heading for such activities as hydrographic surveying, dredging, pile driving, or salvage in enclosed waters.

While there are several different systems available for position fixing for off-shore and medium range hydrographic surveying and allied operations, there is nothing entirely satisfactory for near-shore close range applications where a greater degree of accuracy is often required and where automatic operation without the need of specialized personnel may be necessary.

A considerable variety of short range electromagnetic distance measuring instruments are currently available but are designed specifically for land surveying applications. Some of these have been used for hydrographic positioning operations, but only with mixed success. The main drawback is the necessity to maintain accurate the during the measurement period and invariably this must be done manually. This usually proves difficult to achieve in practice, even with relatively small movements of the vessel.

2. THE ANGLESCAN CONCEPT

The technique of position fixing by resection based on two or more sextant angles observed on board a survey vessel to shore based control is well known to surveyors. It has the advantage that the entire survey process is self contained, being carried out on board without the need for any communication to shore. Anglescan automates this process by measuring horizontal directions to a

M. Kumar and G. A. Maul (directors), Marine Positioning, 59–68.
© 1987 by the Marine Technology Society.

series of control points and computing the position of the vessel as the result
of a least-squares resection. An active line of sight is used to scan the
horizon and detect defined targets. Once intersected, the direction to each
target is measured with respect to the centreline of the vessel by means of a
high precision shaft encoder. From these data both the XY position and the
orientation of the vessel may be determined with respect to the shore control
system.

For many applications the ability to obtain orientation as well as position
is of considerable benefit as it means that the Anglescan instrument may be
situated anywhere on the vessel and be able to provide the position of any other
specified point. Thus in such operations as dredging or pile driving, the
centre of the cutting head or the pile can be monitored, irrespective of the
location of the instrument.

The coherent nature of laser radiation with its very narrow band width and
ability to maintain a closely collimated beam over a considerable distance
coupled with highly sensitive photo-detectors and interference filters have made
the combination an ideal active line of sight. By linking this with suitable
optics and a high resolution shaft encoder interfaced to a micro-computer a
relatively simple yet highly accurate dynamic positioning system is possible.

3.   THE PROTOTYPE INSTRUMENT

Figure 1 shows a schematic layout of the instrument. Figure 2 illustrates
the prototype. Briefly it consists of a 5 mW HeNe laser, mounted vertically
below a rotating mirror assembly which directs the beam to rotate clockwise in a
horizontal plane. The beam passes through a beam expander to ensure a high
degree of collimation. The direction of the beam as it rotates is given at any
time by means of a high precision angle encoder attached to the mirror assembly.
The encoder is capable of resolving angles to 1.5 seconds of arc, though for its
present application a resolution of only 30 seconds of arc is used. The output
consists of a series of pulses as the encoder is rotated, each pulse signifying
a 30 arcsecond interval. These are input to a 16 bit resetable binary counter.
The counter is set to zero at the beginning of each revolution by means of a
vane mounted on the rotating mirror assembly triggering an optical switch set in
the base of the instrument. The position of the switch is such that the counter
is set to zero at the instant the laser beam is pointing along the centreline of
the vessel. The clockwise angle of the beam at any instant with respect to the
centreline direction is thus given by the number of 30 arcsecond intervals
counted.

The targets used for the shore stations are retroreflectors, either of the
type normally used for electromagnetic distance measurement or, where short
ranges only are required, the cheaper plastic reflectors. During the time the
laser beam traverses across the reflector, it will be reflected back along its
outward path. A second larger mirror at right angles to the smaller
transmitting mirror reflects this return signal upwards to an objective lens
which focuses it onto a sensitive photo-detector. Since the return signal
strength from the reflected laser beam may be very small, it is essential to be
able to discriminate this against other light sources scanned by the system. To
achieve this, the field of view of the receiving optics is restricted to within
half a degree of the optical axis of the receiving system by means of a field

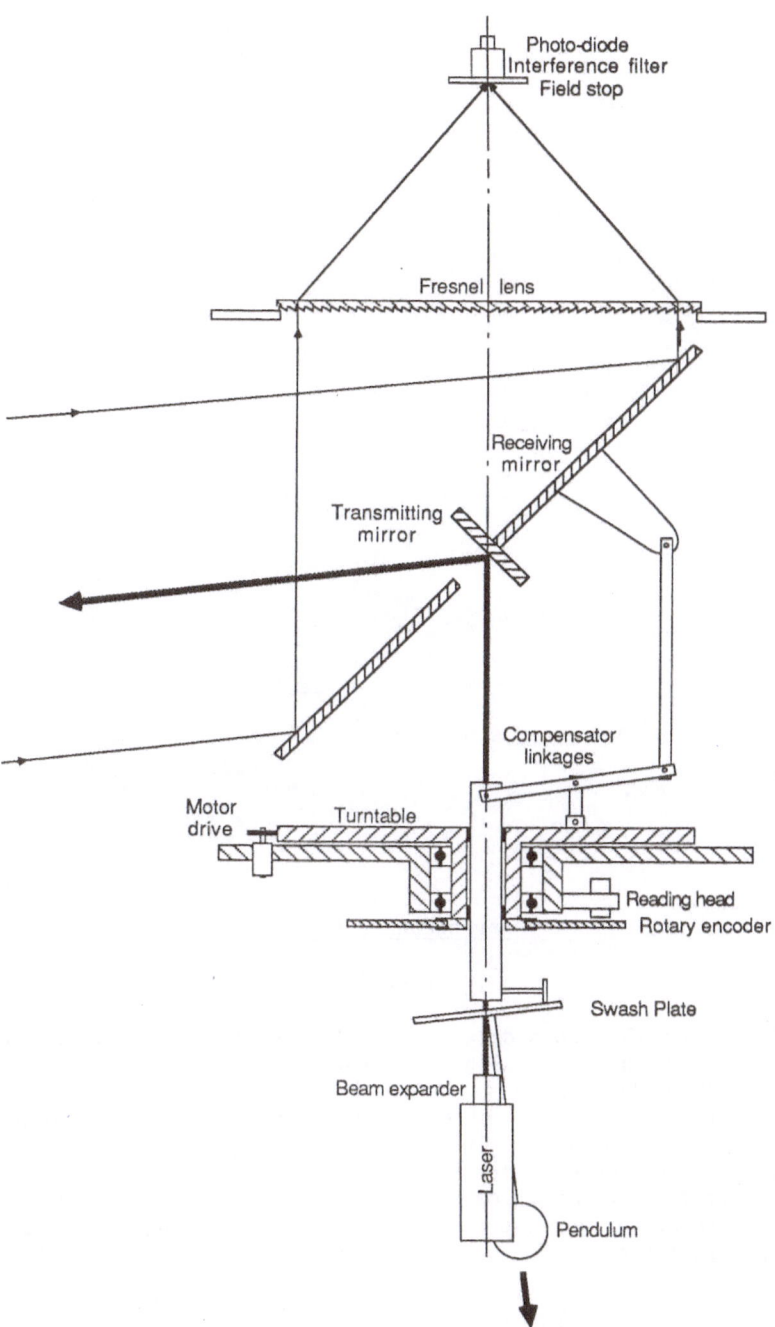

Anglescan MarkII Mechanical Layout

Figure 1

stop, and the photo detector is made sensitive only to the particular wavelength of the laser beam by introducing a narrow band interference filter to the optical system.

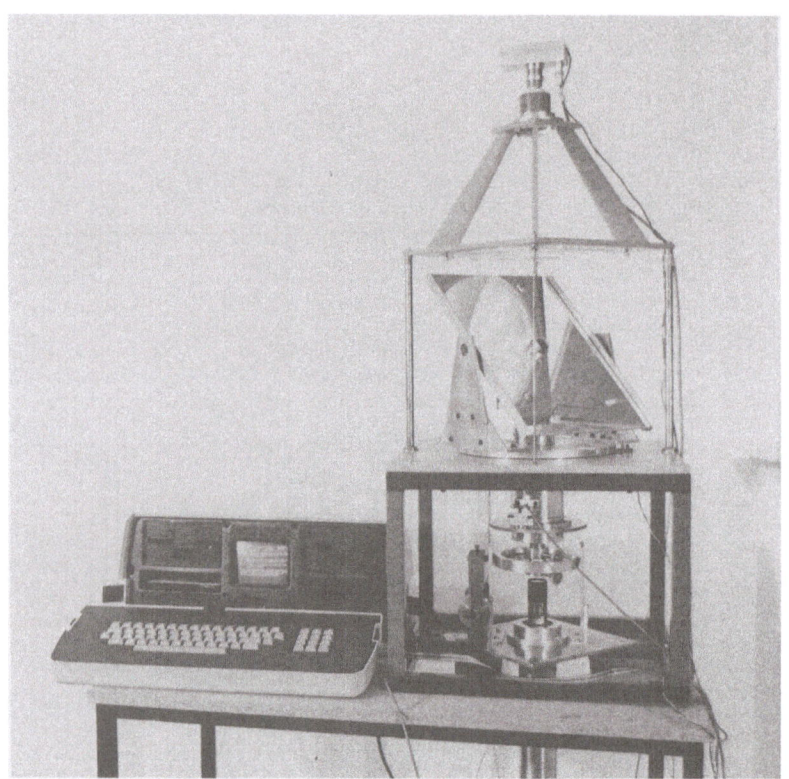

<u>Anglescan Prototype</u>

Figure 2

The output from the photo-detector as the laser beam intersects a reflector appears as a short duration pulse, the length of which depends on the time the beam is actually on the reflector. This is a function of the speed of rotation of the beam, the width of the reflector, the distance to the reflector, and the horizontal width of the beam. The pulse is amplified and then used to trigger the transfer of the current binary counter reading into a holding register for subsequent transfer to the microcomputer. Thus as the beam scans a 360 degree horizon, the direction is measured and stored in the memory of the computer each time a target is intersected. Figure 3 illustrates the data transfer process.

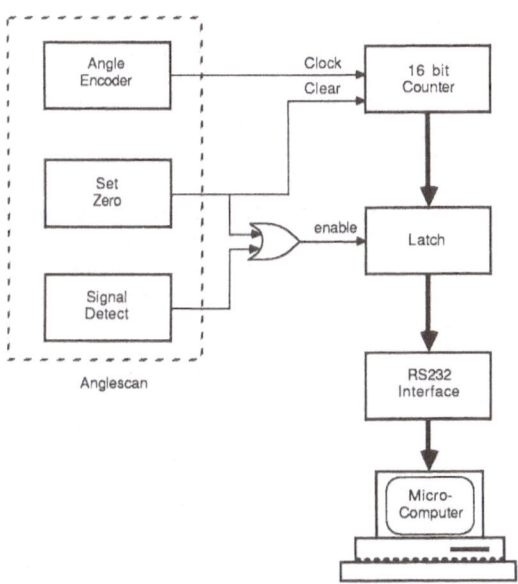

Schematic Diagram of Data Transfer

Schematic Diagram of Data Transfer

Figure 3

4. VERTICAL COMPENSATION

Since the instrument was designed to be mounted on board a moving vessel subject to wave motion, consideration had to be given to the problems involved in ensuring that the laser beam actually intersected the targets as it rotated. Even without the rocking motion of the vessel, it is difficult to place a number of reflectors all on exactly the same plane as that of the rotating laser beam. As a consequence, it becomes necessary either to expand the beam in a vertical direction in the form of a fan by means of a cylindrical lens using a single reflector for each target, or to provide a vertical array of reflectors at each target thus ensuring intersection of the unfanned beam.

Neither of these two approaches is entirely satisfactory in practice. In the first case, the range is severely limited by the considerable loss of return signal as the beam diverges. In the second case, if reasonable ranges in the order of one or two kilometres are desired, the cost of providing a vertical array of retroprisms for each station becomes prohibitive, while the use of plastic reflectors is limited to ranges of 400 metres or less. The best solution to this problem has proved to be a combination of all three. That is, by using a vertical array consisting of a few widely separated high quality retro-prisms with the spaces in between being filled with plastic reflectors.

The laser beam is also expanded slightly in the vertical direction such that its width at 400 metres range is the same as the spacing of the retro-prisms. Thus for the shorter ranges the plastic reflectors will provide sufficient return signal, and at longer ranges the retro-prisms do so. Excluding losses due to atmospheric effects, the return signal strength of this configuration of reflectors is, for all practical purposes, independent of range until the width of the fanned beam exceeds the length of the array.

While this approach eliminates the need to align all of the targets in exactly the same horizontal plane, it does not overcome the effects of the rocking motion of the vessel under wave action. For this, an automatic levelling system is required which compensates for the various tilts of the vessel and enables the laser beam to rotate in a horizontal plane within the vertical extent of the targets. The simplest method of achieving this is to mount the instrument in gimbals as is done with a ship-borne compass. However the main shortcoming of this technique is the difficulty of damping the system. With a compass, it may be achieved by having the whole unit filled with liquid. This is not possible with the Anglescan and, if gimbals are to be used, some other form of damping is necessary. Since the weight and size of the prototype instrument was quite large (about 15 kg) the forces required to damp the motions are excessive and difficult to achieve in practice. As a consequence, instead of trying to keep the whole instrument level, a compensator was developed which merely maintained the beam in a horizontal plane as it rotated.

To achieve this the two mirrors forming the mirror assembly were mounted in bearings to enable them to be tilted up and down. The amount of tilt was controlled by linkages connected to a swashplate mechanism activated by a pendulum. Figure 2 shows the arrangement. The linkage ratios are such that the tilt of the mirror assembly from 45° position is exactly half the component of tilt of the vessel in the direction of the laser beam. The laser beam after reflection thus maintains a horizontal direction as it rotates. Tests of the system indicated that for tilts of up to 10 degrees compensation was better than 0.1 degrees, which was more than sufficient to ensure intersection of the targets with the beam.

It was found that at certain mirror rotation speeds the combined effects of the weight of the mirror assembly and the length of the pendulum would cause the whole system to become unstable and rock wildly in sympathetic motion. By judicious choice of the length of pendulum and its weight, this latter motion of the pendulum was almost completely removed at the desired rotation speed of approximately one revolution per second. No additional damping of the pendulum was found to be necessary as the friction of the linkages was sufficient for this purpose.

A new system is currently being developed that measures the instantaneous inclination of the vessel using two orthogonal electronic inclinometers. The signals from these will be used to control motor driven levelling screws to keep the instrument level continuously. This is shown in Figure 4. Preliminary measurements indicate that levelling within a few minutes of arc will be possible for enclosed waters.

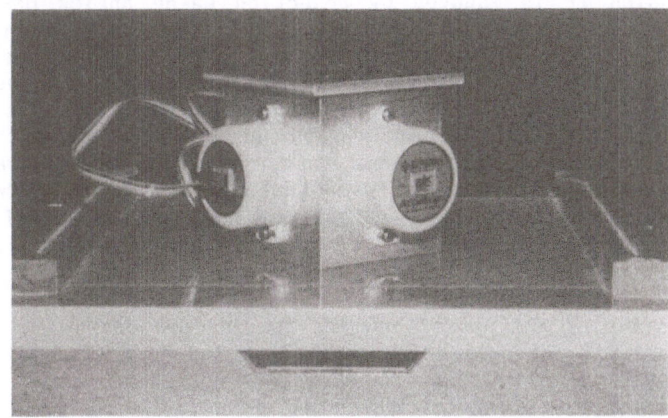

<u>**Electronic Inclinometers for Automatic Vertical Compensation**</u>

Figure 4

5.   COMPUTATION OF POSITION

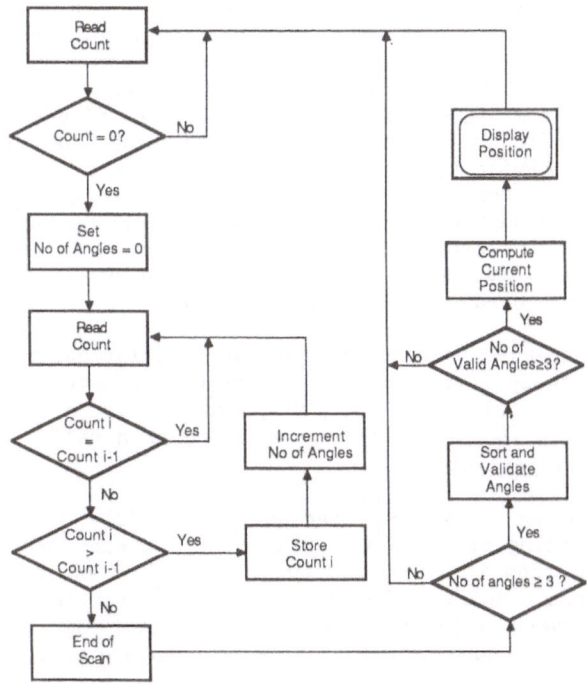

Flowchart: Operation of Anglescan

Figure 5

As described earlier, Anglescan was designed to determine the position of the vessel upon which it is mounted by resection using angles measured to known shore stations. At the completion of a scan of the beam, the information which is passed over to the microcomputer consists solely of a series of counter readings corresponding to angles from the dead ahead position as the receiving system detected light signals of the same wavelengths as the laser beam. There is no guarantee that each of these readings is a result of the beam being reflected from a genuine target, nor that every target has been intersected. Further, there is no way of identifying which target is which from the counter readings alone. A sorting process is thus required to determine firstly which of the readings are genuine returns from fixed shore stations, and then to ascertain which stations have been intersected and in what order (Figure 5).

If the position of the vessel and its heading are known, it is a relatively simple computation to calculate the angle from the dead ahead direction to each shore station and then sort these in order of increasing value. These may then be compared with the observed angles to identify each target sighted, and eliminate any spurious readings which may have been included. At first glance this appears a case of putting the cart before the horse. However, since the time for a complete scan is one second and the vessel is neither able to move very far nor change direction to any extent in this time, the position and heading known from the previous scan may be used to filter the data in the current scan.

The method of filtering the observations involves using the orientation of the previous determination as the starting point of a search routine in which the observed directions are compared to those computed to each control station. A fit between the two is considered to be found if at least three of the observed directions agree with the calculated directions within specified limits.   If no fit is found, the orientation of the observed directions is shifted first clockwise by twice the limit value, then four times anticlockwise. This is repeated backwards and forwards increasing the search angle, each time comparing the observed with the calculated directions until a fit is obtained, or until a maximum search angle has been reached. At this stage, rather than continue to search for a fit, a new set of observations from the scanner is input to the microcomputer and the process repeated.

Some allowance must be made for the small variations in direction due to the shift in position by giving angular limits within which each observed direction may be expected to lie. It is also possible that a control station in very close proximity to the vessel may be excluded as a valid observation due to the large change in direction.

A more rigorous filtering process is possible by first estimating the position and heading at the time of the current scan based on the path of the vessel computed from the previous two or three determinations of position and heading, and then comparing the directions. In practice it was found that this was an unnecessary refinement with the relatively slow speeds involved. Situations may arise however where the more rigorous approach may be necessary.

Having once established which stations have been observed and what are the corresponding directions to them, it is possible to compute the current position of the vessel and its heading with respect to the shore control system. The technique used is a least-squares resection which requires that approximate initial coordinates of the point are known. For each direction observed, an observation equation is derived in terms of corrections to be applied to the approximate coordinates and orientation. These equations are solved using the method of Variation of Parameters. This method of computation is particularly suited to the mode of operation of the instrument, as the position and orientation determined by the previous scan are used as the approximate values in the current determination.

Two sources of error in the determination of position are caused by the motion of the vessel during the actual scan. The first is due to the forward motion of the vessel. This will tend to make all angles to stations to starboard of the vessel larger than they should be, while those to port will be too small. The overall effect of this is to cause the computed position to be shifted to the right of the correct position. The amount of shift is dependent on the speed of the vessel, and the distance to the control stations. The second source of error occurs if the vessel is turning during the scan. If it is turning clockwise, all angles observed will be smaller than they are in reality, while for an anticlockwise turn they will be larger. The error due to this effect is dependent on the amount of rotation and the distribution of control stations around the vessel. Both of these errors can be eliminated in the programme by predicting the motion during the scan calculated from the positions and orientations of the previous two scans. Each angle may then be adjusted by the required amount once the station to which it was observed has been identified.

6. FIELD TESTS

Results of preliminary field tests carried out in King George V Dock by the Port of London Authority with the instrument mounted on board a salvage vessel gave promising results. Six control stations were located around the dock. The maximum range to any station was 300 metres and targets consisted of vertical arrays of plastic reflectors. The laser beam was not fanned in a vertical plane. To test the system, the vessel was held in position in the middle of the test area by means of four anchors, and was shifted from place to place by varying the lengths of the cables.

To ascertain the accuracy of fixation, a series of bearings and distances were observed to the centre of the scanner using EDM and a theodolite. Unfortunately the distances were measured to a reflector hand held above the scanner and as a consequence the coordinates derived do not give a true representation of the accuracy of the system. From the results it would appear that position was determined to an accuracy of better than 0.1 metres.

To obtain further data and ascertain more precisely the limits of the Anglescan prototype, additional field trials were carried out in Hobart on land at the University of Tasmania. The instrument was mounted on a trolley and moved around a flat area surrounded by four control points. The position as determined by the instrument was compared with that determined by EDM. Figure 6 shows the results of the trials. From the results it can be seen that the prototype determined position with a standard error of the order of 0.05 metres which is a better indication of precision than that of the London Docks test.

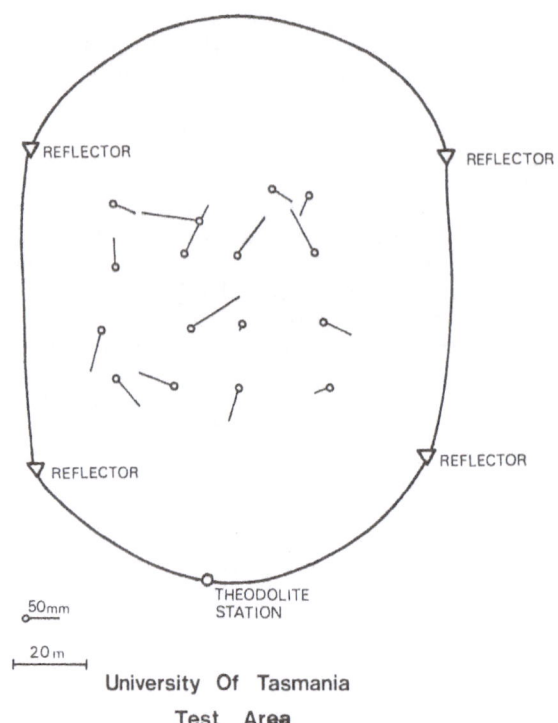

REFLECTOR

REFLECTOR

REFLECTOR

REFLECTOR

THEODOLITE
STATION

50mm

20 m

University Of Tasmania

Test Area

Figure 6

As well as computing the coordinates of the vessel, a graphical representation of its position and heading were also generated on the screen of the microcomputer. The total time for computation and display of position was less than one second. This meant that a new position could be determined for every second rotation of the scanning laser. The first rotation was used to measure the angles, the second to compute position. With some modification of the hardware it should be possible to halve this time and give a new reading every rotation, with the computer processing the results of the previous scan while measuring the angles of the current scan.

7. CONCLUSIONS

The original graphics output was severely limited by the capability of the microcomputer used. However with the availability of fast, graphics oriented micro-computers, it is now possible to generate complex graphical output. It is envisaged that with such a system, a dynamic digital terrain model could be developed enabling the continually updated position of the vessel to be represented in correct relationship with the surrounding features. This could lead to the application of Anglescan to such operations as the navigation of vessels through narrow channels, or the docking of large ships, as well as other hydrographic surveying operations where a high accuracy in position is required. It is hoped that further development of the instrument can take place. The author is actively seeking cooperation with organizations who may be interested in its application to their operations.

POLARIMETRIC RADAR FOR ACCURATE NAVIGATION

Dr. Simon Haykin
Communications Research Laboratory
McMaster University
Hamilton, Ontario, Canada L8S 4K1

ABSTRACT

A new radar system invented by the author and his colleagues provides a novel solution to the problem of automatic navigation along a confined waterway. The system is called "Polarimetric Radar for Accurate Navigtion" (PRAN) as it exploits polarization. The heart of the system is a retro-reflector that changes the polarization of the incident electromagnetic signal through 90°.

The PRAN system uses a dual-polarized antenna (located on board the ship) and a series of reflectors located at strategic points along the shore of the waterway. It operates by transmitting a horizontally polarized signal; each reflector rotates the polarization of the incident signal by 90°; and the resulting echo is picked up by the ship's radar on its vertically polarized antenna. The use of polarization effectively eliminates the clutter produced by natural and man-made objects located along the shores.

An experimental PRAN system (using off-the-shelf components) has been set up at a site located on Hamilton Harbour. Results of experiments conducted there confirm that the system does locate the reflectors, despite the presence of clutter produced by the surrounding environment.

1. INTRODUCTION

How can automatic navigation be provided for ships using a seaway such as the St. Lawrence Seaway so that this valuable resource becomes usable under all weather conditions and for a longer shipping season?

How can a ship navigate in the Arctic, always maintaining a prescribed distance from the shore?

How can the detection of a fishing boat in the presence of sea clutter be enhanced?

A new radar system invented by the author and his colleagues at the Communications Research Laboratory (CRL), McMaster University, provides a novel solution to these navigation problems. The system is called "Polarimetric Radar for Accurate Navigation" (PRAN) as it exploits polarization.

The heart of the PRAN system is a retro-reflector that changes the polarization of the incident electromagnetic signal through 90°. The reflector is patented by Macikunas, Haykin, and Greenlay [1]. The retro-reflector consists of a trihedral reflector with a wire-grid twister positioned on one of the inside faces of the reflector.

M. Kumar and G. A. Maul (directors), Marine Positioning, 69–76.
© 1987 by the Marine Technology Society.

2. THE PRAN SYSTEM

The PRAN system uses a dual-polarized antenna (located on board the ship) and a series of reflectors located at strategic points along the shore of the seaway. It operates by transmitting a horizontally polarized signal; each reflector rotates the polarization of the incident signal by 90°; and the resulting echo is picked up by the ship's radar on its vertically polarized antenna. The use of polarization effectively eliminates the clutter produced by natural and man-made objects located along the shores of the seaway.

An experimental PRAN system has been set up at a site located on Hamilton Harbour. Figure 1 shows a map of the site. The system consists of the following components:

1) Marine X-band radar (DECCA)

2) Mechanically scanned, dual-polarized antenna (Andrew Antenna)

3) Rotary joint (Kelvin)

4) Two reflectors (made by the Engineering Machine Shop, McMaster University)

Figure 2 shows a block diagram of the system. The radar is located on the roof of the Canada Centre for Inland Waters (CCIW), in Burlington. One reflector is located at DOFASCO (at a distance of about 2.3 km) and the other at La Salle Parke (at a distance of about 3.1 km). These two sites are typical of the environment along the St. Lawrence Seaway.

3. RESULTS OF EXPERIMENTS

Figure 3 shows horizontally and vertically polarized sweeps of the reflector at DOFASCO (an industrial site with numerous made-made reflectors). Figure 4 shows the corresponding sweeps for the reflector at La Salle Park (a typical park-setting).

Figure 5 shows the scan-converted image for the horizontally-polarized channel (lower photograph) and the vertically-polarized channel (upper photograph). The radar range is 6.20 km.

The results of the experiments conducted at the site confirm that the PRAN system does locate the two reflectors, despite the presence of clutter produced by the surrounding environment. Indeed, the system locates the reflectors, much like picking out needles in a haystack.

Conclusions

The PRAN system offers an elegant, cost-effective solution to several difficult navigation problems. The idea behind the PRAN system is to exploit a ship's radar for accurate navigation along a confined waterway. The conversion of such a radar requires the

Figure 1

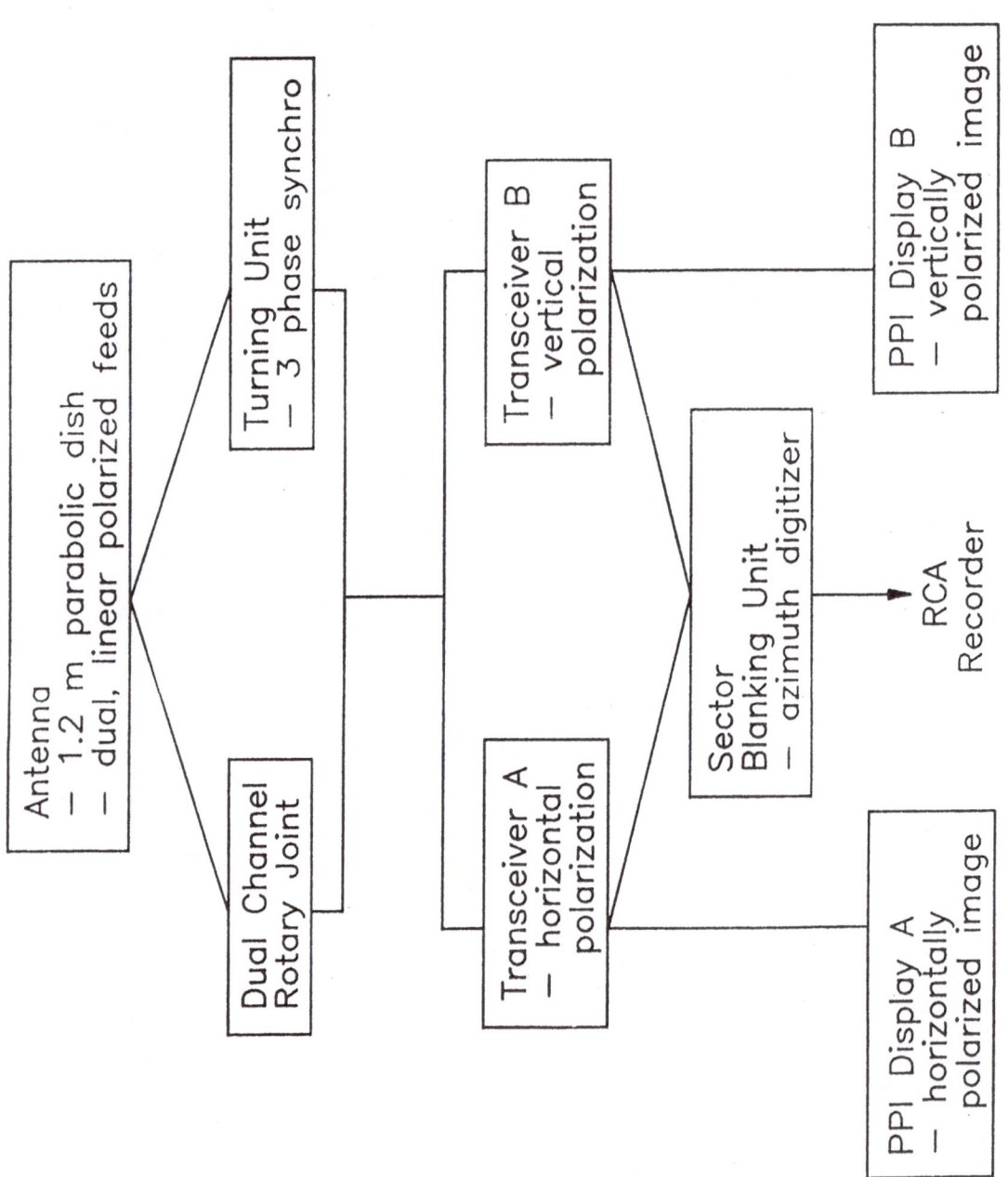

Figure 2   Block Diagram of the
Radar System

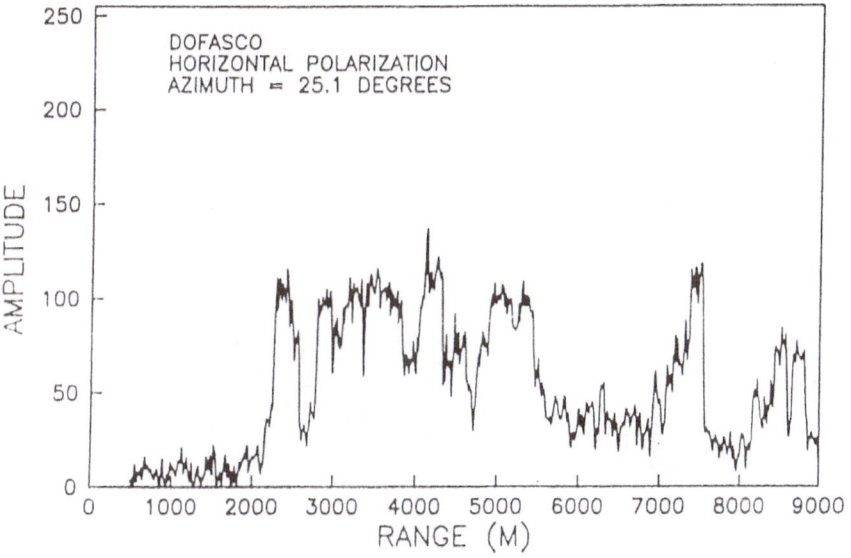

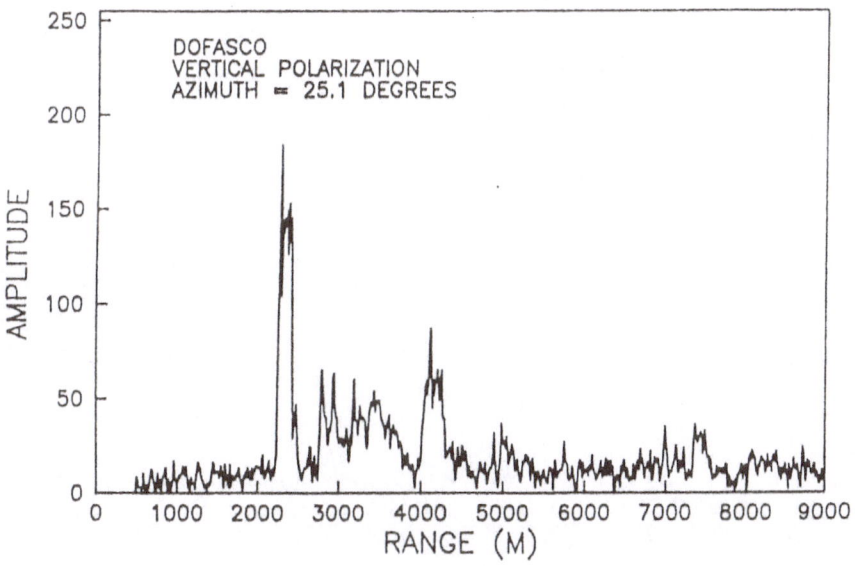

Figure 3   Horizontally and Vertically Polarized
              Sweeps of the Dofasco Reflector

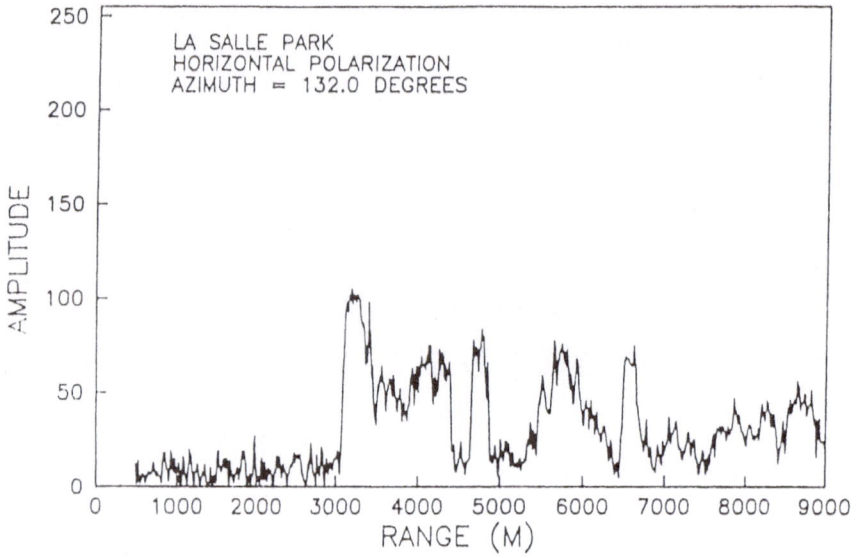

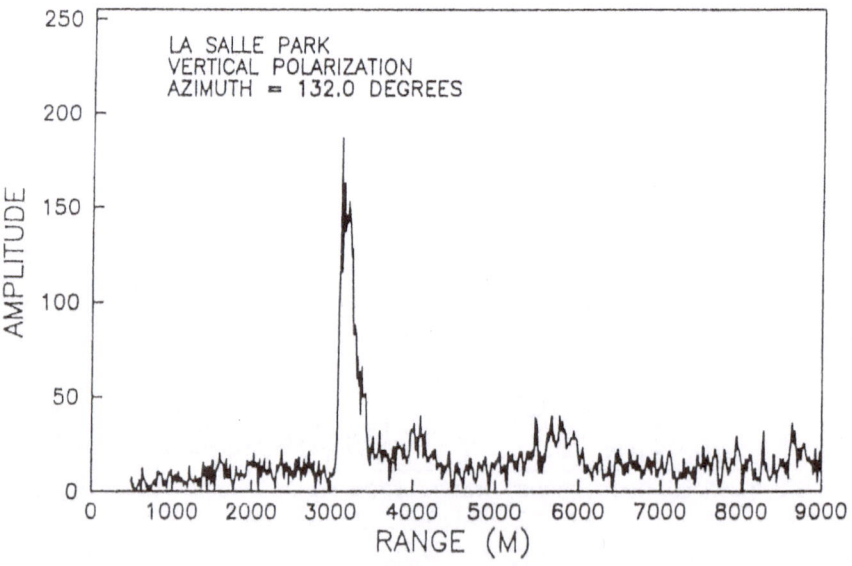

Figure 4   Horizontally and Vertically Polarized
            Sweeps of the La Salle Park Reflector

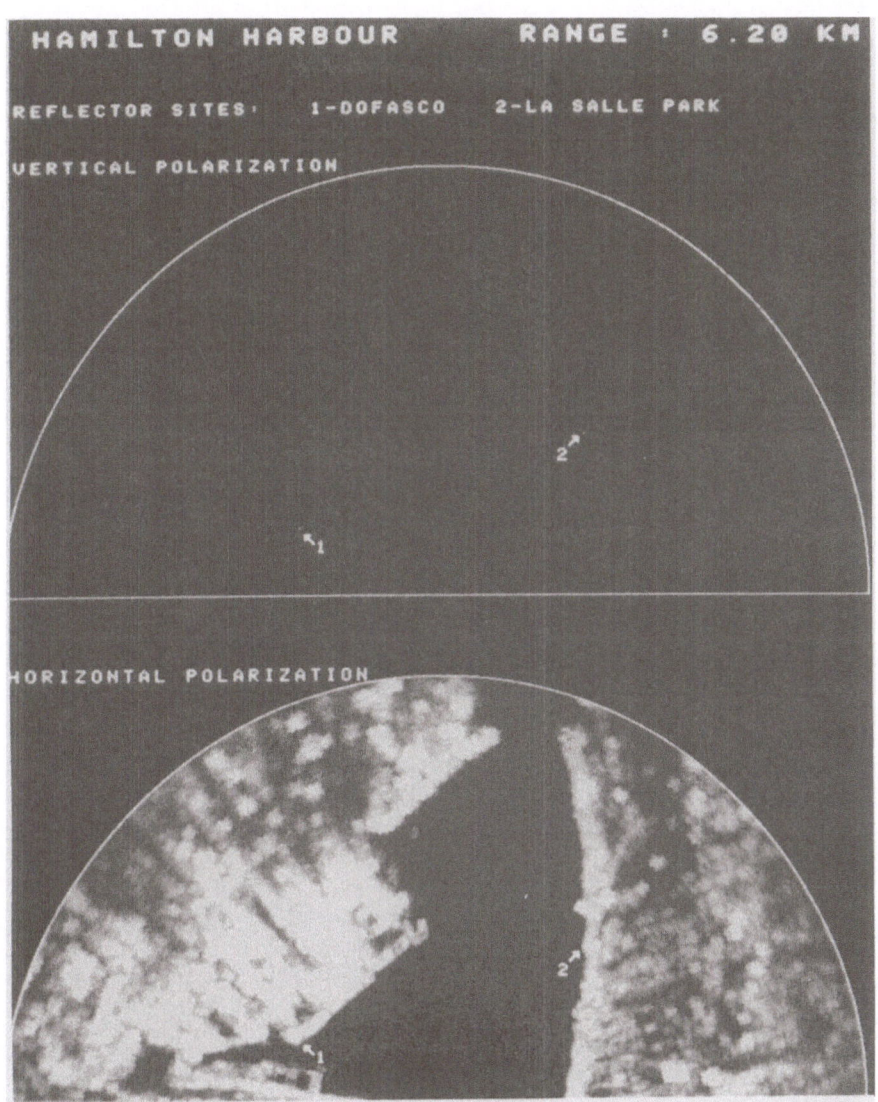

Figure 5   Scan Converted Image of Radar
            Coverage Area (Range=6.20 Km)

addition of a dual-polarized antenna, tracker and digital display. The reflectors are passive, requiring no maintenance once they are installed. The PRAN system is therefore commercially viable.

Acknowledgements

    The author is grateful to his colleagues T. Greenlay, A. Macikunas, and R. Cho for their contributions to the work described herein. He is grateful to the Natural Sciences and Engineering Research Council and Transport Canada for their financial support, and to the Department of Fisheries and Oceans and DOFASCO for providing the site to do the work.

REFERENCES

[1] A. Macikunas, S. Haykin, T. Greenlay, "Trihedral Radar Reflector," Patent applied for November 22, 1984, File No. 265-8007-1.

JAPANESE GEODETIC SATELLITE FOR EXPANSION OF MARINE CONTROL

Minoru Sasaki
Hydrographic Department of Japan
3-1, Tsukiji 5-chome, Chuo-ku
Tokyo 104 Japan

ABSTRACT

The Japanese Experimental Geodetic Satellite "AJISAI" with functions for laser ranging and photographing from the ground was launched. The tracking observation of AJISAI has been made after launch by laser ranging and photographing techniques under international cooperation. According to a simulation the range accuracy of one to two centimeters level is attainable by applying edge detection method. The expansion project of marine geodetic controls around Japan is to be made by the Hydrographic Department of Japan using satellite of Lageos, NNSS and this AJISAI satellite. The field work for precise geodetic purpose by using a transportable laser ranging station with 5 cm accuracy and other equipments will start in late 1987 for this expansion project.

1. INTRODUCTION

The Japanese Experimental Geodetic Satellite (EGS) was launched at the Tanegashima Space Center by using the first H-I rocket of the National Space Development Agency (NASDA) of Japan on August 12, 1986 and the satellite was named "AJISAI" which means Hydrengea of flower in Japanese. The observation project of AJISAI which is and will be conducted by the Hydrographic Department of Japan (JHD) is presented here.

2. FUNCTIONS AND SPECIFICATIONS OF "AJISAI"

The functions of the satellite, "AJISAI", are (1) to reflect input laser light back toward the incident direction by Corner Cube Reflectors (CCRs) and (2) to reflect solar light to the ground by solar reflecting mirrors.

The body of the satellite is hollow sphere made of glassfiber-reinforced plastics The surface of the body is covered with CCRs and solar light reflectors (Fig.1). Twelve pieces of unit CCRs form a set of Laser Reflector(LR) and 120 sets of the LR are distributed on the surface almost uniformly. The effective area for laser light reflection within the full prospect angle of 30 degrees from the center of the satellite is 91.2 cm². The remainder part of the surface is covered with 318 pieces of solar reflectors. The

Fig. 1   Japanese Geodetic Satellite (AJISAI)

M. Kumar and G. A. Maul (directors), Marine Positioning, 77–85.
© 1987 by the Marine Technology Society.

reflectors are mirrors with the radii of the curvature from 8.4 to 8.7 m. The base of the mirror is made of an alloy of alminum and the surface is coated by an oxide silicon for protection from flawing and diminution of quality. The reflective efficiency of the mirror is 0.85. The diameter of "AJISAI" is 2.15 m and its total weight is 685.2 kg.

"AJISAI" was given a spinning of 40 rpm before detachment from the rocket. The spin axis was set in parallel to the earth rotation axis and almost every observers on the ground in the dark can observe the flashing light of reflection from the solar reflectors of "AJISAI" in repetition rate of 2 pps if the satellite is exposed to the solar light. The flashing duration is about 5 msec and the

Table I Brightness of "AJISAI" expressed in star magnitude

elevation	range	transparency		
		k=0.3 (good)	k=0.4 (medium)	k=0.5 (not good)
deg	km	mag	mag	mag
90	1500	1.44	1.55	1.65
80	1519	1.47	1.58	1.69
70	1577	1.56	1.71	1.80
60	1680	1.73	1.86	1.98
50	1841	1.98	2.12	2.26
40	2080	2.32	2.50	2.67
30	2428	2.81	3.03	3.24
20	2931	3.52	3.84	4.15

brightness of the reflective light is from 1.5 to 4.0 star magnitude as shown as followings:

The intensity of the reflected light from a mirror sphere through atmosphere is given by

$$ I = \gamma I_s \frac{a^2}{4 r^2} T $$

and the brightness expressed in star magnitude is

$$ m = -2.5 \log I $$

where I_s : intensity of input light, γ : reflectivity of a sphere, a : radius of a sphere, r : range from observer to a sphere, T : transparency of atmosphere. The magnitude of the brightness is estimated by using some values of specifications and atmosphere as I_s : ($m_s = -2.5 \log I_s = -26.8$: star magnitude of the sun), γ : 0.85, a : 8.5 m , T : exp(-k sec z) for a model transparency of atmosphere, z : zenith distance, k : atmospheric condition (0.3 : good, 0.4 : medium, 0.5 : not good). The results are shown in Table I.

The flash can be taken in a photo with a number of fixed stars by using a camera on an equatorial mount and the direction of the satellite from the observer can be measured referring the star coordinate system.

For the estimation of the ranging accuracy of "AJISAI", a simulation has been conducted by NASDA. The concept of the simulation is given in Fig. 2. In this simulation no atmospheric fluctuation effect is considered. A kind of pulse of the full width at half-maximum(FWHM) of 78 psec (200 psec for total width within three times of the standard deviation of a Gaussian shape) are used. The incident direction of 60 in number are selected geometrically in 30 degrees step from -60 degrees to +60 degrees for latitude and in 30 degrees step for longitude around the spin axis. The resultant change of position (half distance) of the first peak, rising position of half height of the highest peak and rising position of 1/20 height of the highest peak are shown in Table II. The mean value of the position of the first peak and its root mean square(RMS) are :

$$ 1013.3 \pm 11.3 \text{ mm.} $$

The reflection patterns for the three typical cases are given in Fig. 3. The results of the simulation above indicate that the range accuracy to "AJISAI" attains 1 to 2 cm level when a high precision Satellite Laser Ranging (SLR) system with a narrow-pulse-laser-transmitter and with multi-photoelectron-detection or front-

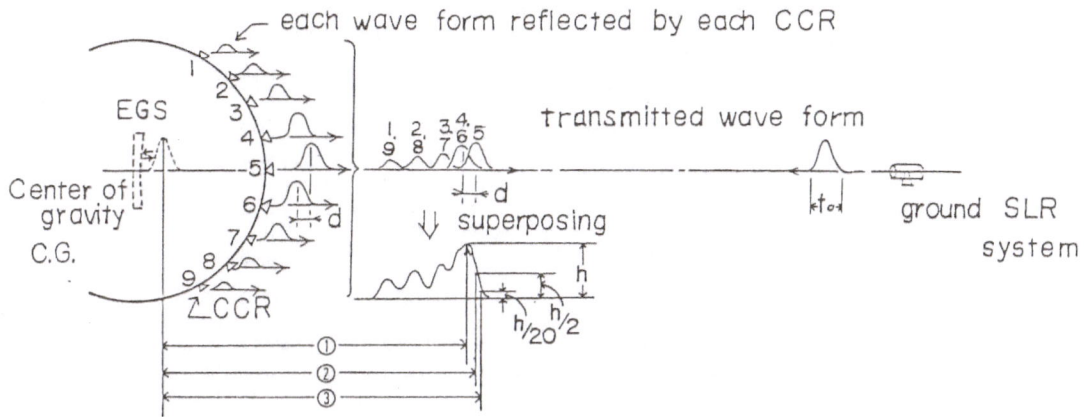

Fig. 2 Concept of simulation to know wave form of return pulse and range bias. Range bias for ① first peak position, ② half height position and ③ rising position.

edge-detection method. The most of Lageos capable SLR systems can be attainable to the front edge detection of return laser light from "AJISAI" since the lower orbit than Lageos(5900 km) gives much stronger return energy of laser light.

The launch of "AJISAI" was made successfully at 20^h 45^m on August 12, 1986 (UT) and the tracking observations by NASDA, JHD and other supporting organization including the National Aeronautics and Space Administration (NASA) of the United States has been continued. The determined orbit is circular with the inclination of 50.0 degrees and altitude from the ground is 1500 km high.

3. METHOD OF OBSERVATION

There are several methods in geodetic use of this "AJISAI" satellite. One is fully geometrical method by using simultaneous observations of distance and direction at some stations. Namely, two kinds of observations are made at a base station (known position, e.g. Simosato) and temporary stations (unknown positions, e.g. isolated islands). The position of the satellite is given by distance and direction observation from the known position and unknown positions are determined from the position of the satellite by similar distance and direction observations at these points. A SLR system and a Satellite Camera of fixed type are necessary at the base station and a transportable SLR system and a transportable Satellite Camera are also necessary at a temporary station in this method. The location of an unknown position is given by only one set of observation in principle.

The other geometrical method is to use several SLR systems simultaneously in a region of a few thousands kilometers. The range correction of this satellite is well determined as shown in a simulation above and simultaneous precision ranging can determine baselines in a centimeter level.

The dynamical method is also useful to determine the locations of SLR stations based on a geocentric coordinate and other geophysical parameters as geopotential coefficients, air drag effect and tidal effects. For each method or combined methods it is effective to hit laser beam well on this satellite because of its brightness.

Table II   Change of range bias for reflective pulse from "AJISAI"

pulse width	measuring position	range bias	
		maximum	minimum
		mm	mm
200 ps	first peak	1026	984
	half hight	1031	1011
	rising	1037	1031
300 ps	first peak	1026	984
	half hight	1034	1019
	rising	1043	1035

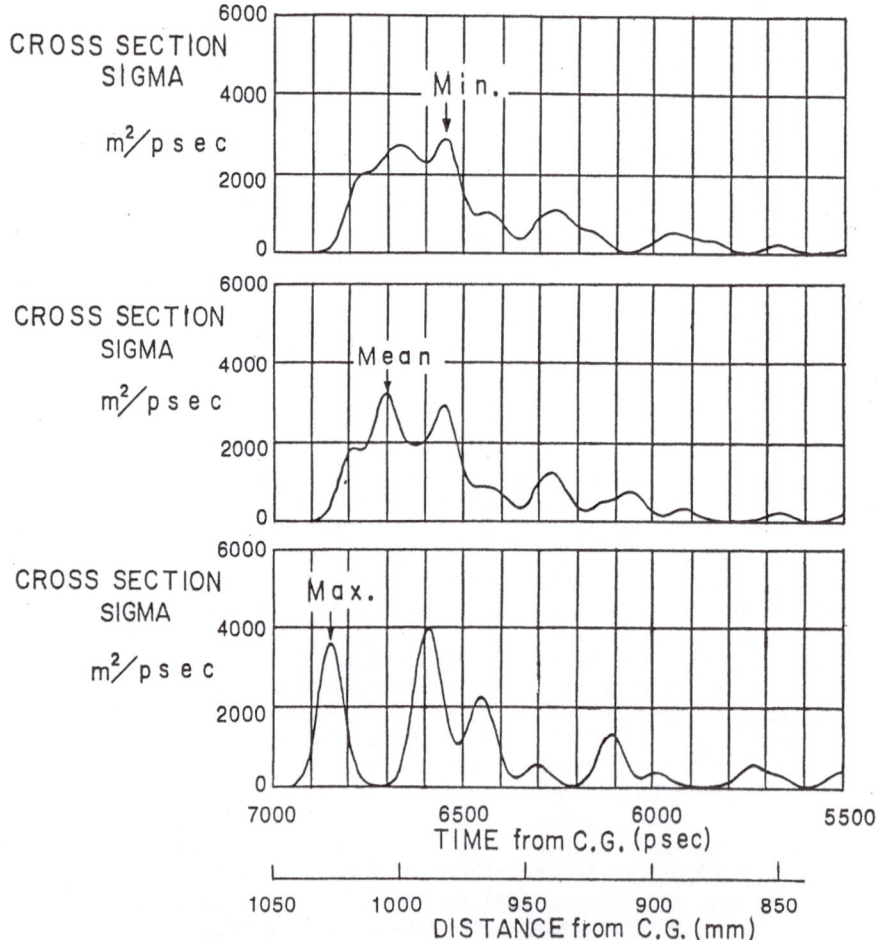

Fig. 3   Examples of laser reflection pattern. For the cases of the
         first peak position of the minimum-, mean- and maximum-range bias.

4. EXPANSION OF THE MARINE GEODETIC CONTROLS AROUND JAPAN USING SATELLITES

The position of the origin of the Japanese Geodetic Coordinate System (Tokyo Datum) and of many isolated islands had been determined by means of astronomical observation. These positions includes errors because of observation errors in old days and the effect of deflection of the vertical for each observation point caused by gravity amomalies. The amount of these positioning errors reaches a kilometer in some cases. To correct these systematic and random errors on geodetic location in Japanese territory, the Hydrographic Department of Japan(JHD) started the Expansion Project of the Marine Geodetic Controls around Japan in 1980. The newly launched Geodetic Satellite "AJISAI" plays a key role for construction of the Controls. The expansion project has been performed as following three stages:

1) The SLR observation to Lageos is made at the Simosato Hydrographic Observatory (SHO) to combine the Tokyo Datum with a global reference geodetic coordinate system under cooperation with worldwide SLR stations. The geodetic position of SHO determined is to be the origin of the marine geodetic controls,
2) The SLR observation and photographing of "AJISAI" is made both at SHO and at the primary stations selected in ten representative islands of each group of

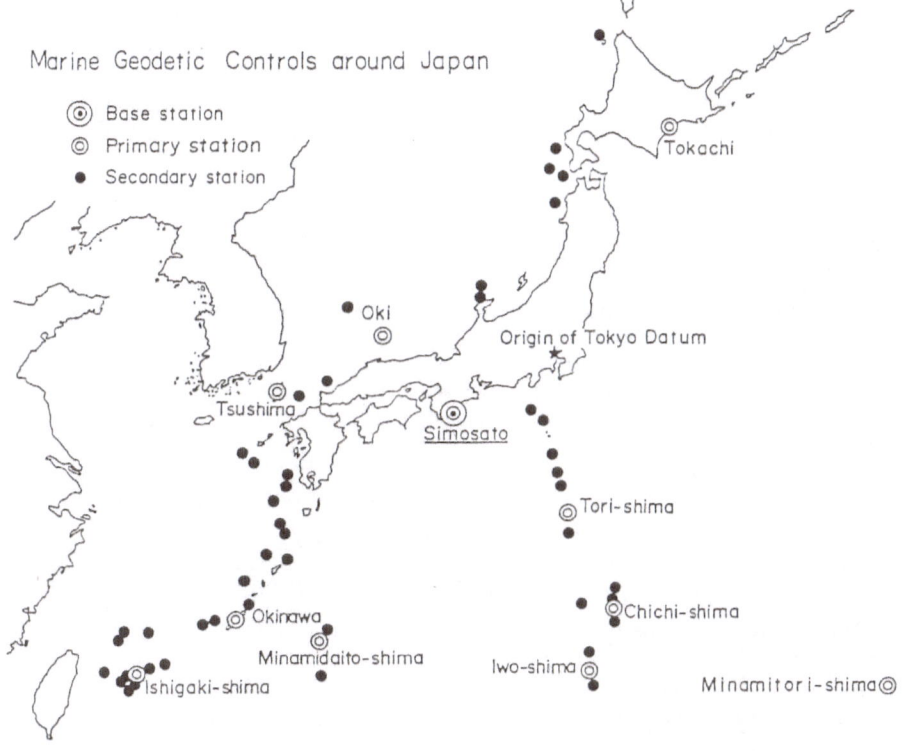

Fig. 4   Configuration of the Marine Geodetic Controls of JHD.

islands around Japan. The geodetic positions of the primary stations are combined with the origin of the marine geodetic controls at Simosato. The field
observation by using this satellite will start in late-1987.
3) The Doppler observation of the US Navy Navigation Satellite System(NNSS) is
made at the primary stations and at the secondary stations selected in some
main islands around the island of each primary stations. The geodetic positions of the secondary stations are combined with the primary stations by means
of the translocation method using three or four Doppler receivers. The Doppler
observation has been continued since 1980. This observation will be replaced
by the observation of the Grobal Positioning System(GPS) in 1990s.

The configuration of the marine geodetic controls is shown in Fig. 4.

According to the project JHD has been conducting SLR observation of Lageos,
Starlette and Beacon-C at SHO since 1982 and the total number of these observations
amounts to 847,000 by the end of August 1986. The observation of "AJISAI" to
determine its orbit and to know geodetic characteristics started at August 13 and
total number of observations in August 1986 is 25,300 of 27 passes and its preliminary range accuracy looks like 7~8 cm level(Table III).
    The Doppler observations by using NNSS have been also made in 25 isolated is-
lands, e.g. Yonakuni-shima, Miyako-shima, Kitadaito-shima, Okinotori-shima, Minami-
iwo-shima and so on. The discrepancies between the ground triangulation network
of Japan and satellite derived positions expressed in Tokyo Datum have been also
discovered in the process of comparisons with other domestic NNSS observations and
surveyed results from the triangulation network. The amount of the discrepancy
reached ten more meters at the north or south edge of the triangulation network.

5. EQUIPMENTS FOR "AJISAI" OBSERVATION

    The SLR system running at SHO will be used for SLR observation of "AJISAI" as
fixed type SLR system when the field work for observation of AJISAI at isolated
islands starts. By the time the SLR system at SHO will finish to equip an additional mechanism for flash timing measurement for photographing by another fixed type
satellite camera. An 25 cm diameter telescope and another PMT will be attached to
the SLR system for flash timing mechanism of AJISAI. As the signal level of flashing of AJISAI is not so high and fairly much background light noise comes to the
detector, both an analog signal recorder and digital photon counting device will be

Table III Data acquisition at Simosato Hydrographic Observatory
and its mean range accuracy

year	LAGEOS		STARLETTE		BEACON C		AJISAI	
	passes	ranges	passes	ranges	passes	ranges	passes	ranges
1982	47	11,000	36	4,700	59	11,900	-	-
1983	137	30,000	116	29,400	199	92,200	-	-
1984	223	93,300	118	37,800	150	56,100	-	-
1985	297	243,800	108	38,800	154	67,500	-	-
~Aug 1986	156	103,900	53	11,700	56	15,400	27	25,300
accuracy	9.0 cm		9.8 cm		9.2 cm		(7~8 cm)	

prepared.

At isolated islands a transportable SLR system named HTLRS(Hydrographic Department Transportable Laser Ranging Station) is to be used. An outline and major specifications are given in Fig. 5. The output laser pulse is raised up through a Coudé path along the azimuth axis and is reflected by a small mirror to the central part of large mirror of 50 cm X 50 cm for transmitter/receiver on the elevation axis. The return signal from a satellite comes to whole part of the transmitter/receiver mirror and goes to the receiver telescope of 35 cm diameter. The detector for return signal is mounted on a bench on azimuth axis behind the receiver telescope. The flash timing of AJISAI is also detected in a part of the detection subsystem. The optical diagram of HTLRS **is** shown in Fig. 6. The system is controlled by multi-micro-processor. HTLRS will be capable to range not only AJISAI but also Lageos in the accuracy of 5 cm. The HTLRS is composed of two sets of separated shelters. The size of each shelter is 2.1 m X 2.3 m X 3.5 m and 2.1 m X 2.3 m X 4.0 m and the total weight is within 5 tons. Whole the station can be transported with a power generator to some isolated islands by ship or a kind of Japanese transport planes.

As for the fixed type Satellite Camera, a refractor telescope with diameter of 20 cm and focal length of 1 m ($F = 5$) on a precision equatorial mount will be newly installed at SHO. A 12.5 cm diameter refractor with $F = 5$ (Nikon Astro Camera) on an equatorial mount is also used as a transportable Satellite Camera. Glass plates of cabinet size are used for both cameras. The flashing dots on the plates are to be identifyed by independently recorded electric timing signals. The neccesary

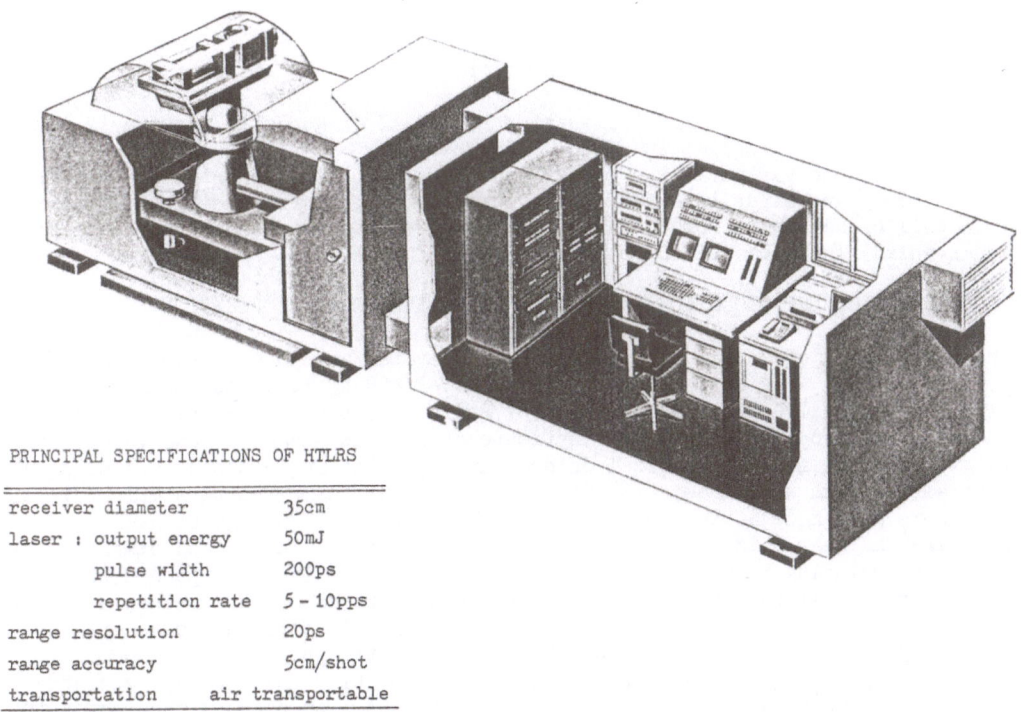

PRINCIPAL SPECIFICATIONS OF HTLRS

receiver diameter	35cm
laser : output energy	50mJ
pulse width	200ps
repetition rate	5 - 10pps
range resolution	20ps
range accuracy	5cm/shot
transportation	air transportable

Fig. 5   Outline of the Hydrographic Department Transportable Laser Ranging Station (HTLRS)

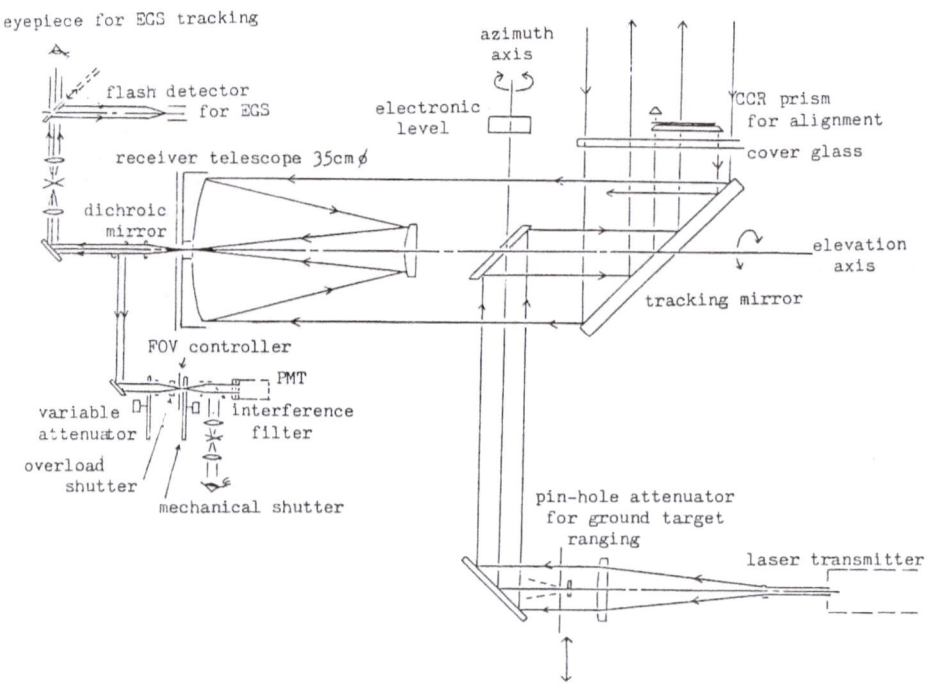

Fig. 6   Optical diagram of HTLRS

timing accuracy for direction determination to AJISAI in geodetic purpose is less than 0.1 msec level.

The field observation at the primary stations will be made twice per year for at least five years after late-1987. The duration of observation at each primary station will be continued for 60 days. The order of observation is Chichi-shima, Ishigaki-shima, Tsushima and so on.

6. SUPPORTING OBSERVATIONS FOR "AJISAI" AND INTERNATIONAL COOPERATION

The observation of "AJISAI" by using the SLR system at SHO has been made from just after launch. The supporting observations of SLR by Tokyo Astronomical Observatory at Dodaira Station, of video by Radio Research Laboratories at Koganei, of photo by the Geographical Survey Institute at Kanozan Geodetic Observatory, the Institute of Space and Aeronautical Science at Kagoshima Space Center and some amateur astronomers were made. The direction observations by using simple Satellite Cameras were also made at four Hydrographic Observatories, Hachijo at Hachijo-shima, Sirahama near Shimoda in Izu Peninsula, Bisei near Okayama and Simosato. The SLR data and pointing information were transferred to the Tsukuba Space Center of NASDA and the Satellite Geodesy Office in the Headquarters of JHD through a micro-computer communication network (PC-VAN). The orbital elements have been created continuously at these places and distributed to domestic observation sites.

JHD has an agreement for cooperation of SLR observation and its data exchange with NASA. The Goddard Laser Tracking Network of NASA also started the supporting observation for "AJISAI" from just after the launch and distributes their own orbital elements of AJISAI to cooperative SLR stations. Observation data and results of research work of AJISAI will be exchanged between JHD and NASA.

Other international SLR stations in England, Switzerland and so on also tracked AJISAI in early stage after launch and more wider cooperation with France, China, West Germany, Australia and Austria is expected. For these international coperation for AJISAI observation JHD with NASA will play a role of data distribution center. The work on collection of SLR and Photograph data, determination of orbit, distribution of orbital elements, data file management and data analysis has started in JHD.

In addition to the work stated above, the efforts to detect plate motions and crustal movements will be made in the SLR observation project and the observations will contribute to estimate the earth rotation parameters and geophysical parameters.

REFERENCES

Ganeko, Y., Komaki, K. and Hashimoto, H., 1983. "On the geodetic satellite GS-1." Proc. Sympo. on Space Techniques in Positional Astronomy, Tsuchiya, A., Ed. Tsukuba, pp.12-18(in Japanese).

Hashimoto, H. and Saito, K., 1985. "Development of the Experimental Geodetic Payload." Proc. 17th Sympo. for engineering results in NASDA(in Japanese).

Sasaki, M., 1979. "Optimum orbit of the Geodetic Satellite GS-1." Rep. Hydrogr. Researches, JHD, no.14, pp.131-144(in Japanese).

Sasaki, M., 1986. "Observation Project of Japanese Geodetic Satellite GS-1." Proc. Sympo. on Application on Space Techniques to Astronomy and Geophysics, Kinoshita, H., Murata, I. and Nakajima, K., Ed. Tokyo, pp.27-31.

Sasaki, M. and Nagaoka, M., 1984. "Satellite Laser Ranging Observations in 1982." Data Rep. Hydrogr. Obs. Series of Astronomy and Geodesy, JHD, no.18, pp.55-67.

Sasaki, M., Sengoku, A., Nagaoka, M. and Nishimura, E., 1986. "Satellite Laser Ranging Observations in 1984." Data Rep. Hydrogr. Obs. Series of Astronomy and Geodesy, JHD, no.20, pp.44-67

Sasaki, M. and Hashimoto, H., 1986. "Launch and Observation Program of the Experimental Geodetic Satellite of Japan." IEEE Transactions on Geoscience and Remote Sensing, to be published.

Takemura, T. and Kanazawa, T., 1983. "Satellite Doppler positioning of off-lying islands in 1980-1981." Data Rep. Hydrogr. Obs. Series of Astronomy and Geodesy, JHD, no.17, pp.61-87(in Japanese).

Yamazaki, A. and Mori, T., 1983. "Marine Geodetic Controls around Japan." Marine Geodesy, vol.7, pp.331-344.

EVALUATION OF A LARGE SHIP GPS SYSTEM:
DIRECT COMPUTATION AND DYNAMIC DIFFERENTIAL

J. Clynch
W. Harper
Applied Research Laboratories
University of Texas at Austin
Austin, TX 78713-8029
and
J. E. French
RCA MTP
Bldg. 989 MU 645
Patrick AFB, FL 32925

ABSTRACT

A GPS system utilizing a TI 4100 and a microprocessor has been installed and tested on two Air Force radar tracking ships. The system has been evaluated using an Autotape radio position system (accuracy at 1m level) as a reference. The system provides both a real time solution via an eight state Kalman filter and records data for post processing. The accuracy of the position and velocity solutions from the real time Kalman filter and from post processing using a shore based reference site are presented.

1. INTRODUCTION

Applied Research Laboratories, The University of Texas at Austin (ARL:UT), has been involved in satellite navigation systems for over 20 years. With the advent of the Global Positioning System (GPS), ARL:UT became involved with the development and exploitation of a geodetic quality GPS receiver. This type of receiver has now been incorporated into a GPS Large Ship System (ARL:UT-GPS-LSS) used for navigation.

This system has been deployed on two radar tracking ships, the USNS Redstone and USNS Observation Island. The system was evaluated for position accuracy off the coast of Cape Kennedy in December of 1985. The system performed very well, giving an average error of 5.7 m over all conditions during 4 days of testing.

2. SYSTEM DESCRIPTION

2.1 Hardware Description

The hardware consists of three principle elements: a TI 4100 GPS receiver, a Zenith Z100 microcomputer, and a nine track magnetic tape drive. The purpose of the system is twofold, 1) to provide real time navigation and 2) to collect data for post processing. The Z100 does the navigation calculation, and tape drive records the raw and computed data.

The TI 4100 receiver is a RAM version receiver running a control software program developed at ARL:UT. This program, called CORE, is designed to interface with an external computer. It has many commands which require long parameter strings which provides a very flexible receiver for use with the external processor. (This software program is used in several applications, hence its name.)

M. Kumar and G. A. Maul (directors), Marine Positioning, 87–96.
© 1987 by the Marine Technology Society.

The Z100 computer has 492 K bytes of RAM, a 8087 co-processor, a 10 Mbyte winchester disk, and a IEEE 488 interface. The operating program requires about 300 Kbytes of RAM. The hard disk is used as a fast means of loading different program elements and the IEEE 488 interface is used to control the tape drive. The 8087 co-processor speeds up operations by about a factor of 20 which is required for the system to function.

The computer has two connections to the TI 4100 receiver, the command line and the data line. The command line is used to pass the various commands to the receiver and receive short answers. The data line, connected to the TI cassette tape recorder port, is used to send binary data to the Z100 in normal operations. It is also used to download the TI 4100 with the CORE program when the system is first brought up.

2.2 System Software

The Z100 is run under the MS-DOS operating system. It was decided to use a standard software environment as much as possible. In line with this decision the vast majority of the code in the Z100 is coded in FORTRAN using the Microsoft 3.2 compiler. There is a small amount of assembly language code, mainly in an interrupt driven I/O driver written to communicate with the receiver.

The process of acquiring this data and analyzing it in real time inherently oriented toward a multi-tasking environment. Because the MS-DOS operating system will not handle multi-tasking, a simulated tasking system was written into the FORTRAN code. The MPM-86 operating system was considered but not chosen.

The navigation software is one large program with a single major loop. This loop always checks for the presence of data and runs the navigation solution. It will also run one of several small subroutines on each cycle. Each subroutine has been coded to take no more than 100 ms and maintain its own internal status.

The entire system software consists of several programs that usually run in an endless loop in a batch file. The main navigation program will exit when it has less than the minimum number of satellites in view or 24 hours has passed. It will pass the current location and an Almanac to the scenario generation programs. These programs will compute the satellites to be tracked in the next session.

The satellites are selected based on the best PDOP available unless the current satellite set has a PDOP less than 3.3. In that case there is no change in the satellites being tracked. This algorithm prevents rapid satellite changes which will not improve the solution. Typically there will be 5 satellite changes in a tracking session of 8 hours.

The scenario generation program writes the tracking instructions to an action file. This file also contains other commands to be sent to the receiver. For example, a new Almanac is collected every 3 days on the direction of the scenario generator via this action file.

In addition there are several parameter files that control the system. A satellite table marks satellites to be used, unused, or used only if less than 4 other satellites are in view. There is also a file containing all the tuning parameters for the navigation filter. A file maintenance program is provided for the operator to aid in changing these files.

2.3 Navigation Filter

The real time navigation is performed with an eight state Kalman filter. This filter is a derivative of one developed at the Naval Surface Weapons Laboratory/Dahlgren Laboratory (NSWC/DL). The filter estimates the 3 position variables and a time bias from range data and the 3 components of velocity and a frequency bias from the phase data.

One of the reasons that the results are better than the JPO specification of 16 meters, is the use of phase smoothed pseudo ranges. This idea, implemented here due to a suggestion of Allen Evans, NSWC/DL, was originally proposed by Ron Hatch [Hatch 1982]. This approach has the advantages of giving effective range measurements with noise levels characteristic of the phase and reducing multipath effects greatly.

Phase smoothed pseudo ranges are obtained by using the integrated phase to obtain the difference in range from an initial point in time. The range at the initial time is estimated from all the pseudo ranges from that time to the present measurement. That is if ρ_0 and ϕ_0 are the pseudo range and phase at the initial time and ρ_i and ϕ_i are the corresponding values at a later time, then

$$\Delta\rho_i = \frac{\lambda}{2\pi}(\phi_i - \phi_0) \ , \tag{1}$$

is the change in range between initial time and the current time. Here λ is the wavelength. At each point after the first, there will also be a measure of this difference from the pseudo ranges,

$$\Delta\rho_i = \rho_i - \rho_0 \ . \tag{2}$$

Equations 1 and 2 can be used to solve for a value of ρ_0 at each time. The average of those solutions will be the current estimate of the range bias

$$\rho_{oi} = \frac{1}{i}\sum_{j=1}^{i}(\rho_j - \sum_{k=1}^{j}\Delta\rho_k) \ , \tag{3}$$

and the current phase smoothed pseudo range will be

$$\rho_{iuse} = \rho_{oi} + \sum_{j=1}^{i}\Delta\rho_j \ . \tag{4}$$

In addition to the use of phase smoothed pseudo ranges, the system makes use of several other facts to improve the solution. Since the system is on a ship, the solution height, adjusted for the distance from the mast to sea level, is constrained to the geoid. This is done with an extra "measurement" from the earth's center. The weight on this measurement is about one tenth that of the true measurements. Since line to the center of the earth from a point on the sea is in the opposite hemisphere from the line to any satellite, this greatly strengths the geometric diversity of the data and the solution. Also the velocity is constrained to be perpendicular to the elipsoid.

The ships using this system have Cesium (Cs) time standards and the Kalman filter is tuned to reflect this standard. This allows the solution to continue with fewer than 4 satellites in view. In fact the solution is continued as long as two satellites are up. The geoid constraint and current clock model provide the extra information. The solution is usually not started till 3 satellites are up, however, as the clock information may be 8 hours old at that time. The clock parameters are estimated only when four satellites are in track.

The Kalman filter is usually cycled at a one second rate. The solutions are recorded to the magnetic tape and a printer at an operator selectable rate, usually 6 sec and 6 minutes respectively.

3.0 EVALUATION TEST

3.1 Test Description

The system was evaluated in December 1985 in a test run just off Cape Kennedy, Fl. with very calm seas. The USNS Redstone sailed around a box 10 Km on a side continuously for 4 days at a nominal speed of 4 knt. The ARL:UT-LSS-GPS system ran automatically during this period. The test was planned and conducted by a Patrick Air Force Base range contractor.

The reference data were obtained from an Autotape system using 3 ranges. The accuracy of data from this system was 1 m. While this reference position was better than GPS, the velocities obtained from this system were quite noisy.

The Autotape system originally had its antenna about 1 m from the TI GPS antenna. It was found that in this location it interfered with the L2 reception causing several dB of increased noise. The Autotape antenna was moved to a location about 1 m below the GPS antenna and this eliminated the RFI.

In addition a second GPS receiver was operated on the shore about 20 km from the ship tracking the same satellites. Its data were used to evaluate the use of differenced data to obtain a differential dynamic position.

3.2 Test Results

Plots of the results from two days in the test are shown in Figs 1-4. In Figs 1 and 2 the error in lattiude and longitude on the first day of the test are shown. The x-axis is time in seconds of the day and the y axis is the position error. Along the top of the plots are indications of the number of satellites in track and of other events. There was one PDOP blow up each day at about 44500 sec.

More typical data are shown in Figs 3 and 4. Note that the error remains mostly within a 10 m band. The GPS solution was quite smooth due to the filter tuning. Therefore few isolated wild points are felt to be due to Autotape errors. The system performed quite well when it came up, even thought only three satellites were in view and continued to have an acceptable error at the end with only two satellites.

A summary of the results is given in Table I. The errors for the two runs

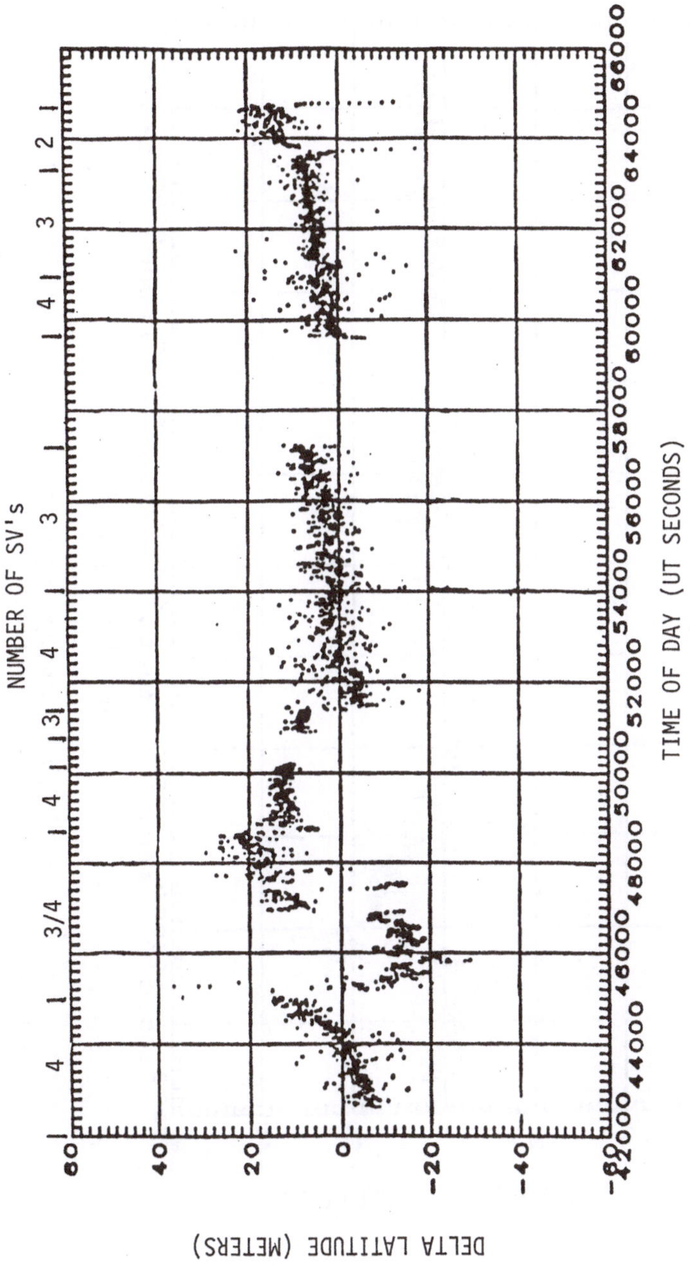

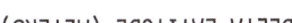

FIGURE 1
ARL:UT-GPS-LSS REAL TIME POSITION
LATITUDE ERROR VS. TIME OF DAY
DECEMBER 10, 1985 USNS REDSTONE

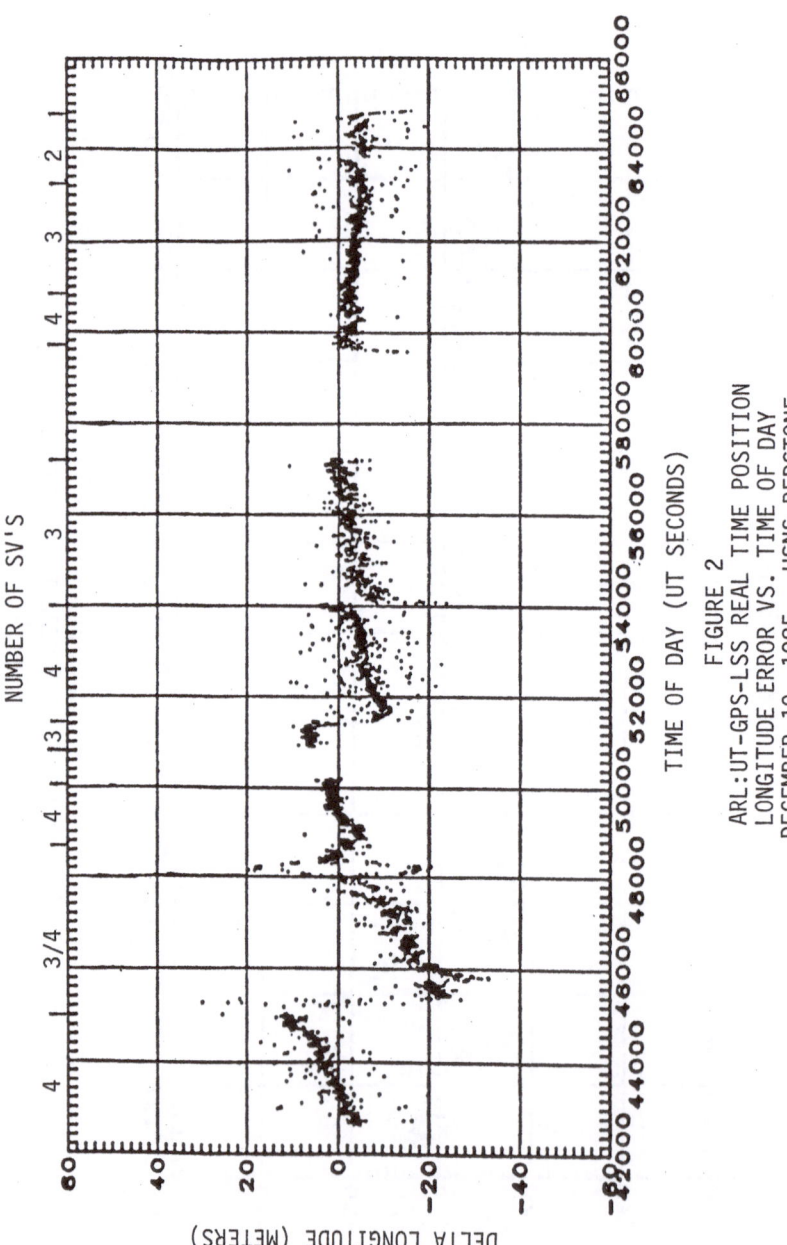

FIGURE 2
ARL:UT-GPS-LSS REAL TIME POSITION
LONGITUDE ERROR VS. TIME OF DAY
DECEMBER 10, 1985   USNS REDSTONE

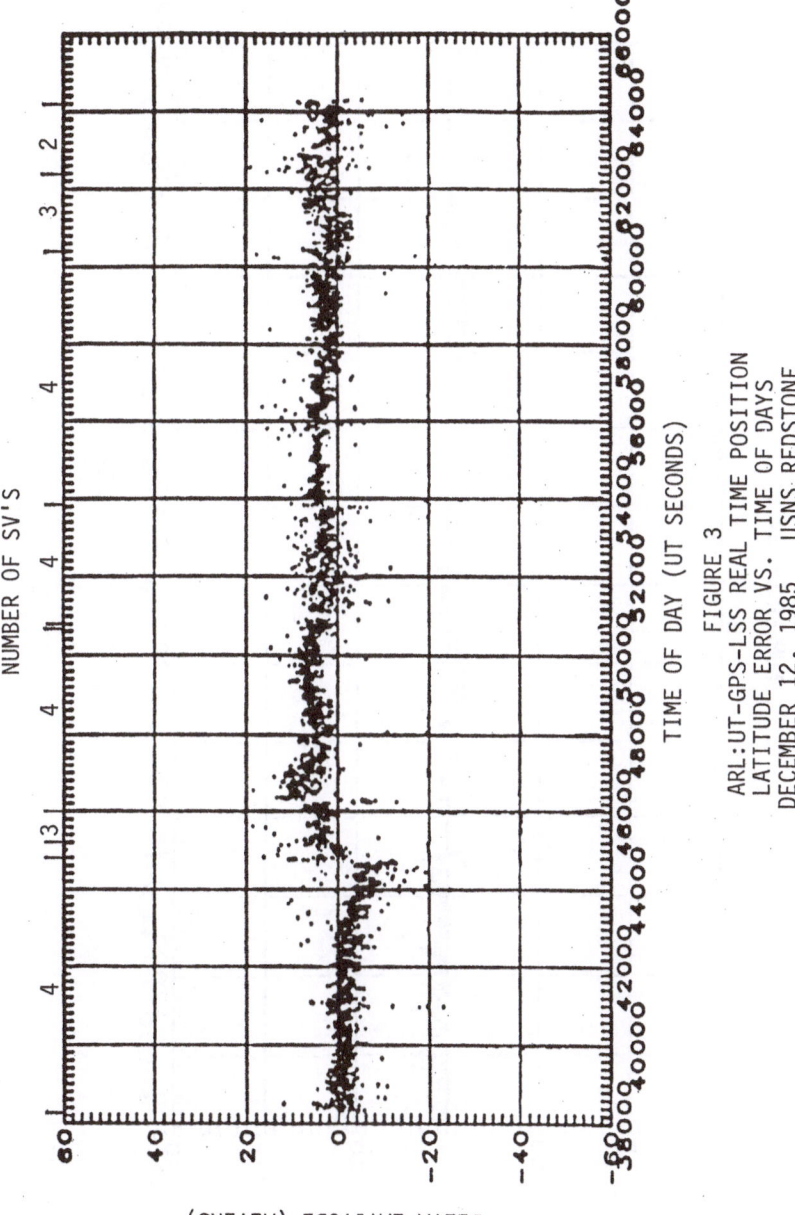

FIGURE 3
ARL:UT-GPS-LSS REAL TIME POSITION
LATITUDE ERROR VS. TIME OF DAYS
DECEMBER 12, 1985   USNS REDSTONE

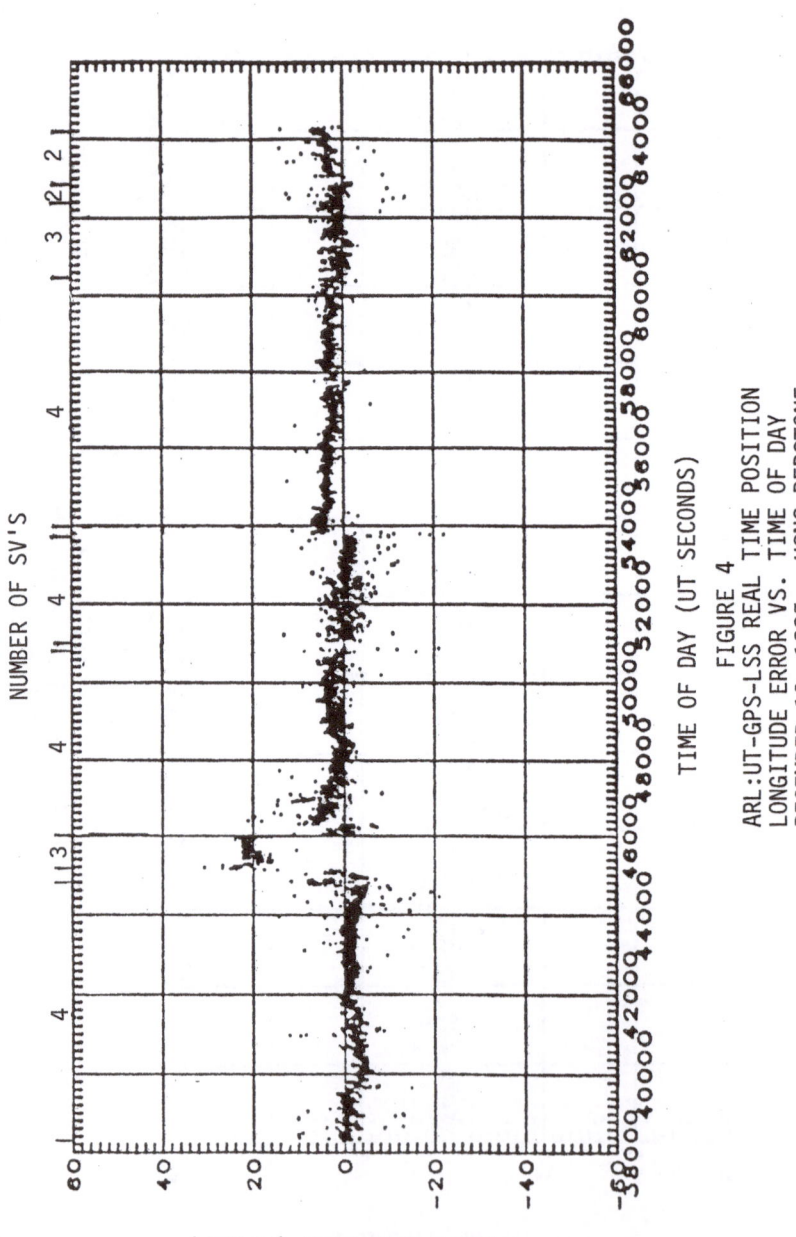

TIME OF DAY (UT SECONDS)

FIGURE 4

ARL:UT-GPS-LSS REAL TIME POSITION
LONGITUDE ERROR VS. TIME OF DAY
DECEMBER 12, 1985   USNS REDSTONE

TABLE I
SUMMARY OF POSITION AND VELOCITY ERRORS

POSITION COMPARISON
(Meters)

RUN	No. SV's	LATITUDE		LONGITUDE	
		Mean	S.D.	Mean	S.D.
2	2	9.4	6.1	-4.6	3.6
	3	3.9	4.0	-3.7	3.4
	4	0.1	5.5	-1.7	6.5
	all	2.7	4.8	-2.9	4.7
4	2	2.5	4.0	2.9	2.9
	3	2.5	2.8	1.4	2.1
	4	1.8	4.4	0.8	3.4
	all	1.9	4.2	1.4	3.2
all	2	1.6	4.1	0.5	3.2
	3	2.0	4.1	-0.9	3.8
	4	1.8	4.3	-0.6	3.9
	all	1.9	4.2	-0.6	3.9

o 5.7 M rms Error Over All Data

VELOCITY COMPARISON

o Autotape Velocity Noisy

o 20 cm/s Comparison Raw Data

o 5 cm/s Comparison Smoothed Data

plotted are given along with the overall statistics.   The results in this table
are divided into sections according to the number of satellites in track. The
rise in error when 4 satellites are in view is   due to the extra clock
parameters being estimated. This can not be avoided as some clock estimates are
needed even with Cs oscillators.

The velocity was also compared to the Autotape data.   A direct comparison
gave a rms error of 20 cm/s.   It is felt that this is a reflection of the noise
in the Autotape data and not the GPS solution. Accordingly the Autotape data
were smoothed with polynominals over periods of 2 minutes and the smoothed data
were used to compute a comparison velocity.   The error in this case was 5 cm/s.
This may be an optimistic estimate of the true real time velocity error as both
systems were producing highly smoothed velocities.   One can only conclude from
this test that the velocity error is between 5 and 20 cm/sec.

In addition to the real time solution, the data from the shore station were
used to obtain a differential position.   The real time filter was modified to
take two data streams and form differences of the data. Using this modified
filter, the solution was found to be only slightly improved on average and
noisier by the square root of two.   The wide swings with PDOP blow ups were
improved however.   The noise increase indicates that the local clock noise is
dominate in these solutions.

4.0 CONCLUSIONS

An automated GPS system aboard a large ship has been evaluated against
Autotape.   The system makes use of the known facts that: the ship is on the
geoid, has a very good oscillator (Cs), and the platform is fairly stable.
Positions can be obtained with only 2 satellites after some 4 satellite
tracking. The average of the error overall condition was 5.7 m (one sigma).

REFERENCE

Hatch, R., 1982.   "The Synergism of GPS Code and Carrier Measurements."
    Proceedings of the Third International Geodetic Symposium on Satellite
    Doppler Positioning, Vol. 2, pp. 1213-1231.

OPTIMUM UTILIZATION OF POSITIONING DATA IN SDS III

Gary C. Guenther
NOAA/National Ocean Service
6001 Executive Blvd.
Rockville, Maryland 20852

and

Robert W. L. Thomas
EG&G Washington Analytic Services Center, Inc.
5000 Philadelphia Way, Suite J
Lanham, Maryland 20706

ABSTRACT

A new, computerized hydrographic data acquisition and processing system, Shipboard Data System III (SDS III), is being designed and built for use by the National Ocean Service. An integrated positioning and navigation system is a critical element of this development. Design features include the ability to benefit from time-deskewed multiple lines of position from mixed sensor types (both electronic and manual), raw data quality evaluation including blunder removal and the use of signal strength data, high precision geodetic calculations, corrections for control and sensor offsets as well as for rare but difficult geometries, and the use of auxiliary speed and heading data in the application of advanced filtering and smoothing techniques for reduction of random noise and recognition of bias errors. Performance has been assessed for a variety of maneuvers via a track simulator which adds both vessel motions and sensor measurement noise. Results are extremely stable and robust. Measurement noise can be reduced by as much as a factor of three without adding significant biases, even on turns, while retaining actual random vessel motions. Operations can continue during complete losses of positioning data for limited but significant periods of time, including during maneuvers.

1. SYSTEM DESCRIPTION

The National Ocean Service (NOS) is the sole agency responsible for the charting of the coastal waters of the United States and its territories plus the Great Lakes. This mission includes the measurement of tides and other oceanographic parameters as well as the locations of obstructions, navigation aids, landmarks, and the like. The bulk of this work is conducted by sonar from small launches in shallow water and larger ships in deep water. Shipboard Data System III (SDS III) is slated as the next generation NOAA hydrographic survey data acquisition, processing, and display system for use on NOS ships and launches as replacement for the venerable HYDROLOG/ HYDROPLOT system (Wallace 1982) which is nearing obsolescence. SDS III (Schiro 1984, Enabnit 1985) is designed to automate much of the work which is presently done manually, to improve accuracy, to significantly reduce the time from survey to printed chart, and to provide a high degree of reliability and maintainability. Interactive color graphics will be used to aid in surveying and field verification of data. The system will use commercially available, general purpose computer hardware and operating system software. Hydrographic applications software which incorporates the knowledge, experience, policies, and procedures of NOS is being developed jointly by the government and a software engineering contractor. An operational capabilities demonstration is planned for 1987, and it is anticipated that these systems will be used beyond the turn of the millenium.

M. Kumar and G. A. Maul (directors), Marine Positioning, 97–111.
© *1987 by the Marine Technology Society.*

Two types of systems, both based on Perkin-Elmer 32-bit computers, are being purchased: the Data Acquisition System (DAS) and the Data Processing System (DPS). Installed on survey ships and 9-meter survey launches, the DAS processes and logs data and produces real-time graphic displays to supply steering guidance to the helmsman and to permit monitoring of the progress and quality of the survey. Industry standard hardware interfaces are provided for current and future electronic positioning and depth sounding equipment. The DPS, located aboard the survey ships and at two Marine Centers, processes data collected by the DAS, provides graphic displays and hardcopy plots to help plan, monitor, and modify the survey, and allows the survey officer in the field to make real-time decisions about the quality, validity, and significance of the collected data.

2. APPROACH

Positioning for SDS III will be handled by a unified "integrated navigation" approach which can simultaneously utilize multiple lines of position (LOPs) from a selection of short, medium, and long range positioning systems, as well as from auxiliary heading and speed sensors. Two LOPs are necessary and sufficient for positioning. More, although not required, are potentially beneficial. Fewer than two LOPs, augmented by speed and heading, can be used for limited periods of time before either increasing the number of LOPs to a minimum of two or terminating operations until at least two reliable LOPs become available. Random noise components are reduced and biases recognized by applying advanced filtering (DAS or DPS) or smoothing (DPS) techniques to restrict position solutions according to selected limitations on vessel dynamics. (The term filtering refers to the use only of present and past data, while smoothing implies the use of future data as well.) The use of these procedures, which are much more ambitious and calculation-intensive than those in current practice, is made possible by the dramatic increase in available computer power.

LOPs can be derived from any combination of ranges, range (phase) differences, sextant angles, azimuths, and latitude/longitude estimates (i.e., processed GPS data) from sensors such as Falcon 484, Miniranger III, Del Norte R03C and 520, Argo, Raydist, Hydrotrac, Northstar LORAN-C, and Texas Instruments 4100 GPS receiver. Auxiliary data can be derived from gyro, digital compass, Doppler speed Log, and engine RPM and propeller pitch pickoffs. Temporally "sparse" data such as theodolite angles (azimuths) and sextant angles will initially be entered manually by the operator, although provisions for automated range/azimuth sensors such as Polar Fix may be included in the future.

All available LOPs and auxiliary data will be calibrated, edited to remove obvious blunders and data with unacceptable signal strengths, and time shifted (deskewed) to common times of interest. Observation equations derived from the LOP data and adjusted for antenna and observer offsets from the vessel "centroid" are calculated with very high accuracy and a minimum of computer burden by Gaussian conformal mapping from a geodetic spheroid to a sphere (Bakker et al. 1985). This ensures that coordinate transformation errors remain much smaller than sensor accuracies regardless of range or latitude. The observation equations developed on the sphere for each of the sensor types are linearized about the previous sensor location as the fiducial.

The linearized LOP observation equations are solved via a standard, weighted least-squares (Cross 1981) adjustment procedure on the sphere to provide an "unconstrained" vessel centroid position solution. Provision is included for iteration of the solution should the LOPs be tightly curved as well as for special cases such as inclined-plane sextant angles and situations of severe vertical displacements at short ranges. Weights for each sensor type are primarily fixed in advance but may be adjusted over limited ranges in response to actual conditions. The acknowledged vulnerability to error of least-squares solutions if a poor quality LOP is included, as noted by Weeks (1984), is mitigated here by the following dynamically constrained filter operation which recognizes sudden biases and deactivates offending LOPs, and by the editor which recognizes and deactivates LOPs with excessive random noise.

The unconstrained solution is then converted to a local easting/northing coordinate system and passed to special-purpose filter or smoother algorithms for reducing the random measurement noise component, recognizing and limiting response to suddenly biased LOPs, and, with the use of speed and heading data, providing the ability to operate for limited periods of time when some or all of the primary positioning data has been temporarily lost. Partitioning the "unconstrained" least-squares LOP solution from the following, dynamically constrained filtering or smoothing operations permits the above solution iteration prior to filtering and saves computer time through the multiplication of smaller matrices and the implementation of a more efficient smoothing procedure. The filtered position is merged with the hydrographic data used to produce plots, error estimates, and diagnostics, and processed in the DAS to provide navigational guidance outputs to the pilot.

A filter can be described as an operation which estimates values of desired output quantities (the state vector) and their uncertainties (the covariance matrix) from a set of noisy measurements of related quantities (the observations vector) based on a model which relates the input and output parameters, the relative accuracy or importance of the observations (the weight matrix), and the pertinent noise factors. "Measurement" noise is that noise associated with the sensors, while "process" noise represents actual variations in the true states of the vessel compared to an idealized model such as a straight line or a mathematical curve. The object of a filtering operation is to reduce the errors in the state estimates caused by measurement noise on the observations without significantly altering the actual process (vessel motions induced by winds, waves, currents, and steering).

The real-time DAS filter is an augmented version of one developed by Houtenbos (1982) who combined it with the least-squares solution. His formalism, properly denoted as a Bayes filter with iterative differential correction, is of the type described by Morrison (1969), but has been extended to include process noise. The outputs or state variables which are estimated are x,y positions and speeds, and speed and heading offsets (between direct sensor measurements and track over the bottom). The Houtenbos approach is unique in that vessel dynamics are limited by invoking a priori pseudo-observations and constraints based on elementary equations of motion. It includes statistical limitations on vessel accelerations as well as incorporating vessel heading inputs and constraints, when available. We have chosen to add speed inputs and constraints in a similar fashion as well. This provides the added valuable benefit of being able to utilize the heading and speed data for dynamically constrained "dead reckoning" when primary positioning data is lost for short periods of time. The Bayes formulation is preferred over the mathematically equivalent Kalman formalism because the time-consuming matrix inversions are done in the smaller state space rather than in the larger observations (and pseudo-observations) space.

This improved filter reduces the standard deviation of the measurement noise and provides the opportunity to optimally utilize heading and speed data for positioning. Water speed and heading offsets are continuously estimated and updated in real time by comparisons with positions derived from the LOPs, when they are functioning.

Consequently, even such indirect "speed" measures as engine RPM and propeller pitch can be used successfully with only a rough initial calibration. The speed offset is set to zero when a Doppler speed (over the bottom) sensor is used. The net effect of the environment (winds, waves, and currents) on vessel motion in a given region, relative to measured speed and heading, is termed "current". A running estimate of the "current" is continuously determined from the offsets, and at signal outages the last value is used to augment dead reckoning. The quality of the fix, as determined in part by the covariance matrix and the closure of the LOPs relative to their standard deviations, is monitored and reported in real time in the DAS. This is one of the means by which biases can be detected. Warning flags are set when unacceptable errors accrue.

Further reduction of random noise can be achieved by additional processing (smoothing) off-line in the DPS because data is then available both before and after each time of interest. A smoother described by Houtenbos requires extremely large matrix inversions and is deemed impractical. An alternate approach, patterned after Mayne (1966), has been developed in which the filtered results from the DAS are combined in a weighted least-squares manner with results obtained by running a predictor (similar in nature to the filter) backwards over the "future" data from the unconstrained least-squares LOP solution. In this way, all available data are used without redundancy. Under typical conditions, the smoother will reduce the random noise component by roughly 30-50 percent over the filtered result. Because processing time for the smoother is roughly double that for the filter alone, the smoother will be optional and invoked only when required based on accuracy considerations.

To reiterate, the key features of the approach are:

- optimal use of overdetermined situations via multiple and hybrid LOPs;
- capability to handle range, range difference, azimuth, sextant angle, and latitude/longitude;
- sensor data "deskewed" to fixed, common times;
- use of auxiliary inputs such as speed, heading, and signal strength;
- data editing to suppress blunders or "fliers";
- geodetic calculations one or more orders of magnitude more accurate than data;
- filtering and smoothing with robust algorithms for random noise reduction and bias recognition;
- position solutions statistically limited by permitted vessel dynamics;
- real-time speed and heading offset and "current" calculation;
- ability to use dead reckoning between "sparse", manually recorded data points and during data outages of limited length, even during maneuvers, with low error;
- real-time digital displays for helmsman and hydrographer including real-time error estimates and diagnostic messages to the hydrographer.

For the remainder of this paper, the topic will be limited to a description of the design, testing, and performance of the filter algorithm.

3. FILTERING

3.1 Algorithms

The primary positioning sensor data for multiple LOPs, reduced via a standard, weighted least-squares algorithm, provides easting and northing (x,y) position estimates based solely on the noisy sensor data and unconstrained by limitations on vessel dynamics. The Houtenbos (1982) method of applying dynamical constraints in the generation of filtered paths involves the representation of elementary equations of straight-line motion as constraints along with the actual input data. The basic assumption or pseudo-observation is that while the helmsman is attempting to maintain a straight-line course at a speed "V", the mean acceleration is zero -- otherwise the course would be curved or the speed varying. The actual accelerations experienced by the vessel (the process noise) are modeled statistically as isotropic with an estimated standard deviation, σ_a, about the zero mean. This value is used in determining the weighting factors for the pseudo-observations. Turns whose centripetal accelerations (V^2/r) are not much greater than the selected value of σ_a are also accommodated by this model.

The "state vector" is composed of the quantities being estimated; i.e., it is the answer. In this case, the state vector is defined as $Y = (x, y, u, v, b_h, b_s)^T$, where x and y are the easting and northing components of the filtered position solution, u and v are their respective speeds, and b_h and b_s are the heading and speed offsets. The transpose notation (T) is invoked simply to write this column vector horizontally on the page to save space. Heading and speed are modeled via observation equations including offsets between measured quantities and the values over the bottom derived from the actual path or "track made good" as determined from the speed components of the filtered solution. The model assumptions or pseudo-observations which enter the constraint equations are that the mean rates of change of the heading and speed offsets are zero with standard deviations, σ_h and σ_s, which are used in determining weighting factors. Based on a straightforward analysis (Thomas 1985), the latter values have been coupled to σ_a via the relations $\sigma_h = \sigma_a/2v$ and $\sigma_s = \sigma_a/2$. In the weight matrix these values are multiplied by heading and speed "variance multipliers" which provide control over the coupling to σ_a in terms which can be related to perceived vessel motions. The heading variance multiplier is calculated from a "turning factor" which expresses the fraction of cross-track vessel movement caused by heading changes. The speed variance multiplier is calculated from a "correlation factor" which relates the response of the speed sensor (i.e., engine RPM) to in-track movements caused by the wave field.

For this filter (Thomas 1985,1986), based on a linearized, iterative differential correction approach, the object is to determine the changes in the state vector, Y, caused by changes in the measurements or observations change vector, ΔX. The "system" coefficient matrix, A, for the linearized observation equations is defined through the relationship $dX = A \, dY$, where the elements of A are $a_{i,j} = \delta X_i/\delta Y_j$. In the Houtenbos approach, the approximate or estimated values to be updated at the present iteration are not predicted formally, but rather, because of the frequent update rate and low speeds, simply set equal to the solution values at the previous iteration. The measurements change vector (observed minus estimated), composed of four true observations and six pseudo-operations, is constructed as follows:
$\Delta X(k)=[x_u(i)-x(i-1),y_u(i)-y(i-1),h_m(i)-h(i-1),s_m(i)-s(i-1),u(i-1)t,v(i-1)t,0,0,0,0]^T$
where i indicates the iteration step number, h is heading, s is speed, the u subscripts denote the unconstrained solution, the m subscripts indicate measurements, unsubscripted quantities at step i-1 are the previously estimated quantities, and t is the time between data updates. The formal,dynamically constrained least-squares solution or state vector update is $Y(i) = Y(i-1) + H(i) \, A(i)^T \, W(i) \, \Delta X(i)$, where W is the combined measurements and constraints weight matrix, and H is the state covariance matrix which is calculated as $(A^T W A)^{-1}$. The constraints weight matrix is updated at each step by adding the state covariance matrix from the previous step to the fixed covariance matrix of the acceleration effects.

There are a number of possible combinations of numbers and types (electronic and optical) of LOPs which are handled as special cases. The most important of these are the cases of either zero or only one active LOP. Included here is the range-azimuth case where the ranges arrive every second, but the "sparse" azimuth data may be entered only once a minute. Because no unconstrained position can be determined directly, each second, from measurement data alone under these circumstances, the speed and heading offsets can no longer be directly determined and must be removed from the state vector, along with the corresponding rows and columns from the system, weight, and covariance mastrices. If this is not done, the system of equations is underdetermined, and errors are partitioned, as rapidly as dynamic constraints permit, into both the offsets and the solution.

In order to compensate for this loss of real-time information, a separate "dead reckoning" formalism is invoked. The effects of wind, waves, currents, etc., on the vessel, as they affect the difference between measured speed and heading and the actual track, are reduced to a single net mean vector termed "current". Its x and y components are constantly updated from the speed and heading offsets, while at least two electronic LOPs are active, by averaging over a given period (30-120 seconds) of preceding data, i.e., "running boxcar" means. The standard deviations about the mean "current" components are used in the weight matrix, W. It can be noted that the lateral accelerations which cause the true path to differ from the planned path lead to a small speed bias, because they increase the path length but are not sensed by the speed measurement. The value of this bias is proportional to $(\sigma_a/V)^2$. Formally, this bias should not be part of the speed offset but rather included as a separate term in the speed equations so that the speed offset which results from the net "current" can be used to calculate an unbiassed estimator of the "current".

When the number of active electronic LOPs drops below two, the value of the "current" remains fixed at its last estimated value for application to the dead reckoning calculations in the filter involving measured speed and heading. For fewer than two active electronic LOPs, the measurement change vector is rewritten in terms of the current components, c_x and c_y. When a two LOP fix is again obtained, such as at the arrival of a "sparse" point, the value of the "current" is recalculated to reflect the observed difference between the new solution and the filtered prediction.

If the sensor complement drops to zero active electronic LOPs, the appropriate matrix elements are zeroed out, and the algorithm will continue to provide positioning and navigation outputs through the use of speed and heading data, the previously estimated net "current", and the dynamic constraints imposed on vessel motions. The algorithm, which is the result of examining the performance of a number of diverse formulations, has been designed specifically to function well even during maneuvers such as U-turns. Errors, of course, grow with time, and operation cannot continue indefinitely without accruing an unacceptable positioning bias. Estimates of the expected positioning error derived from the measured uncertainty in the "current" estimate and the LOP outage time are presented to the hydrographer to aid his judgement as to when operations must be terminated until LOPs can be reactivated.

For the one active electronic LOP case, since no unconstrained solution is possible, the measurements and observation equation coefficients are sent directly to the measurements change vector and system matrix, respectively, in the filter. The processing procedure depends on whether the one-LOP case was arrived at from two LOPs or from zero LOPs. For the former case, the one-LOP data is used in the solution. This mode of operation, which might be termed "augmented dead reckoning" provides superior performance over a longer period of time than dead reckoning with no incoming LOP data, but as with that case, errors grow with time. It is designed chiefly for the range-azimuth mode, but it also permits operational capability while waiting for lost LOPs to be reactivated. As with the zero-LOP case, positioning error estimates are continuously updated and provided to the hydrographer. For the zero-to-one LOP case, the dead reckoning position has already accrued potentially significant errors, and the addition of one LOP would, in general, not be particularly useful. Indeed, it could, for example, cause an apparent reversal in the ship's track. For this reason, the one-LOP procedure, when coming from zero LOPs, will be to simply report the residual (from the dead reckoning position to the LOP) to the hydrographer and continue dead reckoning.

Sparse optical data must be handled differently from electronic data because of the highly diverse data rates, and because the sparse data is entered manually into the computer some time (e.g. 10-30 seconds) after it is measured. At the times of the optical data measurements, the dynamically constrained, augmented dead-reckoning positions estimated at a 1/second rate will not agree precisely with the actual fix obtained including the optical measurements, but this will not be known until the data is entered manually and processed. At that point in the DAS, the unconstrained solution at the time of the measurement is calculated, the estimate of the "current" is updated, and a new position estimate based thereon is calculated at the actual clock time. In the DPS, the corrected values of "current" necessary for closure at the sparse, unconstrained solutions can be accessed from DAS outputs or precalculated (by looking ahead in the data) and applied immediately at the actual measurement time such that the augmented dead-reckoning solution will merge with the sparse, two LOP fix.

The values of σ_a, σ_h, σ_s, and the variance multipliers in the constraints weight matrix thus act as tuning parameters on the filter which regulate the rate at which changes in indicated position, velocity, and heading and speed offsets can occur. Small values of these parameters produce very smooth tracks which approximate the true path along straight lines but which cannot react without overshoot to rapid changes in course and speed. Large values of σ_a yield the ability to follow maneuvers but may not provide sufficient reduction of random measurement noise. Depending on the size of the vessel, the sea state, and the positioning system error, there generally exists a compromise value which provides adequate random noise reduction on lines without causing unacceptable biases on turns. This is true because in typical hydrographic operations, positioning accuracy on turns is not critical, and the only real requirements are to maintain lane count in hyperbolic systems and to shed biases and get back on line quickly after the maneuver. The availability of heading data greatly aids the filter in maintaining a low bias condition during turns.

The primary logistical factor which must be considered in designing a survey is to allow sufficient time on line after major accelerations (turns and speed changes) for the biases to damp out before highly accurate positions are required. The greater the noise reduction (small σ's) the greater the time needed for equilibration. Typical times for launches might be 10 - 30 seconds, while for ships it could take as long as 40 - 60 seconds for very small σ_a. If it were desired to further reduce the biases accrued during planned turns (at the cost of less measurement noise reduction) for some specific application such as lines ending in shoal waters, the values of the tuning parameters could be increased in real time according to navigation requirements. As will be seen, this is unnecessary in most cases because, with speed and heading data available, adequate noise reduction can be achieved without significant added bias.

3.2 Simulator

In order to evaluate performance against known quantities and prior to the availability of actual field data, the algorithms have been exercised first via a "track generator" which simulates the output of the preceding weighted least-squares solution converted to x-y coordinates to exercise the filter/smoother alone, and secondly by a "data generator" which simulates noisy LOP data for testing the observation equations and unconstrained least-squares code as well. Initially, a "planned" ideal path, selected from a menu containing a straight line, a 90-degree turn, a U-turn, an S-turn, or a racetrack, is constructed. The vessel travels at a selectable constant speed through the water. Process noise representing actual wind/wave-induced vessel track and speed deviations for random accelerations of selectable magnitude and five-second duration is calculated in along-track and cross-track components. The along-track magnitude is permitted to differ from the cross-track magnitude by a selectable factor to permit simulation of various wave fields and attack angles. The resulting deviations are interpolated to one-second intervals, converted to x and y deviations with the use of the heading information, summed, and applied incrementally to the x and y components of the planned path to produce the "true" path. The five-second duration, selected to be representative of the yaw rate of a survey launch under moderate sea conditions, provides desired cross-track deviations in the 5-10 m range for a cross-track acceleration of 0.5 m/s².

Instantaneous vessel heading values are calculated from the true path by invoking a selectable coupling factor appropriate for the size of the simulated vessel and the update period. Two types of simulated heading sensor errors are generated. The first is random one-second deviations of selectable magnitude corresponding to roll and pitch effects. The second is compass bias errors (due to its damping and subsequent delayed dynamic response) with magnitudes initially equal to the five-second process noise-induced course changes times a fractional multiplier called the "compass damping factor". These values are linearly damped to zero in five seconds and interpolated to one-second values. Random speed errors of 5-second duration and variable magnitude, interpolated to one-second values, are applied to the assumed speed through the water to simulate speed measurements. Simulated water currents may be applied to skew the path and yield different speeds over the bottom. For ease of computation, the currents are permitted to distort the planned path rather than the more complex case where headings would have to be modified in order to recreate the undistorted original. This shortcut has no effect on subsequent filter analysis or performance.

For the "track generator", random measurement errors with a priori standard deviation, σ_m, representing apparent deviations in the vessel track caused by random noise in the positioning sensor systems, are added to the x and y components of the true path to yield the "measured" path which is the input signal for the filter. The measurement errors consist of two populations: the typical, limiting random noise expected during normal operating conditions, and a "flier" population of selectable probability and magnitude which can introduce the infrequent but much larger spurious responses not uncommon in some systems. Antenna motion caused by vessel roll and pitch, although not totally random, is considered to be part of the measurement noise magnitude. Provision has been made to simulate data dropouts by turning off any combination of x, y, heading, or speed inputs halfway through the run.

For the "data generator", the position increments of the true path are geodetically inverted, according to the appropriate observation equations for each sensor type, to yield the corresponding LOP data values. Appropriate random measurement noise is added to these "true" LOP data values to produce characteristically noisy data. This is sent to the positioning algorithms, beginning with the calculation of observation equations, to test all following code. If varied observation time were added, the editor/deskewer could be exercised in a similar fashion. At this point, neither time nor a mixture of high-rate and sparse data have been simulated.

3.3 Performance

One measure of the performance of a filter or smoother is the ratio, R, of the standard deviation of the output about the true path to the standard deviation of the input (measurement) noise. These so-called "filtering or smoothing ratios" are functions of the ratio, Q, of the assumed process (acceleration) noise to the measurement noise ($Q \equiv \sigma_a t^2 / \sigma_m$). The numerator of Q depends on the size of the vessel and the sea state while the denominator depends on the positioning system. Theoretical performance curves for prediction, filtering, and smoothing with no heading or speed inputs, as reported by Houtenbos for the case of random process noise at the measurement period, are depicted in Fig. 1. These levels of performance have been confirmed with simulated data inputs conforming to the Houtenbos noise model. If the ratio, Q, is small, the measurement noise dominates, and the filter or smoother will be able to reduce its magnitude. If Q is large (i.e., unity or above), the measurement noise cannot be distinguished from the actual ship motions, and the filter and smoother become ineffective (R approaches unity). Note that for large Q the predictor actually degrades performance as R increases above unity.

The theoretical performance of the hybrid smoother (forward filter, backward predictor) for no heading or speed inputs, calculated from the predictor and filter curves, is indicated as a dashed line. It is a distinct improvement over the filter, particularly at small Q, but reflects the poor performance of the predictor at large Q. Although R for the hybrid smoother is slightly larger than for the theoretical Houtenbos smoother, it is probably no worse than for a practical implementation of the Houtenbos technique.

The availability of heading and speed inputs further lowers the filtering and smoothing ratios by amounts depending on the heading and speed error magnitudes via assumed values of σ_h, σ_s, and the variance multipliers. Heading information reduces cross-track positioning errors, while speed data reduces along-track positioning errors. Heading and speed inputs are also valuable in reducing biases accrued during maneuvers and in providing "dead reckoning" information when the electronic positioning system signals are lost for short periods of time.

Figure 1 cannot be directly applied to the more general case where the process noise is not random at the measurement update period. For SDS III, the LOP update rate will be once a second, while it is felt that typical vessel yaw motions are more appropriately represented by a roughly five-second period. The actual performance is expected to lie between the now overly optimistic Fig. 1 value based on σ_a and a pessimistic value obtained by replacing σ_a with $5\sigma_a$ (the value needed to yield the same total five-second track deviation in five summed one-second pieces).

Figure 2 includes plots of the a) "planned", b) "true", and c) "measured" paths from the track simulator for a straight-line case of 60 measurements at one-second intervals starting at the bottom of the figure with a constant 5 m/s speed. Process noise accelerations are isotropic at 0.5 m/s² RMS and yield maximum off-track deviations of about 10 m. This represents, for example, a 9-m survey launch in moderate seas. Simulated compass lag errors modeled for the inability to respond instantaneously to track direction changes are as large as 15 degrees with an RMS of about 8 degrees. Measurement noise in x and y is rectangularly distributed with a 5-m standard deviation and no fliers.

In Fig. 3, for the "true" path (3b) from Fig. 2b, filtered paths for σ_a=0.5 m/s² are compared, without (3a) and with (3c) heading data. For the latter case, the heading variance multiplier is 0.09, and estimated random heading error is six degrees RMS. Filtering ratios in the cross-track direction for the two examples are 0.48 and 0.32, respectively. Figure 3c exhibits significantly reduced random excursions and, as it should, clearly reflects the character of the process noise contaminated "true" path. R = 0.40 can be achieved for σ_a = 0.1 m/s² and no heading data, but typical turns of 0.5 m/s² would not be successfully negotiated at this value. Reducing σ_a to 0.1 m/s² for the case with heading data again yields R = 0.32, the same value obtained at 0.5 m/s². This is because the actual vessel excursions of 0.5 m/s² RMS are now being treated improperly as measurement noise and partially filtered out. Thus, for values of σ_a appropriate for maneuvers, the filter reduces the measurement noise by a factor of two without heading data and by a factor of three with heading data, for this particular parameter set. The use of speed data produces similar improvements in R for the in-track direction, although the percentage improvement is typically somewhat less due to the consideration that a larger variance multiplier is deemed appropriate.

Figure 4 demonstrates the utility of heading data in improving overall performance in a U-turn situation with a 50-m radius and 0.5 m/s² acceleration. Figure 4a depicts the "measured" path with 5-m RMS errors superposed over the "true" path with its 0.5 m/s² RMS process noise. Figure 4b depicts the filtered path for σ_a=0.5 m/s² for the case of no heading or speed data. This situation is an improvement over Fig. 4a, but it retains a large component of the measurement noise character. Reducing σ_a to 0.3 m/s², as seen in Fig. 4c, leads to a slightly smoother but clearly more biased result with maximum deviations of 9 m and 13 m in x and y. Although these are probably acceptable in the field under most conditions, Fig. 4d demonstrates the added gain from heading data. With σ_a remaining at 0.3 m/s², slightly less than the actual turning acceleration and process noise, the filter utilizing heading information has produced a result which nearly removes the measurement noise but adds little bias and clearly retains the character of the process noise -- the actual vessel motions!

This filter algorithm has proven to be very stable and robust. Attempts to cause it to "blow up" on unrealistically extreme maneuvers with mismatched parameters have generally failed. Although large temporary biases may be incurred in such instances, the algorithm continues to function in a reasonable manner and returns as quickly as it can to a satisfactory solution.

In instances where all primary positioning sensor data are lost during operations, the algorithm will use speed and heading inputs and the estimated "current" to predict position. The modified, zero-LOP algorithm is expected to provide useable results with primary data drops for periods of more than 30 - 60 seconds. Useable, here means that errors do not grow large enough to cause lane drops in phase measurement systems. Figure 5 is an example of a U-turn in which data from a 5-m RMS, 2-LOP positioning system are lost halfway through the turn. The filter, augmented with noisy speed and heading data and a "current" estimate (zero in this case), successfully completes the turn, and 30-seconds later, at the end of the simulation, the positioning errors from the true path are only 2 m in x and 5 m in y. These errors, which depend on the speed and heading noise and on the uncertainty in the estimated "current", are considered to be representative of typical operational conditions. The RMS heading errors were 3 degrees, true, and 6 degrees, estimated (to account for a compass damping factor of 0.8); and the speed errors were 0.25 m/s, true, and 0.15 m/s, estimated. Heading and speed variance multipliers were 0.09 and 0.50, respectively. Changing the speed variance multiplier to 0.95 caused no significant change in the final error magnitudes. These results are certainly far more than satisfactory considering the extreme conditions under which they were generated and the performance requirements under such circumstances.

Figure 6 displays the additional increase in accuracy which can be achieved through use of the bidirectional smoother. The parameters were set for less than optimum performance (e.g., use of no heading or speed data, and σ_a larger than necessary) to better demonstrate the effects visually. For this case with 0.5 m/s² process accelerations and an estimated σ_a of 0.5 m/s², the noise reduction ratios, R, for the filter in (x,y) are 0.49 and 0.59, and those for the smoother are 0.24 and 0.47. The theoretical equilibrium values predicted by the model are R_f=0.60 for the filter and R_s=0.45 for the smoother. The differences of the measured values from these theoretical performance predictions reflect the statistics of the particular arbitrarily selected random noise sequences used, and the fact that the model assumes one-second (the measurement rate) uncorrelated noise, while the process noise is really applied on a 5-second basis to more closely resemble physical expectations. The smoother provides an extra increase in measurement accuracy when needed, but at the expense of doubling the processing time over that for the filter alone, because of its bidirectional nature. The visual evidence of performance in Fig. 6 is certainly striking. The smoother has reproduced an amazingly faithful rendition of the true path which was heavily contaminated and made almost unrecognizable by measurement noise.

4. CONCLUSIONS

SDS III will be equipped with an integrated positioning architecture which makes optimal use of information from a wide variety of electronic and manual positioning devices as well as heading and speed sensors. Data from multiple lines of position from a variety of diverse sensors are combined in a least-squares algorithm followed by a sophisticated Bayes filter which utilizes measured speed and heading data and which invokes knowledge of the dynamic limitations of vessel motions. Heading and speed sensor offsets and the effective "current" are continuously estimated and updated. Random measurement errors are greatly reduced without significant alteration of actual vessel motions. Bias errors caused by poor data are detected and automatically corrected to the greatest extent possible. In the event of total loss of primary positioning data due to null zones or other temporary problems, the algorithm will continue to supply reasonable position estimates for times long enough to provide a good chance of reacquiring input data before the occurance of a lane loss or the need to stop due to error growth. Navigation guidance to the pilot and accuracy estimates to the hydrographer are provided in real time. Further improvements in accuracy can be achieved in post-processing by the use of a smoothing algorithm. This software system will provide the hydrographer with higher accuracies and more information, flexibility, and convenience than previously achieved with the same positioning systems.

5. REFERENCES

Bakker, G., de Munck, J. C., and Strang van Hees, G. L.,1985. "Radiopositioning for Marine Geodesy," Printed Course Lecture Notes, Delft University of Technology, Department of Geodesy, p. III-33.

Cross, P.A., 1981. "The Computation of Position at Sea," The Hydrographic Journal, No. 20 (April), 7-16.

Enabnit, D. B., 1985. "Shipboard Data System III," Proceedings of "Hydrography and Technology in the Mid-80s", the First Biennial Conference of the Canadian Hydrographic Service and Canadian Hydrographers Association, Halifax, Nova Scotia, April 1985, 44-49.

Houtenbos, A. P. E. M., 1982. "Prediction, Filtering and Smoothing of Offshore Navigation Data," The Hydrographic Journal, No. 25 (July), 5-16.

Mayne, D. Q., 1966. Automatica, 4, p 73.

Morrison, N., 1969. Introduction to Sequential Smoothing and Prediction, McGraw-
Hill, New York, p. 424.

Schiro, R. A., 1984. SDS III -- "The U.S. National Ocean Service's Next Generation
Hydrographic Surveying System," Proceedings of the National Ocean Service Hydrographic Conference, "Hydro 84," Rockville, Md., April 1984, 177-184.

Thomas, R. W. L., 1985. "A Study of Techniques to Improve the Performance of the
Shipboard Data System III Positioning System," Contractors Report WASC TR-X46-
003(85), EG&G Washington Analytical Services Center, Inc., 131 pp.

Thomas, R. W. L., 1986. "Positioning System Development for SDS III - Observation
Equation Derivation and Solution," Contractors Report WASC TR-8E6-0001(86), EG&G
Washington Analytical Services Center, Inc., 184pp.

Wallace, J., 1982. "Fifteen Years of Automated Hydrography in the National Ocean
Survey," Proc. of the Fourth Int'l. Symposium, "Hydro 82," Southampton, England,
December 1982, Vol. 3, Paper 1.

Weeks, C. G., 1984. "Field Experience with a New Automated Hydrographic Survey
System," Proceedings of the National Ocean Service Hydrographic Conference, "Hydro
84," Rockville, Md., April 1984, 59-68.

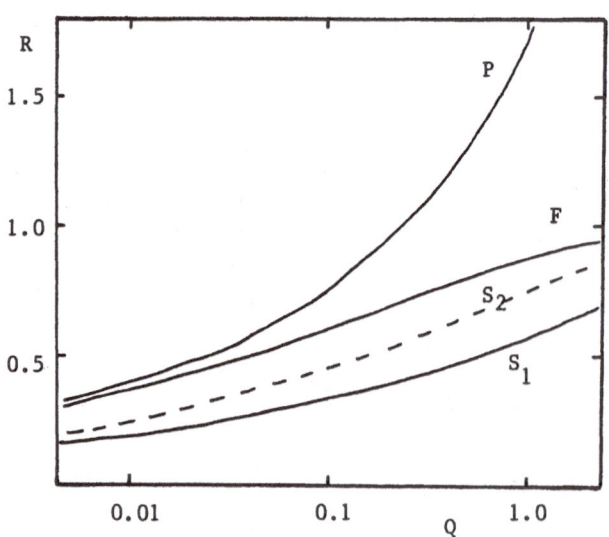

Figure 1. Noise reduction factors for predictor(P),
filter (F), Houtenbos smoother (S_1), and
bidirectional smoother (S_2).

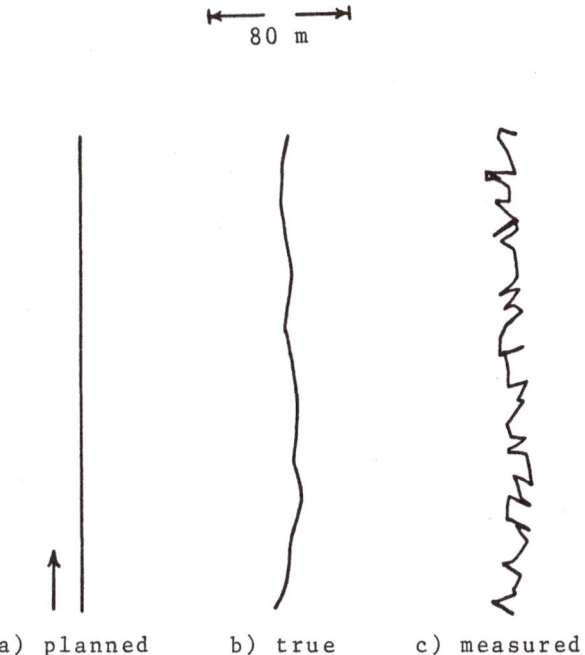

Figure 2. Data simulator output paths.

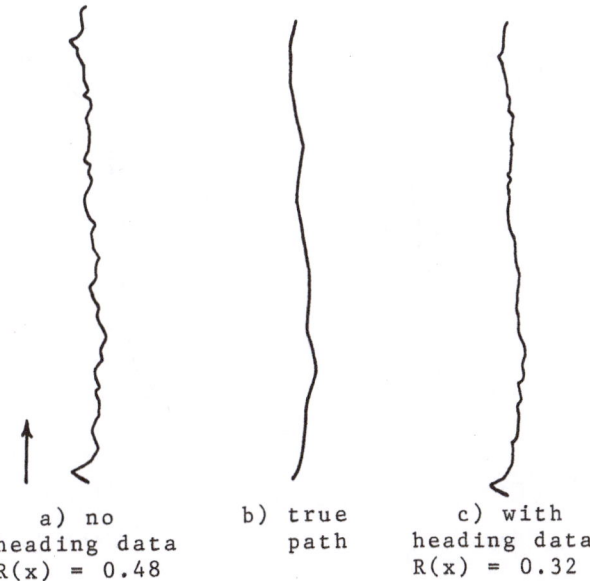

Figure 3. Filter outputs compared to true path for
0.5 m/s^2 RMS process noise and σ_a = 0.5 m/s^2.

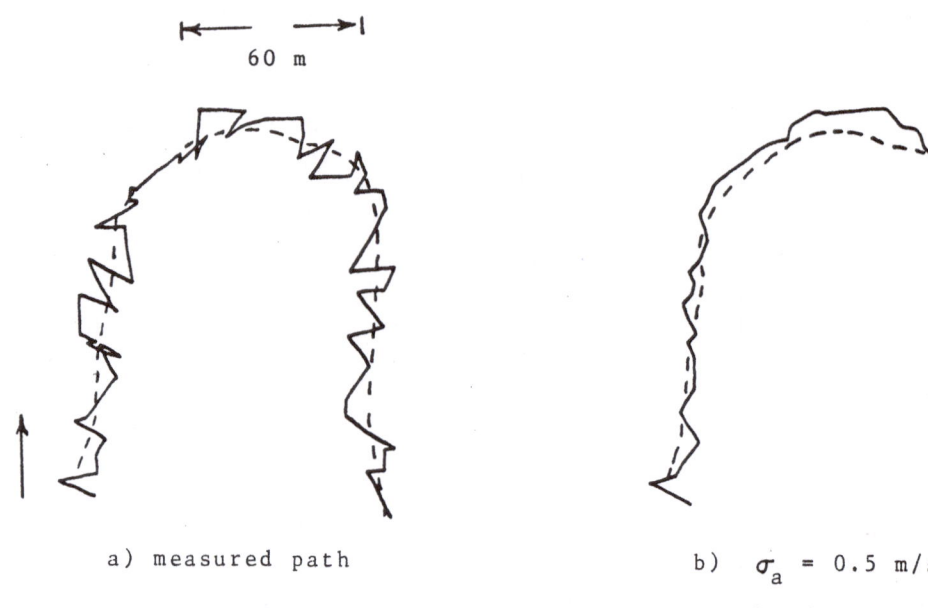

a) measured path   b) σ_a = 0.5 m/s^2

                                    no heading data

dashed line = true path

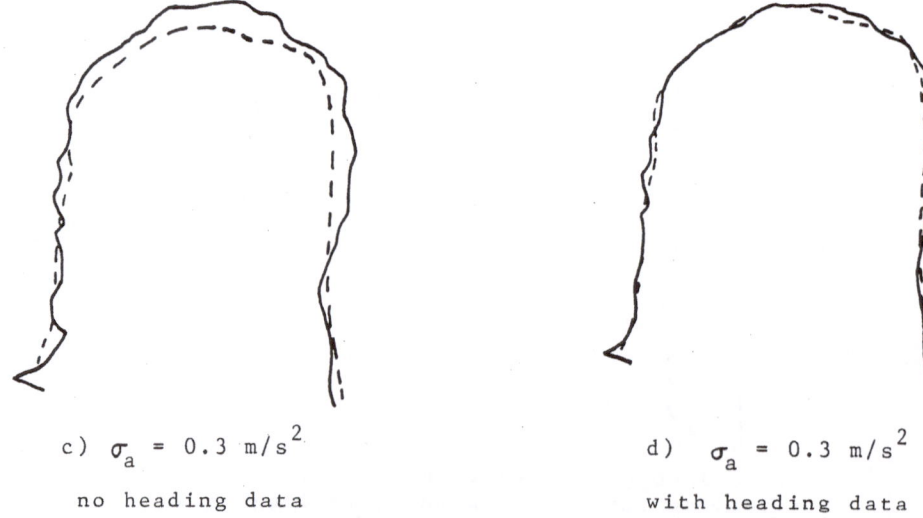

c) σ_a = 0.3 m/s^2 d) σ_a = 0.3 m/s^2

no heading data   with heading data

Figure 4. Filter output for 50-m radius U-turn at 5 m/s.

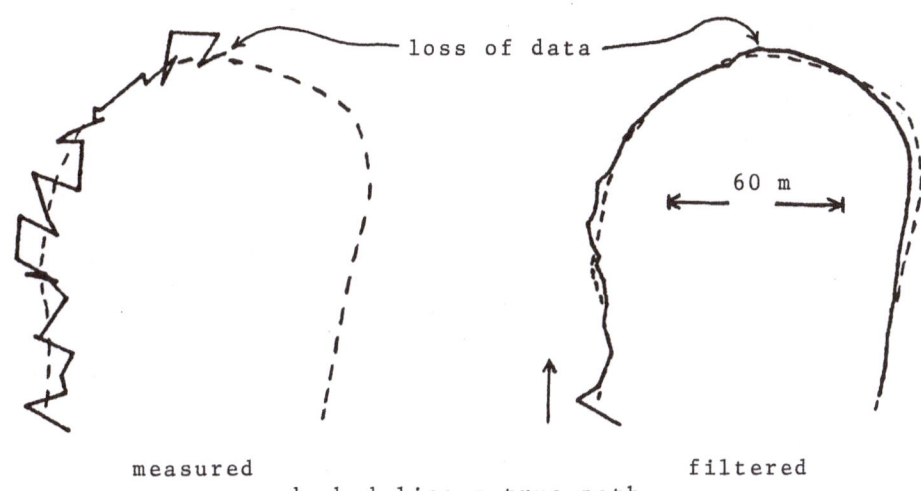

Figure 5. Measured and filtered paths for U-turn
with dead reckoning after loss of positioning data.

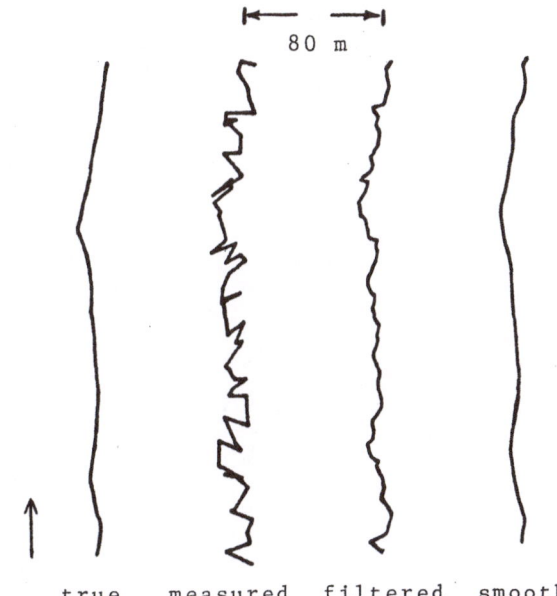

Figure 6. Example performance of the filter and
smoother for 5-m measurement noise, 0.5 m/s^2
process noise, and σ_a = 0.5 m/s^2; no heading, speed.

GPS MARINE KINEMATIC POSITIONING
ACCURACY AND RELIABILITY

G. Lachapelle[1]
M. Casey[2]
M. Eaton[3]
A. Kleusberg[4]
J. Tranquilla[5]
D. Wells[4]

ABSTRACT

Positioning accuracy and reliability are two parameters of major concern to GPS marine users. GPS limitations and error sources are analysed from a user's point of view. The coverage limitations inherent to the (18+3) satellite constellation are discussed and related outages inherent to specific geographical areas are described. Receiver and antenna characteristics are described in relation to their impact on positioning accuracy and reliability. Antenna requirements for shipborne applications are discussed. The effect of the troposphere and ionosphere on GPS measurements is quantified for various cases. Multipath effects on both P and C/A code and carrier phase measurements are discussed. A variety of user accuracy and reliability enhancements are described. These include height and time constraints to limit outage effects, the use of differential methods to reduce satellite and propagation medium errors, the integration of code and carrier phase measurements to increase differential positioning accuracy, and the use of external sensors such as pitch and roll sensors, and inertial navigation systems. Sample differential land kinematic results obtained over a baseline of 1,000 km are described to illustrate the level of accuracy currently obtainable with GPS.

(1) Nortech Surveys (Canada) Inc., 319 - 2 Ave, S.W., Calgary Alberta T2P 0C5
(2) Canadian Hydrographic Service, Department of Fisheries and Oceans, 615 Booth Street, Ottawa, Ontario, K1A 0E9
(3) Canadian Hydrographic Service, Bedford Institute of Oceanography, Dartmouth, Nova Scotia, B2Y 4A2
(4) Department of Surveying Engineering, The University of New Brunswick, Fredericton,New Brunswick, E3B 5A3
(5) Department of Electrical Enginering, The University of New Brunswick, Fredericton, New Brunswick, E3B 5A3.

M. Kumar and G. A. Maul (directors), Marine Positioning, 113–147.
© 1987 by the Marine Technology Society.

1. INTRODUCTION

The 21-satellite GPS constellation will provide, when complete, a powerful tool for multi-purpose marine applications. These will typically range from kinematic positioning for surface hydrographic surveying (e.g., Lachapelle et al., 1984a) and airborne hydrographic bathymetry (e.g., Lachapelle et al., 1984b) to precise static positioning for the monitoring of offshore structures (e.g., Collins, 1986; Lachapelle, 1986). In this paper, the discussion will focus on kinematic applications. The current constellation of seven satellites has already been used very successfully for a variety of research and operational purposes. The complete constellation, which will consist of 18 satellites plus 3 active spares (18+3), is meant to provide U.S. Department of Defense (DoD) users with 10 to 15 m absolute accuracies 24 hours per day on a worldwide basis. The corresponding accuracies available to civilian users will be somewhat less, namely 100 m, due to the use of the one frequency C/A code in a degraded mode.

In order to improve the above 100 m accuracy by 1 to 3 orders of magnitude with a high degree of reliability, the user will have access to a variety of techniques which are a function, among others, of the characteristics of the receiver and antenna used, the state of the propagation medium, and the baseline length over which differential positioning techniques are applied. The reliability of the GPS user's system, i.e., its ability to yield a specified accuracy at a specified location for a specified duration, will be strongly affected by the geometry and proper operation of the satellites. In fact, predictable and unpredictable outages (periods of poor satellite geometry) will be a major parameter affecting adversely the reliability of the user's system. The use of accuracy and reliability enhancement techniques is better understood by a discussion of GPS satellite coverage limitations and errors sources due to user's equipment and the propagation media. This will be the subject of Section 2.   Enhancement techniques are discussed in Section 3. These consist in the use of height and time constraints to provide measurement redundancy, the integration of code and carrier phase measurements to increase accuracy, the utilization of differential positioning techniques to increase both accuracy and reliability, and the use of external sensors such as pitch and roll sensors and inertial navigation systems to provide attitude data required for a variety of shipborne and airborne marine applications.

2. GPS LIMITATIONS AND ERROR SOURCES

2.1 Coverage Limitations and Outages

The (18+3) GPS satellite constellation will consist of 6 planes inclined 55^0. This constellation, although more optimal than a constellation consisting of the same number of satellites distributed over 3 planes (Jorgensen, 1984), will still contain outages, i.e. time intervals in specific areas during which the Dilution Of Precision (DOP) is higher than a specified number, thus decreasing the reliability of GPS as an instantaneous positioning system below an acceptable level. The 3 additional satellites of the (18+3) constellation will be configured to reduce the number of outages (Kalafus, 1984). Outages will occur in specific areas of the world located along 4 parallels of latitude, roughly at 60^0N, 45^0N, 45^0S, and 60^0S (Doucet, 1986). The duration and size of the areas affected by these outages will depend upon the mask angle utilized. Figure 2.1 shows the number of satellites (with a mask angle of 5^0) available at latitude 65^0N and longitude 275^0E (Northern part of Hudson Bay area) as a function of time over a 24-hour period for the (18+3) satellite constellation. The number of satellites available is always greater than 5 except for 3 periods of short duration during which it is 5.

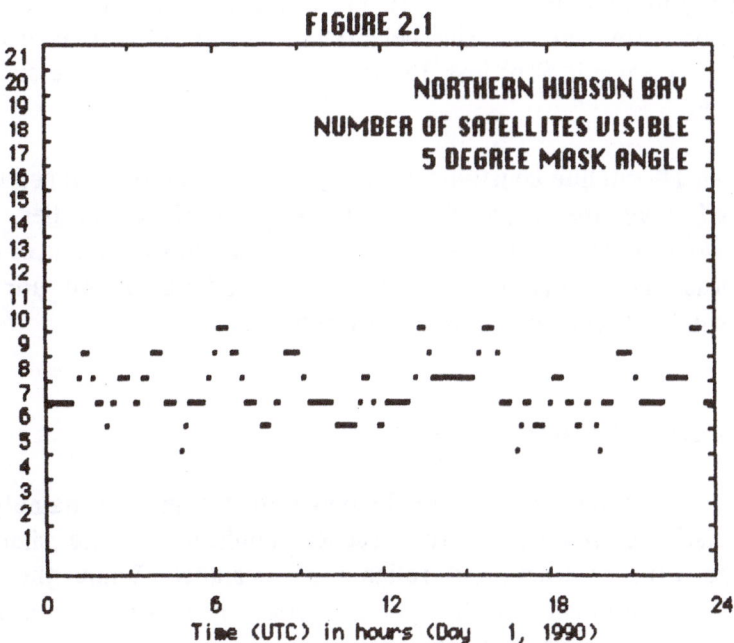

FIGURE 2.1

The corresponding PDOP (Position Dilution of Precision) is shown in Figure 2.2. The DOP's shown in this paper are based on the assumption that all visible satellites can be tracked simultaneously. Outages (PDOP > 5) of less than 10 minutes occur at epochs when only 5 satellites in a poor geometry are available. 3-D positioning would become unreliable during these periods. The use of height and time constraints will improve the situation, as will be discussed in Section 3.1. Figure 2.3 shows the corresponding PDOP when a mask angle of 10^0 is used. The number of outages increases to 6, their duration also increases significantly, and the area over which they extend also increases.

Figures 2.4 and 2.5 show the number of satellites and PDOP for the Amazon, with a 5^0 mask angle. The PDOP value is always better than 3 and no outages occur. The number of satellites available is always seven or greater except for two short periods with only six visible satellites. This shows that greater reliability will be more easily maintained in equatorial regions.

In addition to outages due to the limitations of the (18+3) satellite constellation described above, unpredictable outages will occur due to satellite malfunctioning. The location and size of areas affected by such outages will depend upon the number of simultaneous failures and the position of the failed satellites. Kalafus (1984) showed that no outages will occur over the continental United States due to a failure of a single satellite from the (18+3) constellation, provided that the height-constrained solution is used and/or that the mask angle is 5^0 or less. In other areas of the world, the situation will be different.

For an 18-satellite constellation, the areas affected by outages would still be located along the 4 parallels of latitude mentioned earlier, but their number would increase substantially (e.g., Doucet, 1986). Means of enhancing the accuracy and reliability of GPS during predictable and unpredictable outages will be discussed in Sections 3.1 and 3.4.

2.2 Receiver Characteristics

These characteristics can be divided into two groups, namely physical characteristics of importance for practical applications, and characteristics associated with accuracy and reliable performance. While the discussion herein will concentrate on the second group, it is worth mentioning the characteristics associated with the first group.

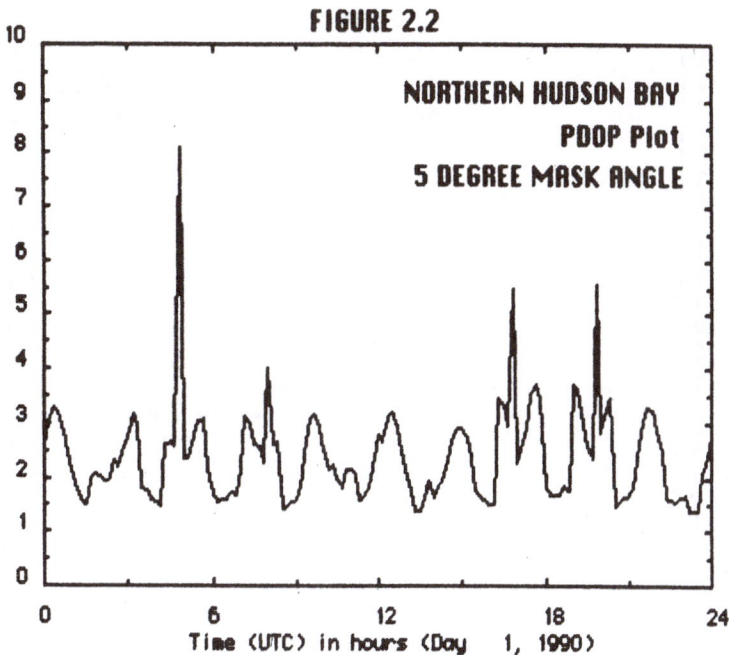

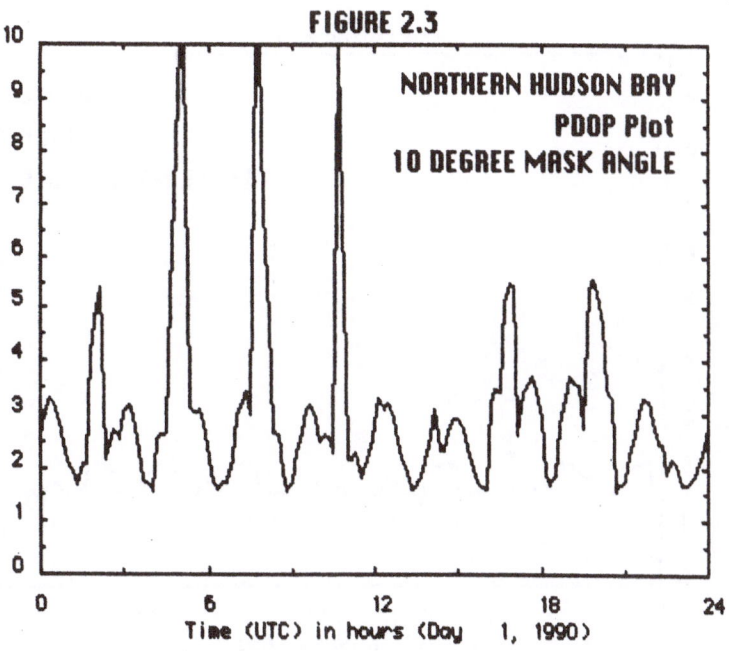

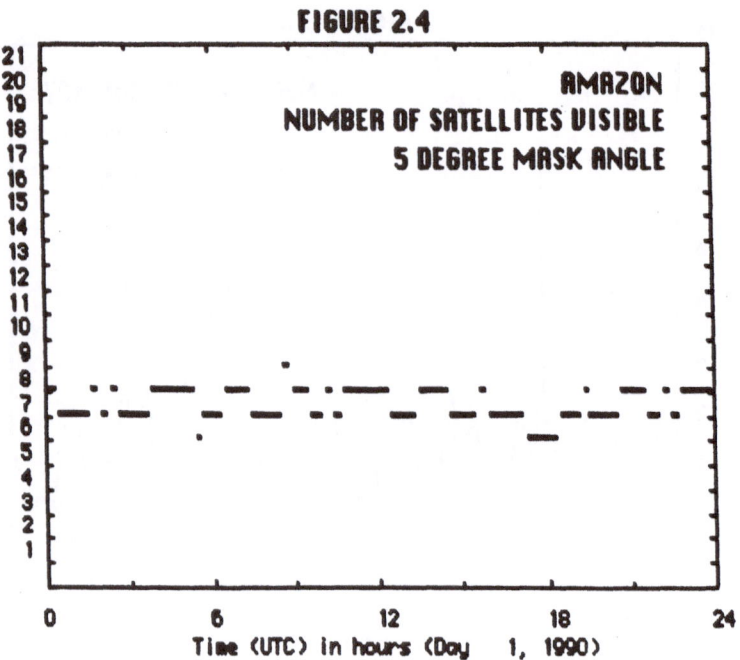

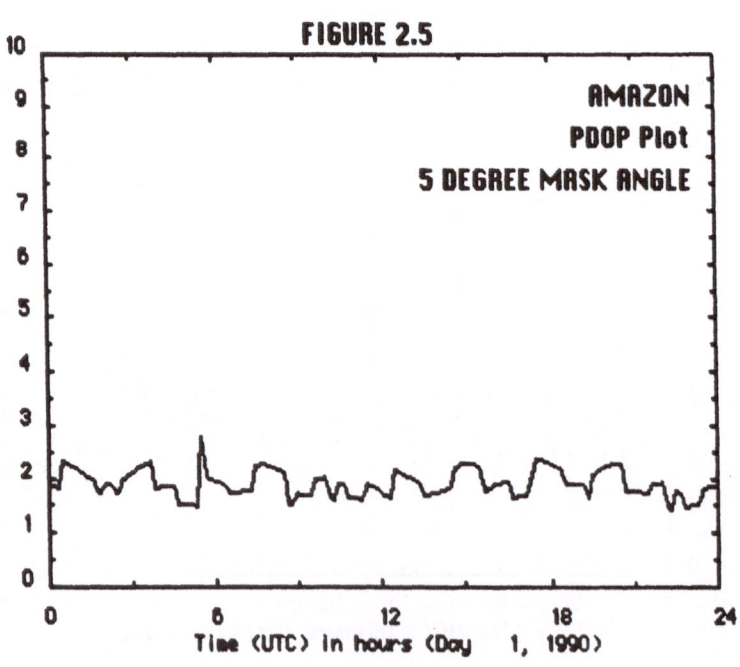

Physical characteristics of special importance for marine applications include small size, flexible electrical power requirements, ruggedness, water resistance, buoyancy, self checking mode, availability of interfaces for external sensors, antenna cable length, antenna characteristics, and cost. The relative importance of each of these parameters is obviously a function of the specific application sought, shipborne applications typically ranging from hydrographic surveying with 10 m survey launches to multipurpose oceanographic missions with 100 m vessels. Airborne marine applications call for additional physical characteristics, such as a flat and rugged antenna shape.

The following characteristics, which are of importance for obtaining good performance in terms of accuracy and accuracy reliability, will be discussed in this Section: (1) number of channels, (2) dual versus single frequency, (3) P code versus C/A code, (4) availability of continously integrated phase measurements, (5) raw data availability, and (6) tracking bandwidth. The antenna characteristics are discussed in the next Section.

2.2.1 Number of channels

Early GPS receivers sometimes consisted of only 1 channel operating sequentially between satellites. If this sequencing was slow compared to the motion dynamics, the GPS pseudo-range measurements were aided with external sensors such as ship's log and gyro measurements. This was the case of the STI5010 (e.g., Lachapelle, 1983), which became available in 1980. With such a sequential receiver, no continously integrated phase measurements are available and positions are derived solely from pseudo- range measurements. A major improvement was the introduction of Texas Instruments' TI4100 in 1982 (Ward, 1982), a P code multiplex receiver which has basically the same advantages as a 4-channel receiver. Many other geodetic type receivers currently available for marine purposes, e.g., Trimble 4000A (Ashjaee and Helkey, 1984), Motorola Eagle Mini-Ranger (Durboraw, 1986), are 4-channel receivers. Figures 2.1 and 2.4 show that a receiver with up to 7 channels will contribute to enhancing positioning reliability through increased redundancy in most parts of the world once the (18+3) constellation is available. Five-channel receivers are already available on a commercial basis (e.g., Trimble 4000SX, Magnavox T-Set). Six-channel receivers are also in use for geodetic, rather than marine, applications (e.g. Macrometer V-1000, Wild-Magnavox WM101). Many manufacturers have no doubt plans for 6 or 7-channel marine receivers.

2.2.2 Dual versus single frequency

Dual frequency measurements are important to correct for the effect of the ionosphere. While the upper limit of this effect will be discussed in Section 2.4, it is important to note at this point that, for sub-metre accuracy, the use of dual frequency measurements is highly desirable. Although the differential method can reduce the effect of the ionosphere very significantly, residual effects of the order of 1 m or more could occur if the differential baseline exceed a few hundred km. Dual frequency measurements can also be useful to detect and correct for cycle slips, as discussed in Section 3.3.

The TI4100 is the only dual frequency receiver currently available commercially for kinematic applications. However, C/A code receivers are being developed which provide code and carrier measurements on L1 and carrier measurements on L2 through a codeless technique (e.g., Counselman et al., 1986). This type of receiver, which can be effectively used for kinematic applications, does not require access to the P code. This is of paramount importance for civilian use.

2.2.3 P code versus C/A code

Although the P code will be denied to civilian users, it is worthwhile to stress some of the differences between P and C/A code from an accuracy and accuracy reliability aspect. The P and C/A code rates are 10.23 MB/s and 1.023 MB/s, which translate into code chip sizes of 29.3 m and 293. m respectively. The ultimate accuracy ratio between P and C/A code measurements is dictated by this chip code ratio of 10. The actual measuring accuracy is also a function of other parameters such as tracking bandwidth (See sub-section 2.2.6 below), carrier-to-noise density ratio, and code mechanization parameter constant, etc (Martin, 1980). For benign GPS applications, the pseudo-range (P code) noise will typically be of the order of 1 to 2 m (1σ). The corresponding C/A code values will evidently be 10 to 20 m. For high dynamics applications, in which case the tracking bandwidth is increased, the noise increases accordingly.

In addition to providing higher accuracy, the P code, in view of its smaller chip size, is more immune to large multipath errors than the C/A code. This will be discussed in more details in Section 2.4.

2.2.4 Availability of continously integrated phase measurements

Continously integrated carrier phase measurements are used to provide a higher positioning accuracy than that available with code measurements only. In view of the accuracy limitations placed upon the single point user by orbital errors, effective accuracy gains with phase measurements are obtained in conjunction with the differential method. If a code receiver is used, carrier phase and code measurements are combined as discussed in Section 3.3. The accuracy of carrier phase measurements is, among others, a function of the carrier-to-noise ratio, and is typically of the order of 3 to 11 mm (on L1) when measurement noise and quantization errors are taken into account (Martin, 1980). This accuracy is 2 orders of magnitude higher than that of code measurements. Phase measurements are also more immune to multipath effects as will be discussed in Section 2.4.

Most geodetic type receivers currently available commercially (e.g., TI4100, T-Set, Motorola Eagle Mini Ranger, Trimble 4000S) provide continously integrated phase measurements.

2.2.5 Raw data availability

Most receivers provide internally a filtered position solution based either on pure code measurements or on a combination of code and carrier phase measurements. While this is an important advantage for real time single point positioning applications, access to raw data is required for high accuracy differential applications and for applications where the integration of the raw measurements with other sensors is sought. Access to the receiver's raw data is therefore important for several types of application. Ideally, this data should be available directly through a standard interface.

The rate at which raw data is measured and made available to the user is important for high accuracy applications if the dynamics of the vehicle are high. Many cycle slip detection and correction methods are based on phase velocity linearity assumption between successive measurement epochs. In a high dynamic case, such as an hydrographic survey launch where antenna dynamics are particularly high, availability of carrier phase measurements at intervals of less than 1 s will be important for maintaining sub-metre accuracy. Receivers currently available provide raw data at intervals of about 1 s, although the actual measurements may be made at much smaller intervals.

Raw data is usually processed through the receiver's internal tracking filter. Such a procedure may produce time delays which could affect adversely the positioning accuracy under high dynamics. The type of delay which may occur under such a situation is illustrated in Figure 2.6, which is extracted from (Lachapelle et al., 1986a). Latitude and longitude differences for periods of 1.2 seconds, which is the data rate of the TI4100 in U.D. 2, are given as a function of time. The test was conducted in land differential pseudo-range mode. From Epoch 40 seconds onwards, the vehicle was moving in an East-West direction at a velocity of approximately 90 km/h (30m/1.2 s). At Epoch A, the brakes were applied, and the vehicle was brought to a complete stop 5 seconds later, namely at Epoch B. The longitude differences did not reach 0 until 3 seconds later. This 3 second time lag is presumably due to the internal tracking filter of the TI4100. Effective countermeasures to such a phenomena are yet to be developed.

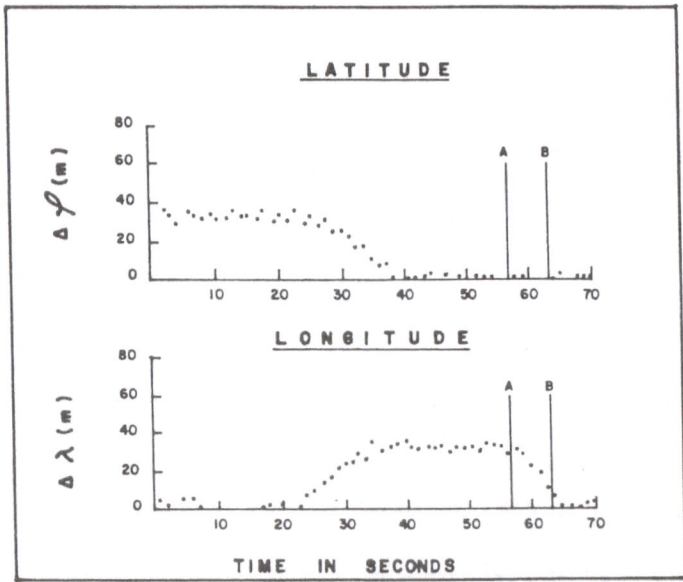

FIGURE 2.6: STOP TEST TO DETECT TI4100 RECEIVER TRACKING FILTER LAG IN U.D. 2. THE VEHICLE IS MOVING IN AN EAST-WEST DIRECTION WHEN BRAKES ARE APPLIED AT EPOCH A. COMPLETE STOP OCCURS AT EPOCH B. A LAG OF 3 SECONDS IS APPARENT AS SHOWN BY THE POSITION DIFFERENCES IN LONGITUDE. THESE POSITION DIFFERENCES ARE BASED ON DIFFERENTIALLY CORRECTED PSEUDO-RANGES AND ARE GIVEN FOR SUCCESSIVE INTERVALS OF 1.2 SECONDS. THE TEST WAS CONDUCTED IN ALBERTA ON DAY 74, 1985, OVER A BASELINE OF SOME 95 KM.

2.2.6 Tracking bandwidth

The tracking bandwidth is one important parameter which determines the ultimate code and carrier phase measuring accuracy as discussed in the previous sub-sections. A narrower tracking bandwidth results in more accurate measurements. However, the tracking bandwidth should be wide enough for the receiver to maintain phase lock under prevailing dynamics. The use of an optimal tracking bandwidth is therefore important. A receiver should ideally have an adjustable bandwidth available for applications ranging from high accuracy static differential positioning to high dynamics applications.

Figure 2.7 shows the carrier phase measurement noise obtained with a TI4100 using successively three of its four User Dynamics mode, which correspond to bandwidths of 5, 8, and 16 Hz. The maximum accelerations allowed for each of these bandwidth values are 6, 15, and 40 m/s^2 respectively. U.D. 1 would be typically used for low dynamics land applications while U.D. 3 would be applicable to a survey launch where antenna accelerations in excess of 4 g are possible (Scavuzzo and Pakstys, 1985).

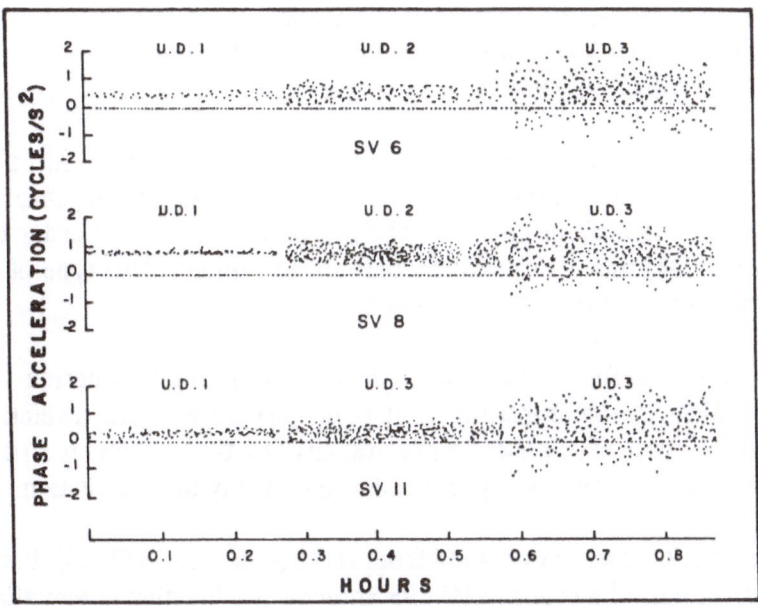

FIGURE 2.7: PHASE ACCELERATIONS, IN CYCLE/SEC^2, MEASURED AS A FUNCTION OF TIME, IN STATIC MODE, WITH A TI4100 RECEIVER IN USER DYNAMICS (U.D.) 1, 2, AND 3. GPS SATELLITES WERE OBSERVED FOR 0.85 HOURS.

2.3 Antenna Characteristics

In addition to physical size, shape, and ruggedness, two important antenna characteristics are amplitude and phase patterns. These two characteristics are a function, among others, of antenna types. At least 4 antenna types are used currently for GPS applications, namely the quadrifilar helix, the conical spiral, the monopole, and the microstrip type. The characteristics of each type have been studied and reported by Tranquilla & Best (1986a) for the quadrifilar helix type, Tranquilla et al. (1986b) for the conical spiral type, Tranquilla & Colpitts (1986) for the microstrip type, and Tranquilla & Best (1986b) for the monopole type. The characteristics are also a function of other parameters such as specific design utilized and quality of construction.

The polar amplitude response pattern is basically the amplitude sensitivity of the antenna as a function of observation angle. The angle cut-off pattern requirements are dictated by the applications sought. For shipborne applications, an antenna with an adequate response down to 20^0 below its horizon is preferable to compensate for ship's pitch and roll when low satellites are tracked. The disadvantage associated with a low angle amplitude response antenna is its increased susceptibility to multipath and other interference. This is why antennas with a minimal amplitude response below angles of 20^0 (above the horizon) are preferred for precise static applications.

Figure 2.8 shows a typical amplitude response pattern for a TI4100 conical spiral antenna at the L1 frequency. A small drop in response is noted at 10^0 above the horizon. At -10^0, the response is still of the order of -5 dB. At angles of less than -15^0, the response drops rapidly. This type of antenna is well suited for shipborne applications.

The phase response pattern of an antenna is used to derive its phase centre, which is the apparent centre of curvature of its radiation field equiphase contour. The phase centre stability of a GPS antenna is a function of parameters such as antenna type, design, and quality of construction.

Figure 2.9, which is extracted from (Tranquilla et al., 1986a), shows phase centre variations of a commercial GPS antenna as a function of elevation angle. Phase centre variations of about 5 cm are noted. A study of three specific GPS antenna types currently available commerically, undertaken by Nortech Surveys (Canada) Inc. and The University of New Brunswick for the Canadian Hydrographic Service, showed that the phase centre stability was of the order

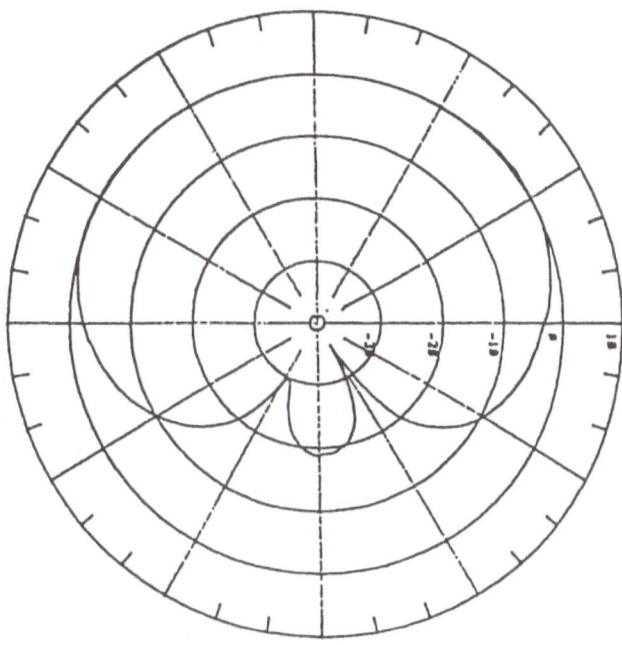

FIGURE 2.8: AMPLITUDE RESPONSE PATTERN OF TI4100 RECEIVER CONICAL
SPIRAL ANTENNA AT THE L1 FREQUENCY.

of 10 cm for anyone of the 3 antennae tested. Two of these antennae were of
the quadrifilar helix type while the third one was of the conical spiral types.
Comparisons between the 2 quadrifilar helix antennae tested showed large
differences between low elevation amplitude response patterns and other
characteristics. Antenna type is therefore not the only criterion which defines
the suitability of an antenna for a specific applications. The particular design
and quality of construction are equally important.

The characteristics of antennae are also affected by the environment. This
is best seen in Figure 2.10 (Tranquilla et al., 1986a) where the polar amplitude
pattern and phase centre variation of a quadrifilar helix antenna are given for
both a multipath free case and for a case when strong multipath interference
is induced. Relative phase centre variations of 8 cm occur between the 2
cases. This illustrates the strong dependence of antenna performance on
prevailing environmental factors.

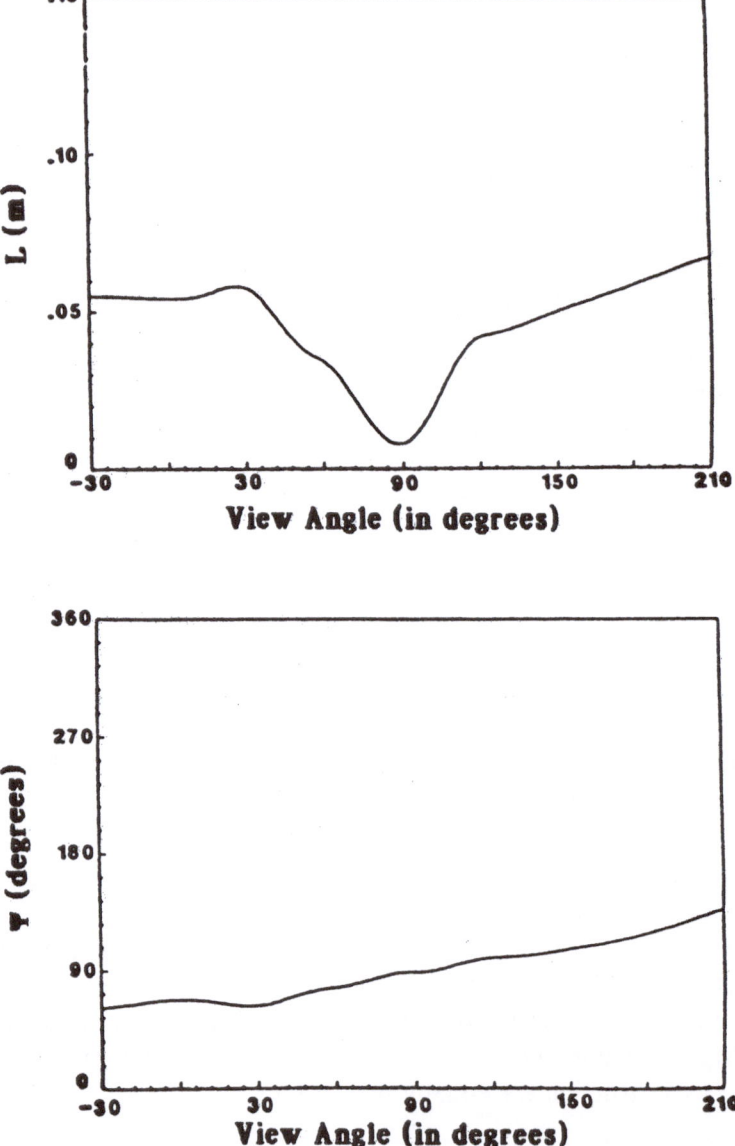

FIGURE 2.9: COMPUTED PHASE CENTRE LOCATION OBTAINED FROM MEASURED PHASE RESPONSE OF A COMMERCIAL GPS ANTENNA. PHASE CENTRE LOCATION IS DESCRIBED BY A RADIAL DISTANCE, L, MEASURED FROM THE COORDINATE REFERENCE ORIGIN, AND AN ANGLE Ψ, MEASURED FROM 0 ELEVATION IN THE OBSERVATION PLANE. MEASUREMENTS WERE MADE IN A POLAR (ZENITH) OBSERVATION PLANE.

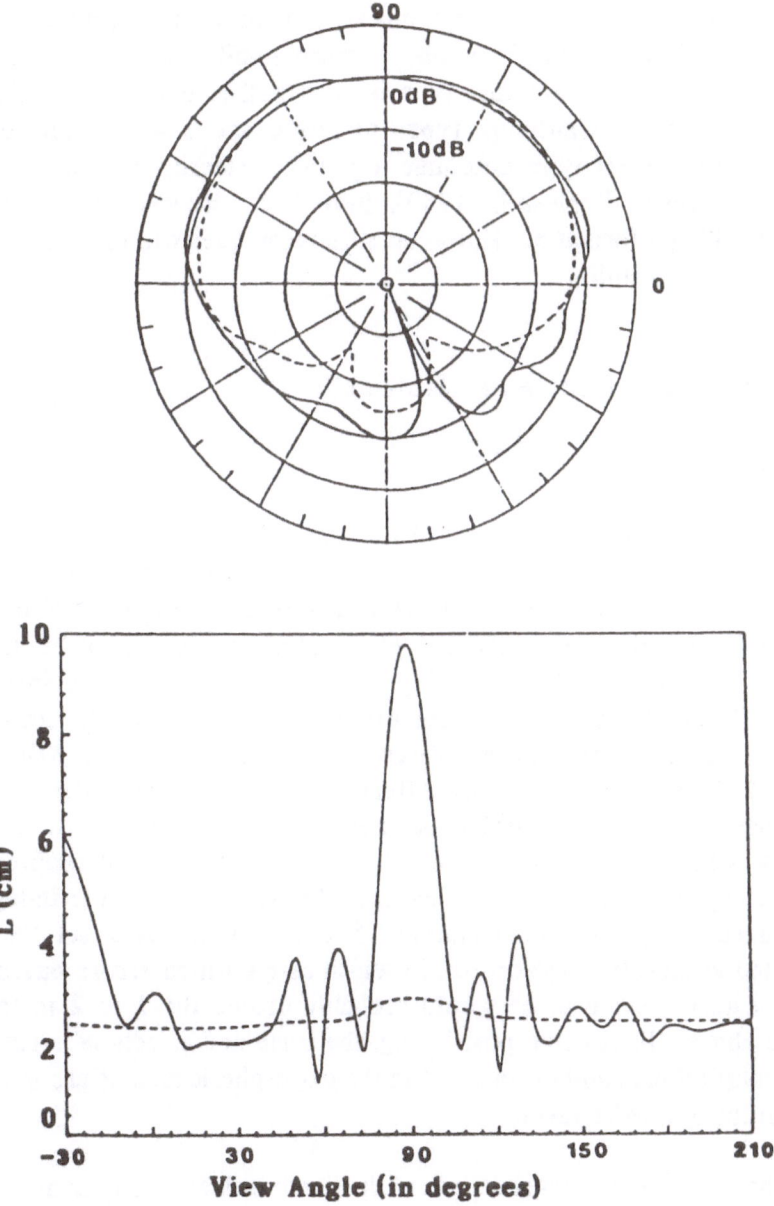

FIGURE 2.10: MEASURED POLAR RADIATION PATTERN AND COMPUTED PHASE CENTRE POSITION FOR A QUADRIFILAR HELIX GPS ANTENNA IN AN ISOLATED ENVIRONMENT (---), AND WITH MULTIPATH INTERFERENCE (___) CAUSED BY A REFLECTING SHEET INSERTED BETWEEN TRANSMITTING AND TEST ANTENNAE.

The microstrip type antenna appears to have the potential of combining the advantages of several other antenna types with few of the inconveniences (Tranquilla & Colpitts, 1986). A phase centre stability in the 1 cm range is likely achievable. This antenna type is also well suited for high kinematic applications. All the antenna types discussed above are omnidirectional. However, directional type antennae, e.g., beam steering and adaptive null steering antennae (Hemesath, 1980), have been proposed as methods of enhancing GPS performance. However, such techniques will inevitably lead to degraded phase stability.

2.4 Atmospheric Effects and Multipath

2.4.1 Troposphere

The effect of the troposphere (h < 50 km) on pseudo-ranges is a function of the satellite elevation and of various tropospheric parameters. The delay varies from a few metres for a 90° elevation to some 25 m for a 5° elevation. Below 15° elevation, the effect is exponential and can reach 50 m at the horizon. Various models (e.g., Hopfield's, Black's, Saasta- moinen's, and Yionoulis') have been developed for estimating the effect of the troposphere based on surface pressure, temperature, and relative humidity. Ashkenazi et al. (1982) found that the use of different models could result in coordinate component differences, especially in height, of the order of 1 to 2 m. This study was based on absolute positioning results with Transit (Doppler) data but also applies to GPS observations since the refractive index is independent of frequency (up to approximately 15 GHz). Their data set was likely truncated at elevation 15° or 20°, in which case position errors based on the use of data truncated at 5° would probably exceed the 1 to 2 m threshold quoted above. In relative positioning, the combined effects of errors in the meteorological data and differences in the tropospheric models are more likely to be at the decimetre level.

The refractivity consists of a dry and a wet component. The dry component is quantatively the most important while the the wet component is the most difficult to estimate. This is due to its dependance on water vapour (h < 12 km) which is difficult to predict from surface measurements. This is why the tropospheric effect is relatively more difficult to predict accurately for low elevation satellites. The variability of the water vapour also means that, for distances larger than a few tens of km, the tropospheric effects cannot effectively be removed using the differential method. This is especially true for marine applications where the monitor station will typically be on land while the master (mobile) station may be a few hundred km at sea. Hence, the troposphere remains an obstacle for sub-metre accuracy marine differential positioning over baselines of several hundred km.

2.4.2 Ionosphere

The ionosphere (h > 50 km) causes a time delay of the signal and a carrier phase advance which are a function of the Total Electron Content (TEC) along the path of the signal and of the frequency utilized. Another important ionospheric effect on GPS signals is amplitude and phase scintillation (Klobuchar, 1983). Ionospheric effects are frequency dependent and can be evaluated accurately using dual frequency measurements. At GPS frequencies (1575 MHz and 1227 MHz), the ionospheric time delay translates into a pseudo-range correction of the order of 5 to 50 m. The TEC along the wavepath is a function of the elevation of the satellite and of the state of the ionosphere at observation time. The state of the ionosphere is affected by various phenomenae such as solar activity and geomagnetic storms (aurorae). In the early 80s, when solar activity was at a maximum, ionospheric effects on pseudo-ranges reached 50 m for low elevation satellites. By 1985, the effect was down to about 20 m. Yet, daily variations can be very significant. For instance, a test conducted in Hawaii (latitude 20°N) during May 1986 resulted in daily variability of up to 28 ppm in single frequency baseline solutions (Henson & Collier, 1986). The baselines were up to 400 km in length.

The optimal method to correct for the ionosphere is obviously through the use of dual frequency user equipment. The development of C/A code receivers performing carrier phase measurements on both frequencies will basically provide non-P code civilian users with this capability. However, single frequency equipment can be used in differential mode to reduce large scale ionospheric effects. A model has also been developed to predict the ionospheric delay on pseudo-ranges (Klobuchar, 1982). The model is broadcast continously to users and its parameters are updated on a regular basis. Its overall rms error in mid-latitude regions is 50% of the actual correction. Errors can be expected to increase in equatorial and high latitude regions.

Scintillation effects, which consist of short term and localized variations in the TEC, are more difficult to deal with. These effects are particularly severe at the Equator and in high latitude regions. In northern regions, effects such as troughs, bubbles and ducts (Muldrew, 1965, 1980a, 1980b), cause rapid changes in the TEC. This decreases the reliability of the differential method for single frequency users. Steep electron density gradients can further result in the necessity of increasing the tracking bandwidth which, in turns, affects the code and carrier measurement accuracy.

Figure 2.11 and 2.12 show the ionospheric variation over a 30 minute period at two sites located in northern Canada, namely Cambridge Bay (69°N, 255°E) and Resolute Bay (74°N, 265°E), as derived from TI4100 integrated dual frequency carrier phase advance measurements on GPS SV 8. The 2 stations are some 700 km apart. The corresponding total group delays, as derived from the L1 and L2 pseudo-ranges, are shown in Figure 2.13 and 2.14. The use of carrier phase measurements results in a better measure of the ionospheric variation than that of pseudo-range measurements due to the higher accuracy of phase measurements (Sub-section 2.2.4) and to the effect of multipath on pseudo-range measurements. The elevation of SV 8 was initially 48° (Cambridge Bay) and 54° (Resolute Bay) respectively, and decreased by some 15° over the 30 minutes time interval considered. The ionospheric variation at each of the 2 stations over this time interval is of the order of 35 cm. The total differential ionospheric effect between the 2 stations is obtained by subtracting the total group delay at one station from that at the other station. However, this total effect is difficult to estimate precisely due to the noise and multipath affecting the pseudo-range data. A partial differential effect is obtained by differenciating the ionospheric changes derived from phase data at the two stations. This partial differential effect reaches 25 cm at 21.8h, as can be seen by comparing Figures 2.11 and 2.12. This effect does not cancel out in single frequency differential positioning mode (Kleusberg, 1986a). The variations shown in Figures 2.11 and 2.12 are considered average for high latitude regions. The total group delays on L1 (Figures 2.13 and 2.14) reach 6 m. The pseudo-range measurements are strongly affected by multipath at Resolute as will be discussed in the next sub-section.

2.4.3. Multipath

Multipath is caused by reflection of the satellite signal on neighbouring reflective surfaces. Metal and certain sea states can produce such reflective surfaces. The code modulation of the GPS signal provides an inherent rejection of signals which do not occur within one code chip size of the direct pseudo-range. Maximum multipath errors on pseudo-ranges are therefore

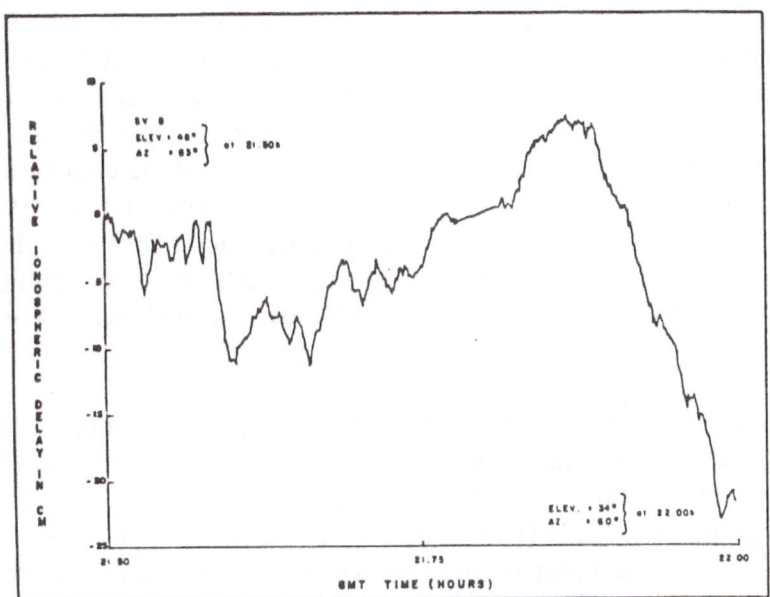

FIGURE 2.11: TIME VARIATION OF IONOSPHERIC DELAY AT THE L1
FREQUENCY, DERIVED FROM L1/L2 CARRIER PHASE OBSERVATIONS ON SV 8
AT CAMBRIDGE BAY, DAY 225, 1985.

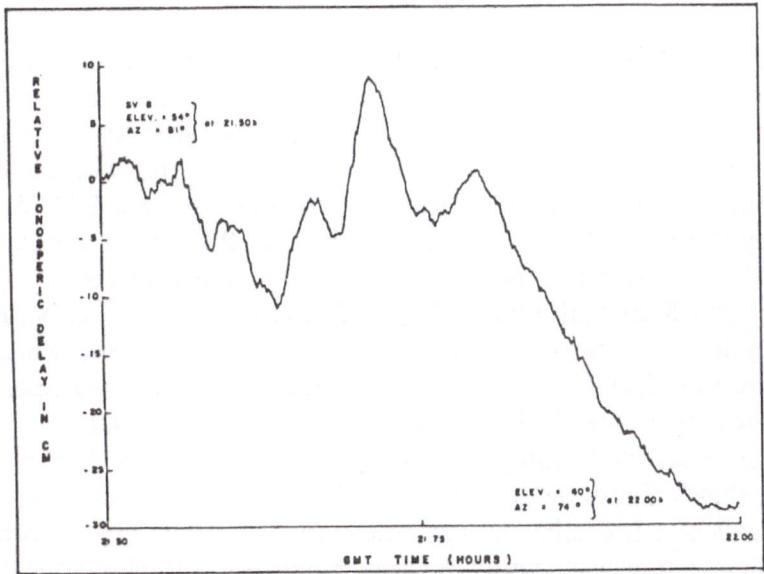

FIGURE 2.12: TIME VARIATION OF IONOSPHERIC DELAY AT THE L1
FREQUENCY, DERIVED FROM L1/L2 CARRIER PHASE OBSERVATIONS ON SV 8
AT RESOLUTE BAY, DAY 225, 1985.

29.3 m and 293.2 m for P and C/A code respectively. Actual errors of up to 20 m using the P code have been observed at the receiver site (Evans, 1985). Multipath effects are also dependent on antenna pattern (See Section 2.3), the geometry of the wavepath, and the type of multipath rejection techniques utilized. Multipath affects measurements not only at the receiver site but also at the satellite antenna. In the latter case, the effect is likely to be reduced to somewhat less than 1 dm using the differential mode (Young et al., 1985). While this remains a problem for cm accuracy static differential positioning, it can be assumed negligible for dm accuracy applications. However, multipath effects on code measurements at the receiver site is a problem of a much larger magnitude.

Carrier phase measurements are less affected by multipath effects than code measurements. The carrier wavelength is 20 cm (L1) and the effect of multipath on carrier phase is 2 orders of magnitude smaller than that on pseudo-range measurements (Thornton et al., 1984; Bishop et al., 1985). The use of phase smoothed pseudo-ranges for kinematic positioning is therefore important to reduce multipath effects to the sub-metre level. Several investigations related to the reduction of multipath are being conducted (e.g., Bishop et al., 1985; Bletzacker, 1985). It is important that effective rejection techniques be developed, especially for shipborne applications, to enhance both accuracy and reliability of GPS positioning.

Figures 2.13 and 2.14 show total ionospheric delays as obtained from pseudo-range measurements on SV 8 at Cambridge Bay and Resolute Bay as described in the previous sub-section. Both data sets are affected by significant pseudo-range noise as can be seen by comparing with the carrier phase-derived differential delays shown in Figures 2.11 and 2.12. At Cambridge Bay (Figure 2.13), the short term noise is of the order of 2 m, which is acceptable considering the measurement noise expected for both L1 and L2 pseudo-ranges. Systematic effects of up to 2 m also occur over periods of 5 or more minutes. The correlation with the carrier phase derived ionospheric delay shown in Figure 2.11 is weak. The pseudo-range derived ionospheric delay shown in Figure 2.13 is possibly affected by multipath effects. The pseudo-range derived ionospheric delay for Resolute Bay (Figure 2.14), however, shows clearly the presence of multipath effects reaching some 10 m over periods of a few minutes. The antenna was located between two large steel buildings which obviously induced multipath. The observations were conducted over 2 consecutive days and further analysis of the data showed strong day-to-day correlation. Multipath characteristically repeats itself under identical conditions (Bishop et al., 1985).

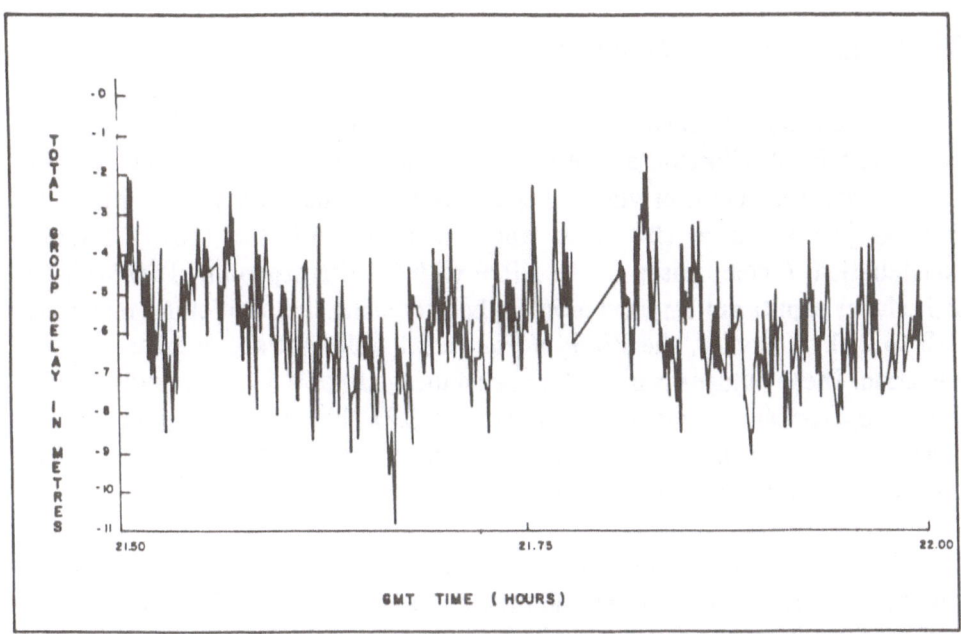

FIGURE 2.13: TOTAL GROUP DELAY AT THE L1 FREQUENCY, DERIVED FROM L1/L2 P CODE OBSERVATIONS ON SV 8 AT CAMBRIDGE BAY, DAY 225, 1985.

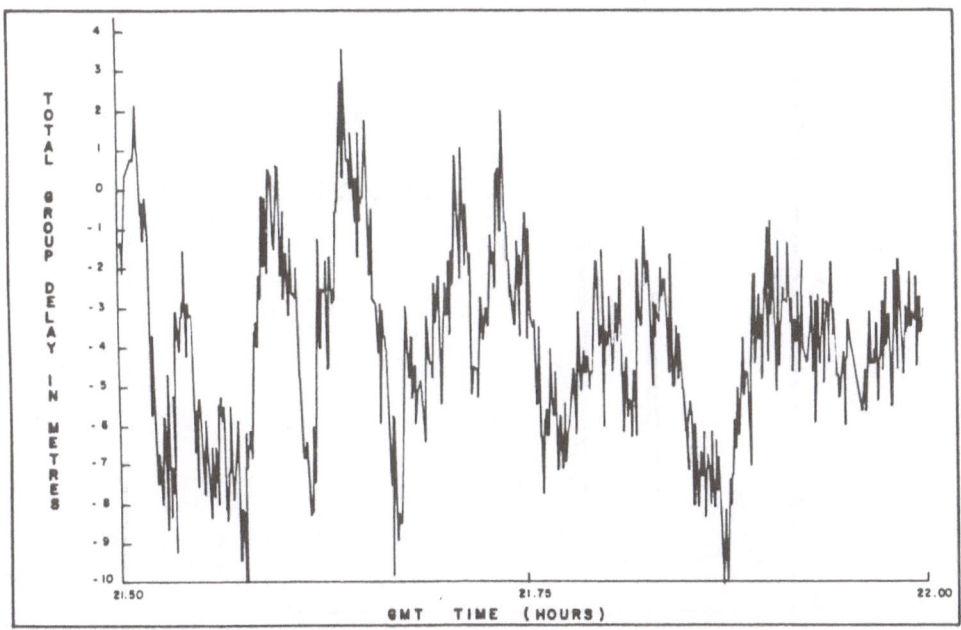

FIGURE 2.14: TOTAL GROUP DELAY AT THE L1 FREQUENCY, DERIVED FROM L1/L2 P CODE OBSERVATIONS ON SV 8 AT RESOLUTE BAY, DAY 225, 1985.

3. USER ACCURACY AND RELIABILITY ENHANCEMENTS

3.1 Height and Time Constraints

In shipborne applications, one is usually interested in 2-D positioning only. The HDOP (Horizontal DOP) becomes the quantity of primary interest from an accuracy point of view. Figure 3.1 shows the HDOP for the Northern Hudson Bay when a 10^0 mask angle is used with the (18+3) satellite constellation. A comparison of the HDOP with the corresponding PDOP (Figure 2.3) shows significant improvements when reducing the positioning task from 3-D to 2-D. However, the HDOP of Figure 3.1 still shows 2 outages of short duration. These outages can be eliminated using either a 5^0 mask angle (Figure 3.3) or a constrained height solution (Figure 3.2). The HDOP of Figure 3.2 was derived using a height constrained with an accuracy of 5 m, which accounts for the uncertainty in the geoid height (2-3 m) and sea state (3-4 m).

The use of a time constrained solution yields an HDOP similar to that of Figure 3.2. Time constrained solutions are especially important for 3-D airborne positioning. A cesium clock is used to maintain accurate time. In shipborne applications, the use of a height constrained solution is generally

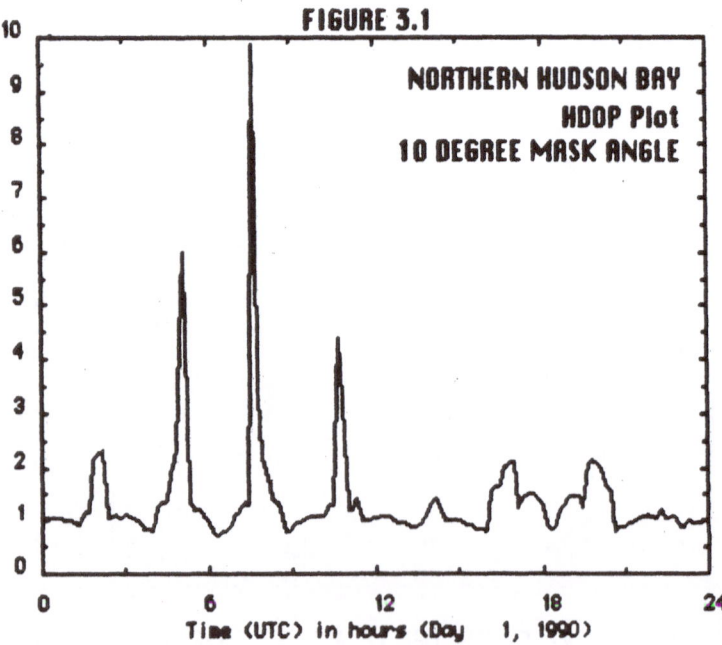

FIGURE 3.1

NORTHERN HUDSON BAY
HDOP Plot
10 DEGREE MASK ANGLE

Time (UTC) in hours (Day 1, 1990)

more attractive in view of its cost effectiveness and greater reliability. The cost of a GPS receiver is likely to drop below that of a cesium clock within the next 3 to 5 years. The use of both height and time constrained solutions will increase reliability further. Such a solution could also increase accuracy during outages.

For high accuracy (< 2 m) shipborne positioning, the use of height and/or time constraints will contribute to improving the reliability within a certain threshold, but not the accuracy sought. The techniques discussed in the subsequent Sections will be more effective to achieve high accuracy during periods of good satellite geometry. During outages, the use of external sensors will have to be considered.

Airborne applications are not necessarily all 3-D. For instance, with airborne bathymetry, a 2-frequency laser is used to measure both ranges to the sea surface and bottom. The height can therefore be constrained with an accuracy of 2 to 3 m, the geoid error still being an accuracy limiting factor. The use of an airborne radar can also be used effectively over water, not only to constrain height, but also to provide highly accurate data on aircraft vertical accelerations (e.g., Brozena, 1984). Aircraft acceleration measurements are of paramount importance for airborne gravimetry.

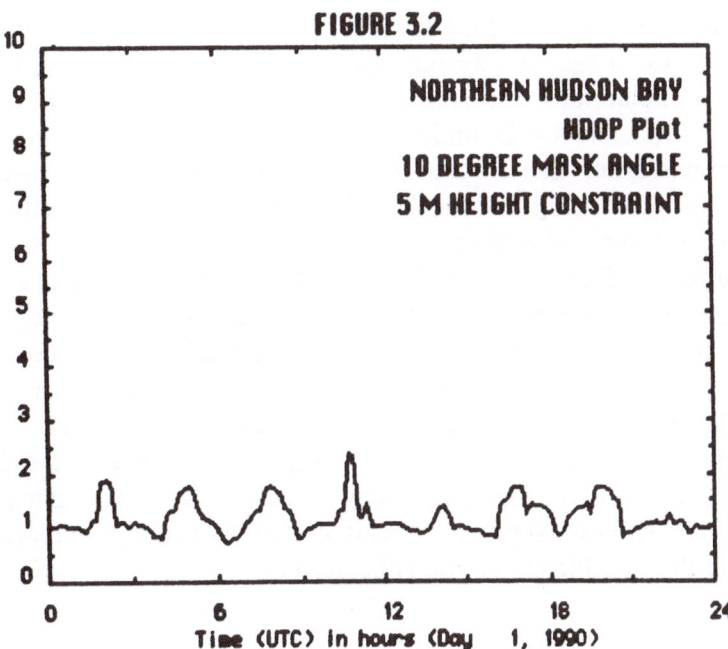

FIGURE 3.2

NORTHERN HUDSON BAY
HDOP Plot
10 DEGREE MASK ANGLE
5 M HEIGHT CONSTRAINT

Time (UTC) in hours (Day 1, 1990)

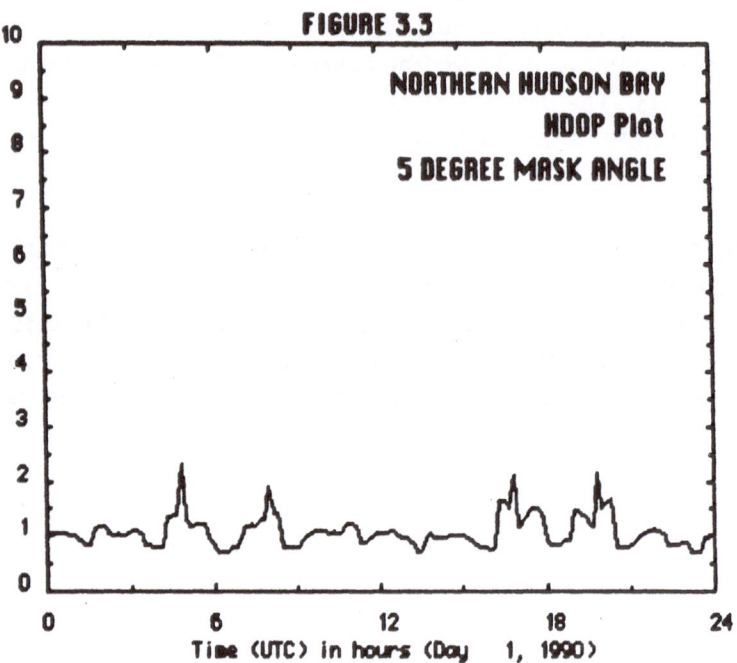

FIGURE 3.3

NORTHERN HUDSON BAY
HDOP Plot
5 DEGREE MASK ANGLE

Time (UTC) in hours (Day 1, 1990)

3.2 Differential Positioning

The differential positioning method is used to reduce error sources due to the orbits and to the propagation medium. The accuracy claimed for the broadcast ephemerides is currently of the order of 25 m. The accuracy of single point positioning is limited by this error source. In addition, C/A degradation implies the introduction of orbital biases which will result in single point positioning biases of 100 m. The differential method is very effective to reduce the above errors and biases. A 25 m orbital error corresponds to a differential position error of 1 ppm on the Earth's surface. If a baseline of 1,000 km is used for differential positioning, e.g., as for the example described in Section 3.3, the effect of orbital errors is therefore of the order of 1 m The biases introduced by the C/A code degradation are likely to be largely eliminated by the use of the differential method over baselines of up to a few hundred km. In addition to improving accuracy, the differential method also improves the positioning reliability and data integrity. The monitor, located at a fixed known point, can detect rapidly unforeseen errors caused by the satellite clocks and ephemerides.

Ionospheric effects are reduced significantly by using the differential method. Large scale ionospheric effects are correlated over distances exceeding 1,000 km. The use of one frequency data in differential mode often yields results which are more accurate than corresponding dual frequency results (e.g., Lachapelle et al., 1986a, 1986b). However, during periods of strong ionospheric scintillation, the use of one frequency data can become unreliable to reduce the effect of the ionosphere as discussed in Section 2.4. In such case, the use of dual frequency data is important to improve the positioning reliability. Tropospheric effects cannot effectively be reduced by the differential method, unless baselines of less than a few tens of km are used. This is due to local water vapour variations in the lower part of the atmosphere as discussed in sub-section 2.4.1.

Several algorithms have been developped and tested for the differential method. The selection of an algorithm should be a function of the accuracy required and of the real time/post mission requirements. If real time differential positioning is needed, the choice of the algorithm will also be a function of the data link characteristics. The maximum baseline length usable will also be a function of the range of the data link. The simplest algorithm is that which uses differential position corrections generated at the monitor. The use of positions generated by the receiver's internal algorithm is then sufficient. Accuracies of the order of 5 m can be obtained if the same satellites are tracked at both ends of the baseline. This requirement, however, constitutes an important limitation from an operational aspect. The differential range correction method overcomes this limitation and yields more accurate results. With existing receivers, an external computer is needed at each end of the baseline to calculate and apply the differential range corrections. However, receiver systems with an internal differential range correction method capability are being developed. Several variations of the differential range correction method exist. A few will be discussed in the next Section.

3.3 Integration of Code and Phase Measurements

The accuracy of a range difference derived from carrier phase measurements is 2 orders of magnitude higher than that of a range difference derived from P code measurements, as discussed in Sub-Sections 2.2.3 and 2.2.4. When C/A code measurements are considered, this ratio is of the order of 3. The use of carrier phase measurements is thus required if sub-metre accuracy is to be obtained. As discussed in the previous Section, the differential positioning method is used to reduce space segment and some of the propagation medium errors to the sub-metre level. It is therefore evident that the use of carrier phase measurements will lead to sub-metre accuracy only when the differential method is used.

The use of carrier phase data will also improve the positioning reliability. In particular, multipath effects on code measurements can be controlled effectively with carrier phase measurements. As discussed in Section 2.4.3. maximum multipath errors on carrier phase data are 2 orders of magnitude smaller than those on code measurements. In view of the high reflectivity environment prevailing in marine applications. the use of phase measurements is therefore advantageous to improve positioning reliability, even if the differential positioning accuracy requirement is not higher than 5 m.

The major drawback associated with the use of carrier phase measurements for positioning is the occurence of cycle slips. If the antenna is temporarily shaded or if a power outage occurs. loss of phase lock occurs on all received signals simultaneously and this will result in a discontinuity in the measured vehicle's trajectory. If loss of phase lock occurs on one satellite only, the geometry of the remaining satellites will dictate the level of accuracy degradation taking place. In order to control positioning errors caused by cycle slips, pseudo-range measurements are integrated to phase measurements. In the case of a loss of phase lock on all satellites simul-taneously, the accuracy with which the next segment of the vehicle's trajectory is recovered is not better than that obtainable by unaided pseudo-ranges in differential mode, namely about 3 m with P code and 10 m with C/A code (e.g., Lachapelle et al.,1986a, 1986b). In order to maintain high accuracy and reliability, cycle slips must thus be detected and corrected. if possible.

Several methods have been proposed and tested over the past few years to combine code and phase measurements. Hatch (1982) proposed smoothing pseudo-ranges with phase measurements. This method was slightly modified and tested extensively in land kinematic mode at speeds ranging from 20 to 100 km/h by Lachapelle et al. (1986a, 1986b). Accuracies at the one metre level could be confirmed. Three-dimensional accuracy confirmation at the sub-metre level in kinematic mode remains difficult to achieve in view of the difficulty associated with providing appropriate reference positions for the vehicle. The method used by Lachapelle et al. (1986a) was initially developed to provide accuracy confirmation at the 1.0 m level.

   Table 3.1, extracted from (Lachapelle et al., 1986b), summarizes
horizontal differential results obtained over a baseline of 1,000 km using
successively unaided C/A code measurements and phase smoothed C/A code
measurements. Unaided C/A code results are accurate to about 10 m in each
of the latitude and longitude components. Phase smoothed results are
accurate to the one metre level. This is considered satisfactory, especially
when the length of the baseline is taken into account. The use of cesium
clocks does not result in significant accuracy improvements in the present
case. Other tests were conducted with P code measurements. Unaided P code
results were accurate to 2 to 4 m while phase smoothed P code results were
generally within 1 to 2 m. See (Lachapelle et al., 1986b) for details.

TABLE 3.1

REAL GPS DIFFERENTIAL KINEMATIC RESULTS OVER BASELINE OF 1,000 KM USING LAND DATA ON DAY 240, 1985[*]

SOL NBR (1)	DIM (2)	CODE & FREQ. (3)	PHASE USED (4)	CLOCK FIXED M R (5)	VEHICLE SPEED (km/h) (6)	MEAN DIFFERENCES LAT LON (7)	ST. DEV. (ABOUT 0) LAT LON (8)	ST. DEV. (ABOUT MEAN) LAT LON (9)
1	4D	C/A L1	Y	N N	100	0.1m -3.6m	1.1m 3.8m	1.1m 1.3m
2					50	-0.1 -3.3	1.1 3.6	1.1 1.2
3					20	-0.4 -3.3	0.7 3.4	0.5 0.9
4	4D	C/A L1	N	N N	100	0.5 -2.9	9.3 13.2	9.4 13.0
5					50	-1.7 -1.1	9.1 13.5	9.0 13.5
6					20	-0.9 0.8	5.9 11.6	5.9 11.8
7	3D	C/A L1	Y	Y Y	100	0.9 -5.1	1.3 5.2	0.9 1.1
8					50	0.7 -5.1	1.2 5.2	1.0 1.3
9					20	-0.1 -3.5	0.5 3.6	0.5 0.9

(1) Solution identification number
(2) Dimension of solution, either 4D (Lat, Lon, Hght, Time) or 3D (Lat, Lon. Hght)
(3) Either P or C/A code pseudo-ranges are always used in the solution
(4) Yes if phase measurements were combined to pseudo-ranges. No otherwise
(5) Cesium clocks were used both at Monitor and Remote stations. Yes if clock was held
      fixed (straight line fit); No otherwise. The cesium was always held fixed at at least one
      end of the baseline when a 3D solution was tested.
(6) Speed of vehicle along the surveyed poles
(7) Mean differences between GPS derived pole positions and surveyed pole
      positions
(8) Standard deviations of GPS derived (assuming kinematic mode) positions
(9) Same as (7) after removing the mean differences
  * Data observed with two TI4100 receivers between Calgary, Alberta, and Brandon,
      Manitoba. Each solution is based on between 24 and 120 pole positions

The phase smoothed pseudo-range method described above is well suited for real time applications. The data transmission requirements are relatively straightforward since averaged differential (phase smoothed) pseudo-range corrections need be transmitted approximately every minute only. More sophisticated methods of combining code and carrier measurements have been proposed (e.g., Cannon et al., 1986; Kleusber, 1986b, Kleusberg et al., 1986; Kleusberg & Wells, 1986) and will likely lead to more accurate results, especially in post-mission mode. The data transmission requirements for real time applications would be more stringent than those associated with the method described in (Lachapelle et al., 1986a). Accuracies better than one metre have been confirmed up to date. Again, a major difficulty is the obtention of the vehicle reference positions at an accuracy level compatible with that sought by GPS. In a recent investigation based on the utilization of the land kinematic data described in (Lachapelle et al., 1986a), Cannon (1987) was able to confirm a sub-50 cm accuracy using a sequential adjustment method of combining code and phase measurements. Confirmation of high accuracy in actual shipborne mode is mode difficult than in land kinematic mode. In a marine experiment involving the use of a POLARFIX system, Seeber et al. (1986) confirmed an horizontal components accuracy of the order of 2 m using phase smootheed P code measurements.

The integration methods described above generally contain tests to detect cycle slips. Two detection methods are currently used, namely the phase velocity trend method and the dual frequency phase ratio method. The phase velocity trend method (e.g., Lachapelle et al., 1986a) assumes phase velocity linearity between phase measurements. While such an assumption is fairly accurate in a benign land kinematic mode, it is hardly adequate in shipborne mode in view of the rapid phase velocity changes caused by the ship's pitch and roll. However, if phase measurements can be extracted at intervals small enough for the linearity assumption to hold, the method can yield adequate results. In this respect, the TI4100 is advantageous in view of the availability of measurements over a 160 msec interval at the code FTF (Fundamental Time Frame). The use of an appropriate bandwidth is also important to increase the reliability of the method (Lachapelle et al., 1986a). Cycle slips can also be corrected for if the phase velocity linearity assumption is acceptable. The dual frequency method (Goad, 1985) of detecting cycle slips is independent from phase velocity linearity assumption and is therefore well suited for harsh dynamics applications. However, it cannot be used reliably for cycle slip correction. A combination of both method using a high data rate dual frequency receiver would likely result in a satisfactory method to detect and correct cycle slips.

3.4 Use of External sensors

The use of cesium clocks to enhance accuracy and reliability of GPS was analysed in Section 3.1. In this Section, other sensors are discussed to improve both accuracy and reliability and provide additional information required in various shipborne and airborne applications.

It was previously shown that the use of phase data will result in highly accurate shipborne and airborne kinematic positioning, provided that phase lock can be maintained. Recent tests by the U.S. National Geodetic Survey (Mader, 1986) have shown that, in the airborne case, phase lock can indeed be maintained for reasonably long (> 30 minutes) periods of time. In the marine case, the high g-forces acting on a ship's mast may result in more frequent losses of phase lock. In any case, satellite switching and/or losses of phase lock, however unfrequent, will always remain a major problem for high accuracy/high reliability unaided GPS positioning. In shipborne applications, the situation is further complicated by relative position variations of typically 3 to 8 m between antenna and sensors, which are typically located in the ship's hold. In airborne remote sensing applications, unaided GPS cannot realistically provide the attitude parameters required for a variety of applications.

Several types of external sensors can be utilized for GPS aiding. These range in complexity from inclination monitors to inertial navigation systems, and in cost, from $1K to $100K. At the lower end of the scale, inclination monitors can provide pitch and roll measurements with an accuracy of the order of 0.1°. An inclination monitor which could deliver this type of accuracy and meet other shipborne utilization requirements would resolve the problem of relating the antenna position to that of the sensors in the ship's hold. The pitch and roll data provided may also be usable to bridge gaps caused by occasional losses of phase lock or satellite switching. The use of such a low cost sensor offers interesting possibilities for high reliability GPS marine positioning at the 2 m accuracy level.

At the upper end of the scale lies the inertial navigation system. The use of such systems for marine and airborne GPS aiding has been the object of numerous investigations (e.g., Cox, 1980; Schwarz et al., 1984; Goldfard and Schwarz, 1985). Different levels of integration have been proposed. Feasibility studies have concluded that sub-metre accuracy is possible for some of these integrated systems. In addition to providing highly accurate and highly reliable positioning information, an integrated GPS-INS system would also provide the attitude parameters with the accuracy and reliability required for most, if not all, shipborne and airborne applications. Inertial navigation systems are available in a wide range of accuracy, cost, and complexity. The next few years will no doubt be see the development of a variety of integrated GPS-INS systems for a wide range of applications.

4. CONCLUSIONS

GPS is revolutionizing marine kinematic positioning. The (18+3) cons-tellation will fulfill many of the civilian needs. However, countermeasures will have to be used to deal with occasional outages and lack of redundancies, if high accuracy and reliability are prime objectives. The use of time and height constraints, together with high gain antennae near the horizon, will enable users to deal effectively with predictable outages. The use of constraints and receivers which can track up to seven satellites will be important to increase positioning reliability. The application of differential methods will result in residual orbital and ionospheric errors below the metre level. Specific receiver and antenna characteristics will enhance positioning accuracy and reliability. External attitude sensors will aid GPS significantly for many applications.

GPS can currently deliver an accuracy of the order of one metre in differential land kinematic mode. The extension of such an accuracy to the marine case requires further development. It will likely be available routinely by the time the full constellation is available in the early 1990s. Decimetre-level accuracies in marine kinematic mode are potentially achievable. However, significant error sources related to satellite orbits, the troposphere, the ionosphere for single frequency users, losses of phase lock, and multipath will have to be reduced further before the above threshold can be attained.

REFERENCES

ASHKENAZI, V., S.A. CRANE, and R.M. SYKES (1982) "The Significance of Various Approaches to the Tropospheric Correction". Proceedings of the Third Intern. Geod. Symp. on Satellite Doppler Positioning, DMA/NOS, Wash., D.C., pp. 463-474.

ASHJAEE, J.M., and R.J. HELKEY (1984) "Precise Positioning Using a Four-Channel C/A Code GPS Receiver". Proceedings of PLANS '84, I.E.E.E., New York, pp. 236-244.

BISHOP, G.J., J.A. KLOBUCHAR, and P.H. DOHERTY (1985) "Multipath Effects on the Determination of Absolute Ionospheric Time Delay from GPS Signals". Radio Science, Vol. 20, No. 3, pp. 388-396.

BLETZACKER, F.R. (1985) "Reduction of Multipath Contamination in a Geodetic GPS Receiver". Proceedings of First International Symposium on Precise Positioning with the Global Positioning System, N.O.A.A., U.S. Department of Commerce, Rockville, Md., pp. 413-422.

BROZENA, J. M. (1984) "A Preliminary Analysis of the NRL Airborne Gravimetry System". Geophysics, Vol. 49, pp. 1060-1069.

CANNON, M.E. (1987) "Kinematic Positioning Using GPS Pseudo-range and Carrier Phase Observations". M.Sc. Thesis, Depart- ment of Surveying Engineering, The University of Calgary.

CANNON, M.E., K.P. SCHWARZ, and R.V.C. WONG (1986) "Kinematic Positioning with GPS - An Analysis of Road Tests." Proceedings of Fourth International Geodetic Symposium on Satellite Positioning, The University of Texas at Austin.

COX, D.B. (1980) "Integration of GPS with Inertial Navigation Systems". Special Issue of Navigation, The Institute of Navigation, Washington, D.C., pp. 144-153.

COLLINS, J. (1986) "Measuring Platform Subsidence Using GPS Satellite Surveying. Proceedings of CPA/CHA Colloquium IV on Land, Sea and Space - Today's Survey Challenge, The Canadian Institute of Surveying and Mapping, Ottawa, pp. 377-382.

COUNSELMAN III, C.C., J.W. LADD, and T. SKERL (1986) "A Dual-Band Interferometric GPS Marine Navigation System". Presented at the International Symposium On Marine Positioning, Reston, Va., Oct. 14-17.

DOUCET, K. (1986) "Performance Considerations for Real-Time Navigation with the GPS". Techn. Rep. No. 122, Dept of Surveying Engineering, The University of New Brunswick.

DURBORAW, I.N. (1986) "Test Results of a Differential GPS Receiver in a Dynamic Environment". Presented at the XVIIIth Congress of the Fédération Internationale des Géomètres, Toronto.

EVANS, A.G. (1985) "Measurements Analysis for signal Multipath (Spring 1985 GPS Precision Baseline Test". Presented at Fall Meeting of American Geophysical Union, San Francisco, December 8-13.

GOAD, C.C. (1985) "Precise Positioning with the Global Positioning System". Proceedings of the Third Intern. Symp. on Inertial Technology for Surveying and Geodesy". Publ. 60005, Department of Surveying Engineering, The University of Calgary.

GOLDFARB, J.M., and K.P. SCHWARZ (1985) "Kinematic Positioning with an Integrated INS-Differential GPS". Proceedings of First International Symposium on Precise Positioning with the Global Positioning System, N.O.A.A., U.S. Department of Commerce, Rockville, Md., pp. 757-772.

HATCH, R. (1982) "The Synergism of GPS Code and Carrier Measurements." Proceedings of Third Intern. Geodetic Symposium on Satellite Doppler Positioning, DMA/NOS, Washington, D.C.

HEMESATH, N.B. (1980) "Performance Enhancements of GPS User Equipment". Special Issue of Navigation, The Institute of Navigation, Washington, D.C., pp. 103-108.

HENSON, D.J., and E.A. COLLIER (1986) "Effects of the Ionosphere on GPS Relative Geodesy". Proceedings of PLANS '86, I.E.E.E., pp. 230-237.

JORGENSEN, P.S. (1984) "Navstar/Global Positioning System 18-Satellite Constellations". Special Issue of Navigation, Vol. II, The Institute of Navigation, Washington, D.C.

KALAFUS, R. (1984) "Service Outages in GPS Associated with Satellite failures". Proceedings of PLANS '84, I.E.E.E., New York, pp. 191-197.

KLEUSBERG, A. (1986a) "Ionospheric Propagation Effects in Geodetic Relative GPS Positioning". Manuscripta Geodaetica, in press.

KLEUSBERG, A. (1986b) "Kinematic Relative Positioning Using GPS Code and Carrier Beat Phase Observations", Marine Geodesy 10, pp. 167-184.

KLEUSBERG, A., and D.E. WELLS (1986) "High Precision Differential GPS Navigation". Proceedings PLANS '86, I.E.E.E., pp. 389-392.

KLEUSBERG, A. S.H. QUEK, D.E. WELLS, G. LACHAPELLE, and J. HAGGLUND (1986) "GPS Relative Positioning Techniques for Moving Platforms". Proceedings of Fourth International Geodetic Symposium on Satellite Positioning, The University of Texas at Austin.

KLOBUCHAR, J.A. (1982) "Ionospheric Corrections for the Single Frequency User of the Global Positioning System". Presented at National Telesystems Conference, Galveston, Texas, November.

KLOBUCHAR, J.A. (1983) "Ionospheric Effects on Earth-Space Propagation". Environmental Research Papers, No. 866, Air Force Geophysics Laboratory, Hanscom AFB, Mass.

LACHAPELLE, G. (1983) "Use of Kalman Filtering for GPS Aided Marine Navigation". Published in "Geodesy in Transition", Publ. 60002, Division of Surveying Engineering, The University of Calgary.

LACHAPELLE, G. (1986) "GPS - Current Capabilities and Prospects for Multipurpose Offshore Surveying". Invited Paper 502.2, XVIII Congress of Fédération Internationale des Géomètres, Toronto, June.

LACHAPELLE, G., J. HAGGLUND, H. JONES, and M. EATON (1984a) "Differential GPS Marine Navigation". Proceedings of PLANS '84, I.E.E.E., New York, pp. 245-255.

LACHAPELLE, G., J. LETHABY, and M. CASEY (1984b) "Airborne Single Point and Differential GPS Navigation for Hydrographic Bathymetry". The Hydrographic Journal, No. 34.

LACHAPELLE, G., J. HAGGLUND, W. FALKENBERG, P. BELLEMARE, M. CASEY, and M. EATON (1986a) "GPS Land Kinematic Positioning Experiments". The Hydrographic Journal, No. 42, pp. 45-55.

LACHAPELLE, G. W. FALKENBERG, and M. CASEY (1986b) "Use of Phase Data for Accurate Differential GPS Kinematic Positioning". Proceedings of PLANS '86, I.E.E.E., pp. 393-398.

MADER, G.L. (1986) "Decimeter Level Aircraft Positioning Using GPS Carrier Phase Measurements". Proceedings of Fourth International Geodetic Symposium on Satellite Positioning, The University of Texas at Austin.

MARTIN, E.H. (1980) "GPS User Equipment Error Models". Special Issue of Navigation, The Institute of Navigation, Washington, D.C., pp. 109-118.

MULDREW, D.B. (1965) "F-Layer Ionization Troughs Deduced from Alouette Data". Journ. Geophys. Res., Vol. 70, pp. 2635-2650.

MULDREW, D.B. (1980a) "The Formation of Ducts and Spread F and the Initiation of Bubbles by Field-aligned Currents". Journ. Geophys. Res., Vol. 85, No. A2, pp. 613-625.

MULDREW, D.B. (1980b) "Characteristics of Ionospheric Bubbles Determined from Aspect Sensitive Scatter Spread F Observed with Alouette 1. Journ. Geophys. Res., Vol. 85, No. A5, pp. 2115-2123.

SCAVUZZO, R., and M. PAKSTYS (1985) "Shock Environment NOAA 20-foot Survey Launch". NOAA Technical Report NOS 113, Charting and Geodetic Services Series CGS 6, U.S. Department of Comerce, Wash., D.C.

SCHWARZ, K.P., R.V.C. WONG, J. HAGGLUND, and G. LACHAPELLE (1984) "Marine Positioning with a GPS-Aided Inertial Navigation System". Proceedings of San Diego National Technical Meeting, The Institute of Navigation, Washington, D.C.

SEEBER, G., A. SCHUCHARDT, and G. WUBBENA (1986) "Precise Positioning Results with TI4100 GPS Receivers on Moving Platforms". Proceedings of Fourth International Geodetic Symposium on Satellite Positioning, The University of Texas at Austin.

THORNTON, C.L., L.E. YOUNG, S.C. WU, and J.B. THOMAS (1984) "GPS-Based Certification for the Microwave Landing System". Proceedings of PLANS '84, I.E.E.E., New York, pp. 256-263.

TRANQUILLA, J., S.R. BEST, and B.G. COLPITTS (1986a) "Selection and Application Criteria for GPS Receiver Antennas". Presented at Annual Meeting of the Canadian Geophysical Union, Ottawa, May.

TRANQUILLA, J.M., and B.G. COLPITTS (1986) "The Microstrip Antenna with Finite Ground Plane for Use with Satellite Positioning Systems". Presented at Antenna Applications Symposium, Winnipeg, August.

TRANQUILLA., J.M., S.R. BEST, and B.G. COLPITTS (1986b) "Phase Centre Considerations for the Log-Spiral Antenna". Presented at I.E.E.E. International AP-S/URSI Symposium, Philadelphia, June.

TRANQUILLA, J.M., and S.R. BEST (1986a) "The Quadrifilar Helix as a GPS Receiver Antenna". Presented at Antenna Applications Symposium, Winnipeg, August.

TRANQUILLA, J.M., and S.R. BEST (1986b) " Phase Centre Considerations for the Monopole Antenna". I.E.E.E. Transactions on Antennas and Propagation, May.

WARD, P. (1982) "An Advanced Navstar GPS Geodetic Receiver". Proceedings of the Third International Geodetic Symposium on Satellite Doppler Positioning, Las Cruses. DMA/NOA, Wash, D.C., pp. 1123-1142.

YOUNG, L.E., R.E. NEILAN, and F.R. BLETZACKER (1985) "GPS Satellite Multipath: An Experimental Investigation". Proceedings of First International Symposium on Precise Positioning with the Global Positioning System, N.O.A.A., U.S.Department of Commerce, Rockville, Md., pp. 423- 432

DEMONSTRATION OF THE COMBINED USE OF GPS PSEUDORANGE AND DOPPLER MEASUREMENTS FOR IMPROVED DYNAMIC POSITIONING

STANLY L. MEYERHOFF AND ALAN G. EVANS
NAVAL SURFACE WEAPONS CENTER
DAHLGREN, VIRGINIA 22448-5000, USA

ABSTRACT

This paper presents the results that can be achieved by using Doppler phase (change in range) measurements to smooth pseudorange measurements. This real-time smoothing procedure reduces the effects of signal multipath and measurement noise associated with pseudorange measurements. Results of a shipboard test are presented. For this high signal reflectivity environment, the combined measurement smoothing procedure improved the test positioning accuracy by about 20 percent.

1. INTRODUCTION

The standard Global Positioning System (GPS) procedure for determining dynamic positioning is to use pseudorange measurements to four satellites to continually perform navigation fixes (Mulliken and Zoller, 1980; Yiu, Crawford, and Eschenback, 1984). These fixes provide the user with corrections to position and the receiver's clock. The corrections are generally obtained using a Kalman filter.

Field testing of the Texas Instruments 4100 Geodetic GPS Receiver has demonstrated that pseudorange measurements may be significantly affected by signal multipath (Evans, 1986). Pseudorange measurement errors have increased from about 1 meter root mean square (rms) in a low reflective desert-type location to over 4 meters rms for high reflective rooftop and shipboard environments. It has also been demonstrated that the phase measurements, Doppler change in range measurements, are far less affected by signal multipath than the pseudorange measurements. The phase measurements can be considered as nearly perfect measurements in comparison to the pseudorange values.

In (Evans, 1986) it was demonstrated that subtracting the ionospherically corrected Doppler range values from the ionospherically corrected pseudorange yields, for the most part, the multipath effects on pseudorange, the receiver measurements noise and a bias. An example of this difference for the test case presented is given in Figure 1. The rms value for the pseudorange multipath effect and receiver measurement noise is about 4 meters. The multipath causes this rms value to be slightly over three times the value expected due to receiver measurements noise alone.

The effects of signal multipath and noise on position solutions in a shipboard environment are demonstrated in this paper. Test description and results are presented. Improvements obtained by using Doppler change-in-range measurements to smooth the Kalman filter.pseudorange observations are shown. A description of the smoothing algorithm is also given. This real-time Doppler smoothing reduces the effects of signal multipath and noise. It is very similar to smoothing algorithms described in (Hatch, R., 1982) and (IBM, 1981), except that it uses ionospherically corrected values of the initial pseudorange in the smoothing process.

Although this Doppler smoothing procedure does not remove all the multipath effects, it does demonstrate a significant improvement over the standard pseudorange positioning procedure. A 90 percent circular error of probability (CEP) of less than 15 meters for the standard solution case was reduced to less than 12 meters CEP when Doppler smoothing was used.

M. Kumar and G. A. Maul (directors), Marine Positioning, 149–156.

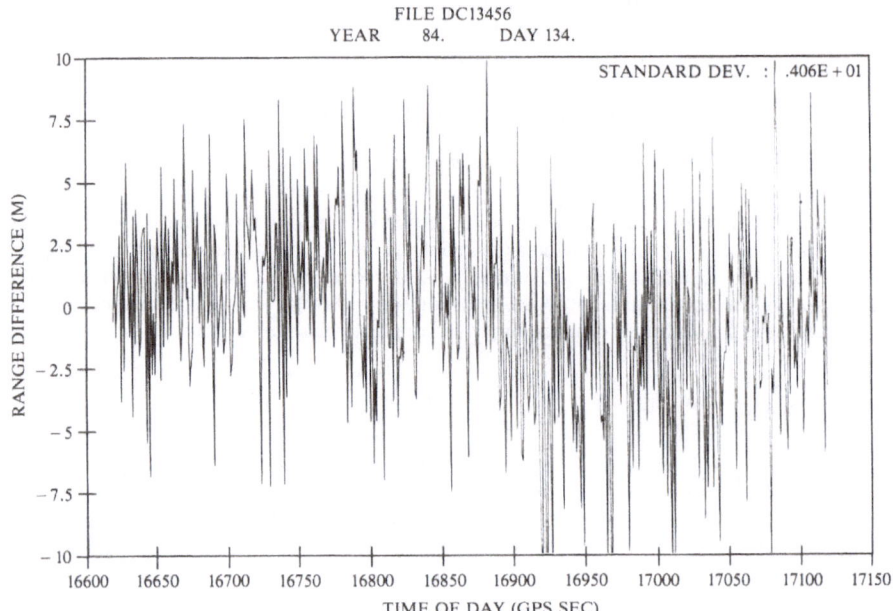

FIGURE 1. PSEUDO MINUS DOPPLER RANGE
(mean value set to zero)

2. TEST DESCRIPTION

The GPS Geodetic TI 4100 Receiver System was tested in a dynamic mode on board a launch on the Mississippi Sound in May of 1984. Data gathered from this test were used to determine whether the receiver system could meet U.S. Naval Oceanographic Office (NAVOCEANO) real-time and postprocessing requirements for absolute positioning on the ocean surface (two-dimensional positioning). The goal of this test was to determine whether the receiver system could provide shipboard dynamic positioning with errors of less than 15 meters CEP (90 percent).

2.1 Receiver System.

The receiver system is a geodetic model of the Texas Instrument 4100 GPS Satellite Receiver System. It can track four GPS satellites (SVs) simultaneously (in a multiplexing sense) in either the P or C/A code. Sixteen primary measurements covering pseudorange and Doppler in the form of continuously counted integrated Doppler on L1 and L2 frequencies are simultaneously provided from the four in-view SVs. Navigation messages from the four SVs are continuously gathered by the receiver system. Time-tagged data are sent, on command, to the navigation processor portion of the receiver system.

2.2 Data Collection.

For this data analysis, nine nights of data were collected and processed. It should be noted that the satellite-receiver geometry was far from optimum for navigation during most of the test. For this test, there were only four satellites (satellite PRNSA numbers 6, 8, 9, 11) available. Figure 2 shows the position dilution of precision (PDOP) versus time plotted for day 134.

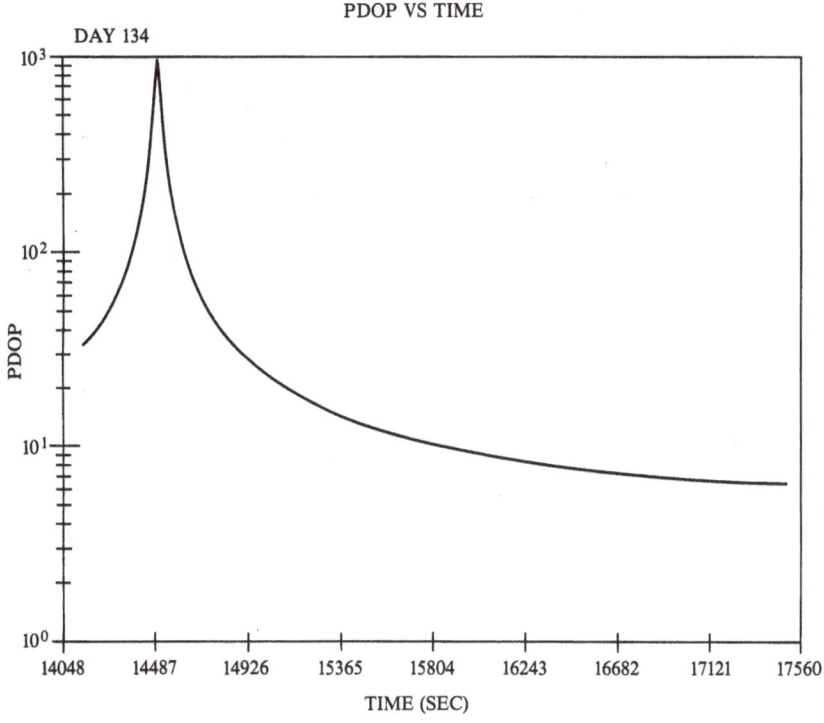

FIGURE 2. PDOP VS TIME

The pseudorange, continuous count Doppler, quality vectors, signal-to-noise density ratio, GPS time of reception, and internal time of reception for every satellite in track were recorded every second. The data from all satellites being tracked had a common time of reception.

The receiver system satellite tracking data were recorded on cassette tapes. The data on these tapes were read, reformatted, and written on 9-track standard exchange tapes by the Applied Research Laboratory of the University of Texas at Austin (ARL/UT). The satellite data on the exchange tapes and the "truth" data were sent to the Naval Surface Weapons Center (NSWC) at Dahlgren, Virginia.

2.3 Truth System.

The navigation processor produces a 1-sec pulse and the accompanying time tag for that pulse. This pulse and time tag were used to time tag the NAVOCEANO truth system. The time error between the truth and the receiver system is less than 20 ms. At a 12-knot velocity, the time tag error leads to less than 12 cm of error in interpreting the truth positions.

On the launch, a Del Norte ranging system that used four land-based Trisponders collected data. When processed by NAVOCEANO personnel, these dataprovided 1-sec truth positions of the launch.

3. DATA ANALYSIS

At NSWC the satellite tracking data were used in a Kalman filter to calculate the positions of the launch. These calculations were done as if they were being done in a real-time environment. These positions were then compared to the truth positioning, and error statistics were calculated.

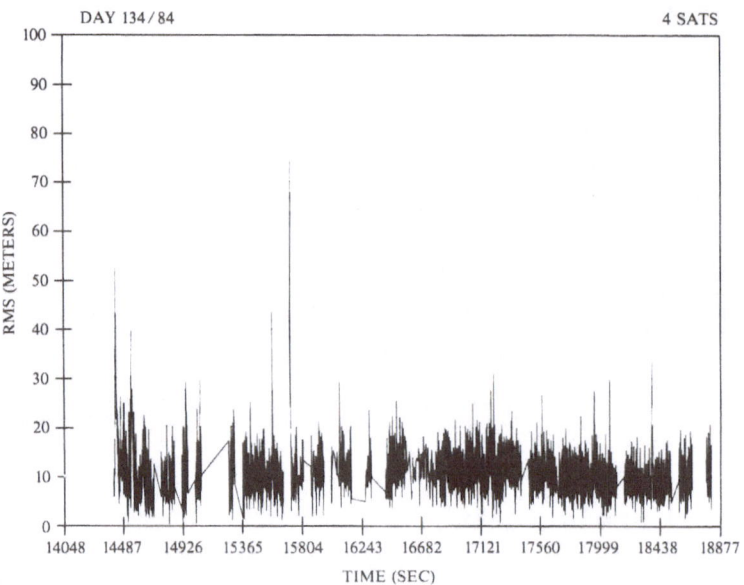

FIGURE 3. RMS VS TIME
(Doppler smoothing not used)

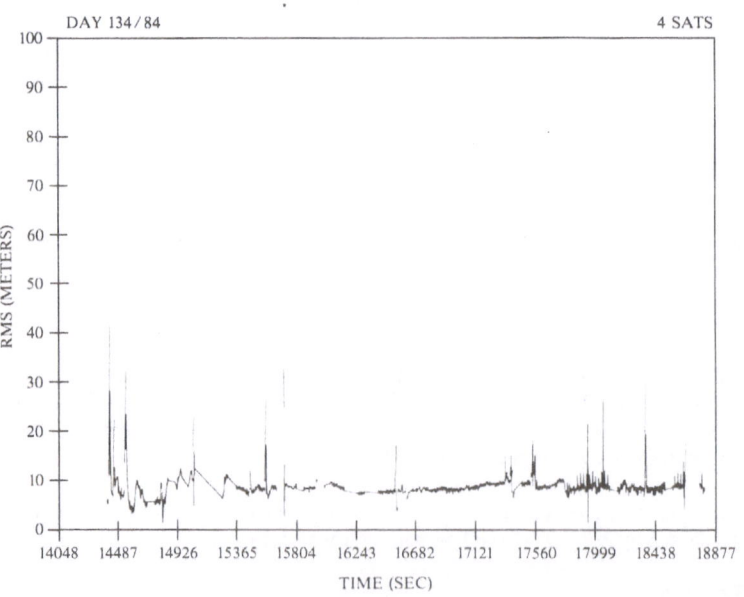

FIGURE 4. RMS VS TIME
(Doppler smoothing used)

3.1 <u>Data Correction</u>.

The following corrections were applied to the pseudorange data.

1. The first-order ionospheric refraction delays were calculated from the two-frequency pseudorange data and the Doppler change in range data.

2. The pseudorange data were corrected for satellite clock errors (Rockwell, 1983). Corrections for clock bias, clock drift, clock aging, and relativistic errors were made by using data from the GPS Broadcast Navigation Message.

3. The Chao Model (General Dynamics, 1978) was used to compute the tropospheric refraction correction.

4. The motion of the satellite in the Earth- fixed coordinate system during the signal propagation was corrected by using the slant range in the calculation for the range residuals.

5. The range data were also corrected for the rotational motion of the coordinate system during propagation. This correction was made by using WGS 72 data values.

3.2 <u>Data Editing</u>.

The quality vector flags for the satellite tracking data were not reliable for determining whether the data were good or bad. Therefore, it was required that the ionospheric delay value be less than 20 meters for pseudorange data to be considered good.

3.3 Pseudorange Smoothing.

This section describes an algorithm that uses the Doppler data to smooth the pseudorange data. The input variables needed for this algorithm are

tt	=	The GPS time tag for pseudorange and Doppler data.
ttfst	=	The GPS time tag for the first good pseudorange and Doppler count of this data span.
R(tt)	=	Ionospheric corrected pseudorange in kilometers.
DR(tt, ttfst)	=	Ionospheric corrected Doppler change in range (delta range) derived from the difference in the continuous count Doppler measurements at time "tt" and "ttfst."

The smoothing span is a span of data during which there are no breaks in the continuous count Doppler data. A smoother span for a satellite will start with the first good data point for that satellite; that is, the first data point for which the quality vector is zero. A new span of data will be started for that satellite any time that there is a break in the data (quality vectors not zero) for that satellite.

To start a new span, "ttfst" is defined as the GPS time tag for the start of the span. The continuous count Doppler data for time "ttfst" must be saved so that an ionospheric corrected change in range (delta range) from this first point to each following point can be calculated by differencing the first Doppler counts with the succeeding counts.

The following steps are done for every data point in a span as the data points become available from the receiver.

The number of data points in the span "N" is incremented by one.

"B_N," the "Nth" estimate of the pseudorange at time "ttfst," is found by subtracting the ionospheric corrected change in range for time "tt" from the ionospheric corrected pseudorange for time "tt":

$$B_N(tt) = R(tt) - DR(tt, ttfst) \tag{1}$$

The average value for the first pseudorange is

$$B_{avg}(tt) = SUM \ Bj(tt) \ / \ N \tag{2}$$

The smoothed pseudorange for time "tt" is then obtained by adding the ionospheric correct change in range for time "tt" to the average value for the first range:

$$R_{sm}(tt) = B_{avg}(tt) \ + \ DR(tt, ttfst) \tag{3}$$

If the absolute value of the difference between the smoothed pseudorange and the raw pseudorange (between $R_{sm}(tt)$ and $R(tt)$) is greater than 20 meters, then this point is considered bad.

3.4 Solution.

The launch position was estimated by using the UDU factorization of a four-state, extended Kalman filter (Bierman, 1977). The four states were receiver system position (X, Y, Z) and Receiver System Clock bias. The state matrix was not formally propagated in time; however, a Q matrix of noise was added to the covariance for propagation. The off diagonal elements of the Q matrix were zero, the diagonal elements for position were set to 750 meters squared, and the diagonal element for clock bias was set to 10.0 meters squared. The solutions were updated every second.

Since the GPS satellite-receiver geometry was poor for most of the test, the estimates were constrained by the distance from the center of the Earth to the launch; that is, the solutions were forced to be on the surface of a sphere whose radius was the initial distance from the Earth's center to the original estimate of the launch position.

The Earth-fixed position solutions were first converted to WGS 72 geodetic coordinates and then converted to UTM (Universal Transverse Mercator) Northings and Eastings.

3.5 Error Statistics.

The NSWC Kalman filter solutions were compared every second to the NAVOCEANO truth positions. At each comparison, a Northing error, an Easting error, and a total error were computed. The Northing and Easting errors were the differences between the filter-solution derived estimate of position and the truth position. The total error (the two-dimensional error magnitude) at each 1-sec mark is the square root of the sum of the Easting error squared and the Northing error squared. The total error versus time was plotted for the time marks with four satellites in track (four satellites in track means receiving good data from four different satellites simultaneously). Figure 3 is the total error versus time plot where Doppler smoothing was not used for day 134. Figure 4 is the total error versus time plot where Doppler smoothing was used for the same day.

A 90-percent CEP was computed for each day and for all days together. The 90-percent CEP is larger than the 1-sec total position errors 90 percent of the time. Table 1 under Standard CEP contains the daily CEP's from the total position error for the case of four satellites in track without Doppler smoothing. The CEP's for total position error when Doppler smoothing was used are listed under Smoothed CEP. The daily CEP improved by at least 2 meters for every day except day 136. There apparently was a problem with the Doppler data for the first half of the four-satellite data on day 136 that corrupted the results. The CEP's for total position error combined for all days were 15 meters for solutions not using Doppler smoothing and 12 meters for solutions using Doppler smoothing.

TABLE 1. FOUR SATELLITES IN TRACK

Day No.	Standard CEP (m)	Smoothed CEP (m)
134	17	11
135	12	8
136	14	16
137	13	10
138	12	8
139	13	11
140	14	11
141	13	9
144	18	15

4. CONCLUSION

Figure 3 shows that the total error plot has a component that remains fairly constant over time and a finer component that varies substantially over time. This finer component of error is due primarily to multipath and noise on the range measurements. Multipath is dependent on the geometric relationships among the satellite, the receiver antenna, and the reflective surface causing the multipath. In the case of a stationary antenna, this geometry varies slowly since its change is usually dependent only on the motion of the satellite. In the shipboard case, some of the angular relationships among the satellite, the receiver antenna, and the reflective surface vary with the roll, pitch, and yaw of the ship. This will lead to multipath error that will rapidly change with ship motion. Since the errors due to multipath and noise are far less for the Doppler data than for the pseudorange data, the method using the Doppler data to smooth the pseudorange was employed to reduce the effect of the multipath and noise on the positioning solution.

A comparison of Figures 3 and 4 shows that the Doppler smoothing significantly reduces the variations in the total errors. Although the smoothing does reduce the variations in the errors, there are still some large excursions. Most of these are due to the poor geometry and temporary loss of lock. Figure 2 shows that the geometry was poor most of the time.

The use of the Doppler smoothing does reduce the combined CEP for all days by about 3 meters. This reduction is due primarily to a reduction in the higher frequency components of the total error. If continuous count Doppler data are available, use of that Doppler data to smooth the range will reduce the total error in a dynamic positioning system and significantly reduce the short-term variations in error for dynamic positioning.

5. REFERENCES

Bierman, Gerald J., 1977. "Mathematics in Science and Engineering, Volume 128: Factorization Methods for Discrete Sequential Estimation," AcademicPress.

Evans, A.G., 1986. "Comparison of GPS Pseudorange and Biased Doppler Range Measurements to Demonstrate Signal Multipath Effects," Paper presented at the Fourth International Geodetic Symposium on Satellite Positioning, Austin, TX.

General Dynamics, 1978. "Computer Program Development Specification for the GPS Master Control Station Ephemeris Computer Programs," CP-CS-304.

Hatch, R. 1982. "The Synergism of GPS Code and Carrier Measurements," Magnavox Technical Paper MX-TM-3353-82, Torrance, CA.

International Business Machines Corporation, 1981. "Appendix A to the Computer Program Development Specification for the Master Control Station Ephemeris/Clock Computer Program of the NAVSTAR GPS Operational Control System Contract: 15O4701-81-0011,CII:, 7939151." Gaithersburg, MD.

Mulliken, R.J. and Zoller, C.J., 1980. "Principle of Operation of NAVSTAR and System Characteristics," NAVIGATION, Vol. 1, The Institute of Navigation, Washington, DC.

Rockwell International, Space Systems Group, 1978. "NAVSTAR GPS Space Segment/Navi gation User Interfaces. ICD-GPS-200."

Yiu, K.P., Crawford, R., and Eschenback, R. 1984. "A Low-Cost GPS Receiver for Land Navigation," NAVIGATION, Vol. 2, The Institute of Navigation, Washington, DC.

Accurate Positioning
of Marine Vessels
Using GPS

by
Dr. I. Newton Durboraw, III

Motorola Inc.
2100 E. Elliot Rd.
P.O. Box 22050
Tempe, Az. 85282

ABSTRACT

Offshore geophysical studies usually require that accurate position of the various observations that are collected on board a moving platform be documented and referenced to absolute coordinates. Shore-based systems such as Mini-Ranger, Raydist, Trisponder, and Microfix have traditionally been used for this purpose. However, the advent of the Global Positioning System (GPS) and its potential for operating with longer distances from shore will certainly change the methods for offshore positioning. The geometry of GPS orbits and the highly structured nature of transmitted GPS signals imply that the system may be used in a precision differential mode to achieve high accuracy at considerable distances from shore using a C/A code receiver that is small, light weight, and relatively inexpensive.

This paper presents results of recent marine vehicle tracking tests using the Motorola Eagle GPS receiver. This receiver, which has only recently been introduced to the market, is a relatively modest cost, digital, 4-channel, C/A code GPS receiver with several features that produce very high performance. The receiver incorporates fully integrated carrier-aided tracking of the independent C/A codes from four satellites to acheive a high level of precision in a dynamic environment. Simultaneous integrated carrier phase measurements and C/A code phase measurements are obtained for four satellites in this receiver and are used in computing a precise estimate of position via an 8-state Kalman filter. Test results indicate that a geodetic level of measurement precision can be obtained which is unprecedented in a receiver of such modest cost, operating in a real-time dynamic environment. The availability of such equipment, especially at this early stage of GPS development, indicates that the technology of marine positioning is very dynamic indeed. In this, and many other realms of positioning technology, GPS will clearly have a very dramatic impact.

INTRODUCTION

There are many applications for the capability to precisely position marine vessels. Denaro and Yoerger describe the application of GPS for survey vessels in reference 1. For example; when drilling a hole in the ocean floor, it is important that the ship be controlled so that it remains stationary over the drill hole. Acoustic stabilizing methods using bottom-mounted transponders is one method of accomplishing this; however, acoustic interference from thrusters used to control the vessel creates a problem which then must be dealt with. Other applications for precise positioning capability include dredging operations and varous marine surveillance missions. In cases where it is necessary to precisely control an area of data collection or to precisely navigate a vessel through controlled waterways, the requirements for position accuracy are much tighter.

M. Kumar and G. A. Maul (directors), Marine Positioning, 157–164.
© 1987 by the Marine Technology Society.

The GPS has promised to serve this broad class of commercial users with the caveat that autonomous receivers will be denied the very high accuracy potential of GPS. High accuracy autonomous performance operation will be reserved for military users who have access to the P code. The solution for commercial users who require high accuracy positioning is the use of a differential mode of GPS, which has been widely discussed. In this mode, with the aid of corrections that are computed at a monitor station, a high level of positioning accuracy can be achieved for dynamic marine applications as well for static surveying. These applications have received much attention in the last year.

Motorola was invited to participate in an experiment conducted by the U.S. Naval Post Graduate School to collect offshore GPS data with its new Eagle GPS receiver (Ref. 2) on board a marine vessel. This experiment, referred to as the Seafloor Benchmark Positioning Experiment Phase II, and reported elsewhere in this meeting, was conducted in August of this year. Falcon range tracking equipment as well as two Eagle GPS receivers and data recording equipment were provided by Motorola. Prior commitments for the GPS receivers prevented their use for the entire test period. However, simultaneous Eagle GPS data were recorded on board the research vessel Pt. Sur and at a shore station on August 12th and 13th. The subject of this paper is the preliminary analysis of this data and an assessment of its utility for positioning a hydrographic vessel in real time.

The Motorola Equipment

The Motorola Eagle GPS receiver is shown in Figure 1. Although under 200 cubic inches, this compact, 4-channel, continuous tracking, carrier-aided receiver packs a lot of performance. Dual RS232 ports are provided: one for a remote control display unit and the other for high-speed data output. Outputs include position at the rate of one independent (three coordinate) positon per second as well as multiplexed ephemeris data and optional raw measurements. This data can be directly downloaded to recording and/or computation facilities for integration with other data. The antenna consists of a microstripline patch antenna mounted together with the RF assembly and measures less than 30 cubic inches. The antenna can be located remotely up to 150 feet from the receiver using a single RG 223 coaxial cable. The total power required for both units is less than 25 watts at 12 volts which is desirable when battery power is a consideration.

The Eagle receiver employs a distinct technique for integrated carrier phase aiding of the C/A code tracking loop to minimize errors as the receiver tracks signals from four satellites simultaneously without the need for multiplexing. The result is very precise tracking and highly accurate position computation with a relatively wide bandwidth to accommodate dynamic platforms. Using differential techniques (i.e., with two receivers) this receiver will support accuracies better than 2 to 3 meters in real time at a 1-second full three coordinate update.

For this test, a monitor GPS station using one Eagle receiver was set up on a hilltop (Ferrier) overlooking Monterey. The test area was approximately 30 miles offshore from this site. GPS shore station data was recorded using an ALGOL cartridge tape recorder at the site. In addition to recording GPS data at the shore station, Motorola operated a UHF data link that permitted the use of shore station GPS data for real-time differential display.

Reference stations for a Mini-Ranger Falcon 492 tracking system were set up at three locations: the one at Ferrier and two others near the shoreline which were configured to provide good geometry for Mini-Ranger Falcon positioning in the test area.

On board the Pt. Sur research vessel, a second Eagle GPS receiver was configured with its antenna mounted on a platform to the rear of the bridge. Real-time plots of the Eagle GPS indicated position were recorded using a plotter provided by the U.S. Naval Post Graduate School. The plotter was driven by an HP 9816 desktop computer that interfaced with the RS232 outputs from

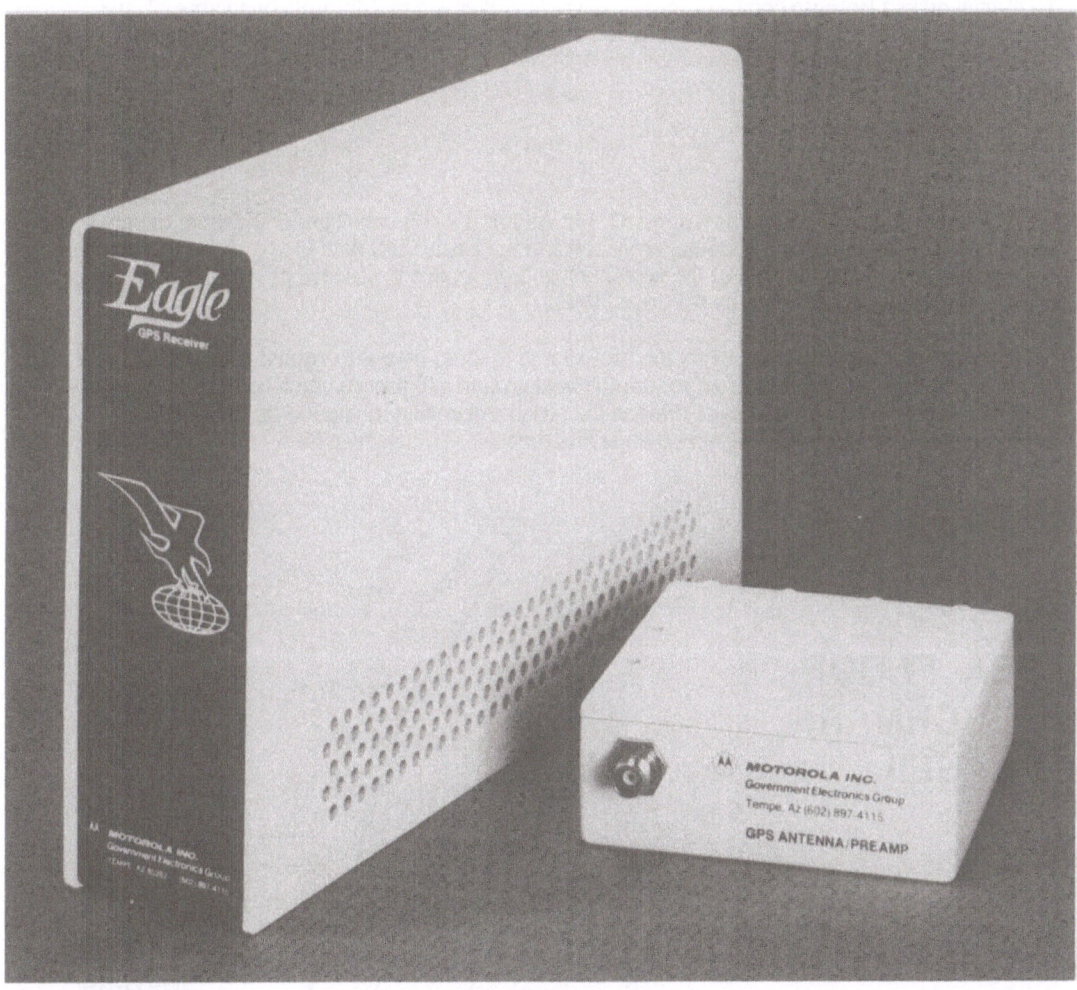

Figure 1. Eagle GPS Receiver and Antenna

the Eagle receiver and the shore-based data link. In addition, the computer was equipped with a Winchester disk drive intended for logging the GPS data at the shore station as well as on the boat. However, software problems in the computer prevented simultaneous display of differential solutions and recording of real-time data from both RS232 ports as planned. For this reason, it was opted to rely on recorded data from the shore station to compute differential solutions on a postprocessing basis. The onboard computer and Winchester drive were used to record the mobile platform data and drive the real-time plotter to display the autonomous GPS indicated position of the boat.

All Eagle GPS and Falcon data were left with Navy personnel at the end of the last day of Motorola participation. This included copies of data collected on the Pt. Sur Winchester drive that had been made on 3-1/2 inch computer disks. (Note: especially for those who use large capacity hard disk drives - this procedure was clearly valuable, for, as it turned out, the Winchester disk drive was damaged during air shipment back to Phoenix and all data resident on this disk was IRRETRIEVABLE! Thus the data shown in this paper is data that was copied from those "backup" 3-1/2-inch disks.) Because copies of the test data were received by Motorola on October 1, the analysis of data for this paper is considered very preliminary. More detailed reporting can be expected in future meetings.

Data Analysis

The data sets are identified as August 12 and August 13. Figures 2 and 3 illustrate the track created by plotting both the Mini-Ranger Falcon and the Eagle GPS data in x/y. Note that arrows are included to indicate the path of the vessel. Data dropout due to absence of satellite coverage or change of satellites is indicated by the dotted lines.

During the August 12 maneuvers the boat started to cross over the original path at 12:12. At this time, the captain was instructed to turn the boat around and proceed on a course that was roughly parallel to the original path. Falcon data is absent during the period of 12:30 to approximately 1:15 for reasons not known at this time.

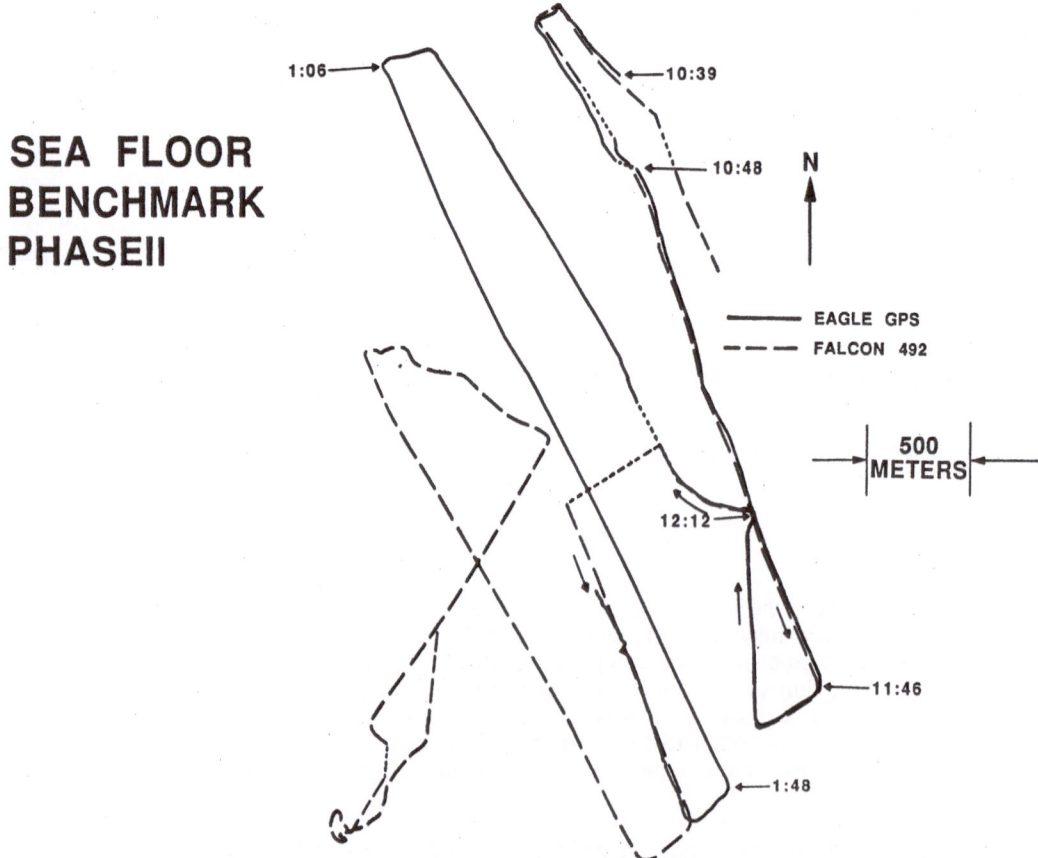

Figure 2. Comparison of Eagle GPS and Falcon 492 Reported Positions, PT. Sur - 8/12/86

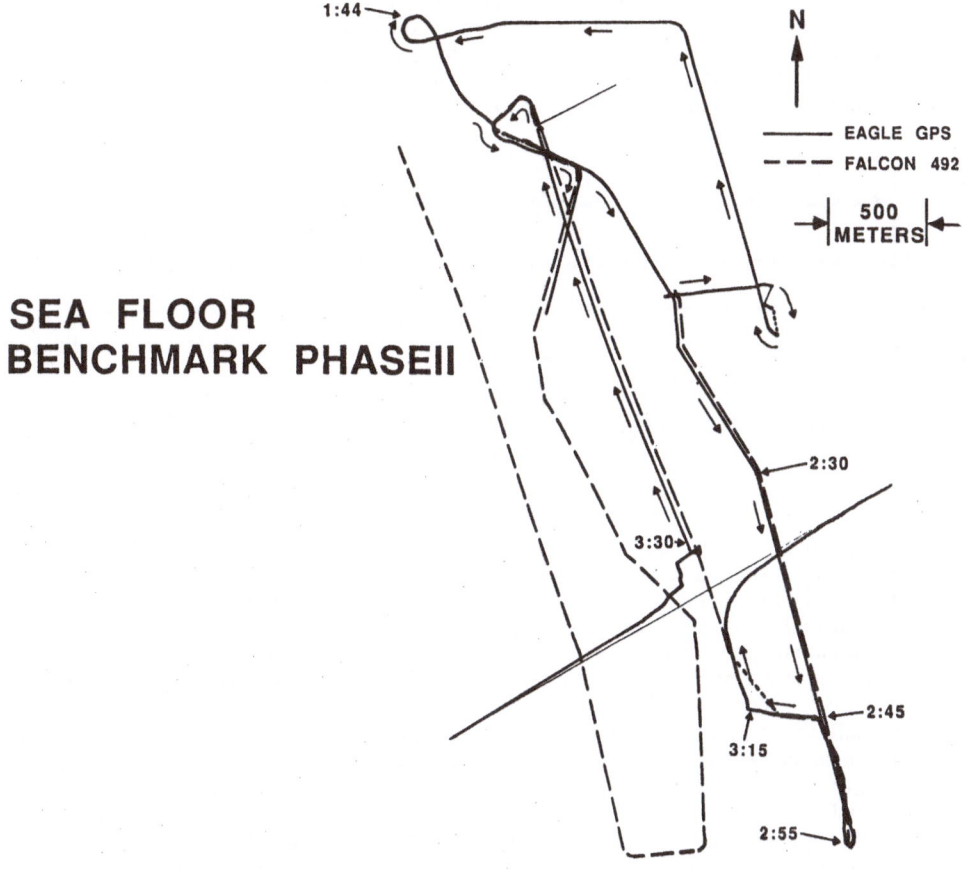

Figure 3. Comparison of Eagle GPS and Falcon 492 Reported Positions, PT. Sur - 8/12/86

Shortly after 1:00, as the Pt. Sur turned from a heading of approxiamtely 260 degrees to a heading of 155 degrees, the Eagle lost track of satellite 13. At that particular time, satellite 13 was rising but still low on the horizon at an azimuth of approximately 200 degrees. At that position, extensive blockage of the satellite was created by the ship's superstructure, especially during the turn. In Figure 2 , the effect of not being able to track satellite 13 is evident in the significant drift that was noted after 1:00 during the August 12 plot. However, by design, the Kalman filter implemented in the Eagle should be able to track three satellites in this situation without such error, as long as the height is constrained to the ocean surface. Subsequent to this test, a subtle error was discovered in the Eagle software which caused the drift of position to be excessive under conditions of tracking only three satellites. Otherwise, quite good comparison was achieved between the Eagle GPS data and the Falcon data.

At approximately 3:30 each day, performance for the satellite combination of 6, 9, 12, and 13 disintegrated due to GDOP. At that time however, very little tracking time remained for any of the alternative satellites because of low elevation angles.

During the second day, more extensive coverage was obtained from the Eagle receiver. Despite the fact that Falcon tracking data were not available during the early part of this period, a considerable overlap period was found between the mobile and shore station Eagle GPS data and the Falcon tracking data as is illustrated in Figure 3, showing the xy plot overlay. Figure 4 illustrates the direct comparison that was made between the differential solution, determined from the mobile and shore station data, and the reference tracking system solution, defined by reducing the Falcon range data from Dome and Ferrier stations. The plots shown are the differences of the differential solution for Pt. Sur and the independently determined Falcon position.

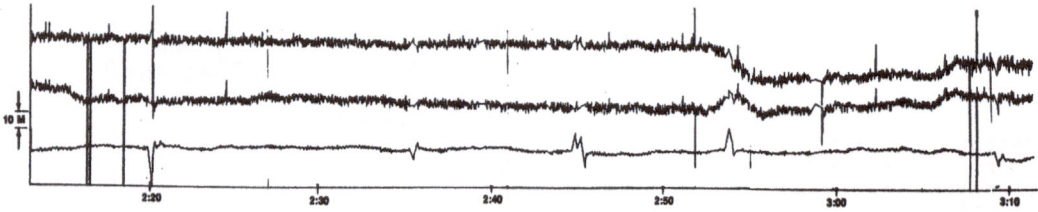

Figure 4. Differential Eagle GPS Comparison with Falcon 492 vs. Time - PT. Sur August 13, 1986

In this comparison, it is difficult to establish an ultimate accuracy that is achievable with the Eagle GPS receiver on board a marine vessel because of the limitations of the Falcon reference instrumentation and the fact that the Falcon and Eagle antennae are not colocated. Because these antennae are separated by approximately 10 meters in height and by a similar distance along the axis of the boat, the motion of the vessel contributes to a relative displacement that is periodic at approximately 6 second intervals. In addition to the harmonic motion that is relatively constant over the data interval, a shift in x and y positions is clearly evident at approximatly 2:54. Analysis shows that this shift is due to the 180 degree turn of the boat at this time and that the Falcon and GPS antennae are not colocated. At this reporting, there is an uncertainty in the zero point on this chart due to confusion of coordinates for the Mini-Ranger reference stations. However, it is clear that the differential solution is stable over a time period of approximately 1 hour. A more detailed analysis of the data is expected in future studies by Naval Post Graduate School personnel, using onboard gyro data as well as other GPS instrumentation which was simultaneously deployed at shore stations and on board the Pt. Sur.

As noted above, the harmonic signature of the boat motion is clearly evident in all of the data. This harmonic behavior is easily observed by expanding the scale of the difference plots shown in Figure 4. However, a more dramatic example of the boat motion observed from the Eagle receiver can be obtained without interference from Falcon-related errors by expanding the scale of the Eagle data using a GPS track reference system of coordinates. This reference system is defined by considering two points on the track of the vessel during a time when a constant heading is being maintained. Specifically, the x coordinate system is parallel to the line formed by connecting the two points. The y axis is perpendicular to the x axis on the horizontal plane and the z axis is the vertical. By simple differencing of the reported GPS solution at various points along the intermediate time interval, a measure of relative motion is obtained that can be plotted with a reasonably high resolution. For example, Figures 5a and 5b illustrate typical plots that are generated and clearly show the characteristic dynamic motion of the vessel in response to sea conditions. These figures were obtained for 14-minute time intervals when the boat heading was relatively constant resulting in minimum crosstrack motion. The figures demonstrate the very high degree of resolution that is achieved with the Eagle GPS receiver and clearly show the 6-second period motion of the boat.

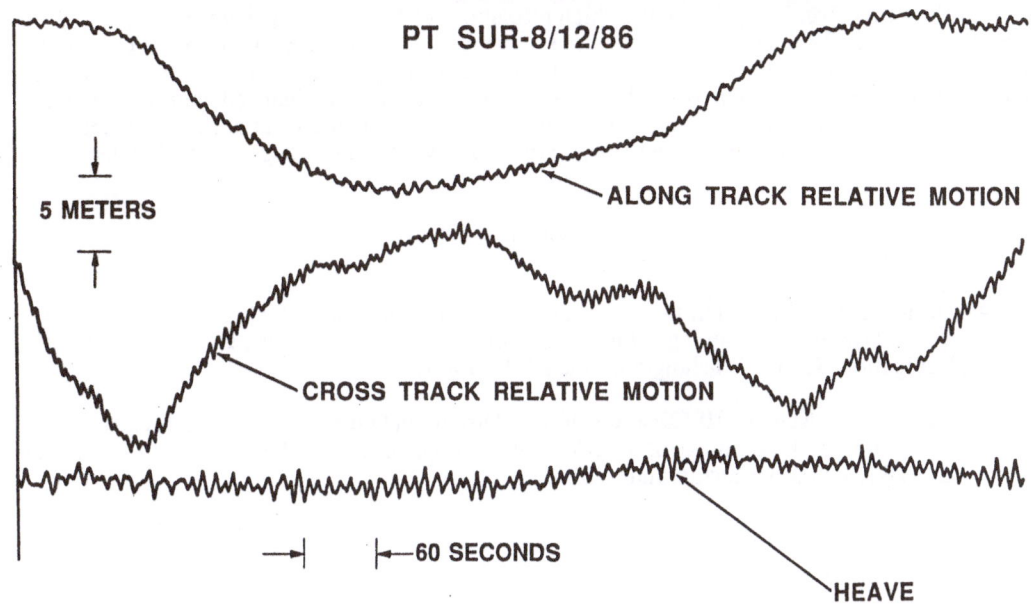

Figure 5. Typical GPS Indicated Boat Motion Relative to Track Line - PT. Sur August 13, 1986

a) 5 Meters Per Tick

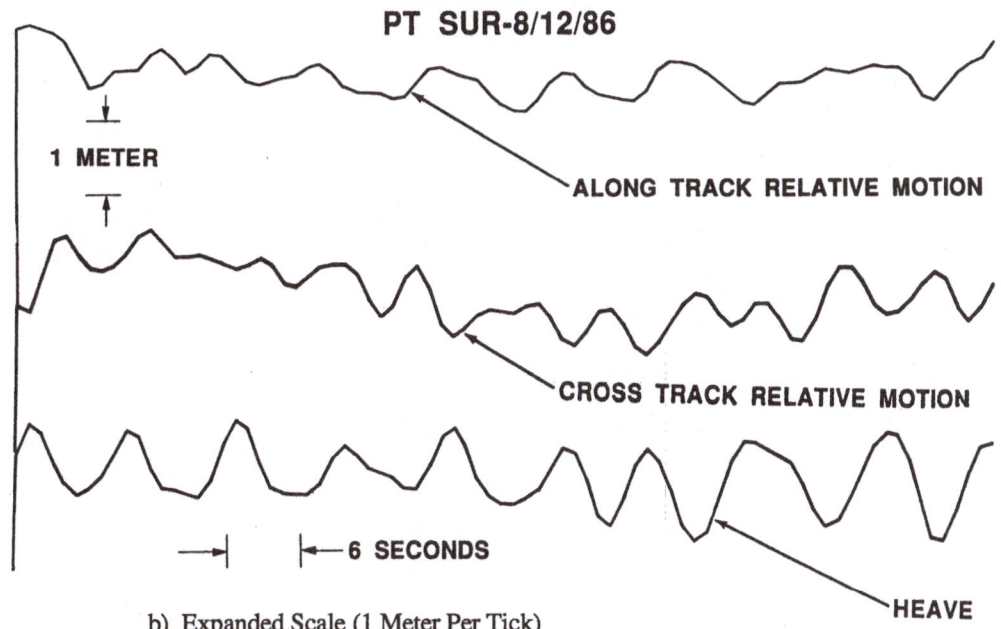

b) Expanded Scale (1 Meter Per Tick)

Conclusion

The Eagle receiver is capable of very high resolution (and, presumably, high accuracy) in reporting position. With position computations at the rate of one per second, this receiver is very suitable for precision instrumentation on board marine research vessels such as the Pt. Sur, where it is necessary to resolve motion of the boat over the period of sea state-induced harmonic motion. Although absolute accuracy of differential positioning was not established at this reporting, it is expected that differential accuracy under these conditions will be in the range of 2 to 3 meters or better.

References

1) Denaro, Robert P. and Dr. Dana Yoerger "The Application of Differential GPS to Marine Vessel Dynamic Positioning". Proceedings of the 42nd Annual Meeting of the Institute of Navigation, Seattle, Washington, June 24-25, 1986.

2) Durboraw, I. Newton, III "Test Results of a Differential GPS Receiver in a Dynamic Environment". Proceedings of the XVIII International Congress, Federation of International Geodesy, Toronto, Canada, June, 1986.

EVALUATION AND CORRECTION OF LORAN-C POSITIONS
IN COMPARISON WITH GPS DATA

Toshio Furuta, Hiromi Fujimoto
Ocean Research Institute, University of Tokyo,
1-15-1, Minamidai, Nakano, Tokyo 164 Japan
Vincent Renard
IFREMER, Centre de Brest, 29273, Brest Cedex, France
Jean-Paul Allenou
GENAVIR, Centre de Brest, 29273, Brest Cedex, France
and
Gerard Riou
IFREMER, Centre de Brest, 29273, Brest Cedex, France

ABSTRACT

Navigation by Loran-C was evaluated in comparison with GPS's data during the Jean Charcot cruise for the French-Japanese cooperative project "Kaiko". Simple overlapping mean of Loran-C positions for five minutes showed a good relative-accuracy of about 50 meters. Loran-C positions, however, showed systematic shift up to 1 kilometer due to propagation delay of Loran-C wave through land area. The value of the shift was measured in the south off the Japanese Islands, and it was found that the shift of Loran-C position can be corrected within an accuracy of about 100 meters in most cases by use of the correction table prepared by the Hydrographic Department of Japan. Corrected Loran-C positions are very close to the positions obtained by GPS system (TI-4100). In such a special case as the Loran-C wave propagates over a high mountain area or along a long coastal line, the recommended correction procedure for propagation delay should be improved to correct lthe local effects (~200m). Dome of the test data obtained by other GPS receivers are also demonstrated.

1. INTRODUCTION

The "Kaiko Project" led by a Franco-Japanese Scientific Party started on the first of June, 1984. The main objective of this project in '84 (Phase I) was to build up the precise bottom topographic and geophysical maps around the Japan Trench and related areas using the multinarrow beam echo sounder (Sea Beam), seismic profiler, gravimeter and magnetometer installed on the French R/V Jean Charcot (Shipboard Scientific Staff of the Kaiko Project, 1985).

The determination of precise ship's position is one of the most important factors for the conduct of oceanographic surveys. In particular precise ship's positioning was essential to make detailed bottom topographic maps that should be used for diving of the manned submersible.

Data processing of Sea Beam of the Jean Charcot previous to the Kaiko cruise had been carried out using the ship's positions obtained by NNSS and dead reckoning between satellite fixes. However, it is problematic to get an accurate fix from NNSS in places where the ocean current is rather fast as it is around the Japanese Islands. For example, if the error of estimation of current velocity is one knot, the position obtained by NNSS includes an error of 0.2 to 0.3 n.m., dependent on the direction of satellite aviation (Fujimoto et al., 1980).

The target areas were in the Loran-C service area (Northwest Pacific Chain: GRI=9970), where the relative accuracy of Loran-C ship's positions is

165

approximately 50 meters. Both of the on-line and off-line data processing of Sea Beam survey on board the Jean Charcot were, therefore, carried out by use of Loran-C positions.

A GPS receiver was also installed on board the Jean Charcot just before the Kaiko cruise. Although the GPS position was obtained for only 4 to 5 hours a day, the reliable navigation data obtained were valuable to calibrate the Loran-C data, and to estimate the accuracy of Sea Beam and other geophysical observations based on the Loran-C positions. Loran-C navigation in the Rho-Rho mode was also obtained by use of the cesium frequency standard from the GPS receiver.

2. POSITIONING SYSTEM OF THE JEAN CHARCOT

The instruments for position determination of the Jean Charcot during the Kaiko cruise were as follows:

```
NNSS        dual sets of Magnavox 7100R
Loran-C    JRC        JNA-760
            "          JNA-902  (Rho-Rho)
GPS        TI-4100
```

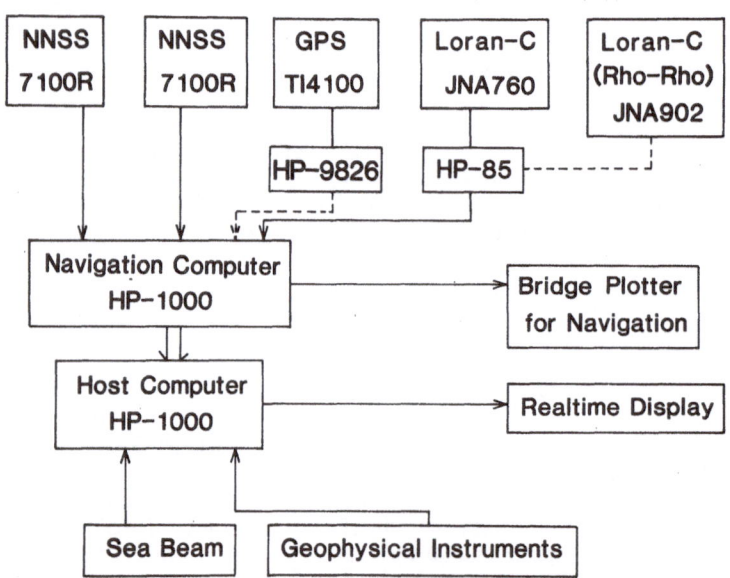

Figure 1. Block diagram of navigation instrumentation of the Jean Charcot during the Kaiko Cruise. Data of GPS and Loran-C were interfaced to microcomputers and were sent to the navigation computer.

All instruments were connected to the navigation computer, and position data obtained by each system were recorded on magnetic tapes. Figure 1 shows the block diagram of positioning instrumentation of the Jean Charcot during the Kaiko cruise. As shown in Fig. 1, these position data were sent to the navigation computer to be used for on-line data processing and real time navigation of the Jean Charcot. The Loran-C data obtained by JNA-760 were sent to the host computer through the navigation computer in real time to process the data of Sea Beam, gravity, total magnetic force and seismic profiler.

At the time of the Kaiko cruise in 1984, only four reliable NAVSTARs were in orbit and the receiving duration of the GPS system with more than three NAVSTARs simultaneously was only four to five hours a day around the Japanese Islands. Moreover it took about one hour for the GPS receiver to be able to lock on four NAVSTARs simultaneously. The positions obtained by the GPS were, therefore, solely utilized to check the NNSS and Loran-C positions. Data processing of Sea Beam and geophysical measurements did not directly utilize the GPS position.

3. RESULTS OF MEASUREMENT AND DISCUSSION

The cruise started on the first of June, 1984 from Kochi. The target areas of the first leg were the Nankai Trough (Box 6), northern part of the Kyushu-Palau Ridge (Box 7) and eastern part of the Nankai Trough and the Zenisu Ridge (Box 5). Figure 2 shows the tracks of the Jean Charcot and survey areas during the first leg. Track intervals in each target area were about 1.5 to 2 n.m.

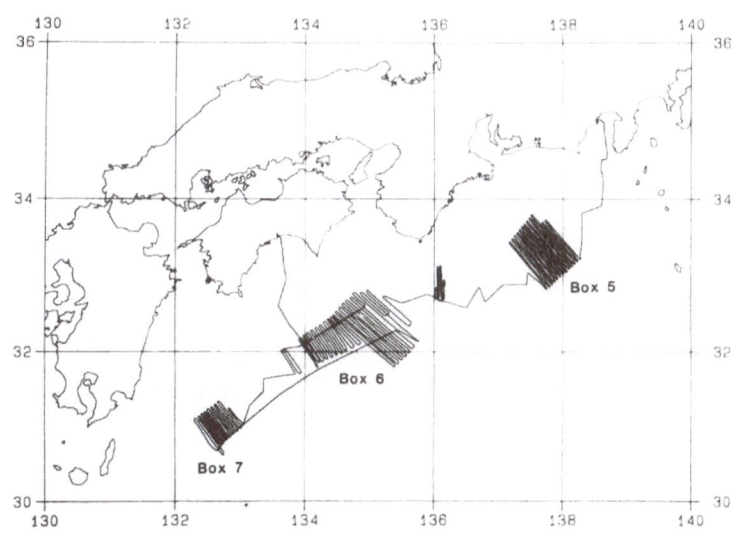

Figure 2.   Tracks of the Jean Charcot during the Leg I of Kaiko cruise.   Box 5 is the northern part of the Nankai Trough and the Zenisu Ridge area. Box 6 is the Nankai Trough area off Kochi. Box 7 is the northern part of the Palau-Kyushu Ridge.

3.1. Fluctuation of Loran-C position

For onboard processing of Sea Beam and other geophysical data, the Loran-C positions were utilized. The navigation instrumentation and its software for the on-line data processing was prepared during the transit cruise from Singapore to Kochi. In the Kaiko cruise, simple overlapping mean of Loran-C positions showed good repeatability, so that the off-line data processing used Loran-C positions which were simply filtered by overlapping mean of 10 points at thirty seconds interval. The error of Loran-C positions obtained in the off-line processing ranges normally from 30 to 50 m, and the worst case is around 200 m judging from the bottom topography of Sea Beam map. The fluctuation of ship's positions obtained by Loran-C solely is quite small and less than expected.

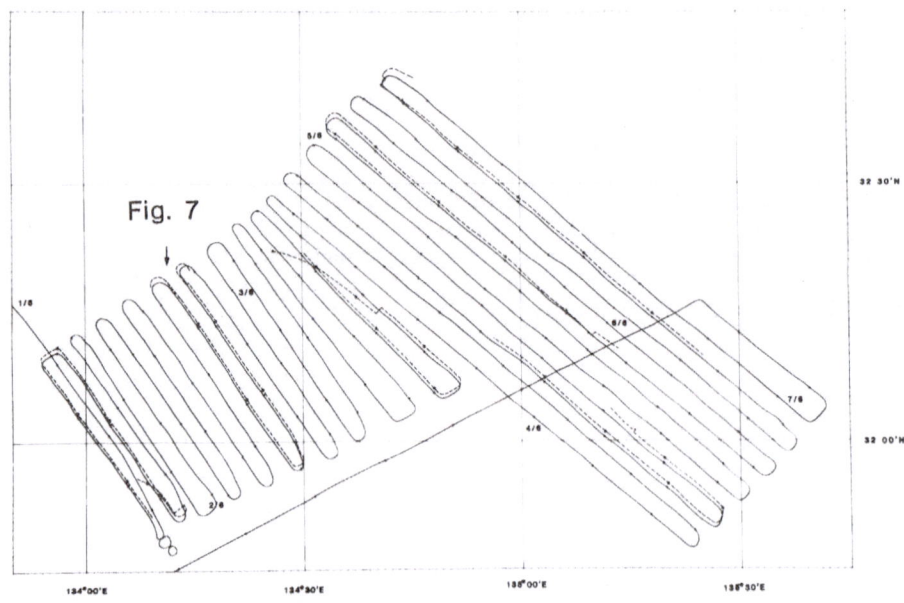

Figure 3.   Tracks of the Jean Charcot in the Box 6.   Solid line denotes the ship's positions obtained by Loran-C, and broken lines denote the positions obtained by GPS.   Numerals in the figure denote day and month as the first June (1/6) at the midnight position.   Positions by GPS were obtained during four to five hours a day.   Time mark is every one hour.

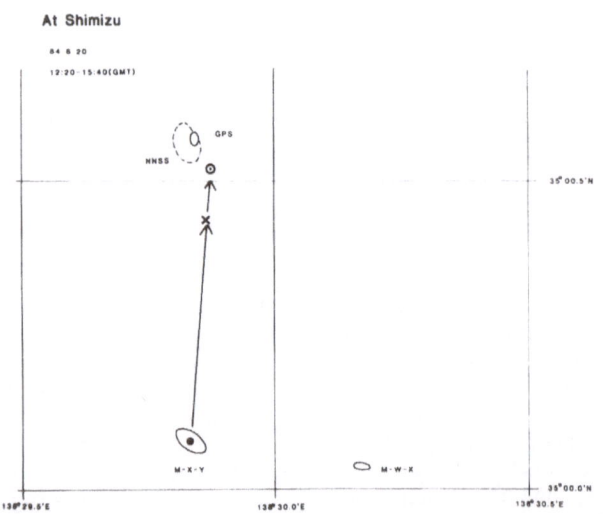

Figure 4.   Position data at the Shimizu Port obtained Loran-C, GPS and NNSS. Loran-C position using M-X-Y station network was shifted south from positions of GPS and NNSS. It is about 900 m, like as observed at the Box 6. Position of cross mark represent the normal correction of propagation of Loran-C wave.   In addition to this correction, correction of local effect of the topography of Mt. Fuji move the position close to NNSS and GPS' position (double circle).

3.2. Systematic shift of Loran-C position

Figure 3 shows one of the examples of the tracks of the Jean Charcot in the Nankai Trough (Box 6) area obtained by both of Loran-C and GPS. The discrepancy between Loran-C and GPS positions is 950 to 1,000 meters, the positions of Loran-C were southward against these of GPS. The position obtained by GPS at Shimizu Port is almost the same as that by NNSS (Fig. 4), and so it is the Loran-C position that suffers a systematic shift as observed in Box 6 area.

Figure 4 shows the position of the Jean Charcot anchored at the Shimizu Port obtained by Loran-C, GPS and NNSS. The position of Loran-C after correction is also shown in Fig. 4. However, there are some differences between corrected Loran-C and GPS positions, because a local effect remains in this area. Mt. Fuji is located close to Shimizu city, and the Loran-C wave from the X-station to the Shimizu Port passes over the summit of Mt. Fuji. The estimated delay of this local effect is 0.8 microseconds, and the corrected position is also shown in Fig. 4 (double circle).

Because Loran-C, NNSS and GPS use the world geodetic system (WGS-72) co-ordinates, a little difference in the parameters of WGS-72 co-ordinates alone cannot account for the difference of about 1,000 meters. The assumed cause of this shift is the variation of the propagation velocity of Loran-C wave on land compared with that on sea. Loran-C stations for the determination of ship's positions during the Kaiko cruise were the Master (Iwo Jima), slave of the X (Hokkaido) and slave of the Y (Okinawa) (Fig. 5). More than a half of whole propagation path from the X-station was through land area (Japanese Islands) when the Jean Charcot was in the area of the Nankai Trough. What should be considered is the delay of propagation of Loran-C wave from the X-station. The propagation path through land area from the Master and the Y-stations was negligible.

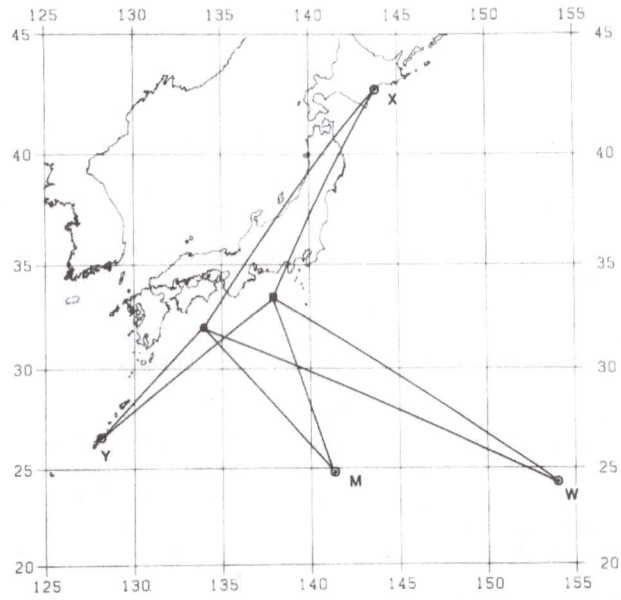

Figure 5.   The great circle propagation paths of Loran-C waves from each station. Solid circle denotes the position on the 2nd of June and solid square on the 16th of June. At these positions, more than half way of Loran-C wave from the X-station propagates over the Japanese Islands.

Figure 6 shows the differences of minutes in geographic co-ordinates between
Loran-C and GPS positions at the several points during three legs of Kaiko. The
values of latitudinal difference around the Nankai Trough area are negative and
significant, that is, observed position of Loran-C is southward against that of
GPS, and the differences decreases toward east being one order small near 140°E.
Ship going northward parallel to the northeast Japan, the values of these
differences are less than 0.1 minute, because the distances of Loran-C waves from
the X- and Y-stations through land area are negligible. While the values of
longitudinal difference in whole are less than those of latitudinal ones. The
variation of these differences both of latitudinal and longitudinal components is
due to the geometry of Loran-C stations.

Recently Ono and Nagamori (1985) of Hydrographic Department, Japan, proposed
the detailed data for the correction of propagation delay of Loran-C wave. They
proposed that the calibration factor of delay time is 0.6 microseconds per a
hundred kilometer of propagating through land part against that of sea-surface, on
the basis of precise observation at the several land and sea stations. We applied
this factor to estimate the delay time. Judging from the results of this
procedures, Loran-C positions in the Box 6 after correction should be shifted by
1000 to 1080 m towards the north from the observed positions, and the positions
after correction were closed to those of GPS.

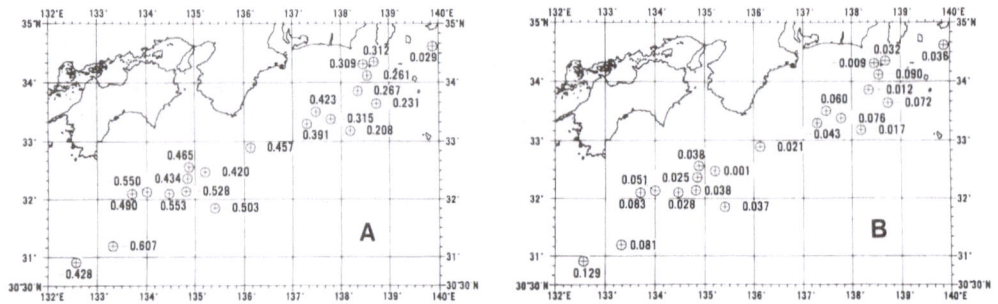

Figure 6. Differences in minute between Loran-C and GPS positions (Loran-C -
    GPS) at several points in Kaiko cruise.
    A: Latitudinal differences off the southwestern Japan. The differences
    show the largest values in the Nankai Trough area, because of the
    longest distance of propagation of Loran-C wave from the X-station (see
    Fig. 5). The values decrease when ship goes toward east.
    B: Longitudinal differences off the southwestern Japan.

Figure 7 shows the example of the difference of ship's positions between
Loran-C and GPS. Bottom profile of Fig. 7 shows the differences of propagation
of Loran-C wave from the X- and Y-stations before correction, and top one after
correction. Loran-C position has a bias of about 1,000 meters in the Nankai
Trough area (Box 6)(Fig. 7). The bias of the latitudinal component is 1,000
meters towards the north, and that of the longitudinal one is 70 meters towards
the east.

In Box 5, the value of latitudinal bias is to some extent smaller than those
of in Box 6, because the distance of propagation of Loran-C wave from the X-
station through land area is shorter than that of at Box 6. In box 5, there
remain several differences after correction between Loran-C and GPS. The cause of
this trend is assumed to be due to <u>path</u> of propagation of Loran-C wave. That is,

Loran-C wave propagates not along a geographic shortest path, but along a fastest one. If the shortest path is on land area near sea coast, Loran-C wave would choose the path at sea, which is a little more distant but faster path. Ono and Nagamori (1985) computed the correction table assuming a Loran-C path on a great circle based on the coastal lines of the Japanese Islands. If a correction table for the propagation delay is computed along the fastest path taking into consideration of radio-wave velocity around the coastal area, the table would give more precise corrections than the one we applied. The fluctuations of relative positions in both Box 5 and Box 6 are around 100 meters peak-to-peak of latitude and 40 meters of longitude components. The fluctuation of large latitude component rather than longitude one may be due to large portion of land wave and/or geometric arrangement of NAVSTARs.

2 June '84

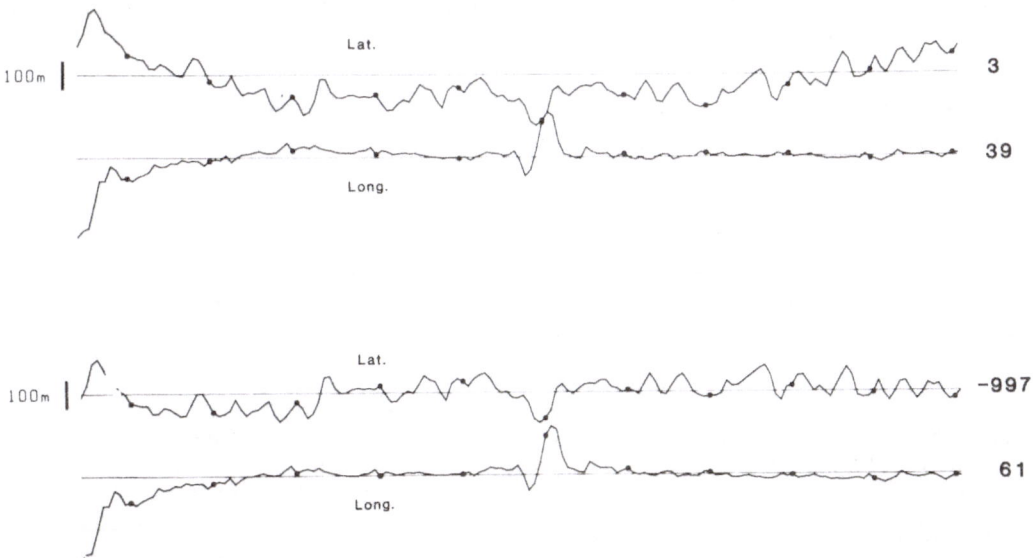

Figure 7. The differences of ship's latitudes and longitudes between Loran-C and GPS on 2nd of June. Bottom; Differences between Loran-C and GPS's positions before correction when the GPS's data are fixed. Top: Differences after correction. After correction, the positions of Loran-C are quite close to those of GPS. Time mark is shown every thirty minute. Numerals on the right side denote the differences in meter.

4. TEST OPERATION OF SEVERAL GPS RECEIVERS

Recently we have operated several kinds of GPS receivers at the stationary point and on the seas. All of them use C/A (SPS) code. Preliminary results are as follows. (1) Several GPS systems show that the fluctuation of position fixing through filtering with time constant of 100 seconds is less than 20 m. (2) Offsets of positions around 30 m were observed by a system when the satellite combination is changed. (3). Repeatability of positions at the stationary point is around 10 m averaging from a hundred three dimensional position data. (4) The motion of antenna on board is one of the serious problem. The ship's pitching and rolling affect directly to determine the range data from each satellite. For example, when the top of antenna is shaking one meter in east-west direction, the fluctuation of positions is around 5 m , rather smaller than that of north-south direction (10-15 m) dependent on the present satellite arrangement above the northwest Pacific. A GPS receiver is a ship's navigation equipment necessary for geological and geophysical cruise.

5. SUMMARY

During the Kaiko cruise, the precise acquisition of ship's position was undertaken by Loran-C system by the aid of other systems and calibration procedures. The repeatability of the Loran-C navigation was fairly good, and the relative accuracy of position of Loran-C was within 50 meters. Comparing the ship's positions of Loran-C with those of GPS, the difference between them is around 1,000 meters in the Nankai Trough area. This discrepancy is due to the delay of propagation of Loran-C wave through land area. After the correction of Loran-C positions, the ship's positions are nearly consistent with those of GPS. Because the acquisition of positioning using GPS cannot be continuous through a day at the present, Loran-C is one of the most useful instrument to continuously acquire precise ship's positions.

ACKNOWLEDGEMENTS

We thank Drs. F. Ono and K. Nagamori for the use of their correction table and also Prof. Y. Tomoda for the critical reading of manuscript and valuable suggestions. Our thanks are also to the Captain, crew and technical staff of the Jean Charcot for making the collection of the data.

REFERENCES

Fujimoto, H., Kakuta, C. and Ganeko, Y. 1980. Accuracy of NNSS doppler positioning. Jour. Geodetic Soc. Japan, 26: 172-179.
Ono, F. and Nagamori, K. 1985. Effect of Loran-C wave propagation on land and preparation of correction chart based on its evaluation. Rep. Hydrographic, Res., 20: 151-166.
Shipboard Scientific Staff of the Kaiko Project. 1985. Japanese deep-sea trench survey. Nature, 313: 432-433.

GPS-AIDED INERTIAL POSITIONING IN IRREGULAR OCEAN GRAVITY FIELDS

Erwin Groten
and
Dieter Keller
Institute of Physical Geodesy
Technical University
Petersenstraße 13
D-6100 Darmstadt

ABSTRACT

Long and short range real and artificial profiles are considered along which gravity changes monotonously and in more or less irregular fashion. Instead of ZUPT's
(zero velocity updates) which are inconvenient under marine conditions mainly coordinate updates at various interval lengths are studied in relation to inertial positioning. The mutual interaction between GPS- and inertial positioning is comtem-
plated: elimination of cycle slips in GPS-data and short term drift elimination in
inertial surveys. The support of inertial (INS) data by direct use of GPS-veloci-
ties was also considered.

1. INTRODUCTION

It seems that GPS-techniques will sooner or later dominate navigation and, to some
extent, also geodetic positioning. There are relatively few gaps where GPS cannot
fulfill the needs even in case of high-accuracy requirements (Hausch, et al.,
1985). Besides military applications the necessity of having self-supporting systems such as in inertial technology is not always of crucial importance. Offshore
as well as underwater technologies are, however, areas where due to irregular surrounding topography, as, e.g., in narrow Norwegian Fjord areas or around structures
such as borehole platforms, difficulties could arise where combinations of GPS-
with inertial technology might be of primary importance, mainly when accuracies of
a few decimeters or better are desidered. Underwater techniques are still dominated
by ranging, particularly by acoustic methods such as sonar. Whenever only relative
positioning is of interest, they still fulfill present needs in many cases. In this
paper, the role of inertial techniques is seen in two aspects mainly as part of
combination approaches: (1) inertial methods can provide attitude, velocity and position in an efficient way when frequent updating is possible; this could be provided by improved acoustic techniques in underwater and by GPS in surface areas;
(2) extreme accuracy requirements can be provided by GPS-techniques if continuous
tracking of satellite signals is possible; if interruptions do occur associated
cycle-slips lead to consequent deterioration of accuracy; whenever inertial techniques can be used in order to bridge the gaps in orbital tracking it can attribute
to high accuracy standards in GPS-methods. The short-term accuracy of inertial
technology can then be optimally used in combination with long-term positioning,
thus providing, in addition, attitude and velocity information.

M. Kumar and G. A. Maul (directors), Marine Positioning, 173–185.
© 1987 by the Marine Technology Society.

The following considerations are based on three concepts:
(1) in sea surface positioning more frequent GPS-updates can imply higher accuracy
of inertial methods and replace or reduce ZUPT's,
(2) in underwater positioning less frequent updates are possible if modern
electronic inertial equipment is available, and
(3) inertial techniques can supplement GPS under special offshore conditions in
high accuracy work.

2. UPDATE PROBLEMS

In marine and, particularly, in underwater applications the conventional terrestrial zero velocity updating (-ZUPT) technique can be useless. It may be replaced
by coordinate updating; doppler veloctiy updating sometimes does not fulfill the
accuracy requirements. The stop- and go-technique associated with ZUPT's is anyway
associated with a loss of accuracy due to the irregular accelerations which occur
in that case; therefore they should anyway be avoided whenever possible! But acoustic ranging in irregular subsurface topography and close to subwater structures is
often perturbed by erratic effects. Consequently, it is not unproblematic. At the
sea surface, frequent velocity or coordinate updating is now possible, in general,
because of the short observation time now available with GPS-receivers even in case
of high-accuracy applications. This may be even more important in the future. The
update problem in inertial geodesy might look different in the future as soon as
optical gyros attain geodetic accuracy and new developments in accelerometers
having smaller and/or more regular instrumental drift become available. Recent developments in gravimetry seem to justify such statements. In that case it might be
possible to avoid the frequent zero velocity (and similar) updates which can
strongly perturb the ideal case of "almost-equal-velocity" inertial surveys necessary in order to attain high accuracy. It should be noted that in many off-shore
surveys today ± 0.1 m accuracy is desired. If more regular drift which can be
modelled over longer (distances and) times becomes available the separation of inertial from gravitational variations over intervals longer than 3 to 5 minutes
could be of practical importance.

3. SOME TECHNICAL CONSIDERATIONS

Contrary to inertial applications on land, strap-down-platforms for marine positioning can be equipped with mechanical gyroscopes such as modern dynamically tuned
gyros. Also compromise solutions, such as gyros mounted on a carousel, can be used
in strap-down-systems and applied in order to improve calibration and alignment.
Advantages of modern manufacturing techniques can be essential. A recent japanese
gyroscope gives significantly improved stability in guidance just because of precisely manufactured ball bearings and similar minor details in manufacturing. They
may substantially contribute to higher accuracy in modern inertial equipment in
comparison to classical one. This could also affect misalignment procedures even
though that type of instrumental error is usually least affected by modern technology. With the availibility of more frequent coordinate updates it is possible to
use even medium accuracy inertial systems in various applications. Wong, (1986)
discusses such applications in the air; we are using a LTN 72 aircraft inertial
platform for terrestrial applications. As we are still at the beginning detailed
unambiguous results are not yet obtained. Nevertheless, with decreasing observation
time of GPS-receivers such approaches appear promising in the future even for high
accuracy marine applications. This is mainly true for geodesy and surveys, contrary
to navigation where immediate or real time information is desired. On the other

hand, if electronic or/and optical hardware (such as optical gyros) lead, at least
to some extent, to less instrumental drift and noise then updating can be even reduced in some cases and the separation of gravitational from inertial effects over
longer distances becomes of interest. Therefore, for the area of the Bermuda islands and Madagaskar the gravity induced effects in coordinates have been investigated, in order to illustrate their impact in relatively irregular gravity fields.

4. GENERAL CONSIDERATIONS

As far as inertial techniques are concerned, maritime applications are one of the
areas where conventional geodetic inertial methods based on zero velocity updates
cannot efficiently be applied. This concerns sea surface as well as underwater applications. It is well known that also in underwater applications (Ashkenazi and
Napier, 1986; Napier, 1985; Napier and Parker, 1985) the applications of zero velocity updates is a crucial obstacle to the optimal application of inertial techniques. A similar statement could be used in case of maritime surface applications
where, however, velocity and coordinate updates using Doppler velocities, GPS-coor-
dinates and velocities etc. are often available. For various applications of INS-
GPS etc. see (Kleusberg, et al., 1986; Rose, 1986; Goldfarb, 1986; Hutcheson and
Grierson, 1986); however, the combination of gravity gradiometrie with inertial
technique is still at its very beginning and deserves further investigation. Since
1971 gradiometry is claimed to be available with sufficient accuracy but only recently useful gradiometers seem to be available which are really useful in modern
geodesy. With increasing importance of the Global Positioning System and its
growing impact on other positioning procedures, the interaction between Global Positioning System and inertial techniques becomes one of the crucial questions in
modern geodesy. The original role of interpolatory inertial methods supported by
long-range stability of other techniques may lead to less inertial applications
with decreasing time of observation in satellite techniques, at least at the ocean
surface. With better gravity gradiometers also the advantages of universality, i.e.
the determination of plumb line deflections, gravity, azimuth, coordinates, of the
inertial methods might loose impact, as gradiometers in combination with "almost
real-time" observations of GPS-type might solve any such problem besides attitude
control etc. On the other hand, new optical gyroscopes and similar recent electronic development led to new concepts such as strap-down-systems. They are still far
from being perfect, at least as far as their high-precision performance is concerned. Such drawbacks are presently investigated and new compromises between classical platforms and new "body-fixed" systems, such as carousel platforms, might improve the calibration techniques and, consequently, also the accuracy. However, in
principle non-linearities in instrumentation, torquing errors, different and time-
varying instrumental drift can be substantially reduced with the new electronic devices. It is still unclear, to what extent similar progress could also imply a substantial improvement of accelerometers. In general, precise inertial geodesy is until now basically related to terrestrial applications but a variety of recent
developments can be used in maritime environment, too. But updating techniques are
mainly different at sea and this problem is far from being solved. However, the
combination of GPS with inertial techniques in terms of more frequent coordinate
and velocity updates affects both marine and terrestrial applications. In general,
it has to be admitted that in spite of a lot of recent progress in detail, mainly
in computational procedures, see, e.g., (Huddle, 1985), the basic principles of inertial geodesy on land and at sea still basically reflect the situation encountered
in 1970. This concerns (1) measurement techniques, (2) observation and data
processing as well as (3) the instrumentation. With the application of new electronics it seems that all three will be affected substantially and significant pro-

gress is ahead. Consequently, we do not focus on the further improvement of
conventional inertial techniques but rather on consequences arising from present
technical developments where combination of GPS and inertial techniques at the sea
surface are seen as an important hybrid system whereas subsurface applications are
supposed to be affected mainly by less gyro-drift etc. thus enabling longer
distances between updates. New inertial techniques could imply more and better use
in local applications of small research diving systems, in off-shore applications
and commerical submersile techniques.

5. NUMERICAL RESULTS

By omitting any clear distinction between inertial techniques on land and at sea
one may state that there are two basically different approaches in investigating
gravity induced effects in inertial positioning where (a) (Forsberg, 1985,
Forsberg, et al., 1985) and others investigate inertial results in terms of complex
signals which are analyzed using filter techniques, updates, drift analyses, external terrain data etc. whereas in this paper, similarly to (Groten, et al., 1985,
1986), the exact integration of gravity disturbance components is carried out by
taking into account several modifications where ZUPT and other updates are partially considered. The results do differ somewhat from polynomial fitting procedures and Kalman filtering as usually applied to inertial data. Tests, however, indicate that integrated results do not differ more from Kalman filtered data or polynomial results than the latter two types of data differ from each other. The
separate investigation of drift, updates, filtering results and gravity induced effects is justified in view of the new tendencies associated with new hardware etc.
as described above. As there are very few ocean areas where deflections of the vertical, i.e. the horizontal components of gravity disturbances, are available we selected the Madagaskar area where however smoothed deflections had to be used
(Rakotoary, 1986)   The vertical components were considered in the area of the
Bermudas where reliable information on an irregular gravity field is available. The
selected profiles of lengths ≤100 km and associated results are shown in the following figures. The data are evaluated for a local level system to which also
strap-down system results could be referred. The data are referred to updates at
intervals of 5, 10, 30 and 60 minutes of time. For underwater applications long intervals should be of interest if (and only if) in the future drift effects can be
substantially reduced or better modelled so that updates are no longer necessary at
very short intervals in evaluating drift components etc.. By comparing bathymetry
in the Bermudas area with free air gravity it is realized that - probably due to
density variations in crustal material, besides numerical effects - the local regression between terrain and free air gravity is not perfect. This illustrates the
limits in using bathymetry for gravity interpolation. Submersile and sea surface
applications are, to some extent, similar to helicopter applications on land because the vehicle does not move along the surface of the terrain as in case of terrestrial applications on land where the vehicle path runs along the terrain itself
and the vertical platform channel records (implicity and) automatically the terrain
heights. Numerical results are summarized in Figs. 1 to 7 and Table 1. As (ξ, η) in
Fig. 1 are smoothed values various types of noise were superimposed, but noise effects were found to be insignificant.

6. DRIFT STUDIES

Our laboratory studies indicate relatively small diffferences between stationary drift and drift in a moving platform of practical field tests. Modelling of drifts of inertial equipment using Kalman filtering and polynomial fitting may indicate significant differences between stationary and "field" drifts but this may not be the case in many marine applications. Drift simulations do corroborate these differences for classical platforms, to some extent. It is planned to supplement these investigations using a LINS-system. Cross (1985, 1986) has recently developed Kalman-filters for oceanic applications which might further complement such investigations at sea. For LTN 72 drifts see Figs. 8 and 9.

7. CYCLE SLIP REMOVAL

When more frequent coordinate GPS-updates become available we might ask to what extent cycle slips could be removed using inertial interpolatory techniques between GPS observations and within very short intervals in very precise off-shore applications. Remondi's (1985) dynamic method can be seen under various modifications where one GPS receiver is located at a fixed station the other in a moving vessel or a situation where the GPS measurement is related to an (a priori known) baseline. In any case, the observation equations of the GPS measurements containing the (unknown) number n of cycles can be resolved with the aid of coordinates obtained from an inertial platform whenever the continuous GPS-measurement is interrupted. Similar ideas were also proposed by C.C. Goad (private comm., 1985); see also (Goad, 1985). There is a variety of alternatives to the concepts of Remondi and Goad such as (Mader, 1986).

8. CONCLUSIONS

The main advantage of inertial techniques is the fact that these methods are basically self-contained, can be used at sea surface and in underwater positioning (in submersiles, with towed fish etc.). Even by assuming that correlation of subsurface terrain, i.e. bathymety, may be used for the separation of gravity from depth there is no such analogue for the separation of horizontal coordinates from deflections of the vertical in areas of strong bathymetric variations and/or deflections. As far as high-precision marine inertial positioning is concerned, (1) the decomposition of complex observed results as well as (2) the separate theoretical investigations using precise gravity and deflections fields, drift and alignement errors etc. are two methods which supplement and complement each other in studying possible improvement of present and future inertial applications.
If ZUPT's could be totally avoided in subsurface applications based on better drift modelling and better knowledge on gravity induced effects in all three coordinates of a local horizon system it would be very helpful.

ACKNOWLEDGEMENT

W. Hausch and T. Kling contributed substantially to the paper by giving valuable hints and carrying out some numerical evaluations. The Bermuda data were provided by U.S. Charting and Geodetic Survey, Rockville, MD.

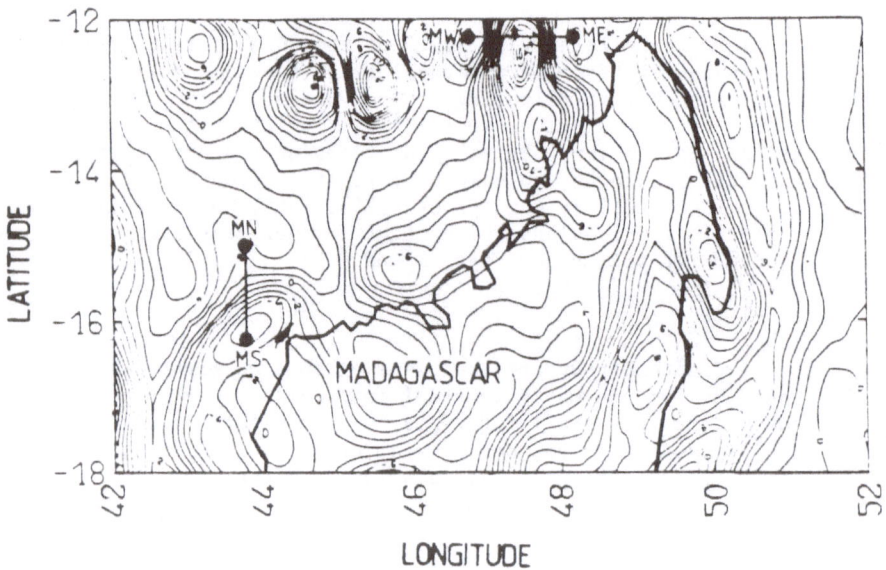

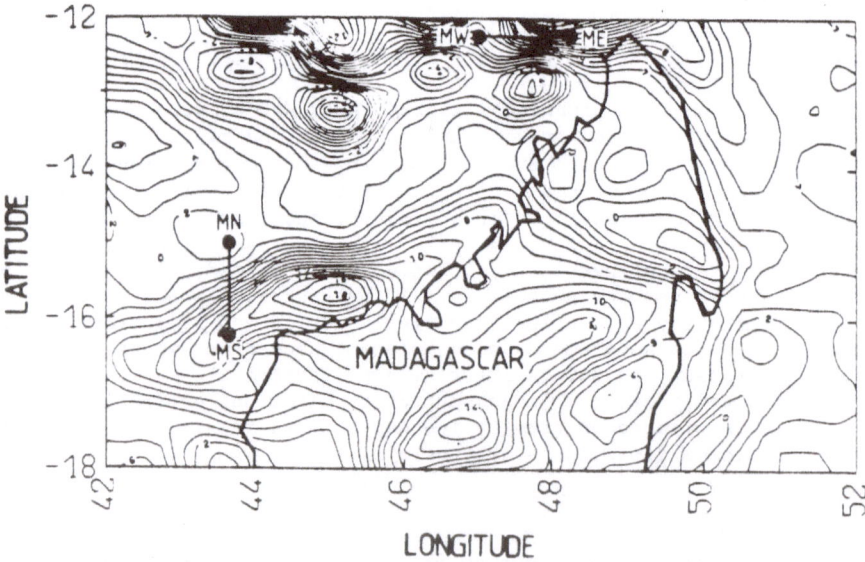

Fig. 1   Madagaskar profiles in (ξ, η) map in evaluating horizontal coordinate perturbations

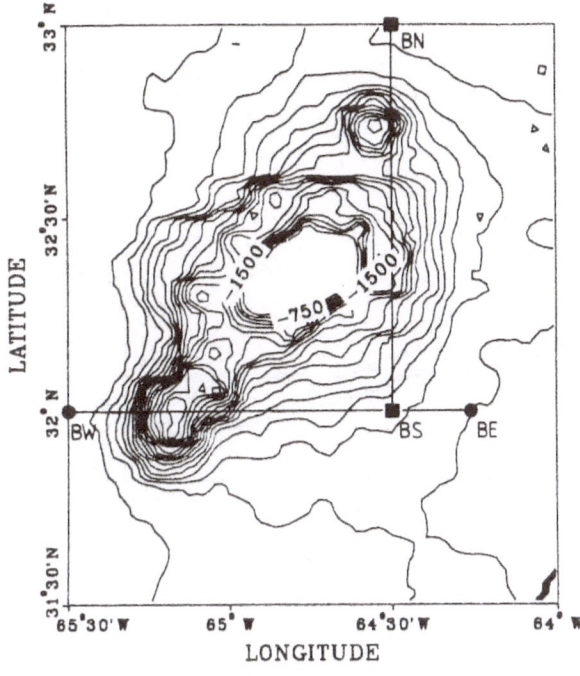

Fig. 2 Bermuda terrain map

(contour interval 250 m)

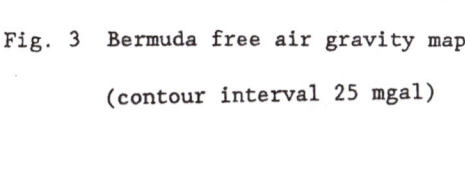

Fig. 3 Bermuda free air gravity map

(contour interval 25 mgal)

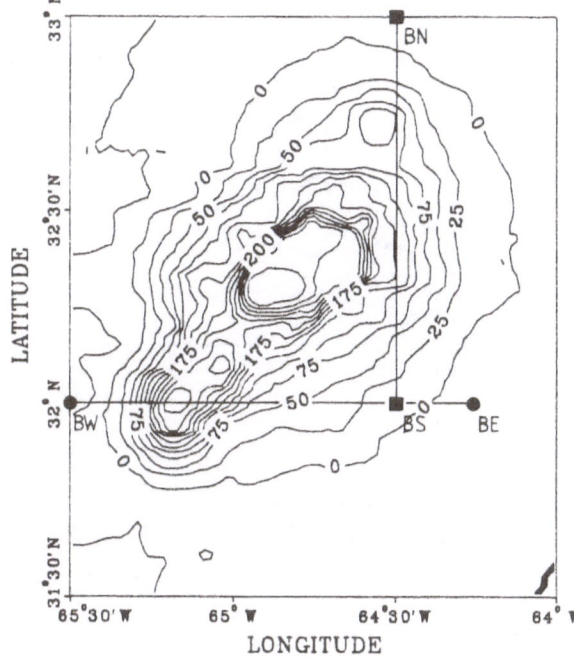

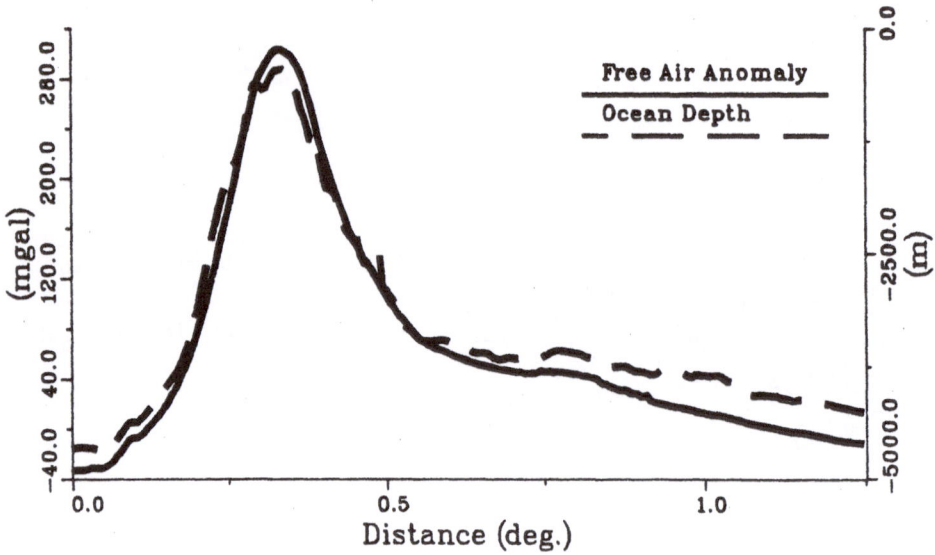

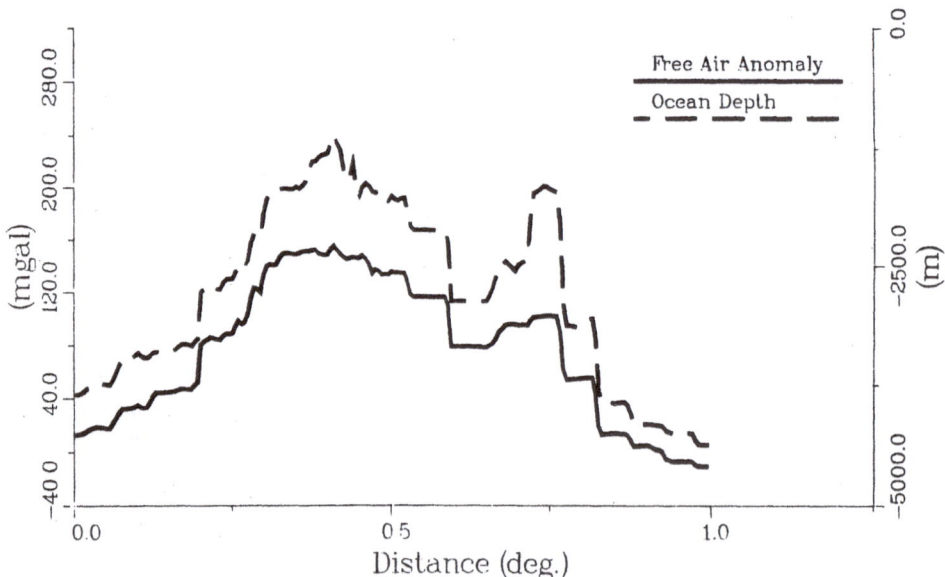

Fig. 4   From top to bottom:
         BW-BE and BS-BN Bermuda profile for the evaluation of vertical coordinate
         perturbations

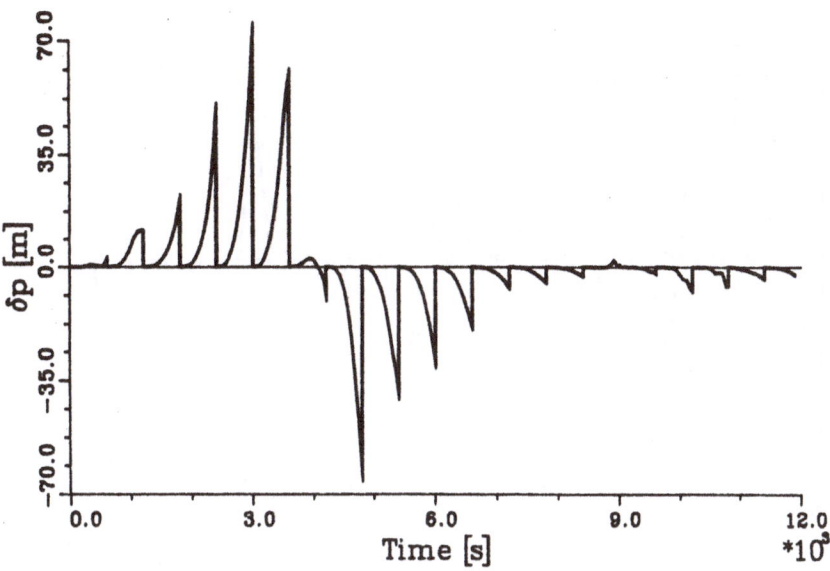

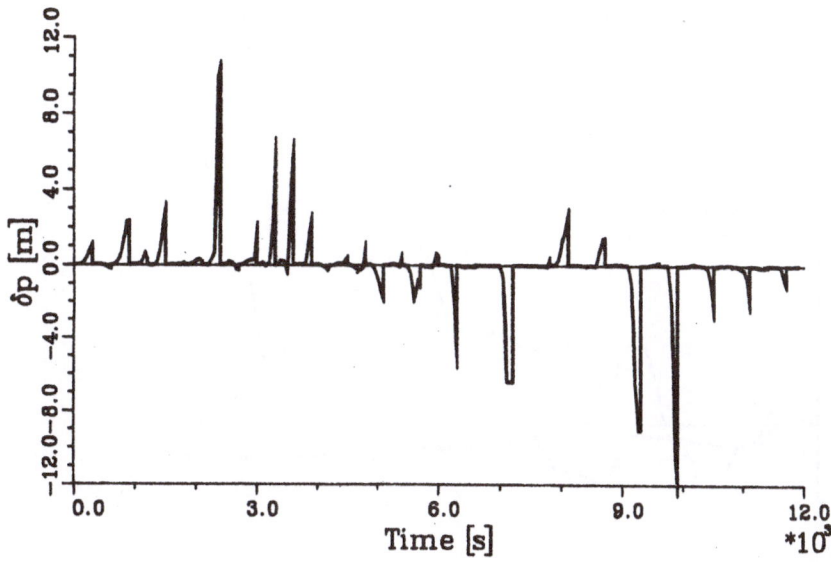

Fig. 5   From top to bottom:
         vertical position error for the BW-BE profile (10 min update interval) and
         the BS-BN profile (5 min update interval) after coordinate updates modeled
         by a step function at the coordinate update points

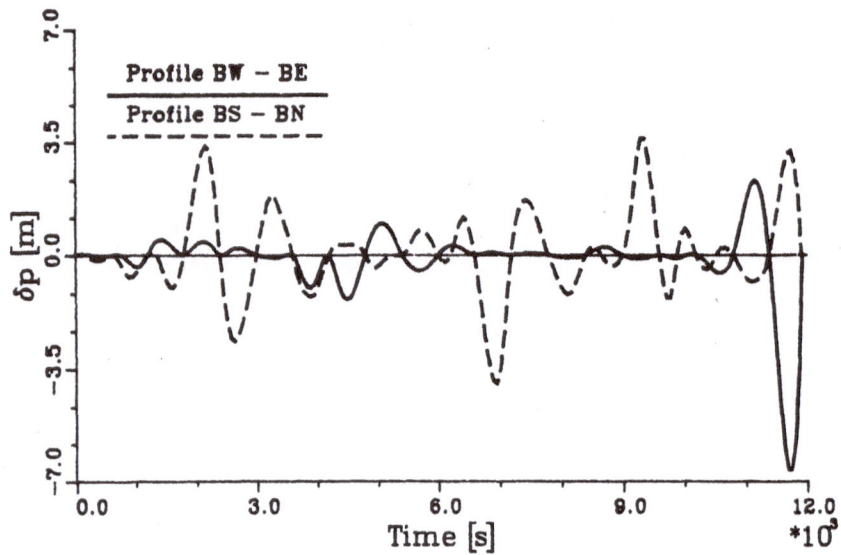

Fig. 6 Residual vertical position error on profile BW-BE after coordinate updates
         at intervals of 10 min modeled by a spline function of the position error
         at the coordinate update points

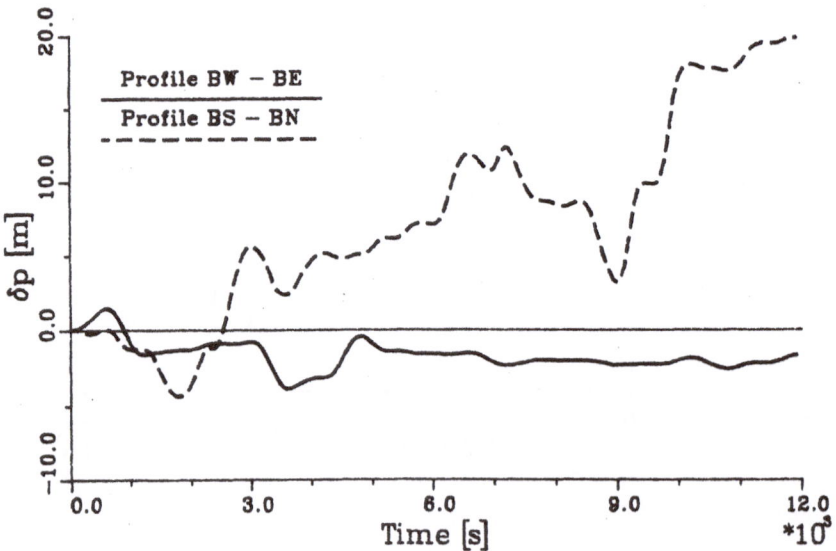

Fig. 7 Residual vertical position error on profile BW-BE after velocity updates
         (v=known) at intervals of 10 min modeled by a spline function of the
         velocity error at the velocity update points

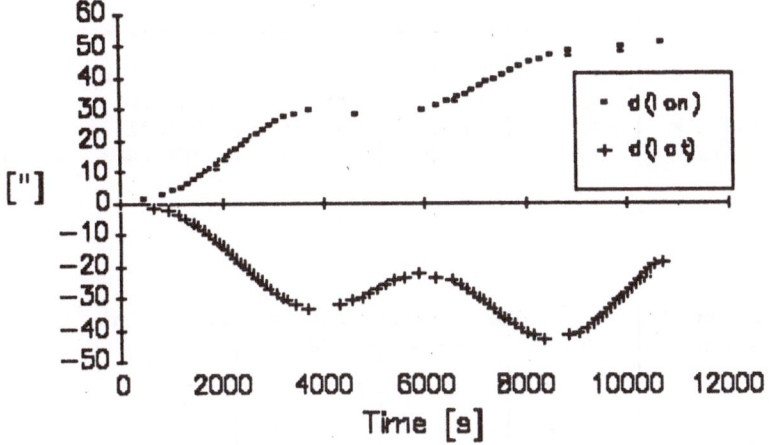

Fig. 8 Stationary drifts (for LTN 72)

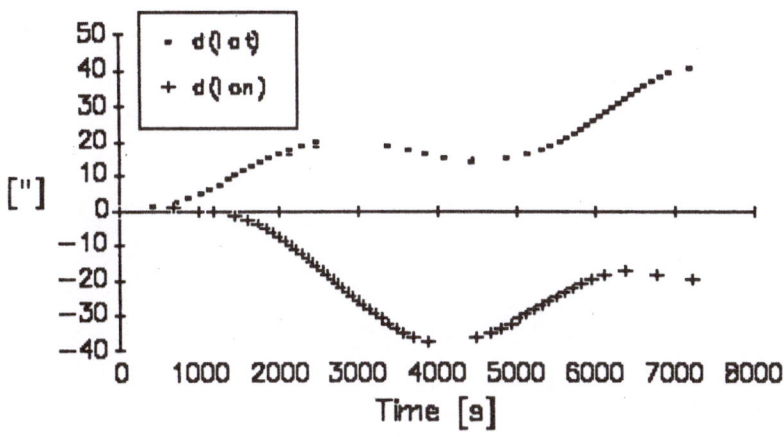

Fig. 9 Simulated drifts for moving vehicle (for LTN 72)

profile	direction of deviation	velocity update interval of [min]			
		5	10	30	60
MN-MS	Cross-Track	4,89[m]	21,25[m]	196,53[m]	750,09[m]
	Along-Track	7,26	31,61	288,32	1104,56
MW-ME	Along-Track	10,90	46,79	397,43	1541,11
	Cross-Track	5,34	23,20	207,70	786,84

profile	direction of deviation	coordinate update interval of [min]			
		5	10	30	60
MN-MS	Cross-Track	0.23[m]	1.75[m]	45.54[m]	339.66[m]
	Along-Track	0.36	2.66	69.47	469.69
MW-ME	Along-Track	0.78	5.86	152.49	1115.22
	Cross-Track	0.25	1.94	50.44	360.86

Table 1   Magnitude of the horizontal position error for the Madagaskar profiles after updates modeled by a step function at the update points

REFERENCES

Ashkenazi, K., and Napier, M.E., 1986. "Modern navigation and positioning techniques." Oceanology, Vol. 6 of Advances in underwater technology, ocean science and offshore engineering, pp.1-9, Graham and Trotman, London

Cross, P.A., 1986. "Kalman filtering and its application to position fixing offshore." Proc. Hydro '86, Hydrographic Soc., Southampton

Cross, P.A., and Pritchett, C.H. 1986. "A Kalman filter for real time positioning during geophysical surveys at sea." FIG-Proc., Toronto

Goad, C.C., 1985. "Precise Positioning with the Global Positioning System." Third Intern. Symp., Proc. Inertial Technology for Surveying and Geodesy, Vol. II, pp.745-756, K.P. Schwarz (ed.), Banff, Canada

Goldfarb, J.M., 1985. "Exposure station control for aerotriangulation with an INS-Differential GPS." Third Intern. Symp., Proc. Inertial Technology for Surveying and Geodesy, Vol. II, pp. 777-789, K.P. Schwarz (ed.), Banff, Canada

Groten, E., Hausch, W., and Keller, D., 1985. "Study of the influence of specific gravity anomaly and deflection fields in inertial surveying." Third Intern. Symp., Proc. Inertial Technology for Surveying and Geodesy, pp. 187-202, K.P. Schwarz (ed.), Banff, Canada

Groten, E., Hausch, W., and Keller, D., 1986. "Some special considerations on gravity induced effects in inertial geodesy." Submitted to manuscripta geodaetica

Hausch, W., Groten, E., Euler, H.J., Strauss, R., and Feltens, J., 1985a. "Three-dimensional geodetic control of regional Macrometer network." manuscripta geodaetica, Vol. 10, pp. 306-316

Hutcheson, W.J., and Grierson, A.D., 1985. "Gravity gradiometer post mission data processing." Third Intern. Symp., Proc. Inertial Technology for Surveying and Geodesy, Vol. II, pp. 687-707, K.P. Schwarz (ed.), Banff, Canada

Kleusberg, A., Quek, S.H., Wells, D.E., and Hagglund, J., 1985. "Comparisons of INS and GPS ship velocity determination." Third Intern. Symp., Proc. Inertial Technology for Surveying and Geodesy, Vol. II, pp. 791-805, K.P. Schwarz (ed.), Banff, Canada

Napier, M.E., 1985. "Application of inertial positioning to high accuracy underwater survey." Ph. D. thesis, New Castle University

Napier, M.E., and Parker, D., 1985. "Simulation of an inertial based integrated positioning system for underwater surveys." Third Intern. Symp., Proc. Inertial Technology for Surveying and Geodesy, Vol. II, pp. 451-463, K.P. Schwarz (ed.), Banff, Canada

Napier, M.E., and Parker, D., 1986. "Application of an inertial based integrated positioning system to underwater surveys." Hydrographic Soc. HYDRO 86, Southampton

Rakotoary, J., 1986. "Geoide Gravimetrique sur Madagaskar, par Rapport au Système de Reference 1980." Bull. d'Information, No. 58, IGC, Toulouse

Remondi, B.W., 1985. "Performing centimeter accuracy relative surveys in seconds using GPS carrier phase." Proc. First Intern. Symp. on Precise Positioning with the Global Positioning System 1985, Vol. II, pp. 789-797, U.S. Dept. Comm. NOAA, NOS, Rockville

Rose, E.J., 1985. "A cost/performance analysis of hybrid inertial/externally referenced positioning/orientation systems." Third Intern. Symp., Proc. Inertial Technology for Surveying and Geodesy, Vol. II, pp. 757-   , K.P. Schwarz (ed.), Banff, Canada

PRECISE GPS-AIDED MARINE POSITIONING DEVELOPMENTS AND RESULTS

Günter Seeber, Andreas Schuchardt and Gerhard Wübbena
Institut für Erdmessung Universität Hannover
Nienburger Str. 6
D-3000 Hannover, FRG

ABSTRACT

Based on two years of practical experiences in GPS research for marine applications, some findings and experiences are communicated. Research at the Institut
für Erdmessung includes software developments, experiments under controlled conditions and practical applications. TI4100 receivers were used in single and
differential mode. Software developments include use of carrier phase observables
in dynamic mode and demonstrate capacity of decimeter accuracy in real time.
Comparison of GPS with acoustic techniques is analyzed with real data. Applications for marine geophysics, marine exploration and near shore hydrography are
demonstrated in various examples.

1. INTRODUCTION

Within this contribution some experiences and results are communicated, related
to the use of GPS for precise positioning at sea. Most of the findings are based
on the use of Texas Instruments TI4100 P-code receivers. The "Institut für Erdmessung" (IFE) at the University of Hannover in relation with the "Special
research center of geodetic and remote sensing techniques in coastal areas and
at Sea" (Sonderforschungsbereich 149) disposes on one TI4100 receiver since early
1984 and has access to additional receivers. Thus it was possible to carry out
several dedicated investigations into the use of GPS for precise positioning on
moving platforms. One rationale behind these investigations was to use original
software developments and to conduct - if possible - the experiments under controlled conditions. Detailed informations on software developments within IFE,
and data lines, realized on different vessels are given in various reports (i.e.
Seeber et al. 1985a, 1985b, 1986). In chapter 2 some of these developments are
summarized.

GPS applications at sea are of interest among others for geophysical research,
prospecting of raw material, mapping of sea bottom and near shore hydrographic
surveying. Examples are given to all of these categories.

With respect to observation strategies the following alternatives may be
distinguished:

- application of the TI4100 user solution on board
- computation of an independent "own" Kalmanfilter solution with onboard GPS
  pseudorange data
- computation of a differential solution on board with pseudorange
  corrections from a landbased reference receiver
- use of GPS in connection with other navigation sensors like acoustic or
  inertial techniques
- use of the carrier phase observables in addition to code-pseudoranges

The examples given within this paper refer to these above mentioned strategies.

M. Kumar and G. A. Maul (directors), Marine Positioning, 187–196.
© *1987 by the Marine Technology Society.*

2. SOFTWARE DEVELOPMENTS FOR MARINE POSITIONING AT IFE

In order to use the TI4100 original data stream, different software packages for data handling, data transfer and decoding had to be developed. Details can be found in Seeber et al. 1985a and 1985 b.

Besides the navigational message, the TI4100 provides two basic types of observables, code phases and carrier phases. The main features of both observables are given in Table 1.

Table 1: Code- and Carrier Phases

	Code	Carrier
Ambiguity	unambigous	ambigous
Wave Length	P: 29,3 m	L1 19.05 cm
	C/A: 293 m	L2 24,45 cm
Measurem. Noise	P: 0.8 m	2-3 mm
	C/A: 10 m	
Propagationeffects	retardation through Ionosphere	acceleration through Ionosphere

It is evident that carrier phases contain a much higher accuracy potential than code phases. This potential is used for geodetic positioning on land. For marine positioning up to now, mainly code phases have been used. It can be shown that under certain conditions also the high accuracy of carrier phases is usable on moving platforms.

2.1 Navigation software GNAV

In navigation with GPS the Kalmanfilter has found a spread use. Typically for positioning of a single moving station an 8-state vector is used which contains one receiver clock error, three position coordinates and their derivatives in time. Vehicle dynamics, i.e. accelerations, are modelled as system noise with zero expectation. Variances for pseudorange measurements were obtained empirically from TI4100 data. For P-code measurements this was derived from ionospheric refraction corrections and found to be $2.7 m^2$ in dynamic mode (Schuchardt 1985). This value was scaled to be $23 m^2$ for C/A-code pseudoranges. The scale factor was obtained from an investigation on the noise of range corrections for differential positioning (Matthes 1986).

2.2 Differential navigation with pseudorange corrections

The various possibilities for differential navigation, with GPS are well known from literature (Kalafus 1985). Pseudorangecorrections, derived from a comparison between predicted and measured pseudoranges to all visible satellites have the advantages that simple algorithms are applicable on the reference station and that the remote station is not forced to use identical satellit selections. One difficulty however is that the reference receiver clock may introduce high correlations between all pseudorange corrections, and that common effects for all pseudoranges exceed by far the individual corrections. In order to avoid

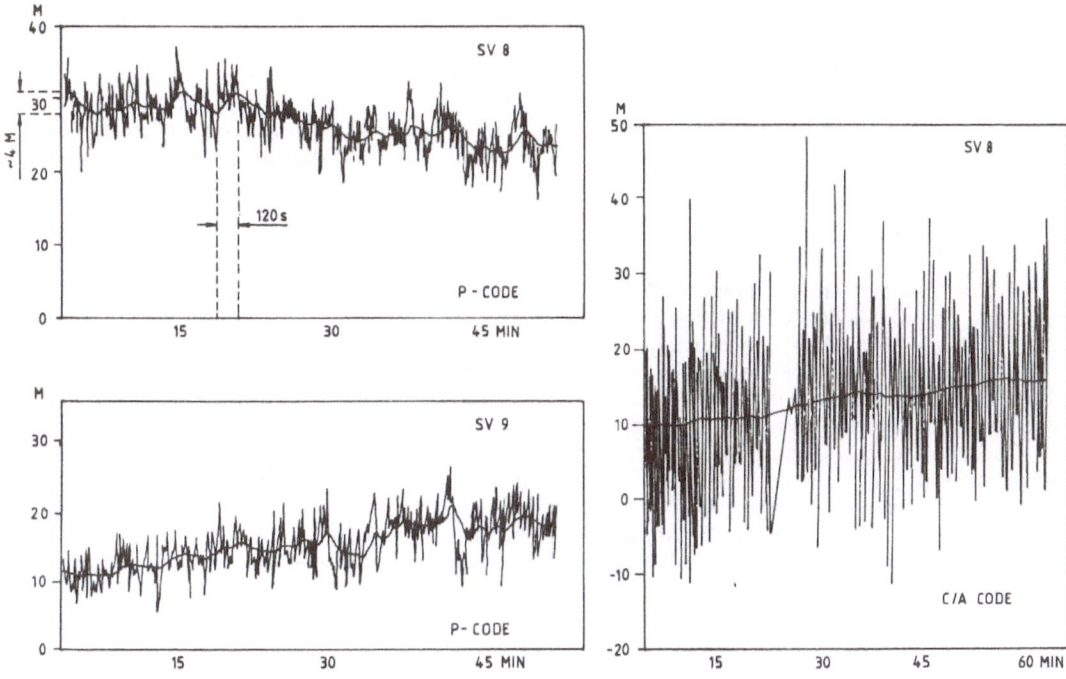

Figure 1: Pseudorangecorrections on a reference station for P-Code (left) and C/A Code (right). Space Vehicles 8 and 9. Full line comes from a simple filter algorithm.

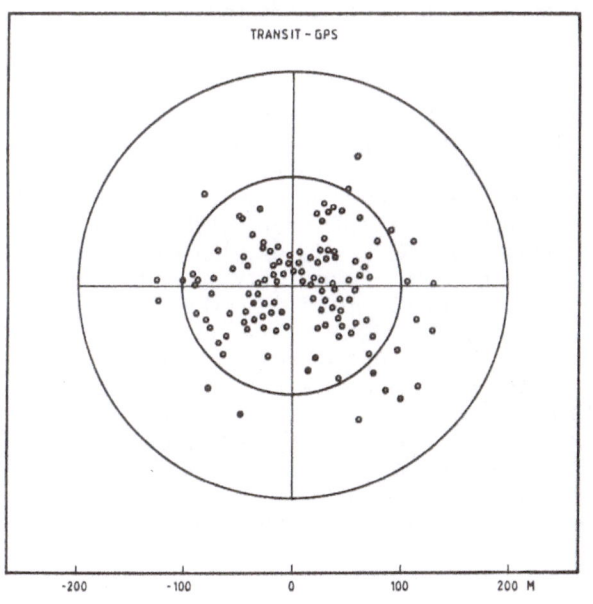

Figure 2: Differences between Transit single pass fixes and GPS results on SONNE cruise East Pacific.

numerical problems, one satellite is selected as reference satellite and only
differential corrections with respect to this satellite are transmitted. The
common effect to all satellites is treated as a receiver clock error on the remote
station. This is done for ranges and velocities.

From empirical data the measurement noise (variances) for range corrections
was found to by 6 m² for P-code and 50 m² for C/A-Code. In order to avoid the
transfer of this noise to the remote station, a simple Kalmanfilter is applied
to the range corrections. Figure 1 shows a sample of unfiltered and filtered
corrections for P- and C/A-Code.

2.3 Batch solution with carrier phases

Table 1 shows an enormous difference in the magnitude of standard deviations
of code pseudoranges and of carrier observations. This can be utilized to improve
successively the code pseudorange results. The basic idea is to use both observation types simultaneously. The solution starts with absolute code results only.
In the sequel, the code-results are improved by the highly accurate relative
positions coming from carrier phase observations.

Details of the algorithm are given in Seeber et al. 1986. The approach shows
some analogy to a batch solution. In comparison to a Kalmanfilter approach, no
assumption have to be made on platform dynamics. Thus the advantages of the very
low measurement noise with respect to the higher system noise can be fully
exploited. The disadvantages of Kalmanfiltering like "overshooting" effects
do not apply. Results with this solution are demonstrated in 5.1.

3. SINGLE TI4100 RECEIVER RESULTS

Comparisons between a TI4100 receiver onboard of a vessel and near shore
navigation aids like HIFIX or SYLEDIS have been reported earlier and are not
to be repeated here(i.e. Seeber et al. 1985b). Dependent from the satellite
configuration, the agreement with SYLEDIS was found to be in the order of
10-30 m after correcting for datum differences. This corresponds to finding
with the high precision near shore positioning system POLARFIX (see Figure 7).

For remote areas this accuracy is much better than with any alternative
navigation system. Figure 2 shows the differences between GPS- and Transit
solutions in the East Pacific. The discrepancies correspond to the accuracy
estimate for Transit fixes including velocity errors. It has to be realized however, that the Transit fix accuracy is deteriorated rather heavily through dead reckoning   errors in standard integrated navigation systems. An example is given
in Figure 3. Here a comparison was made between the TI4100 user solution and
the output of an integrated navigation system INS (NNSS, Log, Gyro)on the German
Polar Research Vessel "POLARSTERN". Update rates after Transit fixes reach 1 km
or more (i.e. at 21.26 hours) and shift the integrated solution to the GPS
results.

The high single receiver accuracy with GPS is already extremely important
for geophysical surveys and the mapping of the ocean bottom with sonar systems
like SEABEAM. This is demonstrated in Figure 4. During a cruise with the German
research vessel SONNE in Central Pacific the positioning of SEABEAM tracks was
partly controlled by GPS respectively INS. The deficiencies in times without GPS
coverage (i.e. after 08:00) are cleary shown. The non-parallelity of tracks
produces gaps between mapped areas, and the weak accuracy in position determination deteriorates the mapping accuracy.

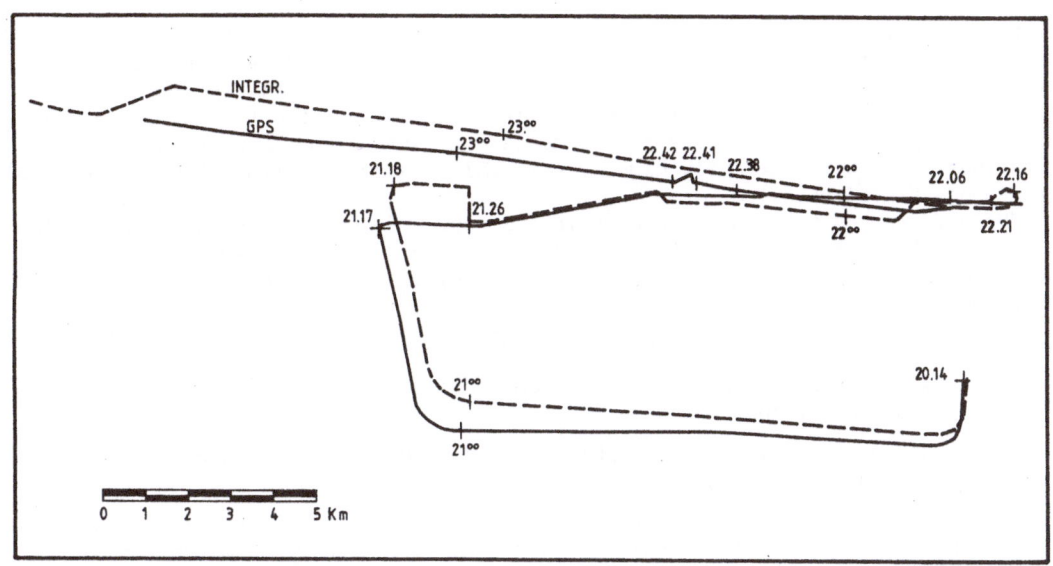

Figure 3: Differences between Integrated Navigation System (NNSS, Log, Gyro) and GPS on RV POLARSTERN in North Atlantic. INS update through Transit fix at 21.26 hours.

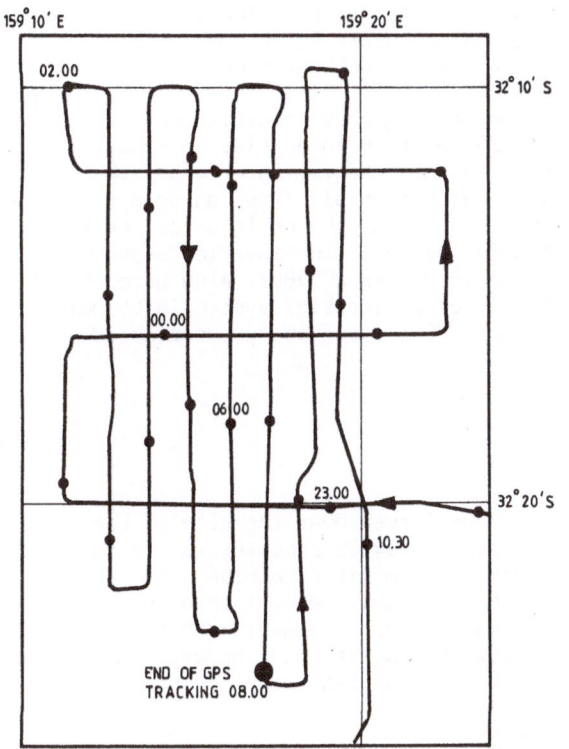

Figure 4: SEABEAM tracks with GPS navigation (before 08.00 hours) and INS (after 08.00) on RV SONNE in Central Pacific.

4. SIMULTANEOUS USE OF GPS AND ACOUSTIC TECHNIQUES

During a cruise with RV SONNE near Galapagos Rift in the second half of 1985, one TI4100 GPS receiver was operated simultaneously with an ATNAV acoustic transponder array and provided valuable insights. Main objective of this cruise was the detailed exploration of a very limited area with respect to hydrothermal sulphide deposits. The accuracy requirements for surface and in particular under-water positions were extremely high. Out of this reason an array of 4-6 acoustical transponders was installed.

The differences between surface positions from GPS and ATNAV on-board solutions turned out to be rather large (Figure 5). Out of this reason an independent soft-ware package for processing acoustic data was developed with the following features

- calibration "on the job" of the transponder array

- determination of datum parameters between local acoustic system and WGS72 via GPS positions

- determination of local and geographical coordinates for transponder- and ships position.

Original observations are the observed signal travel times between vessel and bottom transponder. Details of this software are developed by Heimberg (1986).

The improvement of acoustically determined ship's position through this soft-ware is shown in Figure 6. The correspondence between GPS- and ATNAV positions lies in the order of 10 m.

As a consequence it is recommended, not to integrate GPS-single receiver code results with acoustic results because the relative accuracy of acoustic positions is rather high. GPS however provides absolute positions and is an important means for tying the underwater positions to a global datum. In order to derive reliable datum parameters it is necessary to include ship's heading data into the solution, because the origin of both systems generally is not indentical.

Further investigations have to be done whether GPS phase observations are suitable to improve an integrated acoustic-GPS solution in a relative sense.

A similar situation is given with respect to the integration of GPS with inertial techniques. From a first experiment (Seeber et al. 1986) on an air-plane it could be shown, that GPS is capable to control the drift of an inertial strap-down system. With an update rate of a few seconds, the inertial system can provide an accuracy in the sub-10 m range (Brüggemann 1986). Also here it has to be investigated, to what extend the use of an inertial system is becoming obsolete through the high relative accuracy of GPS carrier phase observables.

5. DIFFERENTIAL GPS

5.1 Use of pseudorange corrections

In July 1985 a three days experiment could be performed near the city of Kiel on the Baltic Sea (KIELNAC 85 campaign), where one TI4100 receiver was installed an board a small surveying vessel "Sturmmöve" and a monitor receiver was operated onshore. The vessel position was simultaneously determined with POLARFIX equipment, a tachymeter which continuously measures range and azimuth to a reflector system onboard. The manufacturer states a positioning accuracy of 0.1 m + 20 ppm (Stednitz et al. 1983). A part of these test runs is shown in Figure 7.

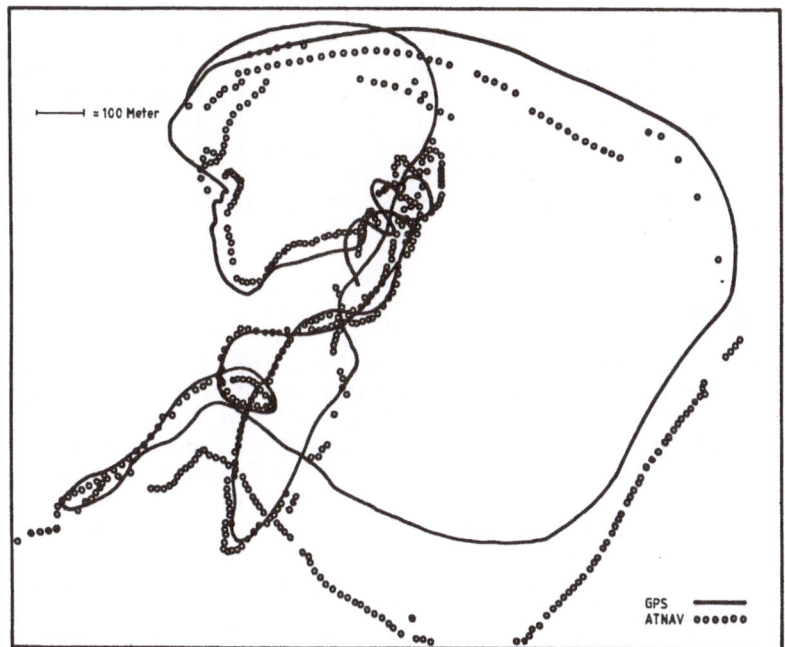

Figure 5: Differences between surface positions from GPS and acoustic transponder System ATNAV (system solution) on RV SONNE cruise East Pacific.

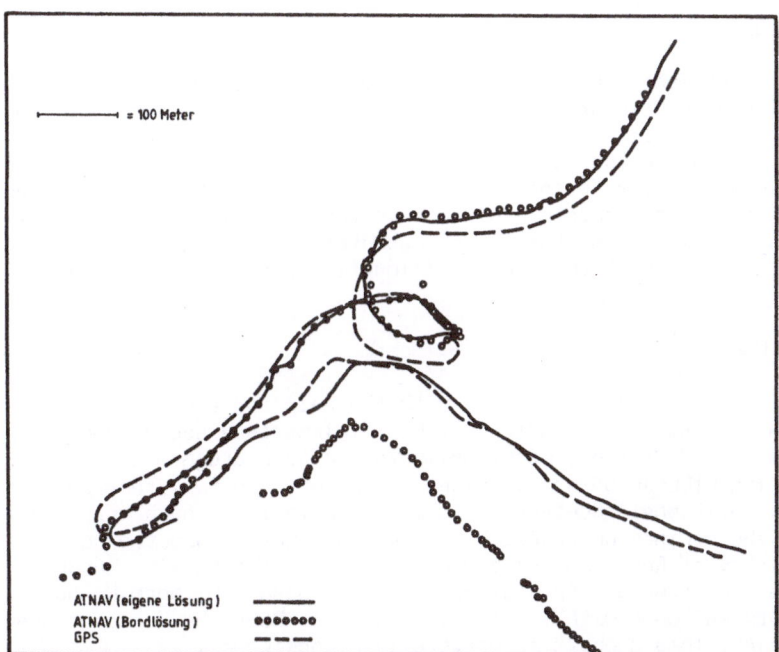

Figure 6: Differences between surface positions from GPS, ATNAV solution (circles)
and independent "own" solution of acoustic measurements (full line).

Pseudorange corrections were determined on the monitor station as demonstrated
in Chapter 2.2 and Figure 1. These corrections were applied to the navigation
solution of the vessel position. Figure 7 shows the differences between Polarfix and
the GPS navigation solution without and with differential pseudorange corrections
for north and east and height component. It is clearly demonstrated that bias
and trend are mostly removed through the differential solution. The remaining
noise however lies in the order of 3-5 m. No significant differences were found
between solutions with P-code and C/A-code. The remaining periodic effects
apparently come from the Kalmanfilter approach.

5.2 Results of the batch solution

During KIELNAC 85 carrier phase were available for both TI4100 receivers and
allowed use of the batch solution as explained in 2.3. Differences of this solution to Polarfix using code and carrier phase data are shown in Figure 8. The improvement against the use of pseudorange corrections is clearly demonstrated. The remaining noise in the solution is partly due to the Polarfix reference data. As a
conclusion we can state, that the utilization of carrier beat phase data together
with corrections from a reference station provides a relative accuracy of better
than 2 m in near shore areas when TI4100 receivers are used. These finding are
also valid for real time applications when telemetric data transfer is available.

Submeter accuracy seems to be a feasible prospect. The analyses demonstrate
the possible use of GPS techniques for continuous height determination at sea
(Figure 8 below) and azimuth determination between two moving platforms. The
latter is of importance in seismic surveys where - as an example - a seismic
streamer has to be positioned high accurately with respect to the vessel.

6. CONCLUSIONS

Various experiments at sea have demonstrated the strong capability of GPS
for providing highly accurate positions at sea. For most tasks the single
receiver accuracy will be sufficient. Together with acoustic techniques, GPS
is capable to improve the calibration of the transponder array and to tie subsurface
positions to a global reference system. Differential GPS with pseudorange
corrections provides accurate positions in the 3-5 m level. By use of the
carrier phase observations the 1 m or submeter accuracy level is achievable in
a batch solution without the need of Kalmanfilter techniques.

7. ACKNOWLEDGEMENTS

The reported work was only possible with the help and assistance of many individuals and institutions. Finding within diploma thesises have been included
coming from Sigrid Matthes, Frank Heimberg and Arno Brüggemann. Logistic
support during KIELNAC 85 was provided by Wasser- und Schiffahrtsdirektion
Nord, Kiel. Additional receivers were made available through Norges Sjökartverk
Stavanger. The cruises on research vessel SONNE were supported partly through
German Ministry of Research and performed in cooperation with Bundesanstalt
für Geowissenschaften und Rohstoffe BGR and Preussag AG, both Hannover. Part of
Research work was done with financial support by Deutsche Forschungsgemeinschaft
within SFB 149. This support is gratefully acknowledged.

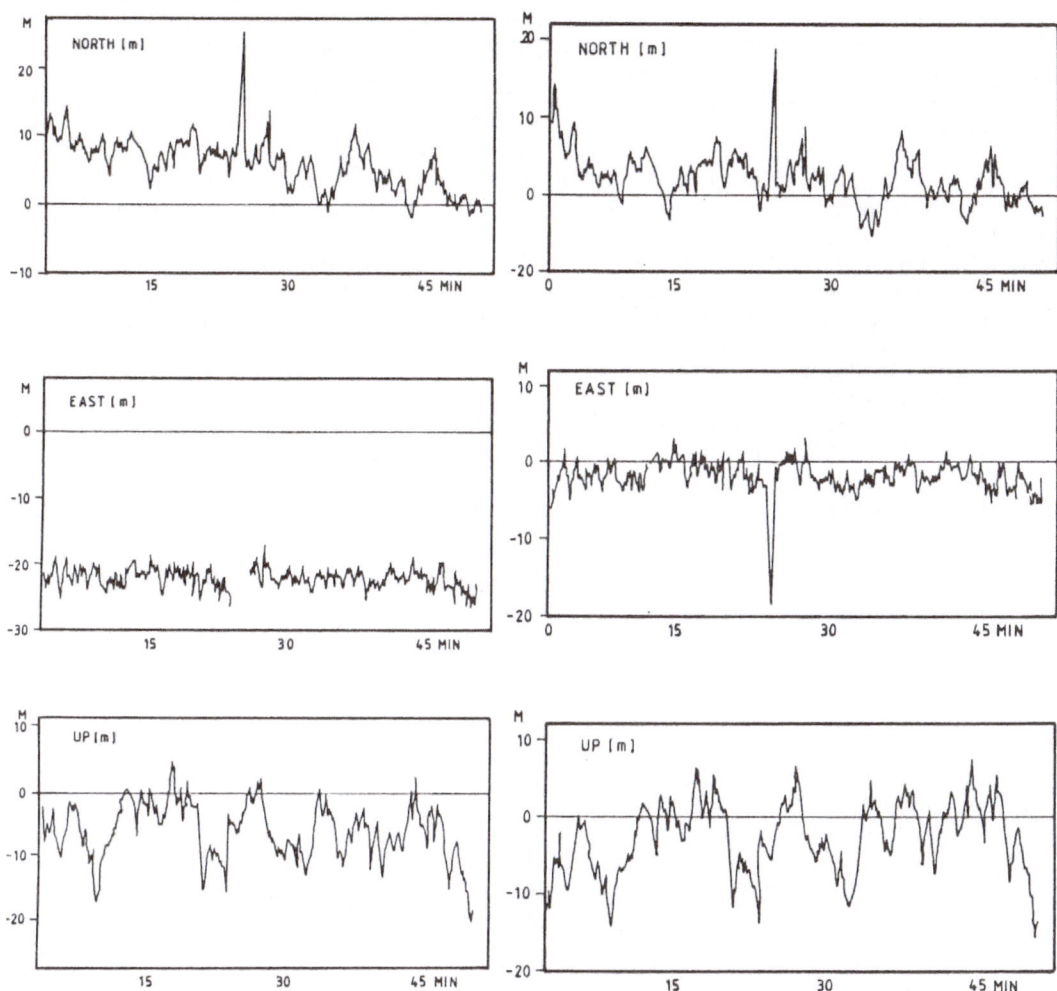

Figure 7: Differences between POLARFIX and GPS during KIELNAC85 experiment.

Left: Single receiver measurements onboard surveying vessel STURMMÖVE
Right: Differential GPS solution with pseudorange corrections.

8. REFERENCES

Heimberg, F., 1986. "Untersuchungen akustischer und satellitengestützter Naviga-
      tionsverfahren zur präzisen Positionsbestimmung im Meeresbereich unter
      besonderer Berücksichtigung des GPS".Diploma thesis,Hannover, not published.
Kalafus, R.M., 1985. "Recommendations of Special Committee 104, Differential
      NAVSTAR/GPS Service." Radio Technical Commission for Maritime Services,
      Washington.

Matthes, S., 1986. "Untersuchungen zur relativen Positionierung von bewegten
      Plattformen mit dem Global Positioning System." Diploma thesis, Hannover,
      not published.
Seeber, G., Egge, D., Schuchardt, A., Siebold, J. and Wübbena, G., 1985a.
      "Experiences with the TI4100 Navstar Navigator at the University Hannover."
      Proc. 1st. Int.Symp. on Precise Positioning with GPS, Rockville USA.
Seeber, G., Wübbena, G., 1985b. "Geodetic Measurements with TI4100 receivers."
      Proc. FIG Joint Meeting on Inertial Doppler and GPS Measurements for
      National and Engineering Surveys
Seeber, G., Schuchardt A. and Wübbena, G., 1986. "Precise Positioning results
      with TI4100 GPS receivers on moving platforms." Proc. 4.Int.Geod. Symp.
      Sat. Positioning, Austin.
Stedtnitz, W., Wentzell, H.F., 1983. "Positionsortung mit Lasergeräten (Polarfix)."
      Ortung und Navigation, No.24.

Brüggemann, A., 1986. "Untersuchungen zur Integration von GPS- und Inertialmessungen".
      Diploma Thesis Hannover, not published.

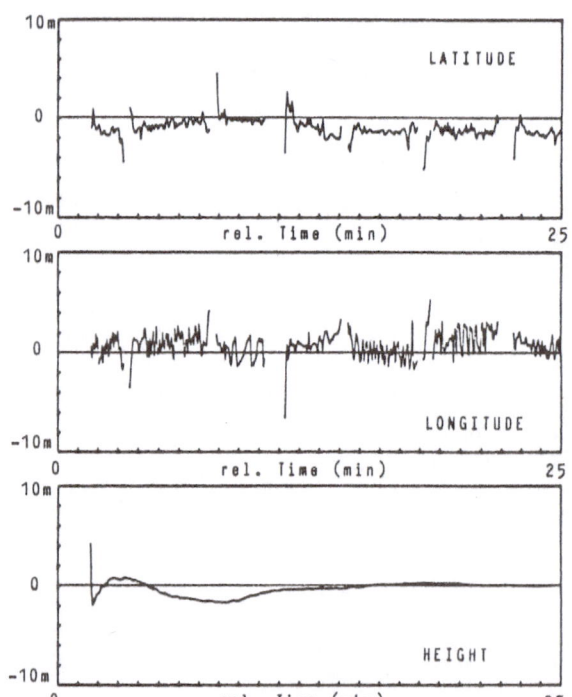

Figure 8: Differences between POLAR
FIX and GPS "Batch" solution
using carrier phases; same
data as in Figure 7.

A DUAL-BAND INTERFEROMETRIC GPS MARINE NAVIGATION SYSTEM

Charles C. Counselman III* and Jonathan W. Ladd
Aero Service Division
Western Geophysical Company of America
8100 Westpark Drive, Houston, TX 77063-6378

ABSTRACT

Since its introduction in 1982, MACROMETRY[sm] technology has consistently delivered the best available accuracy in GPS-based positioning on land. In use on six continents, MACROMETER® Interferometric Surveyors have determined more geodetic positions than any other commercial GPS system. Now Aero Service is extending MACROMETRY technology to marine positioning and real-time navigation.

Aero's new MACROMETER Marine Navigation System uses the full 10-MHz bandwidths of the GPS Precise Positioning Service signals in both the L1 and the L2 bands. Therefore it can outperform any C/A-code, L1-only instrument. The already substantial dual-band advantage will increase greatly within the next few years as we enter the up side of the solar cycle.

Like the dual-band MACROMETER II™ System, Aero's Marine Navigation System operates independently of the GPS P- and Y-codes. Code secrecy and access limitations will have no effect on the performance of Aero's systems. Moreover, since Aero has an independent satellite tracking and orbit determination capability, system performance will be unaffected by degradation of the accuracy of the broadcast orbital data.

Aero's Marine Navigation System will be capable of operating in both stand-alone and real-time differential modes. Differential operation will be needed if immunity to satellite clock and oscillator dither is required. In differential mode, Aero's system will operate as far as 1000 kilometers away from a reference station without degradation of accuracy above the 1-meter level. Because of this ability, it is not necessary for the user always to establish a nearby shore station. This is clearly an economic advantage. Improved reliability and operational flexibility are also gained if marine operations are not dependent on nearby shore stations.

1. INTRODUCTION

The NAVSTAR Global Positioning System (GPS) is not officially operational but its accuracy-improvement and cost-reduction advantages are so powerful that its use is already the preferred method for precise positioning on land. When GPS is fully operational it will replace many presently used methods of offshore positioning and navigation as well. In critical applications such as seafloor mapping, three-dimensional (3-d) seismic surveying, pipeline and cable

---

*also Professor of Planetary Science, Massachusetts Institute of Technology, Cambridge, Massachusetts 01239.

M. Kumar and G. A. Maul (directors), Marine Positioning, 197–214.
© 1987 by the Marine Technology Society.

surveys, rig positioning and borehole relocation, GPS accuracy will soon become the standard.

GPS is capable of providing several-meter accuracy in stand-alone or "single-station" operation, and meter-level or sub-meter accuracy in a relative-positioning or "differential" mode (Kalafus et al, 1985). However, these levels of performance will not automatically be available to all GPS users. For national security reasons the signals transmitted by the Block II GPS satellites will be modified in two ways which will prevent many users from achieving these levels of accuracy (Rockwell International Corporation, 1983; Weinberger and Dole, 1984).

1.1. Selective Availability. One such modification, called "Selective Availability", involves the deliberate variation, or "dither", of the time delay and the rate of change of delay of transmission of the signals from each satellite. The satellite clock correction data which are included in the "navigation" message broadcast by each satellite and which normally enable a user to correct for epoch and rate variations of the clock governing the transmissions, will not be completely accurate. In addition, the transmitted data describing the satellite's orbital position may be inaccurate. Fully accurate information will be made available only to selected users -- thus the name "Selective Availability".

Which, if any, of these Selective Availability measures will be taken, when, and what magnitude the dither will have, are uncertain. According to recent government statements, when 24-hour-per-day, worldwide, two-dimensional (2-d) satellite coverage becomes available, the positioning accuracy available to non-military users operating in a stand-alone mode will be limited to about 100 meters (2-d r.m.s.).

1.2. Differential GPS. In any case, whether artificial effects (Selective Availability) or the various natural sources of error (such as unmodeled forces which perturb the satellite orbits, and random fluctuations of the atomic frequency standards in the satellites) are more important, much better accuracy should be available in a relative-positioning, or "differential GPS" mode (Kalafus et al, 1985). In differential GPS, a receiver at a known location observes the same satellites at the same time as the receiver whose position is to be determined. The observations from both receivers are combined to determine the unknown position with respect to the known position. The combination may be through a real-time communication link, or may be performed in post-real-time data processing. Use of the relative-positioning mode tends to cancel common-mode errors, including not only satellite-related errors but also the common effects of the propagation medium.

1.3. Differential Ionospheric Refraction Error. Unfortunately, the error cancellation in differential GPS is far from exact. The residual, or non-cancelling, error due to ionospheric refraction especially is non-trivial. Ionospheric refraction increases the apparent length of the radio signal propagation path from a satellite to a point on the earth's surface by as much as 20-30 meters during the day, and 3-6 meters at night. Use of the relative-positioning mode fails to cancel this effect exactly because the density of the ionosphere varies significantly, rapidly, and unpredictably over short horizontal distances.

The ionosphere is so inhomogeneous and unpredictable that the differential path length may exceed 10 parts per million (10 ppm) of the horizontal distance between the receivers. For example, if a survey ship is 1000 kilometers from a reference station, the differential ionospheric path length can be 10 meters. In this case the error in determining the ship's position could be much greater than 10 meters, due to geometric dilution of precision (GDOP) effects, or horizontal dilution of precision (HDOP), as applicable

Exacerbating the problem for many applications is the fact that differential ionospheric effects can be rapidly time-varying. Variation approaching 100 percent may occur within less than an hour, sometimes within just a few minutes. This rapid variation results from turbulence in the ionosphere and is quite independent of the time-variation of GDOP which, of course, also occurs.

1.4. <u>Use of Dual-Band Observations to Eliminate Ionospheric Error</u>. Ionospheric error can be virtually eliminated by the use of a suitable dual-band (L1 and L2 band) GPS receiver. Such a receiver can measure the delays of the signals received from the satellites in the lower-frequency L2 band with respect to the high-frequency, L1 band signals. From these measurements and the known inverse-square frequency dependence of ionospheric propagation delay, the effects on the L1 signals can be computed and corrected for.

The quantitative importance of observing the GPS signals in both the L1 and the L2 bands is not universally appreciated. Perhaps because most receivers in use today are single-band (L1 only), and because the vendors of these receivers tend not to advertise the ionospheric effects which degrade the performance of their systems, a false notion has spread that ionospheric effects are virtually completely cancelled when differential, or relative-positioning, techniques are used.

Little data have been published to indicate the typical magnitude of the residual, or non-cancelling component of the ionospheric error as a function of user-station separation. Relatively few civilian users of GPS have had experience with dual-band equipment and observations. Except for Aero's dual-band MACROMETER II™ (Ladd et al, 1985), MINI-MAC™ (Ladd et al, 1986), and the Texas Instruments' TI 4100 (Henson and Collier, 1986) GPS geodetic surveying systems, available civilian GPS equipment does not receive the L2 signals.

Aero's worldwide geodetic survey operations with dual-band MACROMETER II systems háve generated a substantial body of ionospheric refraction data. We present some of these data in Section 4 of this paper in order to demonstrate the importance of being able to observe the GPS signals in both the L1 and the L2 bands.

1.5. <u>Anti-Spoofing: Encryption of the L2 and Wideband L1 Signals</u>. The GPS satellites transmit signals in the secondary, L2, band, for the specific purpose of enabling ionospheric errors to be eliminated. However, the L2 signals will be encrypted and classified when the Block II GPS satellites are launched (Rockwell International Corporation, 1983; Weinberger and Dole, 1984). The 10-MHz wide "Precise Positioning Service" (P-code) signals transmitted in the L1 band will also be encrypted. (The encrypted version of the P code is called the Y code.)

The reason for encrypting these signals is to provide immunity to jamming and spoofing. "Spoofing" is the transmission of counterfeit signals by an enemy. Anti-Spoofing, referred to in the GPS literature as "A/S", is often confused with Selective Availability, or "S/A". It is important to understand the differences. Whereas Selective Availability (e.g., satellite clock dither) may be relaxed or not exercised at all depending on political considerations and the availability of other precise positioning and navigation systems (e.g., the Soviet GLONASS), Anti-Spoofing is virtually certain to be employed. Without Anti-Spoofing our own U.S. defense forces could not rely on the GPS, because the extremely low received signal strength of GPS would make it too vulnerable to jamming and spoofing.

1.6. Codeless Dual-Band Technology. The certainty that the GPS Precise Positioning Service L1 and L2 signals will be encrypted and classified, coupled with the lack of any unclassified component (e.g., a C/A code modulated component) in the L2 transmissions, means that a user without access to the classified Precise Positioning Service codes and keys must use codeless technology to realize the full accuracy potential of GPS.

A codeless MACROMETER positioning and navigation receiver is intrinsically no less accurate than a GPS Precise Positioning Service receiver which employs the classified "Y" code (the encrypted version of the P code). In fact, since 1982 the best available accuracy in GPS positioning on land has consistently been provided by codeless MACROMETRY technology.

1.7. A Wide-Band, Dual-Band, Civilian GPS Marine Navigation System. Now Aero Service is applying dual-band codeless MACROMETRY technology to marine positioning and navigation. We have developed a unique GPS-based Marine Navigation System capable of operating in a stand-alone mode (i.e., without direct reference to observations made simultaneously at another site) and in either real-time or non-real-time differential modes.

Aero's system is unique in three important respects: (1) It uses the GPS signals in both the L1 and the L2 bands without requiring any classified keys or codes. (2) It utilizes the full 10-MHz bandwidth of the GPS Precise Positioning Service signals, not just the 1-MHz C/A signals. (3) Aero's system does not need to rely on the orbital data which are broadcast by the satellites.

Aero's marine navigator operates independently of the GPS P- and (classified) Y-codes which military dual-band systems require. Thus, GPS code secrecy and access limitations will have no effect on the performance of Aero's system. System performance will be unaffected by degradation of the accuracy of the broadcast orbital data since Aero has an independent satellite tracking and orbit determination capability (Ladd, 1986b).

Aero's GPS Marine Navigation System utilizes the proprietary codeless method of interferometry which was originally used in the MACROMETER Interferometric Surveyor, introduced nearly five years ago at the Third International Geodetic Symposium on Satellite Doppler Positioning (Counselman and Steinbrecher, 1982) and now in use throughout the world (Cain, 1986). Aero's Marine Navigation System also uses the unclassified C/A codes which modulate the GPS signals in the L1 band. As we shall explain, this combination of dual-band, full-bandwidth codeless and C/A code capabilities enables Aero's system to outperform all other commercial, unclassified GPS systems.

In stand-alone operation, the positioning and navigation accuracy of Aero's system is determined by the accuracy of the real-time or post-real-time source of satellite orbital and clock data. In differential operation, Aero's system has the ability to operate as far as 1000 kilometers away from a reference station without degradation of accuracy above the 1-meter level. Because of this ability, it is not necessary for the user always to establish a nearby shore station. This is a tremendous operational advantage and will reduce operational costs. Improved reliability and operational flexibility are also gained if the marine operation is not dependent on a nearby shore station.

2. MACROMETRY CODELESS TECHNOLOGY

Aero's proprietary MACROMETRY technology derives precise position information from the NAVSTAR Global Positioning System (GPS) signals without knowledge of the pseudo-random codes which modulate these signals. MACROMETRY technology has been used on a production basis in geodetic surveying since January 1983. Its accuracy and reliability have been demonstrated in official government tests in several countries (e.g., Hothem and Fronczek, 1983; Valliant et al, 1984). MACROMETRY technology is used by U.S. federal and state government agencies and large and small companies throughout the United States and Canada, in South America, Europe, Asia, Africa, and Australia. Current applications include establishment of three-dimensional (3-d) high-order control networks on local, regional, and national scales; establishment of shore control for navigational beacons; monitoring of land and offshore-platform subsidence; and extremely precise engineering and cadastral surveys.

2.1. <u>Proven Millimeter-Level Precision</u>. Of all the GPS systems commercially available, only Aero's MACROMETER systems have demonstrated millimeter-level precision in the determination of three-dimensional relative position vectors between fixed points on land (Ruland et al, 1985; Ladd, 1986b).

2.2. <u>Unmatched Productivity in High-Precision Applications</u>. Aero's MACROMETER II land-surveying instruments, the only GPS systems commercially available which use both the L1 band and the L2 band signals without the GPS codes, have demonstrated 1-part-per-million (1 ppm) precision measurements with as little as 15 minutes of satellite observing time (Ladd et al, 1985). The ability of the MACROMETER II system to achieve high precision in such a short time enables Aero Service's GPS operations group to establish even high-order geodetic networks in a very cost-efficient manner. Recent MACROMETER II system project results affirm that horizontal and vertical measurement precision are currently limited by available satellite orbital information (Ladd, 1986a).

3. A COMPLETE GPS MARINE NAVIGATION SYSTEM

As a service company, Aero Service Division of Western Geophysical Company is committed to providing its customers with complete GPS positioning services--not just receivers. Aero's established services include the provision of accurate, reliable, and timely satellite orbital data. To this end, Aero maintains a network of satellite tracking stations, a central facility for processing the tracking data to determine the satellite orbits, a quality control system for verifying the accuracy of the resulting orbital information, and a data communication system for rapid dissemination of the information to its customers (Figure 1).

Experience has proven Aero's orbital information to be both more reliable and more accurate than the information contained in the "navigation message" broadcast by the satellites themselves. MACROMETRY survey results worldwide are typically accurate within 1 ppm in horizontal coordinates and better than 2 ppm in vertical coordinates. Aero's tracking network and orbit-determination software are being upgraded with the goal of providing orbital information accurate at the 0.1- to 0.2- ppm level.

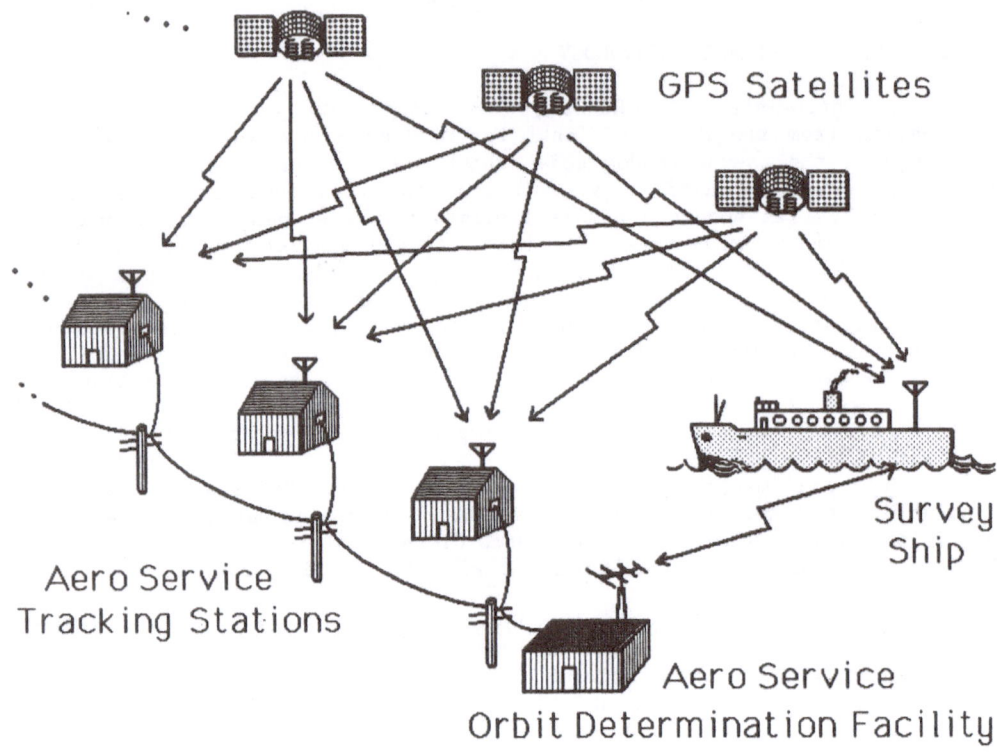

Figure 1. Aero's complete GPS marine navigation service and system includes shore-based tracking stations and a data processing facility to determine the satellite orbits accurately, in addition to equipment carried by a survey ship.

Aero also provides comprehensive data post-processing services to determine geodetic positions from data collected by MACROMETER Interferometric Surveyors in the field. For offshore navigation applications, Aero will be able to provide both processed satellite orbital data and raw satellite tracking data from its onshore stations. MACROMETER marine navigation systems will be able to use these data in real time to perform navigation in a differential mode, with respect to reference points whose positions are accurately known. However, the MACROMETER marine navigation system can also operate in a standalone mode using the information broadcast by the satellites and/or extrapolated orbital information from the Aero Service orbit determination facility in Houston, Texas. Extrapolated orbital information accumulates error typically at a rate of about 1 ppm per day.

3.1. Block Diagram of the Shipboard System. A block diagram of the shipboard portion of Aero's marine navigation system appears in Figure 2.

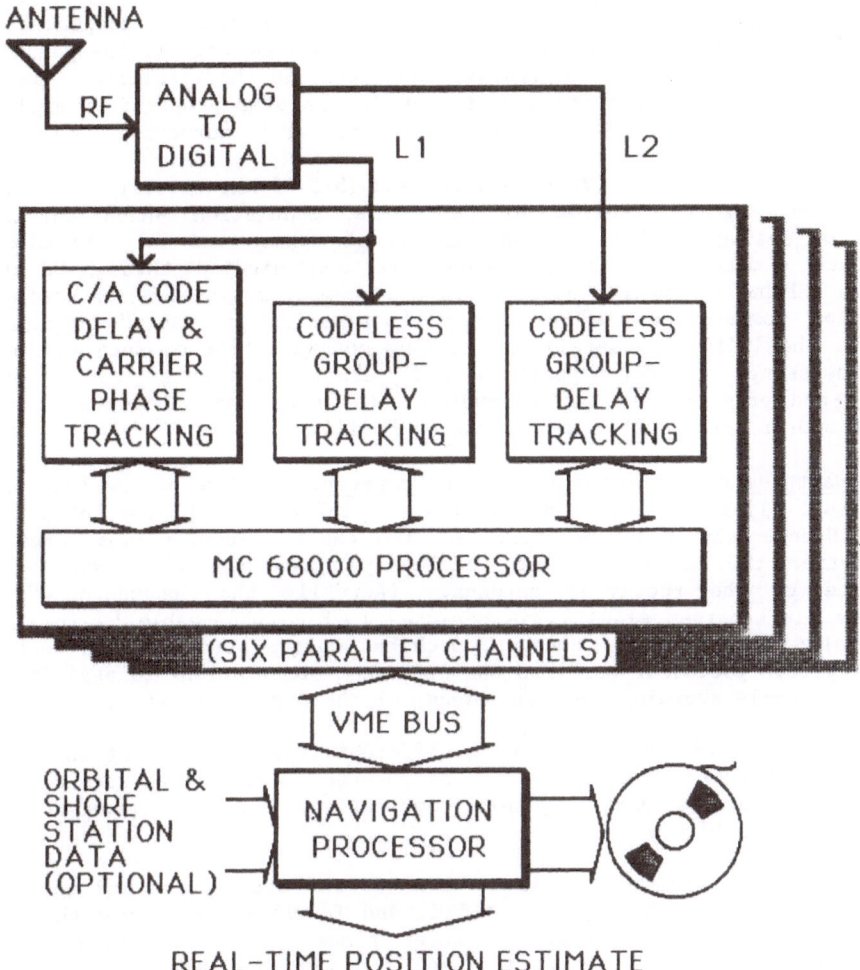

Figure 2. Block Diagram of the Shipboard System

The antenna, atop a mast, receives the GPS signals in both the L1 and the L2 bands. The signals are amplified and filtered, then converted from analog to digital form. The digital signals are applied to a set of parallel satellite-tracking channels. (Our prototype system has six channels, but the production system will have eight.) Each tracking channel includes a Motorola 68000 microprocessor, within which many of the necessary signal-processing functions are performed. Each channel includes hardware and software (or "firmware") for tracking and accurately measuring the delay of the C/A code component of the modulation of the signal received from a selected satellite; for tracking and accurately measuring the phase of the 1575.42-MHz L1 carrier wave of this signal; and for tracking and accurately measuring the group delays of the approximately 10-MHz bandwidth, P- or Y-code-related components of the signals received in both the L1 and the L2 bands.

The group delay, in other words the modulation delay, or loosely speaking the "pseudorange" of the wide-bandwidth P- or Y-code-related component of the signal received in each band, is measured without knowledge of the pseudorandom P or Y code which was used to generate the signal in the satellite. The method used to perform this codeless measurement is the same proprietary method used in Aero's MACROMETER and MACROMETER II interferometric surveyors.

A byproduct of the L1 C/A code delay and 1575.42-MHz carrier phase tracking is the decoding of the 50 bit-per-second modulation which carries the navigation message. This message and all the phase and delay tracking data from each satellite tracking channel are communicated through a standard Motorola VME bus to another 68000-based microcomputer system which performs the navigation computations. The latter, "navigation processor" computer also controls the individual satellite tracking channels, stores data for possible post-processing, delivers position and velocity estimates to a real-time display and printer, and may communicate (by means not shown) with a shore station and/or Aero's orbit determination facility.

3.2. Advantages of Combining Different Observables. Each of the four observed quantities, (1) the C/A code delay, (2) the 1575.42-MHz L1 carrier phase, (3) the L1 P-code-related group delay, and (4) the L2 P-code-related group delay (the latter two being measured codelessly), measures the range from the satellite to the receiving antenna. (Actually the "pseudorange", which includes clock-offset effects, is measured.) Each observable by itself would be useful for navigation. However, each has unique strengths and weaknesses. The navigation processor combines the different observations to exploit all the strengths, while avoiding the weaknesses, of the different types.

The notion of optimally combining different types of observation is old and well known. However, the combination of observable quantities formed in Aero's marine navigator is unique. This combination, and its consequences, are illustrated in Table 1.

In the first column of Table 1, the various observables and combinations thereof are listed. The second, third, and fourth columns show the relative sensitivity of each observable to three kinds of error: due to multipath interference, ionospheric refraction, and bias, respectively.

Multipath interference refers to interference between the desired signal which arrives at the antenna by a straight-line ray path, and undesired signals which arrive after having been scattered or reflected from objects or surfaces near the receiving antenna -- for example, from the deck of the ship or the sea surface.

Ionospheric refraction effects have already been mentioned, and will be discussed further below. By "bias" error in Table 1 is meant any error which is substantially time-invariant. Such errors may stem from the integer-cycle ambiguity of a phase measurement, and from a variety of more-or-less subtle effects which may distort the signal waveform and thus affect a group-delay measurement. Multipath interference due to reflection of a signal from a surface both nearby, and stationary with respect to, the receiving antenna can bias a group-delay measurement, for example.

Table 1. Sensitivities of different observable quantities to various error sources.

OBSERVABLE QUANTITY	RELATIVE SENSITIVITY TO		
	MULTIPATH INTERFERENCE	IONOSPHERIC REFRACTION	BIAS
1. L1 C/A CODE DELAY	HIGH	1	MED.
2. L1 CARRIER PHASE	LOW	-1	HIGH
1. & 2. COMBINED	LOW	1	MED.
3. CODELESS L1 GROUP DELAY	MEDIUM	1	LOW
4. CODELESS L2 GROUP DELAY	MEDIUM	$(77/60)^2$	LOW
3. & 4. COMBINED	MEDIUM	0	LOW
1., 2., 3. and 4. COMBINED	LOW	0	LOW

Of the four observables listed in Table 1, the L1 C/A code delay has the highest sensitivity to multipath interference because the bandwidth of the C/A code-related modulation is relatively small, about 1 MHz. This observable is also affected by ionospheric refraction. As discussed, the ionosphere delays the modulation of the L1 signal by an amount equivalent to lengthening the signal propagation path by, typically, 20-30 meters during daytime. In Table 1 this ionospheric sensitivity is denoted by "1", standing for one unit of path-length change.

Observable number 1, the L1 C/A code delay, has "medium" sensitivity to bias error in the sense that the ten-times-wider bandwidth codeless group delay observables (numbers 3 and 4 in Table 1) have about an order of magnitude less bias error, whereas the L1 carrier phase observable has virtually infinite potential for bias due to its integer ambiquity.

The L1 carrier phase observable, number 2 in Table 1, has extremely low sensitivity to multipath interference. This low sensitivity is exploited by MACROMETER interferometric surveyor systems in order to perform millimeter- and centimeter-level geodetic positioning, as discussed in Section 2 of this paper.

The sensitivity of the L1 carrier phase observable (#2) to ionospheric refraction happens to have exactly the same magnitude (unity) but the opposite sign as observable number 1, the L1 C/A code delay. Thus, the ionosphere appears to shorten, rather than lengthen, the signal propagation path indicated by observable #2.

With the use of a properly designed algorithm observables 1 and 2 may be combined, as indicated by the next line in Table 1, to obtain low sensitivity to multipath, unit sensitivity to ionosphere, and medium sensitivity to bias error. Thus, the best properties of each observable may be retained in the combination.

The most serious deficiency of this L1-only, "1 & 2" combination is its ionospheric error sensitivity, whose magnitude is not reduced in the combination. The opposite signs of the sensitivities of the individual #1 and #2 observables do not cause cancellation in the combination. This unpleasant fact is a consequence of the high bias-error sensitivity of observable #2.

Aero's marine navigation system solves the ionospheric error problem, and gains a major reduction in multipath error to boot, by utilizing the wide-bandwidth, codeless, L1 and L2 group delay observables, numbers 3 and 4 respectively.

As indicated in Table 1, each of these observables has "medium" sensitivity to multipath -- more than the L1 carrier phase but less than the L1 C/A code delay -- and low bias error, the least of all the observables. Both of these advantages stem from the wide bandwidth, about 10 MHz, of the P-code-related signal component which these observables measure.

Observable #3, the codeless L1 group delay observable, has exactly the same sensitivity, +1, to ionospheric refraction as observable #1, the L1 C/A code delay. This follows because both of these observables are of the group-delay type and both the C/A- and the P-code-related signal components have the same center frequency, 1575.42 MHz.

The sensitivity of observable #4, the codeless L2 group delay observable, to ionospheric refraction is greater than that of observable #2 by a factor equal to the square of the ratio of the respective band-center frequencies. The value of this factor is precisely equal to $(77/60)**2$, or approximately 1.647. Since the ratio of the ionospheric sensitivities is known and neither of the codeless group delay observables is substantially biased, it follows that these two observables may be linearly combined to determine separately the ionospheric refraction and the ionosphere-free path length. As the "3 & 4 COMBINED" line in Table 1 indicates, such a combination has a medium level of multipath sensitivity and low bias, in addition to being insensitive to the ionosphere.

However, the optimal combination of observables uses all four: numbers 1, 2, 3, and 4. The optimal combination has low sensitivity to multipath, thanks to the immunity of the L1 carrier phase observable; it has zero sensitivity to the ionosphere, thanks to the dual-band codeless group delay combination; and it has low bias, thanks to the wide bandwidth of the codeless group delay observations.

4. SIZE OF THE DUAL-BAND ADVANTAGE

Because the quantitative importance of observing the GPS signals in both the L1 and the L2 bands is not widely appreciated, we present in this section a typical sample of ionospheric measurements collected in the course of geodetic survey operations with dual-band MACROMETER II systems. However, a few introductory comments should be made first.

As mentioned in Section 1 of this paper, ionospheric refraction increases the apparent length of the radio signal propagation path from a satellite to a point on the earth's surface by as much as 20-30 meters during the day, and 3-6 meters at night. Use of the relative positioning mode fails to.cancel this error exactly because the density of the ionosphere varies significantly, rapidly, and unpredictably over short horizontal distances. The ionosphere is so inhomogeneous and unpredictable that the differential path-length effect often exceeds 10 parts per million (10 ppm) of the horizontal distance between the receivers. The differential effect can also be rapidly time-varying.

The mean density of the ionosphere varies somewhat systematically with time of day, latitude, season of year, and phase of the 11-year solar cycle. Mathematical models of the systematic variation exist, and some such models are claimed to be able to account for as much as 50 percent of the variance of the real ionosphere (and thus to reduce the standard deviation by 29 percent). However, it is practically impossible to model differential effects.

The standard deviation of the unpredictable differential effect on GPS L1 signals, expressed as a fraction of the distance between two observers, varies from a minimum of the order of a few parts in 10 million during pre-dawn hours at solar minimum, to several parts per million during the day at times of average solar activity, to 10 ppm and higher when the sun is more active.

The ionosphere varies on practically all time scales, with important fluctuations known as traveling ionospheric disturbances (TIDs) having characteristic timescales of the order of three-quarters of an hour or less, and having effects on GPS signals of the order of several centimeters to several decimeters, for inter-observer distances as small as a few kilometers to a few tens of kilometers. TIDs occur at all times of day, all seasons of the year, and apparently everywhere on earth (although there is evidence that they are more common in auroral regions).

Near the time of the last solar activity maximum, around 1981, the typical effect of the ionosphere on a relative-position determination from single-band GPS observations, even after time-averaging for two to three hours, was up to several parts per million of the length of the relative-position vector. On several mornings in December 1982, we observed differential ionospheric effects of 10-20 ppm across a 6-km distance in Massachusetts, in two-hour averages.

Even now, near solar minimum, 10 ppm effects are not uncommon. As an example we present, in Figure 3, measurements of ionospheric path length variation made with dual-band MACROMETER II instruments on April 25, 1986. These data were collected in the course of a geodetic control project in Cameroon, done for Photosur-Lavalin, and are reproduced with the kind permission of Photosur-Lavalin. The measurements span nearly four hours in the late afternoon and early evening, a time of day when the ionosphere is usually quiet (as opposed to the hours from dawn until mid-day, when the most rapid variations usually occur).

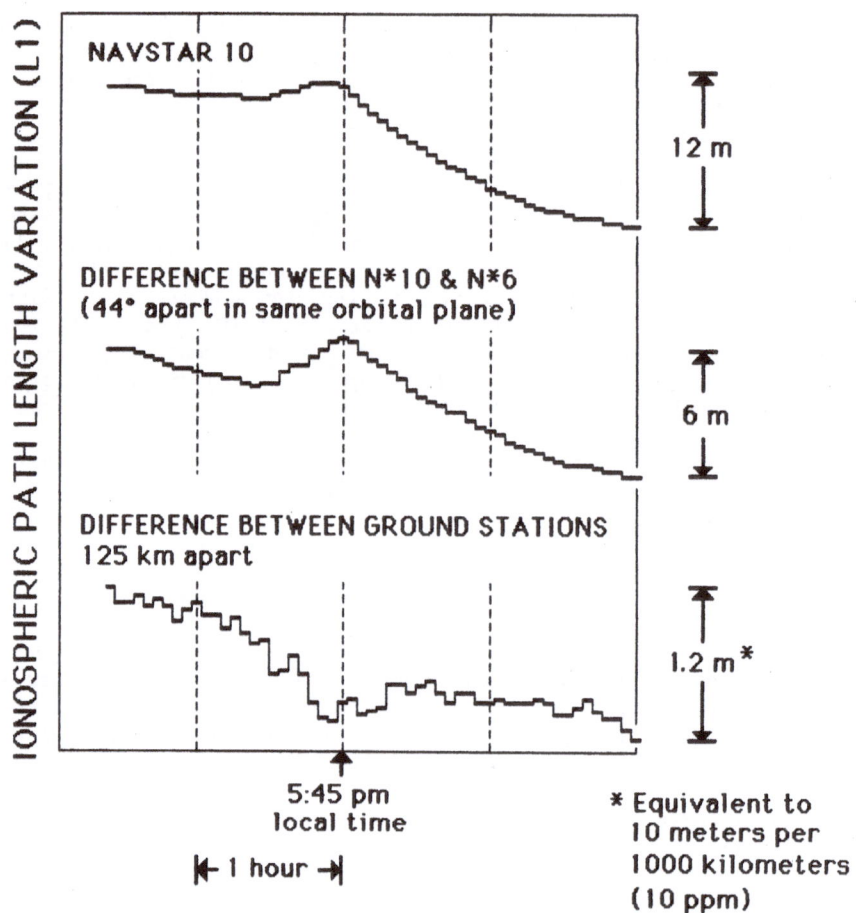

Figure 3. Ionspheric path length variations for L1 band signals
as functions of time, measured by MACROMETER II dual-band interferometric surveyors on April 25, 1986 (solar minimum conditions).
The ionosphere was relatively quiet.

At the top of Figure 3 is a graph showing the variation, as a function of
time during this early evening span, of the ionospheric effect on the L1 signal
propagation path length from one GPS satellite, NAVSTAR 10, to one MACROMETER
II instrument on the ground. The positive-signed effect, for the L1 band group
delay or "pseudorange", has been computed from observations of the phase
difference between the L1 and the L2 band center-frequency carrier waves. The
MACROMETER II observations were continuous, but were sampled and recorded at
four-minute intervals. This coarse sampling is responsible for the
quantization which is evident in Figure 3.

The NAVSTAR 10 path length changed rather little for nearly two hours, then
suddenly began dropping, by about 8 meters between 6 and 7 p.m. local time.
The path length had decreased by about 12 meters by 8 p.m., when the
observations were discontinued. Since NAVSTAR 10 was rising and moving away
from the sun, a decrease would be expected. However, the cusp in the graph

just before 5:45 p.m. represents a random occurrence.

In the center of Figure 3 is a similar graph, of the difference between the ionospheric effects on the paths from NAVSTAR 10 and NAVSTAR 6 to the MACROMETER II instrument. (Note the change in vertical scale.) NAVSTAR 10 and NAVSTAR 6 orbit the earth in the same plane, remaining about 44 degrees apart. Thus, their angular separation in the sky is nearly constant. The ionospheric density disturbance which caused the cusp to appear in the NAVSTAR 10 graph appears to have produced a "doublet" --a negative, then a positive bump--in the graph for the difference between the two satellites. A westward-traveling density wave seems to have passed first through the line of sight from the ground to NAVSTAR 6, then through the line of sight to NAVSTAR 10. The differential path-length variation associated with this event, which took place within less than an hour, was about 3 meters.

At the bottom of Figure 3 is a similar graph, of the difference between the effects observed at two ground stations separated by 125 kilometers roughly north-south. (Again, note the change in vertical scale.) Between about 4:45 p.m. and 5:45 p.m., a drop of about 1 meter occurred. The variation over the four-hour period, 1.2 meters, was about 10 ppm of the distance between the stations.

Since these MACROMETER II observations were of the L-band carrier phases rather than the (relatively unbiased) group delays, the total values of the ionospheric path length effects could not be determined from them; only the variations, or changes occurring within the time span could be observed. The total ionospheric path length, including the "d.c." value, almost certainly exceeds the magnitude of any change. In other words, there is more ionosphere than meets the eye in Figure 3.

Figure 3 showed an example of a relatively slow variation of ionospheric path length, probably caused by the passage of a relatively long-wavelength density disturbance. An example of more rapid variation is shown in Figure 4. In Figure 4 we have plotted the path-length difference between two stations 145 kilometers apart, in the same geographic area and at about the same time of day, but four days later than the observations of Figure 3.

Here one sees trains of quasi-periodic oscillations with several-decimeter amplitude, with "periods" shorter than an hour. Such behavior is typical of the phenomenon commonly known as a "traveling ionospheric disturbance", or TID. Because of their relatively short wavelength and high speed, TID's are most likely to cause trouble in relative positioning over short distances, of the order of 100 kilometers, in applications where sub-meter accuracy is desired.

It is important to keep in mind that both the mean density and the level of turbulence in the ionosphere are known to increase with increasing solar activity. The data which we have presented in Figures 3 and 4 were collected in 1986, at a time of near minimum solar activity.

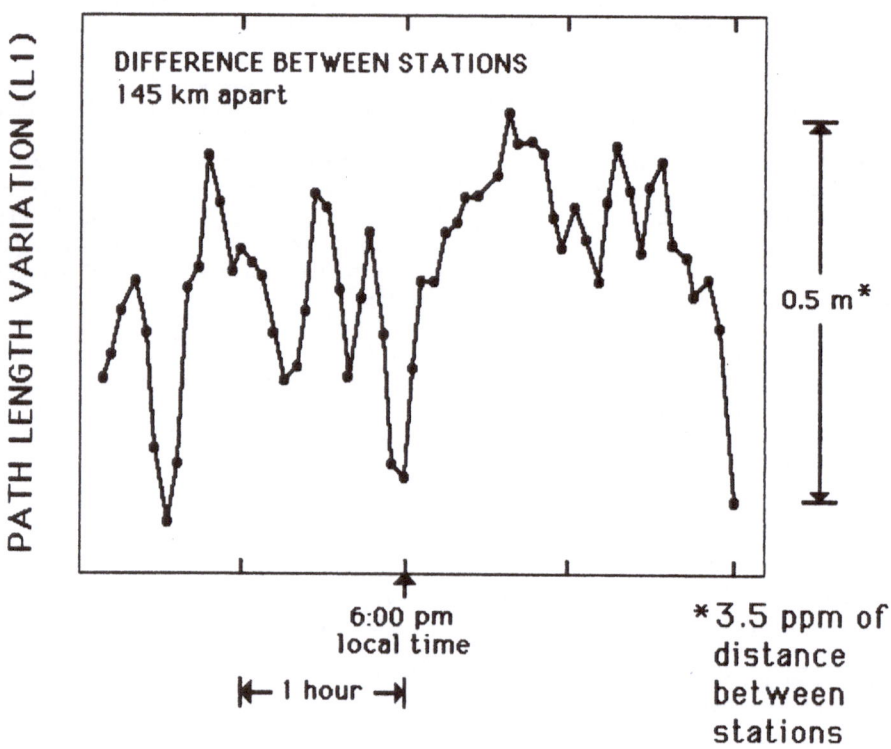

Figure 4. Ionospheric path length variations for L1 band signals
as functions of time, measured by MACROMETER II dual-band interferometric surveyors on April 29, 1986 (solar minimum conditions).
The rapid, quasi-periodic variations seen here are typical of
mild "traveling ionospheric disturbances", or "TIDs".

5. IONOSPHERIC DETERMINATION FROM GROUP DELAY DIFFERENCE

As mentioned in Section 3, group delay observables have more sensitivity to
multipath error than phase delay observables. In Aero's MACROMETER Marine
Navigation System, as opposed to the MACROMETER II system, L1 and L2 group
delay observations are used to determine the ionospheric path length in order
to remove this effect from the L1 delay and phase observations. The accuracy
of this removal is limited by the multipath errors in the L1 and L2 group delay
observations. The size of these errors can be seen in Figure 5.

In Figure 5 we have plotted, as a function of time for 60 minutes, the
difference between a MACROMETER II System's and a MACROMETER Marine Navigation
System's determinations of the L1 band ionospheric path length. The two
receivers were connected to the same antenna at the same time, and should
therefore have sensed the same ionosphere. Since the precision of the L1 and
L2 carrier phase observations made by the MACROMETER II system is much finer

than that of the L1 and L2 group delay observations by the MACROMETER Marine
Navigation System, virtually all of the variation seen in Figure 5 represents
measurement error in the group delay observations of the Marine System.

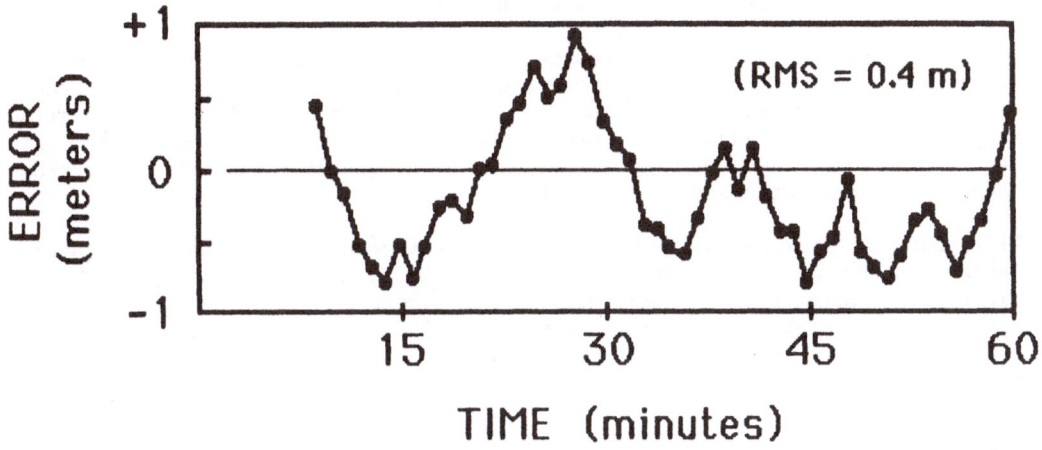

Figure 5. Error in the determination of ionospheric path
length for L1 band signals, from difference between MACROMETER II
System dual-band carrier-phase and MACROMETER Marine Navigation
System dual-band group delay determinations.

Each system made an independent observation once per minute. In Figure 5,
one point is plotted each minute. Thus, receiver noise should appear as
"white" noise, statistically independent from point to point. The obviously
high point-to-point correlation, and the correlation time of between 5 to 15
minutes, reflect the fact that multipath error was dominant. The antenna was
stationary on a flat metallic roof, and the time-variation of the multipath
error resulted from the slow motion of the satellite across the sky. In an
actual marine application, the time-variation might be much more rapid, so that
the multipath error would tend to average out. The magnitude of the multipath
error would be about the same, of the order of 0.5 meter, before averaging.

6. PERFORMANCE SUMMARY

The projected accuracy performance of the MACROMETER Marine Navigation
System, when operating in a stand-alone mode, is summarized in Table 2.

Table 2. Approximate Error Budget for Stand-Alone Operation

ERROR SOURCE	UNCERTAINTY (in meters)
Satellite Orbits	5
Satellite Clocks	3
Daytime Ionosphere	0.5
Multipath	0.5
ROOT-SUM-OF-SQUARES	6

In Table 2 we give estimates of the horizontal position uncertainties due to each of the major error sources. Errors in the orbital information are assumed here to be about 1 ppm. This is the level of error that 24-hour extrapolated orbits would typically have. Given this level of orbital uncertainty, orbit-related error would probably dominate other errors.

If more accurate, post-processed orbital data were used, the Marine Navigation System's positional uncertainty in stand-alone mode would probably be dominated by satellite clock instability. But in post-real-time, the Marine Navigation System observations could probably be processed in combination with observations from some shore stations, in which case differential-mode (rather than stand-alone) accuracy would be obtained.

Differential operation is much more efficient with Aero's dual-band Marine Navigation System than with an L1-only system because dual-band observation virtually eliminates ionospheric error. Thus there is little need, with Aero's System, for a reference station to be located near the ship's operating area.

Note that we are assuming in Table 2 that the GPS clocks are not dithered. If they were, then we would operate in a differential mode. Note also that a non-zero uncertainty is shown for ionospheric error despite the use of dual-band observations. This uncertainty reflects the fact that the dual-band observations contain errors, which are due primarily to multipath. The "Multipath Interference" entry in Table 2 represents the multipath-related error which would be present even if dual-band observations were not employed to eliminate ionospheric effects.

6.1. <u>Differential Operation</u>. The projected accuracy performance of the MACROMETER Marine Navigation System when operating in a differential mode with a reference station 1000 kilometers away, is summarized in Table 3.

Table 3. Approximate Error Budget for Differential Operation with Reference Station 1000 Kilometers Away

ERROR SOURCE	UNCERTAINTY (in meters)
Satellite Orbits	1 (0.5*)
Satellite Clocks	0
Daytime Ionosphere	0.5
Multipath Interference	0.5
ROOT-SUM-OF-SQUARES	1.2 (<1*)

*with post-survey orbit improvement

The orbit- and clock-related uncertainties in Table 3 differ from those in Table 2 due to the common-mode error cancellation which occurs in the differential mode. Clock error is cancelled virtually perfectly, but a portion of the orbital error survives. If 1-ppm orbital uncertainty is assumed, as in 24-hour extrapolated orbits, then the positional uncertainty at 1000-km distance is 1 meter. With post-real-time orbit improvement, the orbit-related uncertainty would certainly be reduced.

7. CONCLUSIONS

In differential mode, Aero's MACROMETER GPS Marine Navigation System can operate as far as 1000 kilometers away from a reference station without degradation of accuracy above the 1-meter level. Because of this ability, it will not be necessary for the user always to establish a nearby reference station.

This is an operational advantage and reduces field costs, particularly in difficult operating areas. Improved reliability and operational flexibility are also gained if marine operations are not dependent on nearby shore stations.

Aero's GPS system is more accurate than all other civilian systems, for three main reasons: (1) Aero's satellite-orbit data are more accurate than the data broadcast by the satellites. (2) Aero's codeless receivers utilize the 10-MHz bandwidth Precise Positioning Service (PPS) signals, not just the 1-MHz Standard Positioning Service (SPS, or C/A code) signals. (3) Aero's codeless receiver is dual-band (L1 and L2, not L1-only).

Use of the wide-bandwidth, dual-band, Precise Positioning Service signals is essential for high-accuracy positioning and navigation. In the next few years as solar activity increases, the performance gap between dual-band PPS and single-band SPS (C/A code, L1 only) receivers will grow much wider.

The PPS signals will be encrypted in the Block II GPS satellites. The Department of Defense is unlikely to allow any civilians to have the classified codes and keys, in part because codeless technology allows civilian users to obtain full accuracy without them.

REFERENCES

Bock, Y., Abbot, R. I., Counselman III, C. C., and King, R. W., 1986. "A Demonstration of 1-2 Parts in 10^7 Accuracy Using GPS," Bulletin Geodesique, in press.

Cain, J., 1986. "MACROMETRY^{sm} Surveys - A Summary of Results," Proc. Fourth International Geodetic Symposium on Satellite Positioning, 28 April - 2 May 1986, University of Texas at Austin, in press.

Counselman III, C. C., and Steinbrecher, D. H., 1982. "The MACROMETER™ compact radio interferometry terminal for geodesy with GPS," Proc. Third Intl. Geodetic Symp. on Satellite Doppler Positioning, vol. 2, pp. 1165-1172.

Counselman III, C. C., Abbot, R. I., Gourevitch, S. A., King, R. W., and Paradis, A. R., 1983. "Centimeter-level relative positioning with GPS," Journal of Surveying Engineering (ASCE), vol. 109, pp. 81-89.

Henson, D., and Collier, E. A., 1986. "Effects of the Ionosphere on GPS Relative Geodesy." IEEE 1986 Position Location and Navigation Symposium Record, pp. 230-237.

Hothem, L. D., and Fronczek, C. J., 1983. "Report on Test and Demonstration of MACROMETER® Model V-1000 Interferometric Surveyor." Federal Geodetic Control Committee: FGCC-IS-83-2, Rockville, Maryland.

Kalafus, R. M., Van Dierendonck, A. J., Stansell, T. A., Pealer, N.A., 1985.

"Recommendations of Special Committee 104, Differential NAVSTAR/GPS Service," Radio Technical Commission for Maritime Services, P.O. Box 19087, Washington, D.C. 20036.

Ladd, J. W., Counselman III, C. C. and Gourevitch, S.A., 1985. "The MACROMETER II™ Dual-Band Interferometric Surveyor." Proceedings of the First International Symposium on Precise Positioning with Global Positioning System, vol. 1, pp. 175-180, National Geodetic Survey, Rockville, Maryland.

Ladd, J. W., 1986a. "Establishment of a Three-Dimensional Geodetic Network Using the MACROMETER II™ Dual-Band Surveyor." Proc. Fourth International Geodetic Symposium on Satellite Positioning, 28 April - 2 May 1986, University of Texas at Austin, in press.

Ladd, J. W., 1986b.   "Three-Coordinate Positioning Within 1 Part in 10 Million Without the GPS Codes." Proceedings of the IEEE Position Locations and Navigation Symposium (PLANS '86), 4 - 7 November 1986, Caesar's Palace, Las Vegas, Nevada, in press.

Ladd, J. W., Welshe, R. G., Brown, A., and Sturza, M., 1986.   "MINI-MAC™ -- a New Generation Dual-Band Surveyor." Proc. Fourth International Geodetic Symposium on Satellite Positioning, 28 April - 2 May 1986, University of Texas at Austin, in press.

Rockwell International Corporation, 1983. "Interface Control Document, NAVSTAR GPS Space Segment/Navigation User Interfaces," ICD-GPS-200, Downey, California.

Ruland, R., and Leick, A., 1985.   "Application of GPS in High Precision Engineering Survey Network." Proceedings of the First International Symposium on Precise Positioning with the Global Positioning System, vol. 1., pp. 483-493, National Geodetic Survey, Rockville, Maryland.

Valliant, H.D., 1984. "Canadian Evaluation of the MACROMETER® Interferometric Surveyor." Open file report 84-4, Department of Energy, Mines, and Resources, Ottawa, Ontario, Canada.

Weinberger, C., and Dole, E., 1984.   "DOD/DOT Policy for the Future Radio-navigation System Mix." Co-signed memorandum available from Offices of the Secretary of Defense and Secretary of Transportation, Washington.

CODELESS GPS POSITIONING

P. F. MacDoran
J. H. Whitcomb
R. B. Miller
L. A. Buennagel
J. H. Lower

ISTAC, Inc.
444 N. Altadena Drive, Suite 101
Pasadena, CA 91107
818/793-6130

ABSTRACT

The Global Positioning System (GPS) offers significant promise for applications to positioning needs but the often stated DoD policy (i.e., W.H. Taft, IV, 22 May 1985) makes it clear that the operational (Block II) phase of GPS will not be as openly available as the current prototype (Block I) satellites. Thus, user organizations and manufacturers of GPS code-dependent equipment have invested substantial efforts aimed at changing that established DoD policy so that their GPS equipment will function in the future as well as they presently perform.

However, the ISTAC-SERIES codeless technology is immune to changes within the satellites and ISTAC has already developed receiver hardware and orbit determination software to enable an independent source for satellite ephemerides and assure future viability. In this manner, ISTAC technology users can conduct their civilian business activity without the vulnerability to GPS changes required by DoD. National security implications of ISTAC technology have already been addressed and countermeasures exist to prevent its applications to tactical/strategic missions. ISTAC codeless GPS technology is now routinely performing commercial differential positioning missions in static environments, achieving accuracies of 2 to 5 parts per million.

GOAL OF ISTAC'S CODELESS GPS TECHNOLOGY

The Goal is to innovate technological methods that accommodate established DoD policy to degrade Block II Operational GPS satellites relative to the present Block I performance, while providing civilian users with high accuracy positioning and navigation.

The approach is to exploit physics in order to: i) derive codeless ranging to GPS satellites; ii) synchronize the receiver clocks to UTC and to each other; iii) derive the orbits of the GPS satellites from codeless observations at known locations; and iv) calibrate ionospheric delays to each satellite.

Immunity to Block II changes is demonstrated by ISTAC codeless technology's current performance with the anomalies of the Block I satellites. The following charts and graphs provide information on results during normal operations and during periods of "unhealthy" performance of the Block I satellites.

M. Kumar and G. A. Maul (directors), Marine Positioning, 215–221.
© 1987 by the Marine Technology Society.

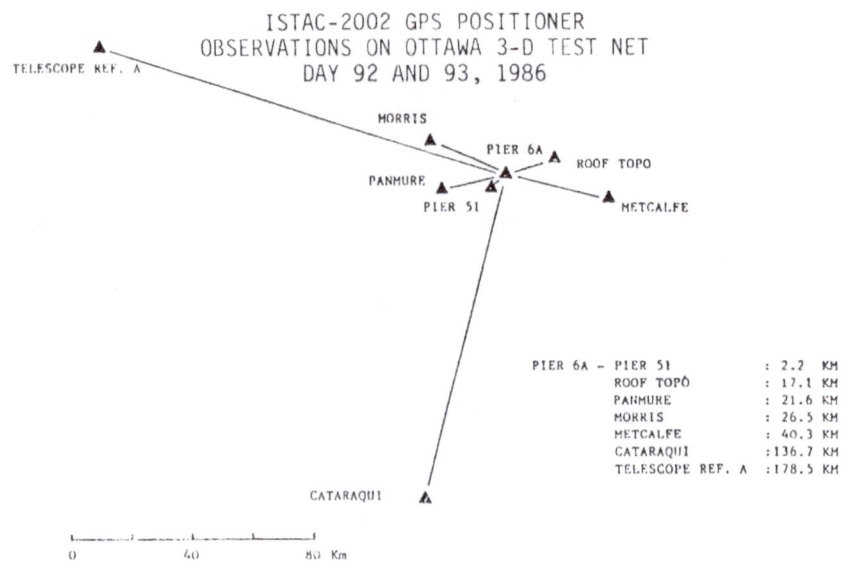

ISTAC-2002 GPS POSITIONER
OBSERVATIONS ON OTTAWA 3-D TEST NET
DAY 92 AND 93, 1986

PIER 6A – PIER 51	: 2.2 KM
ROOF TOPO	: 17.1 KM
PANMURE	: 21.6 KM
MORRIS	: 26.5 KM
METCALFE	: 40.3 KM
CATARAQUI	:136.7 KM
TELESCOPE REF. A	:178.5 KM

CLOSURE ERRORS FOR INDEPENDENT BASELINES

TRAVERSE SEQUENCE (SESSIONS)	X (M)	Y (M)	Z (M)	RSS M (PPM)	PERIMETER LENGTH (KM)
6A-PA-ME-6A (1A-1B-2B)	0.225	- 0.574	0.080	0.622 (5.2)	119.815
6A-PA-ME-6A (1A-1B-2A)	0.298	- 0.429	- 0.125	0.537 (0.5)	119.815
6A-ME-CA-6A (1B-2A-2B)	0.587	- 0.594	- 0.529	0.988 (3.1)	317.498
6A-ME-ARO-6A (1B-2B-2A)	0.010	- 0.168	0.701	0.721 (2.3)	317.498
6A-ME-ARO-6A (1B-2A-2B)	0.635	- 0.670	- 0.287	0.935 (2.2)	437.516
6A-ME-ARO-6A (1B-2B-2A)	- 0.155	0.490	0.205	0.553 (1.3)	437.516
6A-PA-ME-CA-6A (1A-1B-2A-2B)	0.576	0.812	0.449	1.091 (3.1)	356.723
6A-PA-ME-CA-6A (1A-1B-2B-2A)	- 0.001	- 0.412	0.781	0.888 (2.5)	356.723

MEAN RSS   0.792 M ±0.212 M

MEAN RSS   1.2 PPM ±1.4 PPM

Ref., Delikaraoglou, D., and Duval, R., 1986. "Report on Test Results with the
ISTAC-MODEL 2002 GPS POSITIONER." Canadian Geodetic Survey,
Surveys and Mapping Branch, Ottawa, Ontario, Canada.

THE ISTAC DIFFERENCE

ISTAC 2002 systems do not depend upon acquiring their timing from the space vehicles. Thus bad clocks in orbit are no problem. For example, on 9 September 1986, SV7 and SV11, and on 23 September 1986, SV3 indicated that it was 31 January 1993.

Time is introduced into the ISTAC 2002 system by use of civilian timing assets (i.e., WWV, GOES UHF Satellite, Naval Observatory) or even the operator's best guess from a wristwatch. The ISTAC software determines the actual value to within a few milliseconds by exploiting the physics of orbiting GPS satellites which is sufficient to support 1 part per million surveying.

The ISTAC 2002 high accuracy baseline measurement mode is accomplished by forming an explicit phase differential data-type so that space vehicle transmission instabilities (by accident or intent) are identically cancelled out.

Precision orbit element information is introduced as a separate module which in the future will come from an ISTAC independent orbit monitor/determination service. Presently, a 24 hour per day dial-in computer modem service is provided for ISTAC 2002 users. The modem service allows access to the fresh Air Force ephemeris uploads over North America but can also accommodate NGS orbit data when it becomes available. Although they will be considerably behind real-time.

GPS SATELLITE PROBLEMS OF AUGUST AND SEPTEMBER HAVE PRESENTED NO PROBLEM TO ISTAC 2002 USERS

August 1986. Global Surveys, Ltd. operating in U.K. performed sub-decimeter (6 to 8 cm) surveying using three ISTAC 2002 systems even though SV13 broadcast ephemerides indicated it was orbiting 25 km from the earth center while indicating SV13 was healthy. Code dependent receivers are in serious problems when such conditions exist.

September 1986. SV8 and partially SV9 are shown as unhealthy. Sunrise International operating in New Mexico and Montana and Global Surveys operating in Turkey acquire data in the normal manner and derive typically good baseline results.

UNHEALTHY GPS SATELLITES
SEPTEMBER 1986

SUN	MON	TUE	WED	THU	FRI	SAT
7	8	9	10	11	12	13
		SV8-3C SV9-3C	SV8-3C SV9-3C	SV8-3C	SV8-3C	SV8-3C
14	15	16	17	18	19	20
SV8-3C	SV8-3C	SV8-3C	SV8-3C	SV8-3C	SV8-3C	SV8-3C
21	22	23	24	25	26	27
SV7-2A SV8-3C SV11-2A	SV8-3C	SV3-2A SV8-3C	SV8-3C	SV8-3C	SV8-3C	SV6-20 SV8-80

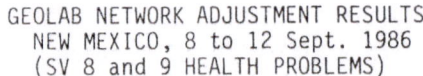

GEOLAB NETWORK ADJUSTMENT RESULTS
NEW MEXICO, 8 to 12 Sept. 1986
(SV 8 and 9 HEALTH PROBLEMS)

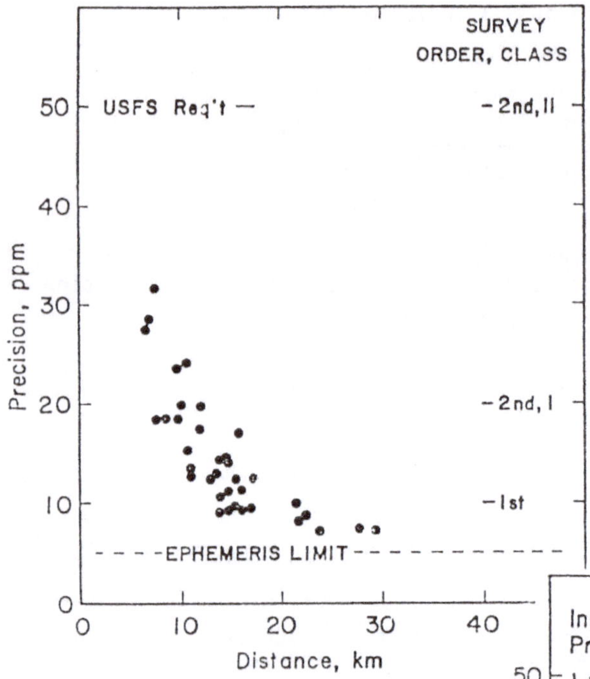

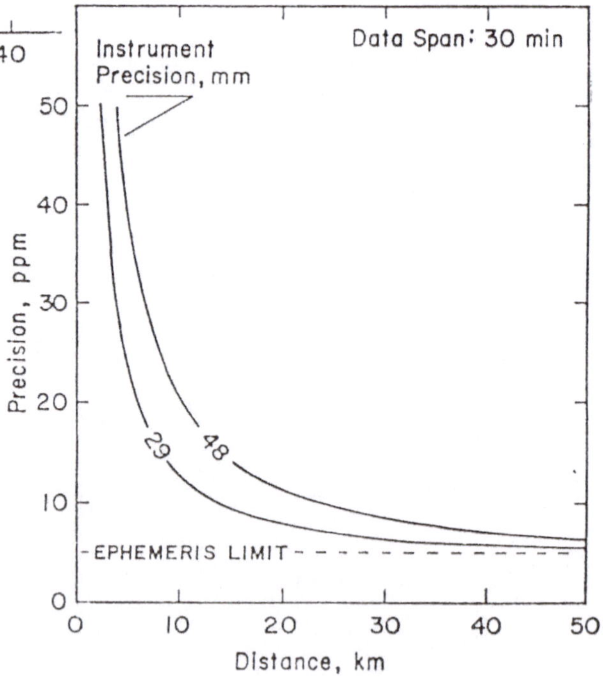

DATA SUPPLIED BY: C. MUNCY
                   SUNRISE INTERNATIONAL
                   MESA, ARIZONA

SPACE VEHICLE 8 (NAVSTAR 4) OSCILLATOR PHASE RELATIVE
TO REMAINING HEALTHY GPS CONSTELLATION
(SINGLE ISTAC 2002 STATION OBSERVATION AT PASADENA, CA 19 SEPT. 1986)
C/A PHASE, 1 DEG. = 0.81 M

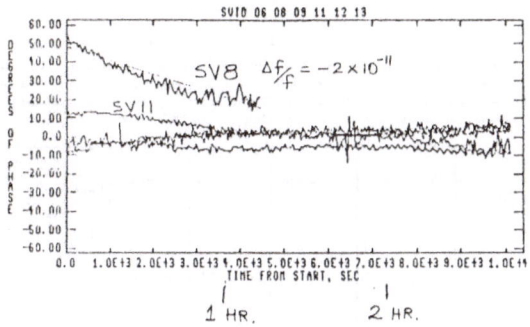

OCCASIONAL INSTABILITIES IN GPS TRANSMISSIONS
(SINGLE ISTAC 2002 STATION OBSERVATION AT PASADENA, CA 19 SEPT. 1986)
C/A PHASE, 1 DEG. = 0.81 M

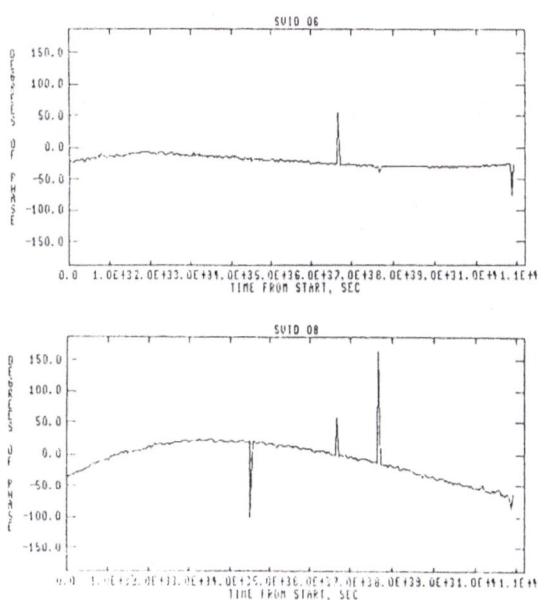

GLOBAL POSITIONING SYSTEM CHARACTERISTICS (SIGNAL-IN-SPACE)

1. Report No. DOD – 4650.4 DOT–TSC–RSPA–84–8	2. Government Accession No.
4. Title and Subtitle FEDERAL RADIONAVIGATION PLAN	

GLOBAL POSITIONING SYSTEM (GPS)

System Description: A space-based radio positioning navigation system that will provide accurate three dimensional position, velocity and time information to suitably equipped users anywhere on or near the surface of the earth. The space segment wil consist of 18 satellites plus 3 operational spares in 12 hour orbits. Each satellite will transmit navigation data and time signals on 1575.4 and 1227.6 MHz.

Accuracy			Availability	Coverage	Reliability	Fix Rate	Fix Dimension	Capacity	Ambiguity Potential
Predictable	Repeatable	Relative							
PPS* Horz - 17.8m Vert - 27.7m Vlcty - 0.2m/sec Time - 48ns max offset 120 ns	Horz - 17.8m Vert - 27.7m	Horz - 7.6m Vert - 11.7m	Approximately 100%	Worldwide Continuous	** Design Life of the Satellite is 7.5 years	Essentially Continuous	3D + Velocity + Time	Unlimited	None
SPS Horz - 100m Vert - 156m Vlcty - ** Time - **	Horz - 100m Vert - 156m	Horz - 28.4m Vert - 44.5m							

* For US and Allied military, US Government, and selected civil users specifically approved by the US Government.

** To be determined.

If millimeter levels of accuracy impress you, YOU DON'T UNDERSTAND GPS ORBIT ERRORS

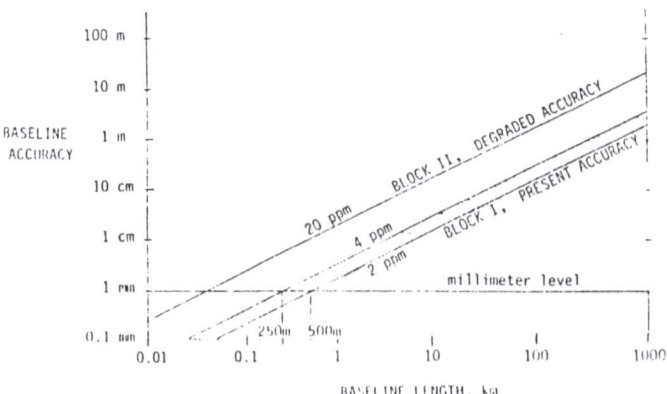

REAL-TIME   EXPLICIT   DIFFERENTIAL   NAVIGATION

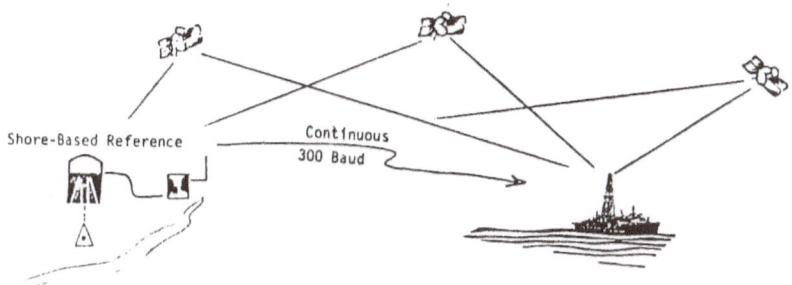

- Using continuous 300 baud communications link, the shore-based reference sends raw time-tagged phase and ephemeris parameters.

- Offshore vessel computes the baseline vector relative to the shore based reference.

- Baseline accuracy is 2 m at 1 second updates at 500 km (dual band receivers).

- Explicit phase differential technique removes errors of GPS Block II signal dithering.

HAZARDS   SURVEYS

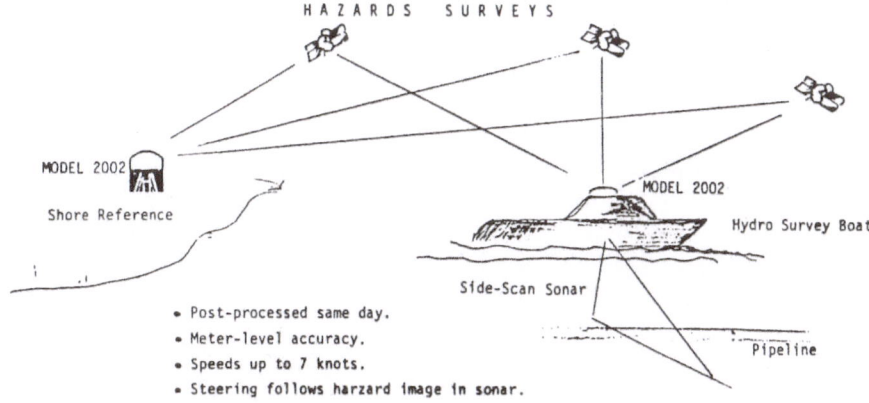

- Post-processed same day.
- Meter-level accuracy.
- Speeds up to 7 knots.
- Steering follows harzard image in sonar.
- Shore-based reference station may be 400 km away (requires dual-band receivers).

DECIMETER GPS POSITIONING FOR SURFACE ELEMENT OF SEA FLOOR GEODESY SYSTEM

Lawrence E. Young
Sien C. Wu
Timothy H. Dixon
Jet Propulsion Lab
California Institute of Technology
4800 Oak Grove Drive
Pasadena, CA 91109

ABSTRACT

This paper describes the results of an analysis of several GPS based positioning systems, which have been designed to allow accurate positioning of the surface element of a seafloor-to-surface acoustic array. The results of covariance analyses are discussed. The precision, accuracy, and data rates required from the GPS receiver are given.

1. INTRODUCTION

Knowledge of the overall rates of motion among the earth's lithospheric plates, as well as a more detailed picture of the strain field in the vicinity of plate boundaries, is important for improved understanding of the kinematics and driving forces of plate motion. To this end, geodetic measurements have been performed on the continents for some time using both conventional and space-based techniques. However, most plate boundaries occur beneath the ocean, and it is clear that the extension of some form of geodetic technique to the marine environment is highly desirable (Space Science Board, 1982; Panel on Crustal Movement Measurements, 1981; Committee on Geodesy, 1983; Walter, 1984).

A design to allow these measurements has been proposed by F. N. Spiess (Spiess 1984). The ocean going instrumentation would consist of two parts. (See Fig 1) The first part would be an acoustic system used to measure the relative location of a reference point at the sea floor with respect to a floating platform. This paper describes alternative approaches for the second part of this approach, consisting of a GPS based system. The GPS system would determine the location of the floating platform relative to a reference point on land.

2. ACOUSTIC SYSTEM AND REQUIREMENTS FOR GPS SYSTEM

The acoustic system design consists of a transducer assembly located on a surface vehicle, which measures ranges to 3 or more precision transponders located at the sea floor (Spiess, et al., 1984). Additional instrumentation will probably be required in order to measure the sound velocity profiles along the acoustic paths. In operation, the surface transmitter emits a sonic signal, consisting of a pseudo-random code, at a time referenced to an on-board clock. The sea floor transponders, which define the geodetic reference mark, receive this signal, and transmit their replies. After a delay equal to the two-way travel time through the water columns, plus the transponder delay, these replies are received back at the surface. The set of delays from all

M. Kumar and G. A. Maul (directors), Marine Positioning, 223–232.
© 1987 by the Marine Technology Society.

transponders is used to determine the horizontal offset of the surface platform from the theoretical point directly above the geodetic reference mark.

To aid this solution, the GPS system can be used to determine the kinematics of the surface platform during the transmission and reception of the acoustic signal.

Because the phase center of the GPS antenna is not collocated with the delay center of the acoustic transducer, attitude information is required for the surface platform. This can also be provided by the GPS system.

The overall error budget for the horizontal components of the land to acoustic transducer baseline is 10 cm. GPS data would be acquired continuously during the few hour time span of the acoustic measurements to determine platform position, attitude, and kinematics. While system demonstrations could be made on baselines of several hundred km, the ultimate use of sea floor reference marks in mid-oceanic regions implies the eventual use of baselines several thousand km in length.

LOCATION OF A SEA FLOOR REFERENCE POINT WITH GPS/ACOUSTIC TECHNIQUES

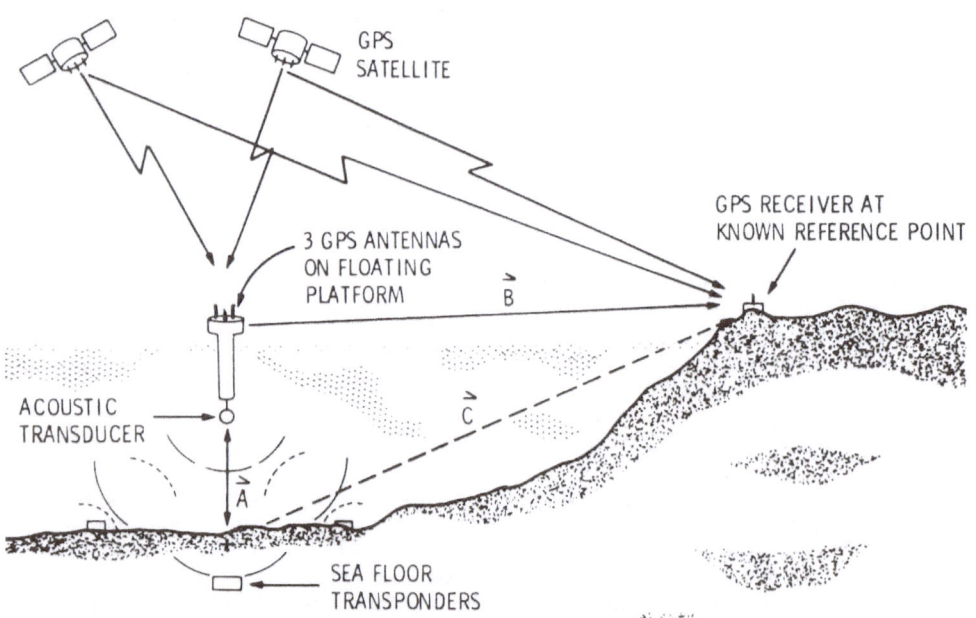

FIGURE 1: This figure schematically represents the determination of the horizontal (at the sea floor mark) components of vector $\vec{C}$, from a geodetic mark (the center of a transponder array) on the sea floor, to a known point on land. This is done in two steps. First, the horizontal components of vector $\vec{A}$ are determined by acoustic ranging techniques. Second, the horizontal components of vector $\vec{B}$ are determined using a differential GPS solution, and $(\vec{A} + \vec{B} + \text{offset of acoustic transducer from GPS antenna}) = \vec{C}$ is formed.

3. OVERVIEW OF GPS SYSTEMS

3.1. GPS Orbit Improvement. Improved satellite ephemerides are required for 10 cm accuracy over baselines of several thousand km. All varieties of GPS system design considered here rely on the simultaneous operation of a network of GPS receivers at known sites on land, called fiducial sites. For our simulations, 6 fiducial sites were assumed, spanning the sea floor measurement area. Data from the fiducial sites is used to reduce the errors in the satellite ephemerides (Davidson, et al., 1985).

3.2. Attitude Measurement. All GPS systems we have considered would use a three antenna interferometer receiving the GPS carrier phase at a single frequency to determine platform attitude.

3.3. GPS Systems The GPS system must provide three types of information; the position of the GPS antenna, the kinematics of the floating platform, and the attitude of the platform. The five GPS system configurations considered in this study are shown in Table 1.

TABLE 1: GPS SYSTEM DESIGN ALTERNATIVES

SYSTEM	POSITIONING MEASUREMENT	KINEMATIC MEASUREMENT
I	Dual frequency carrier range. (Cycle ambiguity removed by momentarily placing the sea going antenna on a known land reference mark near one of the land receivers.)	Dual frequency carrier range change
II	Dual frequency carrier range change	Three axis accelerometer
III	Dual frequency P code range	Three axis accelerometer
IV	Dual frequency carrier range change	Dual frequency carrier range change
V	Dual frequency P code range	Dual frequency carrier range change

3.4. System I. System I uses continuously tracked carrier phase to translate positioning knowledge from a known mark (Remondi, 1985). For example, before going to sea, the GPS antenna would be placed over a reference mark at a known, short distance from one of the land-based receivers. This allows the carrier cycle bias to be fixed for the satellites in view. After the research vessel has gone to sea, and new satellites come into view, the two receivers will no longer be close, and the cycle bias on these new satellites cannot be fixed, unless their orbits are known to much less than a carrier wavelength. This will cause some loss of accuracy with time, as more new satellites are acquired. Another difficulty for this technique, when used for mid-oceanic surveys, is that it requires the carrier phase to be tracked with no unrecovered cycle slips for mission durations of several days. Given the

costs of operating a research vessel at sea, it is desirable to avoid a system
design that could allow a momentary signal loss to destroy the utility of data
collected over several days. No covariance analysis was performed for this
system. This technique may, however, be very appropriate for surveys of
shorter duration.

It may also be possible to resolve carrier cycle ambiguities without
returning to a known location, even on a long baseline, with the use of the L3
carrier frequency (Melbourne, 1985). Due to the uncertain availability of
these signals, this possibility has not been pursued in this study.

3.5. <u>System II</u>. This implementation combines accelerometer data with data
from a GPS receiver. The GPS data is dual frequency carrier range change. The
solution proceeds in three steps. First, the accelerometer data is used to
determine the kinematic motion of the platform relative to its position at an
initial epoch. Second, the effects of the kinematics on the carrier
observables are calculated for each data epoch, and are removed from the
measured history of carrier range change. Third, the carrier range change
data are used to determine a solution for the platform position at the initial
epoch.

Given the position at an initial epoch, the accelerometer data data can be
used to determine the platform position at an arbitrary time during the
experiment. The failing of this technique is the accuracy requirement that is
placed on the accelerometers. It should be noted that the accelerometer and
carrier range change solution errors are correlated. In order to allow an
overall platform solution accuracy of 10 cm, the accuracy required for the
accelerometer is estimated to be about 4 cm over a carrier range change
solution, which will take at least 15 minutes. A 4 cm error would arise from
the accelerometer data, over 15 minutes, if there were a bias in the measured
acceleration of $10^{**}-8$ g. This level of accuracy is not currently available
in a sea going accelerometer, however, and no covariance analysis was
performed for this system.

3.6. <u>System III</u>. This approach is similar to system II, with P code range
data substituted for carrier range change, and it suffers from the same
deficiency. Since at least 15 minutes of data is required for an accurate P
code range solution, this approach is also prevented by accelerometer
inaccuracies, and no covariance analysis was performed for this system,
either.

3.7. <u>System IV</u>. The fourth system uses dual frequency carrier range change
data to determine both position and kinematic information. Simultaneous
solutions for position and clock at N_t epochs are possible if N_s satellites
are continuously tracked over all epochs, and if $N_s \geq 4 N_t /(N_t - 1)$. Since
at least 5 GPS satellites can be observed at any time, a solution exists for
$N_t \geq 5$. Tracking over longer times will strengthen the solution. The
results of covariance analysis for this GPS receiver system are shown in
Section 4.4.1.

3.8. <u>System V</u>. System V uses a GPS receiver which measures the L1 and L2
carrier range change, a high precision, but ambiguous, phase measurement for
tracking platform kinematics. P1 and P2 pseudo range, a lower precision, but
unambiguous, delay measurement, is used to determine position. The combined
data types are used to solve for receiver positions at periodic epochs. With

this technique, the precise carrier phase change data is primarily used to determine platform motion relative to the initial epoch, although the carrier data do add some strength to the position solution. Once the kinematics have been determined, the effects on the P code ranges of motion away from the initial position are calculated and removed from the P code data. The unambiguous P code range data, with kinematic effects removed, are then averaged to reference epochs over the length of the experiment, and these averages are used to solve for the platform position at the initial epoch (Thornton, et al., 1984). Covariance analysis results for this system are described in Section 4.4.2.

4. SYSTEM PERFORMANCE

4.1. General In this section the positioning accuracies for systems IV and V are discussed. Both use the carrier phase data to remove platform and GPS kinematics. Once the instantaneous platform position is fixed at a reference time, its position at other times during the experiment can be determined from the history of GPS carrier phases. In this section, we analyze the effects of different error sources on the determination of platform attitude, kinematics, and position.

4.2. Attitude Determination (Common to all Systems).

4.2.1. Attitude Determination Errors. Three antennas are placed in an L configuration, with separations of 4 m. The differential Ll carrier phases between pairs of these antennas are used to determine the platform orientation. This observable is available at intervals of 0.020 second, allowing kinematic tracking of orientation. Each 0.020 second differential phase measurement has an error, due to system noise, of 0.1 cm.

The expected accuracy of such a system is about 0.0003 radians. If the antennas are 100 m above the acoustic transducer, this would contribute a positioning error of 3 cm.

4.2.2. Ambiguity in Attitude. A complication in using a carrier interferometer is the angular ambiguity that results if the phase difference is off by an integer number of cycles. For a satellite overhead, and 400 cm antenna separations, the ambiguity is about 0.05 radians. When multiple satellites are in view, however, this ambiguity should be easily resolved. This is so because the satellites at different geometries will have ambiguities of different values. For example, consider a satellite at 20 degrees elevation, in the plane defined by a pair of antennas and the zenith. The ambiguity for this satellite is about 0.12 radian (in the same plane).

In order to resolve any large uncertainties in orientation that may remain, knowledge that the long term average platform orientation is horizontal could be invoked. Alternatively, a conventional clinometer could be used.

4.3. Kinematics (Common to Systems I, IV, and V).

4.3.1. Data Types. For systems I, IV, and V, dual frequency carrier range change data are used to determine platform kinematics. These data consist of carrier ranges measured at epochs separated by 0.020 seconds, with all data for one satellite track having the same range ambiguity, or bias, consisting of an integer number of carrier cycles. During an experiment, these data would be averaged over one minute, with a single biased range recorded on 1 minute epochs.

In addition, the 0.020 second data are buffered for 2 seconds. If a message is received from the acoustic package indicating a signal has been transmitted, the buffered data, and all 0.020 second data until the reception of replies from the sea floor transponders, are sent to a microprocessor which is integrated into the GPS receiver system.

4.3.2. On Board Position Interpolation. In near real time, this processor uses the 0.020 second range change data, along with the epochs of signal transmission/reception provided by the acoustic package, to determine the integrated carrier range change from the most recent minute epoch to the transmission/reception times. If these times do not coincide with a 0.020 second epoch, a linear interpolation is performed between the two nearest 0.020 second points. Assuming three sea floor transponders are in use, this means that the platform position must be determined for the time of acoustic signal transmission, and for three reception times.

4.3.3. Data Storage. The requirements on data storage are that a set of carrier data, as well as averaged P code data, must be recorded each minute for all of the satellites in view. In addition, four more sets of carrier data are recorded for each interrogation completed by the acoustic system.

4.3.4. Error Sources for Kinematic Solution. The errors in the determination of kinematics come mainly from four sources. The first error source is the noise on the biased carrier range. For the proposed receiver design, which allows up to 8 GPS satellites to be tracked in parallel, the error for a 0.020 second biased range measurement is 0.3 cm. This number assumes a dual frequency correction is applied, removing ionospheric effects. This noise error is uncorrelated for the four positions determined per acoustic interrogation, leading to an overall differential error of 0.6 cm.

The second error source is caused by acceleration of the platform. Since the interpolation is done in first order, any acceleration causes errors in the position solution. For example, a constant acceleration of 2 g over the 0.020 second interval between carrier data would lead to a maximum interpolation error of 0.1 cm. This error would likely be correlated among the four positions determined per acoustic interrogation, leading to a differential error of 0.4 cm.

The third source is the presence of errors in the GPS velocities. These velocity errors are assumed to be 25 cm per hour, based on orbit accuracies determined from the covariance results. In a differential baseline, this error is scaled by approximately the ratio of the baseline length to GPS altitude. For a 2,000 km baseline, this leads to an accumulated error of about 2.5 cm during 1 hour of tracking.

The final error source for the kinematic solution is the presence of an unmodeled variability in the tropospheric delay. This is estimated to amount to a 1 cm error over 1 hour intervals.

4.4. Position Determination

4.4.1. Covariance Results for System IV. The GPS orbits used for the

covariance analysis are those for the operational system with 18 satellites.
The receiver is assumed to be capable of tracking up to 8 GPS satellites in
parallel channels. The error models used as inputs to the analysis are given
in Table 2.

TABLE 2: ONE SIGMA ERROR MODELS FOR SYSTEM IV

Random errors:

Data noise	<0.1 cm	(dual freq carrier, T=5·min)
Multipath	0.5 cm	(dual freq carrier)
Instrumental phase	<0.1 cm	(dual freq carrier)
Troposphere	1.0 cm	

Systematic errors:

GPS ephemerides	solved for (2 m apriori, each component)	
Coordinates of fiducial stations	5.0 cm	each component
Troposphere	10.0 cm	zenith delay at ocean platform platform
	1.0 cm	at fiducial sites (assumed to have water vapor radiometer at fiducial sites)
Kinematics	2.7 cm	RSS of errors from 4.3.4.
Attitude	3.0 cm	0.0003 radian over 100 m
Sat multipath	<1.0?cm	(magnitude of multipath at satellite antenna is unknown)
Sample time tag error	0.1 cm	(range error caused by 1 microsecond time tag error)

The results of the analysis for the carrier range change only system,
system IV, are given in Table 3.

TABLE 3: COVARIANCE RESULTS FOR SYSTEM IV

Length of data arc	RSS of North and East location errors for the acoustic transducer located beneath the surface platform.
1 hour	268 cm
6 hour	19 cm

4.4.2. Covariance Results For System V. The GPS orbits used for the
covariance analysis are those for the operational system with 18 satellites.
Again, the receiver is assumed to be capable of tracking up to 8 GPS
satellites in parallel channels. The error models used for the carrier data
are the same as for system IV (Table 2). The additional error models for the
P code data which is used, in addition to the carrier data, in system V, are
given in Table 4.

TABLE 4: ONE SIGMA P CODE ERROR MODELS FOR SYSTEM V

Random errors:
    Data noise   1.0 cm (P code, T = 1 hr)
    Multipath   4.0 cm (P code, use 1 hr averaging
                                    to reduce effect)
    Instrumental delay <1.0 cm
    Troposphere   1.0 cm

Systematic errors:
    GPS ephemerides   solved for (2 m apriori, each component)
    Coordinates of
    fiducial stations   5.0 cm each component
    Troposphere   10.0 cm zenith delay at ocean platform
                       1.0 cm at fiducial sites (assumed to
                                have water vapor radiometer
                                at fiducial sites)
    Kinematics   2.7 cm RSS of errors from 4.3.4.
    Attitude   3.0 cm 0.0003 radian over 100 m
    Sat multipath   1 to 10 cm (magnitude of multipath at
                                satellite antenna is unknown)
    Sample time tag   0.1 cm (range error caused by 1
    error   microsecond time tag error)

The results of the analysis for System V, which uses a combination of carrier range change and P code pseudo range data, are given in Table 5.

TABLE 5: COVARIANCE RESULTS FOR SYSTEM V

Length of data arc	RSS of North and East location errors for the acoustic transducer located beneath the surface platform.	
	if 1 cm sat multipath	if 10 cm sat multipath
1 hour	11 cm	15 cm

5. CONCLUSIONS

5.1. GPS System Choice. Of the five GPS systems considered in this work, only system V, using the combined carrier and P code data types, appears capable of 10 cm positioning of an acoustic transducer fixed under a floating platform in mid-oceanic regions. Even system V may not produce the required accuracy, if unmodeled 10 cm magnitude multipath effects are present at the GPS satellites (Young, et al., 1985).

5.2. GPS Receiver Requirements. In order to meet the precision goals, the GPS receiver will need parallel channels for each satellite. In order to allow mutual visibility with a large (in spatial extent) network of fiducial stations, the receiver should be able to track 7 or 8 satellites simultaneously. The operational achievement of time tags with 1 microsecond accuracy will require a receiver that can interpret the GPS data message, allowing accurate clock synchronization.

In order to keep receiver phase and delay errors to an acceptable level, it will probably be necessary to have either a highly digital data path, or careful phase and delay calibrations, or both. A powerful digital processing capability contained within the receiver would facilitate on-board data compression, as described in section 4.3.2. In turn, the reduction in data volume would allow the use of a rugged, compact, data storage device.

For this sea floor geodetic system to become operational, it is important that the GPS receiver system be capable of operating for hours at a time, in the sometimes hostile marine environment, with no loss of carrier cycles.

5.3. Further Work Needed on GPS Receiver System. Work remains in order to support the sea floor geodetic system. Some of this work is described below.

5.3.1. Receiver Development. A GPS receiver to meet the specifications given above must be developed. Its performance, particularly the P code data accuracy, must be demonstrated.

5.3.2. Antenna Development. Antennas must be developed that can meet the multipath error budgets given in Tables 2 and 4.

5.3.3. Satellite Multipath. The effect of multipath from the GPS satellite's antennas must be determined. If it is large (10 cm), models must be generated to allow the reduction of its effect on the baseline solution.

6. ACKNOWLEDGEMENTS

The research described in this paper was carried out by the Jet Propulsion Laboratory, California Institute of Technology, under contract with the National Aeronautics and Space Administration.

REFERENCES

Committee on Geodesy, Panel on Ocean Bottom Positioning, National Research Council, 1983. "Sea floor Referenced Positioning: Needs and Opportunities", National Academy Press, Washington, D.C.

Davidson, J. M., Thornton, C. L., Vegos, C. J., Young, L. E. and Yunck, T. P. Yunck, 1985. "The March 1985 Demonstration of the Fiducial Network Concept for GPS Geodesy: A Preliminary Report", Proceedings First International Symposium on Precise Positioning with the Global Positioning System, pp. 603-611.

Melbourne, W. G., 1985. "The Case for Ranging in GPS-based Geodetic Systems", Proceedings First International Symposium on Precise Positioning with the Global Positioning System, pp. 373-386.

Panel on Crustal Movement Measurements, Committee on Geodesy and Committee on Seismology, National Research Council, 1981. "Geodetic Monitoring of Tectonic Deformation - Toward a Strategy", National Academy Press, Washington, D. C. (1981).

Remondi, B. W., 1985. "Performing Centimeter Accuracy Relative Surveys in Seconds using GPS Carrier Phase", Proceedings First International Symposium on Precise Positioning with the Global Positioning System, pp. 789-798.

Space Science Board, Committee on Earth Sciences, National Research Council, 1982. "A Strategy for Earth Science from Space in the 1980's Part1: Solid Earth and Oceans", National Academy Press, Washington, D.C.

Spiess, F. N., Lowenstein, C. D., McIntyre, M. O., 1984. "Analysis of a
    Method for Precisely relating a Sea floor Point to a Distant Point
    on Land: a Report under NASA Grant NAG 5-320", MPL-U-31/84, Marine
    Physical Laboratory of Scripps Institute of Oceanography, Univ. of
    CA, San Diego
Thornton, C. L. Young, L. E., Wu, S. C. and Thomas, J. B., 1984. "GPS-Based
    Certification System for the Microwave Landing System", Proceedings
    PLANS 1984, IEEE, pp. 256-263.
Walter, L. S., 1984. Proceedings of the Airlie Workshop, NASA Conference
    Publication 2325 (1984).
Young, L. E., Neilan, R. E. and Bletzacker, F. R., 1985. "GPS Satellite
    Multipath: An Experimental Investigation", Proceedings First
    International Symposium on Precise Positioning with the Global
    Positioning System, pp. 423-432.

Performance Appraisal of an ARGO Calibration System Using GPS

Lt. Timothy D. Rulon
and
Knute A. Berstis

NOAA/NOS
6001 Executive Boulevard
Rockville, Maryland 20852

ABSTRACT

The ARGO and Raydist positioning systems are used as a primary means of horizontal control for the conduct of bathymetric surveys of the Exclusive Economic Zone (EEZ) using multibeam sonar systems. An ARGO calibration system using a Texas Instruments TI 4100 Global Positioning System (GPS) P code receiver is described.

1. INTRODUCTION

The Office of Charting and Geodetic Services, a component of the National Oceanic and Atmospheric Administration's (NOAA) National Ocean Service, performs detailed bathymetric surveys of the U.S. EEZ which extends out 200 nautical miles (nm) from our shoreline. These surveys are typically conducted at a scale of 1:50,000 using General Instrument's Sea Beam and Bathymetric Swath Survey multibeam sonar systems. The accuracy criteria for these surveys conform to International Hydrographic Organization standards, i.e., depth to 1 percent, horizontal position to within 50 meters (90 percent).

The ARGO and Raydist medium range (less than 250 nm) positioning systems are used as the primary means of horizontal control for these surveys. Mini-Ranger, a short range (less than 20 nm) microwave ranging system has been used exclusively for the electronic calibration of these systems in the past. System calibrations are required to determine the integer and partial range bias correctors at the beginning of each survey and at frequent (3-5 day) intervals to verify that no integer range jumps or system drifts have occurred.

The efficiency of a survey operation is directly dependent upon the number of calibrations and distances involved. Because of the offshore nature of NOAA's EEZ surveys, vessels must transit considerable distances to reacquire Mini-Ranger signals used for calibration making the impact of calibration on efficiency very significant for these surveys. Consider a 12 knot vessel surveying for 17 days, 150 nautical miles from the calibration area. Theoretically, this vessel could survey 4,896 linear nautical miles during this period. If five calibrations were to be performed only 3,396 linear nautical miles would have been surveyed, i.e., 4,896 nm less 5 trips x 300 nm/round trip. Therefore, if calibrations could be conducted in the survey area rather than by Mini-Ranger the result would be an increase in efficiency of 44 percent in this case. With ship costs ranging from $10,000 to $18,000 per sea day, the impact of using GPS is substantial.

To increase the efficiency of future survey operations prior to GPS becoming fully operational, a Texas Instruments TI 4100 P code GPS receiver was evaluated as a calibration system. Tests were conducted in the static mode and using the Mini-Ranger Falcon 484 and ARGO systems as a reference. This has led to routine operational use of the TI 4100 GPS receiver for calibration of the ARGO system.

M. Kumar and G. A. Maul (directors), Marine Positioning, 233–242.

2. HARDWARE/SOFTWARE

The data acquisition system used for this series of tests was a Hewlett Packard 9825T desk top computer with 64K memory, real-time clock and two RS-232 interfaces. The real time clock was manually synchronized to GPS time in order to correlate the GPS and Falcon/ARGO data sets. (See figure 1)

The data acquisition program acquires the relative navigation record from the TI 4100 GPS receiver and can be configured to acquire either Falcon or ARGO data. The GPS and Falcon/ARGO data is used to calculate and display calibration parameters in real time. The data is also recorded on tape for later analysis. Each 9825T tape cartridge holds approximately 2 hours of data. Although the update rate for the TI 4100, Falcon and ARGO systems is once per second, these data are displayed on the 9825T and recorded to tape at an average rate of once every 12 seconds due to processing delays internal to the 9825T.

The acquired data is post-processed to graphically display performance para-meters and provide summary statistics. Plots have been developed to determine the performance of the TI 4100 using the Falcon or ARGO systems as a reference, or conversely, to determine Falcon/ARGO range bias correctors using the TI 4100 as the calibration standard.

Software routines within the processing program rejected data with bad flags, poor GPS receiver convergence, poor Position Dilution of Precision (PDOP > 10), time skews (< 1.2 seconds) or points which were clearly fliers. The flier rejection limit was set at 25 meters for GPS position errors and GPS predicted Falcon range correctors. GPS position errors were not computed when the Falcon or ARGO intersection angle between the range arcs was less than 30° or greater than 150°. It was necessary to incorporate these criteria in order to limit the data analysis to points with valid GPS receiver performance, acceptable geometry and to eliminate data points which were clearly recognizable as fliers.

Due to a software bug in the data acquisition program, data points less than 10 seconds apart contained anomalous GPS time tags and were rejected by the processing program. A few data points with updates greater than 10 seconds contained these anomalies and these are visible as spikes in some of the data plots, e.g., point A in Figure 3.

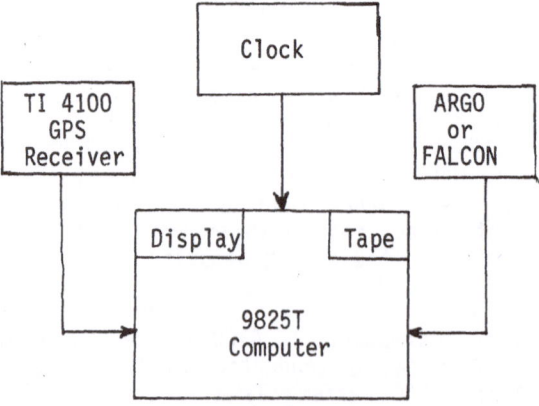

Figure 1

3. STATIC TESTING

The results obtained during the static test were typical for GPS P Code receivers and at times were much better than expected. The average horizontal error observed during static testing was 8.66M Root Sum Square (RSS^2=AVERAGE^2+STD DEV^2). Although the data set is limited, several conclusions were drawn regarding the performance of the TI 4100 and these performance characteristics have been noted repeatedly at other times when data were not being recorded in both the static and dynamic modes. Figure 2 is typical of the performance observed using 4 satellites in the static mode.

It was observed that there usually occurred a shift of 5-15 meters when switching satellite vehicle (SV) combinations. This error tends to decrease with time providing PDOP is not increasing significantly. The linear correlation between horizontal error and PDOP is masked by this effect at low values (<6) of PDOP. Therefore, it would appear to be advantageous to hold a particular SV combination rather than switching continually based solely on PDOP.

When the elevation angle of one of the SV's was less than 5° the effect of the troposphere was clearly visible. There appears to be no significant degradation above 5°, so this mask angle appears reasonable for GPS operations.

The use of the actual mean sea level altitude appeared to significantly degrade the performance of the TI 4100 when operating in the 3 SV, altitude hold mode when compared to a GPS determined altitude. The GPS determined altitudes and the actual altitude were seen to vary by as much as 40 meters from each other. During all testing of the TI 4100, 4 SV operation generally outperformed 3 SV operation independent of the altitude selected for 3 SV operation.

Tape #	GPS POSITION ERROR
1	12.77 M
2	9.87 M
3	14.93 M
4	4.25 M
5	1.49 M
6	7.62 M rej., low angle
AVERAGE	8.66 M RSS

TABLE 1
STATIC TEST RESULTS

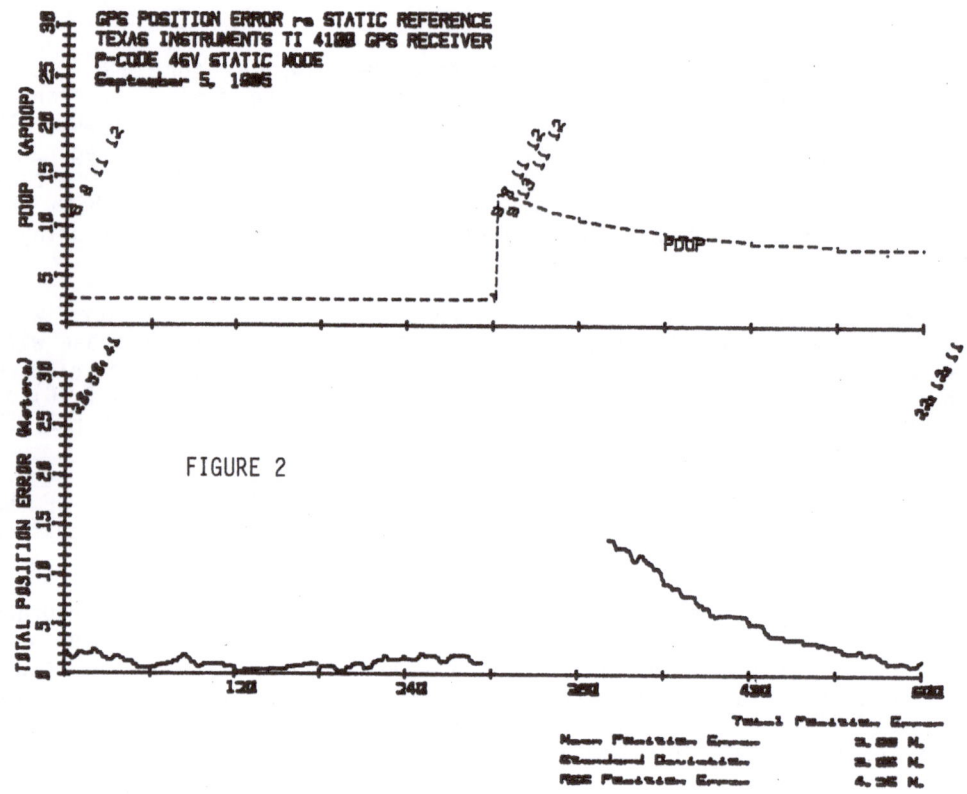

FIGURE 2

4. DYNAMIC TESTS USING FALCON

Evaluation of the dynamic performance of the TI 4100 GPS receiver using the Mini-Ranger Falcon as the reference took place at Solomons Island, MD, aboard the NOAA vessel LAIDLY during the period 19-22 October 1985. Three Falcon reference stations were located 1 nautical mile apart, and the LAIDLY operated at speeds to 13 knots and well within the limits of the triangle formed by the stations.

Prior to the tests the Falcon was calibrated over a known baseline and the range bias correctors were set to zero. The filtered X-Y output of the Falcon was used in determining GPS Position errors and Falcon range bias correctors. The filtered X-Y position was assumed to be accurate to better than 4M 2drms under the conditions of the test.

Testing of the TI 4100 receiver was performed in both the 3 and 4 SV modes. The position error exceeded 25 meters at times, (50 meters) and the data were recognized by the processing program as being fliers and were rejected. The observed performance in the 3 SV mode would have been significantly degraded had the errors in excess of 25 meters been included in the analysis.

The results obtained during the dynamic testing of the TI 4100 followed a pattern similar to the static testing with the exception that the data were considerable noisier. Several meters of additional noise are to be expected in the dynamic mode from the TI 4100, Falcon reference system and time skew errors.

The average RSS horizontal error observed during this series of tests was 9.92 meters, see Table 2. The actual GPS error was on the order of 8 meters after taking Falcon positioning errors and time skew errors into account. This compares very favorably with the static test results. Figure 3 is typical of the performance observed in the 4 SV dynamic mode.

Range bias correctors were computed for range 1 and range 2. Since the Falcon ranges had previously been calibrated, these GPS computed corrector values should be near zero assuming no GPS or Falcon range errors. The Falcon range correctors were determined to an average RSS error of 7.78 meters, see Table 2. This can be extrapolated to determine the performance of the TI 4100 as an ARGO calibration system. At a nominal frequency of 1650 kHz the ARGO lane width is 90.8 meters. Therefore, the ARGO range bias correctors could have been determined to an RSS accuracy 0.086 lanes.

TAPE #	GPS POSITION ERROR	GPS PREDICTED RANGE ERROR	
		R_1	R_2
1	16.20 M	11.39,	13.72 M
2	6.42 M	5.14,	4.39 M
3	5.49 M	4.28,	3.27 M
4	11.57 M	8.64,	11.43 M
AVERAGE	9.92 M RSS	7.78 M RSS	

TABLE 2
DYNAMIC TEST RESULTS USING FALCON

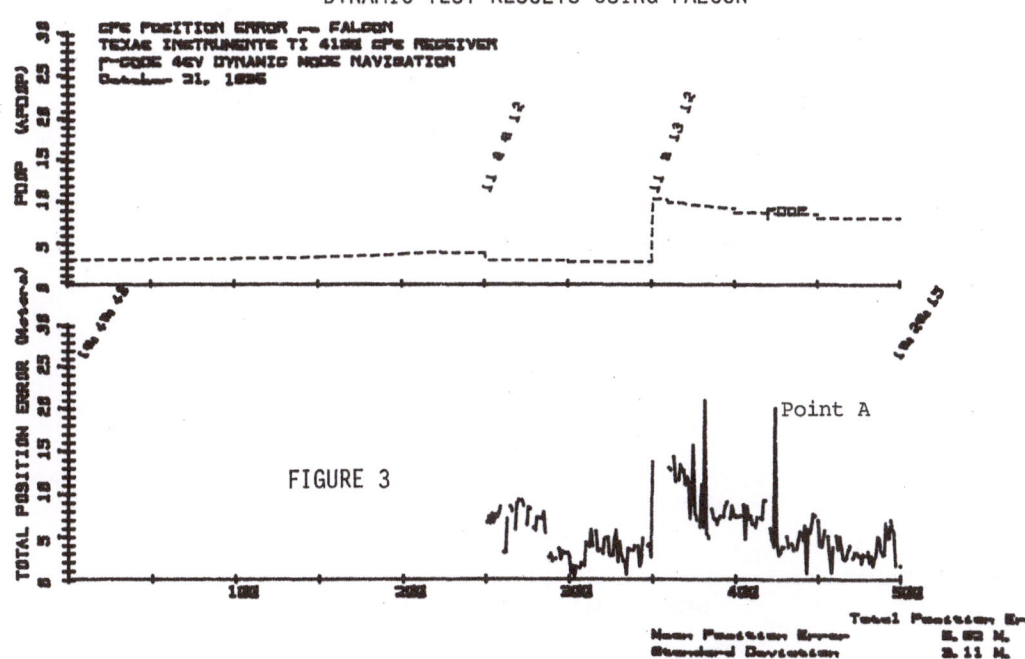

FIGURE 3

5. DYNAMIC TESTS USING ARGO

Evaluation of the TI 4100 GPS receiver and data acquisition system as a calibration system took place aboard the NOAA survey ship SURVEYOR which uses the ARGO system for positioning control. Testing took place 50-75 NM offshore California during the period 7 November to 15 November 1985. The system was operated by personnel who were not totally familiar with the operating characteristics of the TI 4100 receiver and difficulties were experienced at times in acquiring the satellites and obtaining a converged solution.

ARGO performance during the test period was excellent and accuracy to better than 25M 2drms was expected. Three ARGO calibrations were performed during the test period using a Mini-Ranger III system. The Mini-Ranger III determined ARGO range bias correctors were used for comparison with GPS determined ARGO correctors and for determining GPS position error using the ARGO system as a reference.

TAPE #		AVERAGE MINIRANGER III DETERMINED CORRECTORS	AVERAGE GPS DETERMINED CORRECTORS	DIFFERENCE AVERAGE MR III-GPS	RSS ERROR MR III-GPS
1	R1	.46	.55	.09	.12
	R2	-.06	-.07	.02	.06
3	R1	.46	.32	.14	.18
	R2	-.06	.04	.10	.39
5	R1	.46	.51	.05	.10
	R2	-.06	-.10	.04	.08
6	R1	.46	.51	.05	.10
	R2	-.06	-.07	.01	.10
7	R1	.46	.47	.01	.08
	R2	.01	-.02	.03	.10
8	R1	.46	.43	.03	.26
	R2	.01	-.04	.05	.15
AVERAGE				.05	.14

All units in lanes
ARGO Frequency 1640.3 kHz

TABLE 3
DYNAMIC TEST RESULTS USING ARGO

The results obtained from the intercomparison of the GPS and ARGO systems were considerably better than predicted. The average position difference between the GPS and ARGO fixes was 11.8 meters RSS. This error was less than predicted for the ARGO system alone.

The average GPS computed ARGO range correctors agreed to within 0.05 lanes from those computed using the Mini-Ranger III system. The larger RSS error of 0.14 lanes primarily results from the inclusion of anomalies in the data set which were not removed by the processing software. The standard deviation of the GPS predicted range correctors during periods of stable GPS operation can be seen to be only a few hundredth's of a lane. Had these anomalies been rejected during processing, the RSS error between the GPS and Mini-Ranger III computed range correctors would have been less than 0.1 lane. This compares very favorably with the dynamic results using the Mini-Ranger Falcon system.

A good example of these anomalies may be found in Figure 4. The large 0.2 lane spikes are the result of the software bug in the data acquisition program which were not removed by the processing program. The large excursions are easily recognizable to an operator as extraneous since the correctors are significantly different than previous values obtained and are changing rapidly. The cause of these excursions is unknown; however, the phenomenon has not reoccurred during subsequent operations using experienced operators.

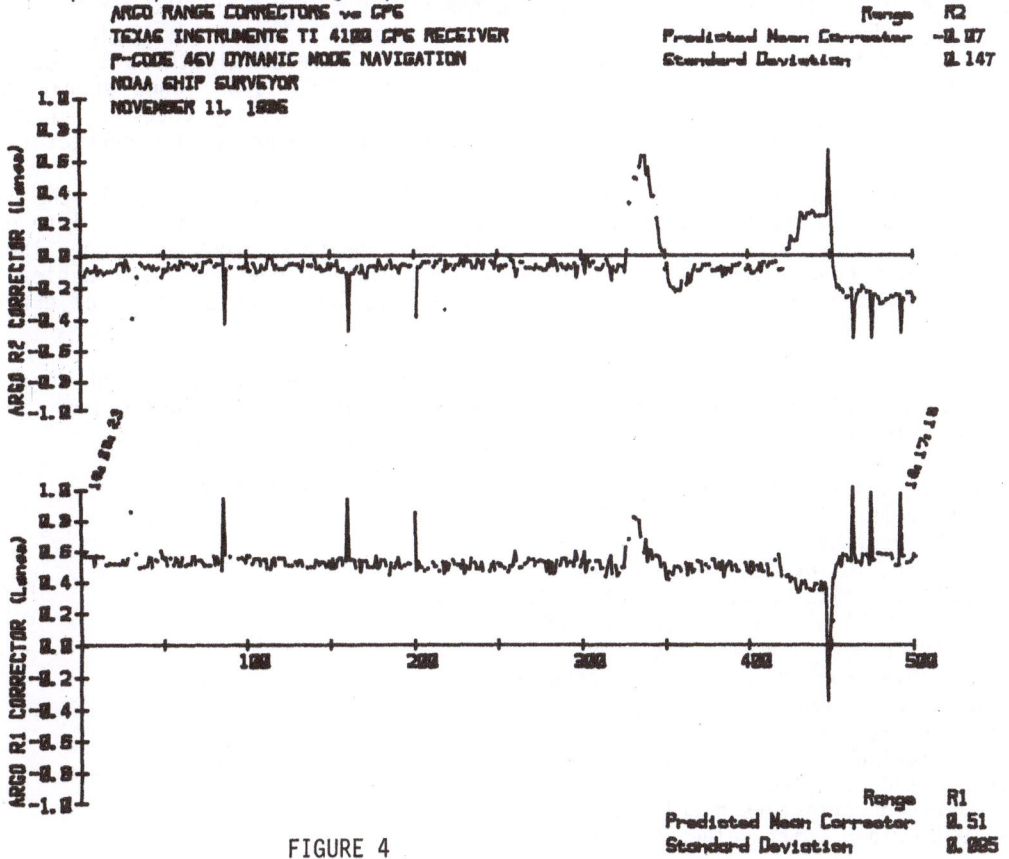

FIGURE 4

6. FINAL RESULTS AND CONCLUSIONS

Texas Instruments specifies the 3D accuracy of the TI 4100 GPS receiver to be 14 meters (one sigma), for a nonaccelerating user in a dynamic environment. Assuming that the error in the X, Y, and Z components are equal, this equates to a horizontal accuracy of 11.4 meters, one sigma. For the purposes of this discussion, the RSS errors will be considered a one sigma error.

In tests using the TI 4100 in the 4 SV mode, a horizontal accuracy on the order of 8 meters RSS was achieved after errors in the reference system and the data acquisition system are taken into account. Therefore, when operated properly, the TI 4100 may exceed its specified performance in the 4 SV mode.

In tests using the TI 4100 in the 3 SV mode, the horizontal error was found to vary from as little as 1 meter to as much as 50 meters. This error was dependant upon whether a GPS determined or the actual mean sea level was selected for the 3 SV mode. The better performance was observed at the end of a GPS window, when the altitude was determined using 4 SV's. An RSS error of 25 meters was typical during 3 SV operation during data collection and at other times when data were not being recorded. Therefore, the TI 4100 did not meet its specified performance in the 3 SV mode during this test.

Operated within the proper constraints in the 4 SV mode, the TI 4100 can be used to determine ARGO range bias correctors to better than 0.1 lanes. Averaging these GPS determined range bias correctors over a period of several minutes will reduce the effect of random errors and accuracies to better than 0.05 lanes should be achievable. Averaging data also serves as an additional quality check to insure that extraneous data is not included in determining the ARGO range bias correctors.

Under less stringent criteria, i.e., using 3 SV's or for 6< PDOP < 10, the TI 4100 can be used to identify the ARGO whole lane count. At a nominal lane width of 90 meters, accuracy to better than 45 meters is required to resolve the proper lane count. It has been found that this level of accuracy is nearly always achieved when the TI 4100's PE data quality indicator displays less than 5 meters of error and the satellite residuals are < 10 meters.

Based upon the performance observed during this series of tests and during several months of operating the TI 4100, the following set of guidelines are recommended for using the TI 4100 to calibrate an ARGO positioning network.

o 4 SV's locked in with signal strengths > 39 (3 SV's lane ID only)
o SV residuals < 10 meters and changing
o All SV's healthy
o PDOP < 6 (< 10 for lane identification)
o PE Quality indicator displays < 5 meters error
o Altitude (MSL) within 30 meters of actual value
o All SV's above 5 degrees elevation
o GPS computed course and speed reasonable
o TI 4100 in NAV mode 1 (2 for 3 SV)
o * Warning is off
o Vessel laying to or at constant course and speed
o Dynamic mode 2 selected
o GPS derived correctors stable ±.1 lane > 10 minutes
o Use DMA origin shift values for the local area

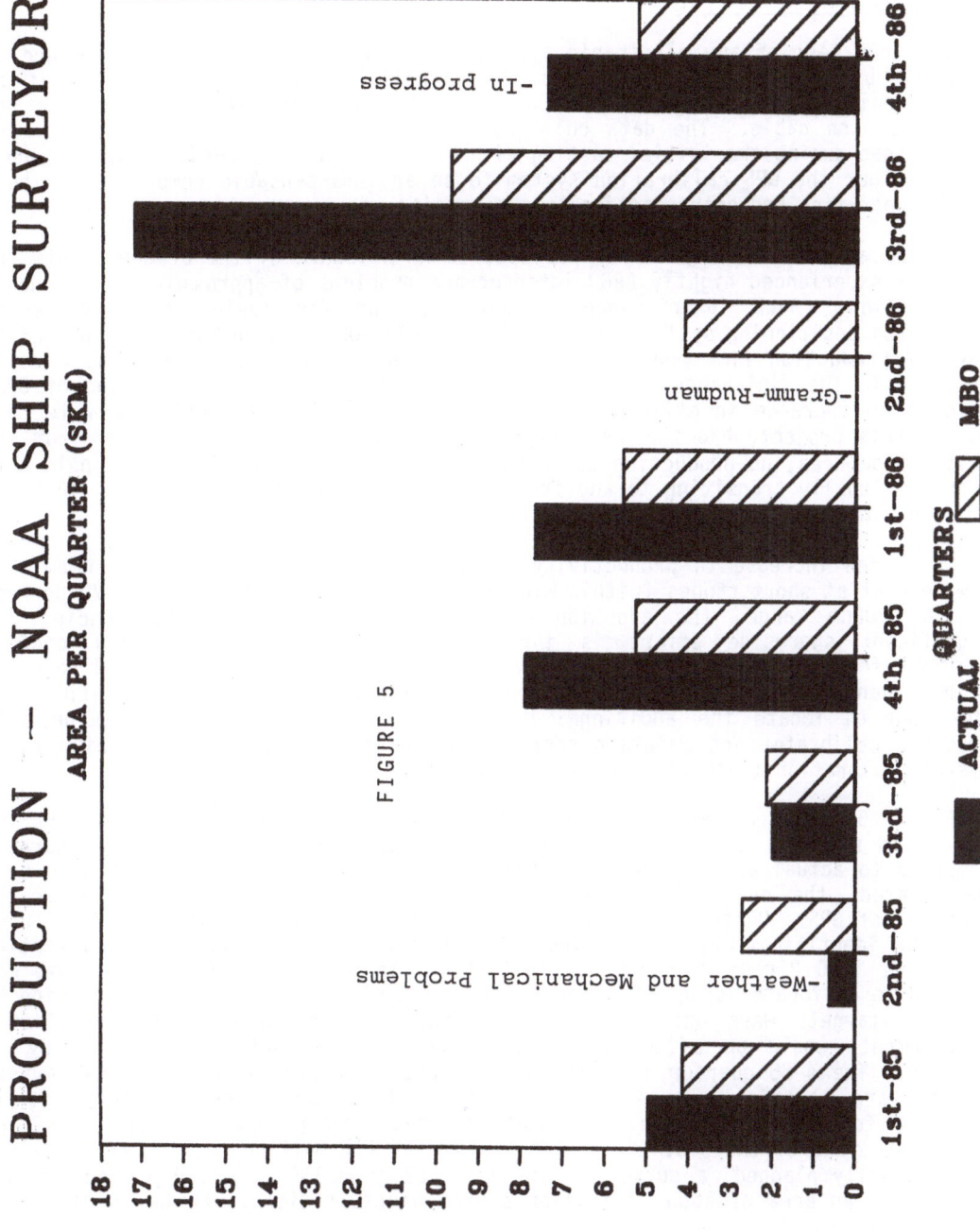

7. OPERATIONAL DEPLOYMENT

The true test of any electronic system is how it performs in the field and is accepted by the users. TI 4100 has been installed aboard the SURVEYOR in an operational capacity since APRIL 1986. To date the only electronic failure has been a broken cable. The data collected indicate that the system is performing better than during the initial testing with few anomalies observed. Ship personnel now consider the GPS calibration system to be an indispensable component of their survey operation and enthusiastically support its use.

During a recent 52 day survey project, 65 nm offshore of the Oregon coast, the SURVEYOR experienced nightly ARGO interference problems of approximately six hours in duration. Such interference is not uncommon for medium frequency phase comparison positioning systems such as ARGO. SURVEYOR personnel calculated that 14 days of production time would have been lost had it been necessary to make daily transits to the Mini-Ranger calibration site rather than taking advantage of GPS. This is an increase in efficiency of 37 percent. It is interesting to note that during this project, had the SURVEYOR been surveying greater than 144 nm from the calibration area, no production could have been accomplished. The ship would have been continually transiting to and from the calibration site. Using GPS, 18 hours of surveying could be accomplished daily.

While the increase in productivity described above varies from a minimum of a few percent at short ranges (within Mini-Ranger coverage) to better than 70 percent at the 200nm range, the adoption of GPS results in other efficiencies. A significant shoreside effort is involved in establishing horizontal control stations for use as Mini-Ranger calibration sites and for deployment and recovery of Mini-Ranger equipment. The adoption of calibration of ARGO by GPS eliminates the need to locate the additional Mini-Ranger calibration sites; the need to procure, calibrate, and maintain short range systems for electronic medium range system calibrations; and releases survey personnel for other tasks.

Figure 5 shows the NOAA Ship SURVEYOR ship production in square kilometers for FY 85 and most of 1986. The graph shows Management By Objectives (MBO) targets compared to actual accomplishments. While other management efficiencies have been implemented, the overriding improvement in productivity has resulted from the adoption of GPS. During the 3-85 quarter, a GPS C/A code receiver was first used for lane identification. During the 1-86 quarter lane identification and testing of the TI 4100 P code receiver was initiated. In the 3-86 quarter, the ship was allowed to calibrate using GPS after an initial Mini-Ranger calibration. During FY 85, MBO targets were exceeded by 23 percent. In recognition that additional improvements would be affected during FY 86, MBO targets were raised by 21 percent. The 4-86 quarter is still in progress at this writing, but the revised targets are already being exceeded by about 25 percent. As a result, it is possible for NOAA EEZ surveys to meet the Departmental goal of characterizing important areas of the U.S. EEZ in ten years. If six multibeam ships are assigned as originally planned, a survey of the U.S. EEZ from 150 meters depth out to the EEZ limit, an area of about 3 million square nautical miles, can be completed in about 20 years.

REFERENCE

1. Thomas O. Sepplin., "Department of Defense World Geodetic System 1972" International Symposium on Problems Related to the Redefinition of North American Geodetic Networks, May 1974.

NAVGRAV, A COMPREHENSIVE COMBINED NAVIGATION AND GRAVIMETRY EXPERIMENT ON THE NORTH SEA; OBJECTIVES AND FIRST EXPERIENCES

M.A. Salzmann
G.J. Husti
G.L. Strang van Hees
P.J.G. Teunissen[x]
Department of Geodesy
Delft University of Technology
Thijsseweg 11
2629 JA Delft
The Netherlands

ABSTRACT

During spring 1986 a unique NAVigation GRAVimetric experiment (NAVGRAV) took place on the North Sea. An overview of the objectives, the experiments performed and some experiences are given.

1. INTRODUCTION

The NAVGRAV project, a combined NAVigation and GRAVimetric experiment, was carried out successfully in the period April 23 - May 13 1986. Summarizing the main objectives of the project are:
- to establish the perspectives of the Global Positioning System (GPS) for positioning at sea.
- to investigate the quality of terrestrial navigation systems in the North Sea region.
- to establish the achievable resolution and (internal and external) accuracy of the gravity field characteristics, derived from sea gravimetry, in comparison with results obtained from satellite altimetry.

The oceanographic vessel H.M.S. Tydeman of the Royal Netherlands Navy was made available for the experiments.

The experiment was carried out in two parts. The first period (April 23 - April 29) was mainly directed at the navigation experiment (see fig. 1). The second period (April 30 - May 13) concentrated on the sea gravimetry experiment (fig. 2).

The NAVGRAV project is the result of a joint effort of the Dutch survey community and several German universities.

x) supported by Z.W.O, Netherlands Organization for the Advancement of pure Research

M. Kumar and G. A. Maul (directors), Marine Positioning, 243–248.

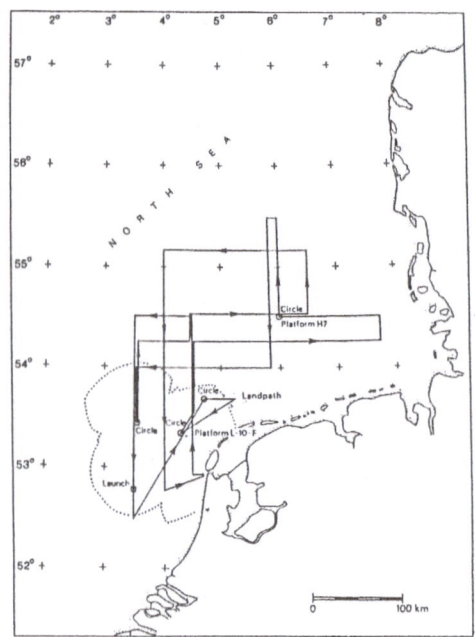

figure 1.   figure 2.

Figure 1. Navigation experiment NAVGRAV. The dotted line indicates the Syledis
          three way fix area.
Figure 2. Gravity survey NAVGRAV.

    A list of all participants in the project is given in the appendix.
The NAVGRAV project was sponsered by the Dutch Council of Ocean Sciences
(Nederlandse Raad voor Zeeonderzoek, NRZ). Additional financial and material
support was obtained from participating universities, government agencies and
private companies.

2. OBJECTIVES OF THE NAVGRAV PROJECT

    The objectives of the navigation and gravimetric experiments will now be viewed
in more detail.
First of all the perspectives of GPS for positioning at sea have to be established.
-Comparison of GPS with classical radio positioning systems (e.g. Syledis).
-Absolute (single point) positioning for the purpose of navigation and
 and velocity determination.
-Relative positioning between the vessel and a reference point on land.
-Relative positioning between two receivers on board.
-Relative positioning between two moving objects.
    In addition the limitations and perspectives of the existing classical radio
positioning systems, concentrating on the physical aspects of these systems,
are investigated.

The purpose of the gravimetric experiment is to investigate to what extend
satellite altimetry can replace sea gravimetry in the future (see e.g. Rummel et al,
1983).

Part of the region was already surveyed in 1979 (Strang van Hees, 1983).
By partly re-measuring and densifying this gravity network, it becomes possible
to investigate the spatial resolution, external accuracy and systematic biases
of sea gravimetry. The measurements were carried out by two gravimeters to
investigate the internal accuracy and instrumental characteristics.

3. EXPERIMENTS AND FIRST EXPERIENCES

The experiments concerning GPS, radio positioning systems and gravity
measurements will be reviewed seperately.

During most of the navigation experiment Syledis was used as a reference system.
As a consequence of ideal weather conditions first-class data were obtained.
The sea gravimetry measurements were mainly carried out on the northern part of
the Dutch Continental Shelf. An overview of the available navigation systems
during the first week of the project is given in table 1.

Available navigation systems during NAVGRAV, April 23 - April 29.

on board	on land
GPS receivers	the Netherlands
2 TI 4100	1 TI 4100 Delft and Haarlem
1 Sercel TR5S	1 Sercel TR5S Delft and Haarlem
1 Trimble 4000S	
1 Cesium standard	1 Rubidium standard
	Norway
	1 TI 4100 Stavanger
Radio positioning systems	
Syledis	
Hifix-6	
Hyperfix	
Loran C	
Other systems	
Gyro	
Log	
Pitch and Roll sensor	
Echosounder	

table 1.

3.1 <u>GPS experiments.</u> The availability of three types of receivers on board
enabled a comparison of the different receivers. During the entire navigation
experiment always at least one GPS antenna was installed in the front mast of the
ship.
   The following experiments were performed:
-Single point positioning and velocity determination with a single receiver.
 A velocity estimate of 0.1 m/s was achieved during good coverage with the Trimble
 4000S. Using a cesium standard it was investigated how long GPS observation
 intervals could be enlarged during unfavourable satellite coverage.
-Relative positioning between the vessel and reference points on land. For this
 purpose receivers were installed ashore. One receiver was installed in Stavanger
 (Norway) about 600 km. from the test area. Two receivers were installed at
 Haarlem and Delft (The Netherlands) much closer (+ 100 - 200 km.) to the test area.
 Both pseudo range and carrier phase were observed. These experiments for longer
 baselines supplement the already performed experiments (e.g. Seeber et al.,1986).
 First analysis of the data of the shorter baselines shows that an accuracy,
 which is equal or better then the reference system Syledis (3 m.) can be obtained.
-Short distance differential GPS in order to determine the ships dynamics.
 During the experiment antennas were installed in the main and front mast of the
 H.M.S. Tydeman at a relative distance of 40 m.
-Relative positioning between two moving objects.
 A TI 4100 receiver was installed in a launch, which was also equipped with Hifix-6.
 The distance ship-launch varied between 0 - 3 km.
   Up to now mainly single point solutions have been computed. The quality of the
data is such that good differential positioning results can be expected.

3.2 <u>Experiments concerning radio positioning systems.</u> These experiments mainly
relate to the physical aspects of medium frequency systems (Hifix-6 and Hyperfix)
and on the possible future use of interchain Loran-C in the North Sea area.
   Hyperfix proved to be very unstable during dawn and dusk. Hyperfix was also
highly affected by skywave during night. Part of the efforts in the first week
have been directed to the calibration of the relatively new Hyperfix chain.
   The following experiments were carried out:
-In order to investigate the effects of landpaths on Hyperfix positioning some
 special lines were surveyed.
-On medium and low frequencies the combination antenna-ship may cause heading
 dependent additive errors to the measured (pseudo-)distances. De Jong and
 de Munck (1986) propose to develop corrections in a fourier series. To obtain
 these corrections gyro and (pseudo-)distance observations, uniformly spaced
 over the horizon, have to be made. For this purpose several circles were steamed.
 First computations show that the effect occurs with periods of π and 2π radians,
 which is in accordance with the expectations. Harmonics higher than order 2
 do not seem present.
 These results should be handled with care, because the total effect (for the
 H.M.S. Tydeman) was of the same magnitude as the accuracy of the reference
 system. Also the antenna lay-out on board could have caused some interference.
-The reflection of signals of all frequencies caused by oil platforms was
 investigated by circling around two platforms at close range.
-The effects of skywave on medium frequency systems were monitored at night by
 observing lane values from distant stations.

-Loran-C interchain measurements. Pseudo ranges were measured to two station in the Norwegian, and one in the French chain. A cesium standard was used to synchronize the chains. This feasibility test for the use of Loran-C in the North Sea area proved that the system is reliable. An assessment of the accuracy of the measurements cannot be made yet.
Despite the 100 kHz frequency, during the experiment no interference of importance was encountered, even in the busy North Sea region.

3.2 Sea gravimetry experiments. 3500 Nautical miles of gravity lines were surveyed. The Delft gravimeter (an Askania Bodensee sea gravimeter KSS-5) was supplemented during the second period by the gravimeter of Hamburg University (Bodensee sea gravimeter KSS-30).

This set-up enables the assessment of mean instrumental biases of the gravimeters, since external effects like the Eötvös correction (vertical component of the Coriolis acceleration) and errors in positioning are eliminated in the difference between the two measured gravity values.

During a preliminary comparison about 200 points were used with a maximal range in gravity anomalies. The root mean square (rms) of the difference was 0.9 mgal in good weather conditions. During a few lines, measured under slightly worse weather conditions, the rms of the differences was about 2 mgal.

The external accuracy of the gravity measurements at sea is strongly influenced by the quality of the navigation; especially the velocity in East-West direction has to be determined very accurately. An uncertainty in the East-West velocity of 0.1 m/s (0.2 knot) corresponds to an uncertainty of 1 mgal in the Eötvös correction. As the determination of the Eötvös correction is seen as one of the main error sources in sea gravimetry, velocity measurements with GPS could help to bring sea gravimetry accuracy down to 1 mgal worldwide.

4. CONCLUSIONS

The NAVGRAV project yielded a large amount of data. All data have been catalogued and stored in a data-base. Preliminary computations show that real time differential GPS at sea can bring position accuracy down to the meter level. Sea gravimetry will benefit from the velocity determination capabilities of GPS.

Phase errors caused by the ship-antenna combination seem present, but need further investigation.

Actual processing of the data is presently underway. Interest is mainly focussed on differential GPS with a reference station on land and sea gravimetry. The gravity data will be processed independently at Delft University of Technology and Hamburg University. Further detailed investigations (e.g. concerning the radio positioning systems) will be performed in the near future.
All research results will be published.

REFERENCES

De Jong, C.D. and de Munck, J.C., 1986. "Computing the heading effect at M.F. and L.F. position fixing." Paper presented at the INSMAP International Symposium on Marine Positioning, Reston, USA.
Rummel, R., Strang van Hees, G.L. and Versluijs, H., 1983. "Gravity Field Investigation in the North Sea", ch. 23 in:"Satellite Microwave Sensing", (ed. T.D. Allen). John Wiley & Sons, New York.

Seeber, G., Schuchardt, A. and Wübbena, G., 1986. "Precise Positioning Results with TI 4100 GPS Receivers on moving Platforms." Paper presented at the Fourth Symposium on Geodetic Satellites, Austin, TX.

Strang van Hees, G.L., 1983. "Gravity Survey of the North Sea." Marine Geodesy, Vol.6, No. 2, pp 167-182.

APPENDIX

List of participants in the NAVGRAV project

-Delft University of Technology, Department of Geodesy	Delft	the Netherlands
-Hydrographic Service of the Royal Netherlands Navy	Den Haag	the Netherlands
-Rijkswaterstaat, Survey Department, section Marine Geodesy	Delft	the Netherlands
-Shell Internationale Petroleum Maatschappij	Den Haag	the Netherlands
-Intersite Surveys	Haarlem	the Netherlands
-Osiris Seaway	Heemstede	the Netherlands
-NeSA	Rotterdam	the Netherlands
-NHS	Leidschendam	the Netherlands
-Radio Holland	Amsterdam	the Netherlands
-Oretech	Vijfhuizen	the Netherlands
-Utrecht University, Institute of Earth Sciences	Utrecht	the Netherlands
-Hannover University, Institute of Geodesy (IFE)	Hannover	FRG
-München University of Technology, Institute of Astronomical and Physical Geodesy	München	FRG
-Hamburg University, Institute of Geophysics	Hamburg	FRG
-Norwegian Hydrographic Service	Stavanger	Norway

MAPPING NUCLEAR CRATERS ON ENEWETAK ATOLL, MARSHALL ISLANDS

John C. Hampson, Jr.
U. S. Geological Survey
Woods Hole, MA 02543

ABSTRACT

In 1984, the U.S. Geological Survey conducted a detailed geologic analysis of two nuclear test craters at Enewetak Atoll, Marshall Islands, on behalf of the Defense Nuclear Agency. A multidisciplinary task force mapped the morphology, surface character, and subsurface structure of two craters, OAK and KOA. The field mapping techniques include echo sounding, sidescan sonar imaging, single-channel and multichannel seismic reflection profiling, a seismic refraction survey, and scuba and submersible operations. All operations had to be navigated precisely and correlatable with subsequent drilling and sampling operations.

Mapping with a high degree of precision at scales as large as 1:1500 required corrections that often are not considered in marine mapping. Corrections were applied to the bathymetric data for location of the echo-sounding transducer relative to the navigation transponder on the ship and for transducer depth, speed of sound, and tidal variations. Sidescan sonar, single-channel seismic reflection, and scuba and submersible data were correlated in depth and map position with the bathymetric data to provide a precise, internally consistent data set. The multichannel and refraction surveys were conducted independently but compared well with bathymetry. Examples drawn from processing the bathymetric, sidescan sonar, and single-channel reflection data help illustrate problems and procedures in precision mapping.

1. INTRODUCTION

As navigation techniques and technologies improve, other sources of error in marine mapping become more important components of the total positioning error. A task force assembled to study two nuclear craters at Enewetak Atoll, Marshall Islands, encountered many of these sources of error while doing a full-spectrum marine survey in a tightly-controlled miniranger navigation net on the atoll.

At the request of the Defense Nuclear Agency (DNA), the U.S. Geological Survey (USGS) completed an intensive study of OAK and KOA craters at Enewetak Atoll (fig. 1) by using marine geophysical surveys and bottom sampling. The work was conducted during summer 1984 in three phases: single-channel acoustic-reflection and sidescan surveys; scuba and submersible sampling; and multichannel-reflection and refraction surveys. The results were released as USGS Bulletin 1678 (Folger, 1986). DNA also requested that the USGS analyze samples collected during a related study, an extensive drilling program carried out in 1985. Only the data collection and reduction techniques for the marine survey are discussed here.

A case history of the first phase of the marine survey illustrates the problems encountered and how they were handled. These are problems typical of precision mapping with several different data collection systems and high data density.

M. Kumar and G. A. Maul (directors), Marine Positioning, 249–258.
© *1987 by the Marine Technology Society.*

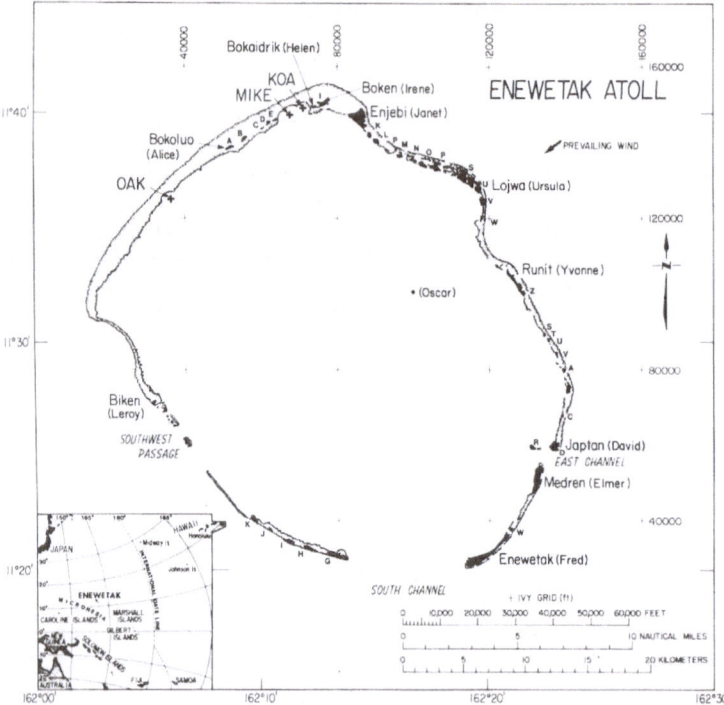

Figure 1.   Index map of Enewetak atoll. Craters are shown with crosses.
             Native names for larger islands are shown. Military names are in
             parentheses.

2. ENEWETAK MAPPING

     The object of the study was to depict in great detail the surface
morphology and geology and the subsurface structure of two nuclear test sites,
each within survey areas less than 3 km by 3 km (1.6 nm by 1.6 nm).   The
survey required horizontal precison in the range of 3-4 m (9.8-13.1 ft) for
the high-resolution geophysics and vertical accuracy of less than 1 m (3.3 ft)
for the bathymetry.

     The three phases of the field work were each about 1 month long. The first
phase included echo sounding, sidescan sonar imaging, and single-channel
reflection profiling; the second phase, scuba and submersible observations and
sampling;   and   the   third   phase,   multichannel   reflection   profiling   and
refraction.   Data collected during phase one were processed quickly to provide
preliminary base maps for the next two phases.

     Geophysical operations were carried out aboard three different vessels: a
41-m (135-ft) mudboat, MS EGABRAG II; a 15-m (50-ft) landing craft, the MIKE
boat; and a 6.4-m (21-ft) Boston whaler.   EGABRAG II was the primary survey
platform, with the MIKE boat and the whaler providing shallow-water surveying
to about 2 m (6.6 ft).

2.1. _Navigation_. Navigation for the Enewetak survey was provided by Meridian Ocean Systems (MOS). A Hewlett Packard[1] HP85 computer, linked to a Motorola Falcon IV Miniranger, calculated positions at 20-second intervals. Fixes were based on four transponder stations located around the atoll on towers up to 21.5 m (70 ft) high. Latitudes and longitudes for the towers were established with Motorola Satellite Surveyor systems using Transit satellites. One tower, located on a preexisting survey marker, was used as a master station, and the other towers were translocated.

To maintain consistency with earlier pretest and posttest surveys, positions were reported in coordinates based on the Enewetak IVY grid. The IVY grid was introduced during pretest surveys in the early 1950's. It is an origin-sourced geographic rectangular grid not a projection grid. Its origin is a large pinnacle reef (point OSCAR in figure 1) in the Enewetak lagoon. The X and Y coordinates at the origin are designated as 100,000 ft E by 100,000 ft N, permitting all positions in the atoll to be positive values. The grid is oriented by aligning the Y axis at point OSCAR with astronomic north. For this project, a modified Transverse Mercator grid on the Hough spheroid, Wake-Enewetak 1960 datum, is used to approximate the IVY grid and to permit transformations between IVY coordinates and latitude and longitude.

An independent survey by the Defense Mapping Agency (DMA) (Woodworth, 1985) yielded latitudes and longitudes and IVY coordinates for the miniranger transponder towers within 4 m (13 ft) of those used for this survey. In 1985, the DMA also processed satellite navigation data gathered by MOS to position each tower. They recalculated positions based on the World Geodetic System 1972 (WGS-72) datum. Changing from Wake-Enewetak 1960 datum to WGS-72 datum results in a shift of all latitude-longitude pairs for tower locations of 118 m (388 ft) to the southwest. After that vector offset, residual position corrections to the towers are less than 0.6 m (2 ft) in all cases.

The IVY grid, being origin sourced, is not shifted by updates to the geoid. The transformation between IVY coordinates and latitude and longitude is updated simply by changing the latitude and longitude assigned to point OSCAR (the origin). On maps with the IVY grid, mapped features and the IVY grid remain fixed while the latitude-longitude graticules are moved. For this survey, all maps were printed with the IVY grid and with latitude and longitude based on the Wake-Enewetak 1960 datum.

2.2 Bathymetric Survey.

2.2.1. _Field work_. Most of the echo-sounding data were acquired from the MS EGABRAG II during phase one of the marine work. Echo-sounding data were gathered during a dedicated survey, the sidescan survey, and all seismic profiling. Shallow-water data were acquired with the Boston whaler and during a shallow-water sidescan survey from the MIKE boat. Additional bathymetry was acquired during the multichannel profiling in phase three. Average trackline spacing was about 25 m (82 ft).

The bathymetric data were collected by using a Raytheon DE-719B Survey Fathometer with a 200-kHz transducer. Soundings were recorded on the fathometer paper record. They also were recorded and processed digitally by

---

[1]The use of trade names does not constitute endorsement by the U.S. Geological Survey or the Defense Nuclear Agency.

the navigation computer, but the need to integrate bathymetry from the whaler and the great detail we were seeking led us to digitize the paper records in the laboratory. The same fathometer was used with a portable transducer on the MIKE boat and on the whaler.

Tidal variations were recorded with Sea Data tide gauges in both craters to make tide corrections to the bathymetry. Barometric pressure was recorded every 10 min by an Atmospheric Instrumentation Research digital barometer to correct the pressure changes recorded by the tide gauges. Barometric pressure changes proved to be insignificant relative to the 1.5-m (5-ft) tidal variation during this study, and no barometric corrections were needed.

2.2.2. Field corrections. The fathometer permitted a field correction for transducer depth. This was set at 4 ft (1.2 m) for the EGABRAG II, and at 2 ft (0.6 m) for the MIKE boat and the whaler. The fathometer also allowed for calibration based on sound velocity in water. Depth data were displayed in feet. Conversion from the two-way travel time of the acoustic pulse to depth in feet was based on a speed of sound in water of 4800 ft/s (1463 m/s). This calibration could be changed, but we left it at the standard conversion and made velocity corrections in the laboratory.

The recorded position was corrected from the location of the navigation transponder to the location of the depth transducer. On the EGABRAG II this was a correction of 71 ft (21.6 m), 13 degrees off the ship's heading. As a result, the ship's heading was entered into the calculation at the beginning of each new line and at major course changes. The navigation transponder was located in close proximity to the depth transducer on the small boats, and no correction was required.

2.2.3. Laboratory processing. Figure 2 is a flow chart showing the procedure used to improve accuracy in the bathymetry. The bathymetry was digitized visually to record all maxima and minima in their precise locations, and the digital data were corrected for tidal variations and the velocity of sound in water and merged with editted navigation.

Column 1 of figure 2 shows the procedure used to create the raw digital depth files. Data were digitized on a digitizing table by using a USGS package of seismic data analysis programs. Position along the track was recorded as Greenwich Mean Time (G.M.T.), to be correlated with the ship's position at that time. Depth was recorded as raw soundings in tenths of a foot. To capture small topographic features that also might be recorded by the sidescan, soundings were digitized every few meters, or every 1 to 4 seconds clock time for a ship's speed of 4 knots (2 m/s). The digitizing programs, designed for larger scale surveys, required modifications to their levels of precision to accommodate these vertical and horizontal scales. Digitized profiles were plotted for comparison with the paper records.

The tide data were recorded as changes in pressure versus G.M.T. every 20 minutes. A standard conversion of 1 mbar/cm of seawater was applied. The tidal maxima and minima were correlated with tidal maxima and minima relative to Mean Lower Low Water (MLLW), predicted by the National Oceanic and Atmospheric Administration for Enewetak during the same time period. This permitted conversion of the relative tidal variations into deflections from MLLW. Tidal variations in each of the two craters were nearly identical, and the tide corrections were based on data from KOA crater, where a complete record was obtained.

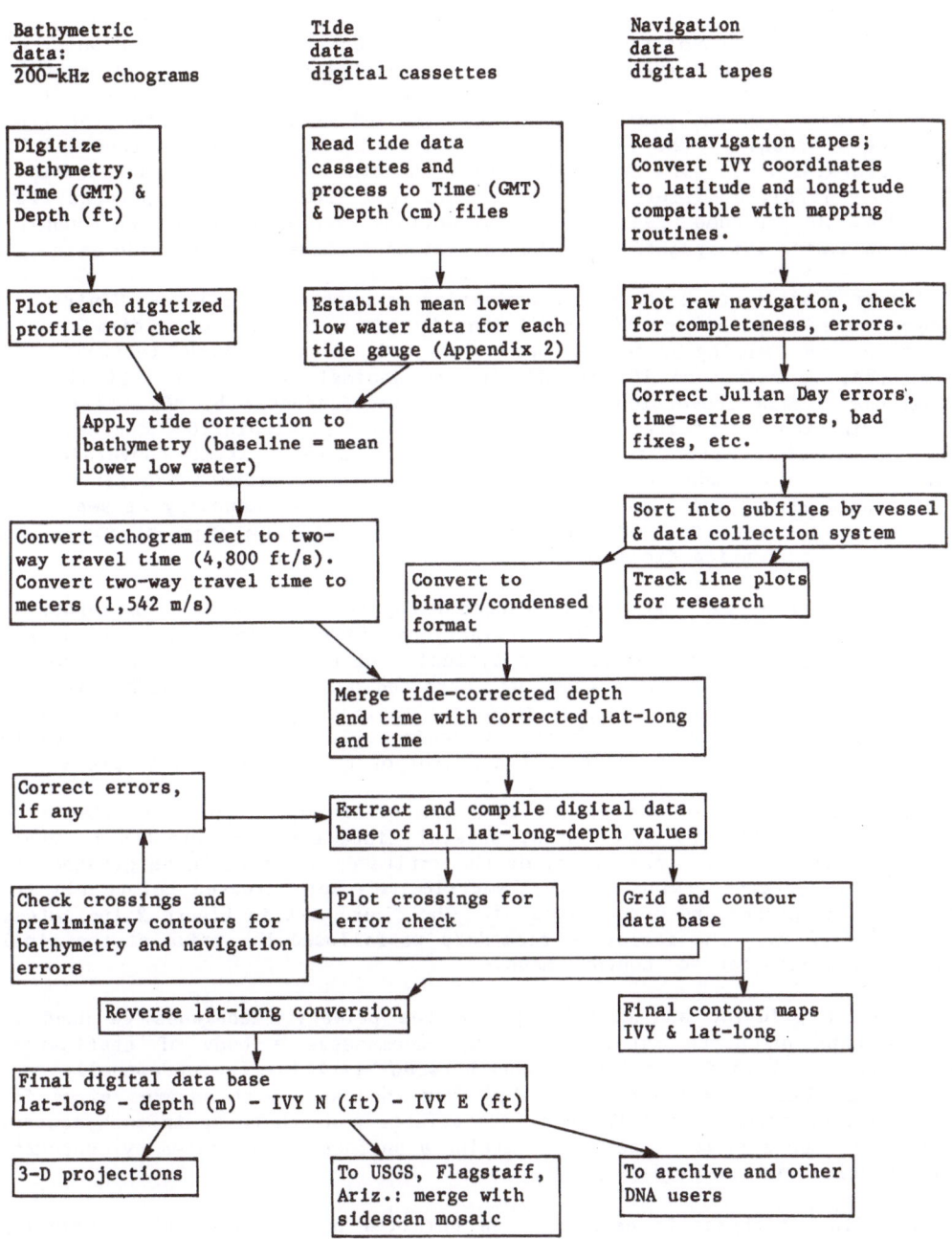

Figure 2. Flow chart showing depth data processing.

The digitized bathymetry was converted to tide-corrected true depth in one program. The factor for converting echogram feet to true meters uses the assumed-velocity on the fathometer recording of 1463 m/s (4800 ft/s) to calculate two-way travel time. For conversion back to depth, the velocity of sound in seawater initially was assumed to be 1500 m/s (4921 ft/s), a standard oceanographic conversion. A difference of 1-m between our depth for the floor of OAK crater and the measured depth (60 m) led us to correct for the true velocity of sound in the lagoon, derived from water temperature recorded by the submersible. In warm tropical waters, the true velocity proved to be 1542 m/s (5059 ft/s), a difference of 3 percent from 1500 m/s and 5 percent from 1463 m/s. In the same program, tide corrections were applied to each sounding based on linear interpolation of the correction factors from the tide data.

Column 3 of figure 2 shows the navigation processing procedure. Miniranger ranges were computed in IVY coordinates by MOS. Because our seismic data analysis and mapping programs were designed to work with latitude and longitude, we converted IVY coordinates to decimal degrees of latitude and longitude. This conversion was calculated independently by MOS and by the USGS navigation and gravity team. The results agreed within 10^{-6} degrees (0.1 m), the level of precision used in the data base. After conversion to latitude and longitude, the navigation was plotted to locate bad fixes, and typographic errors such as Julian day which was updated manually at sea. The navigation also was run through a program to locate any adjacent times and positions that implied excessive speeds.

The navigation and the tide-corrected bathymetry were merged in a routine that interpolated between the 20-s navigational fixes to derive a position for each sounding. As an example of a typical problem encountered in improving the precision of a mapping system, we corrected a subroutine embedded in this program that did the interpolations in radians carried to six decimal places. Calculating in radians to six decimal places, instead of degrees to six decimal places, yields more than a factor-of-fifty reduction in precision.

Once the navigation was merged with the water depth, and the files were organized on a line by line basis, several lines could be plotted to check correlations at profile crossings, or the entire data set could be gridded and contoured. Several preliminary bathymetric maps were created before all the data were processed and errors located. The flow chart in figure 2 indicates, in a general way, the iterative procedure we followed to arrive at the final bathymetric data set and contour maps.

Computer gridding and contouring are the primary techniques we used to create maps of depth data. Gridding encompasses a body of statistical techniques used to interpolate depths between tracklines. Conceptually, one selects a piece of graph paper to lay over the data, with squares sized for the level of detail desired, then interpolates a depth to go in each square. Contouring the grid is a matter of fitting a surface, represented by contours, to the values in the grid.

The final bathymetric data set was compiled on magnetic tape, including both IVY coordinates and latitude and longitude. The data were presented as 1-m contour maps, as shown in figure 3, and some preliminary comparisons were done with bathymetry gathered before nuclear testing at the OAK crater site. Further processing will be done by the Flagstaff, AZ, office of the USGS to compare postevent bathymetry with preevent bathymetry and sidescan sonar data.

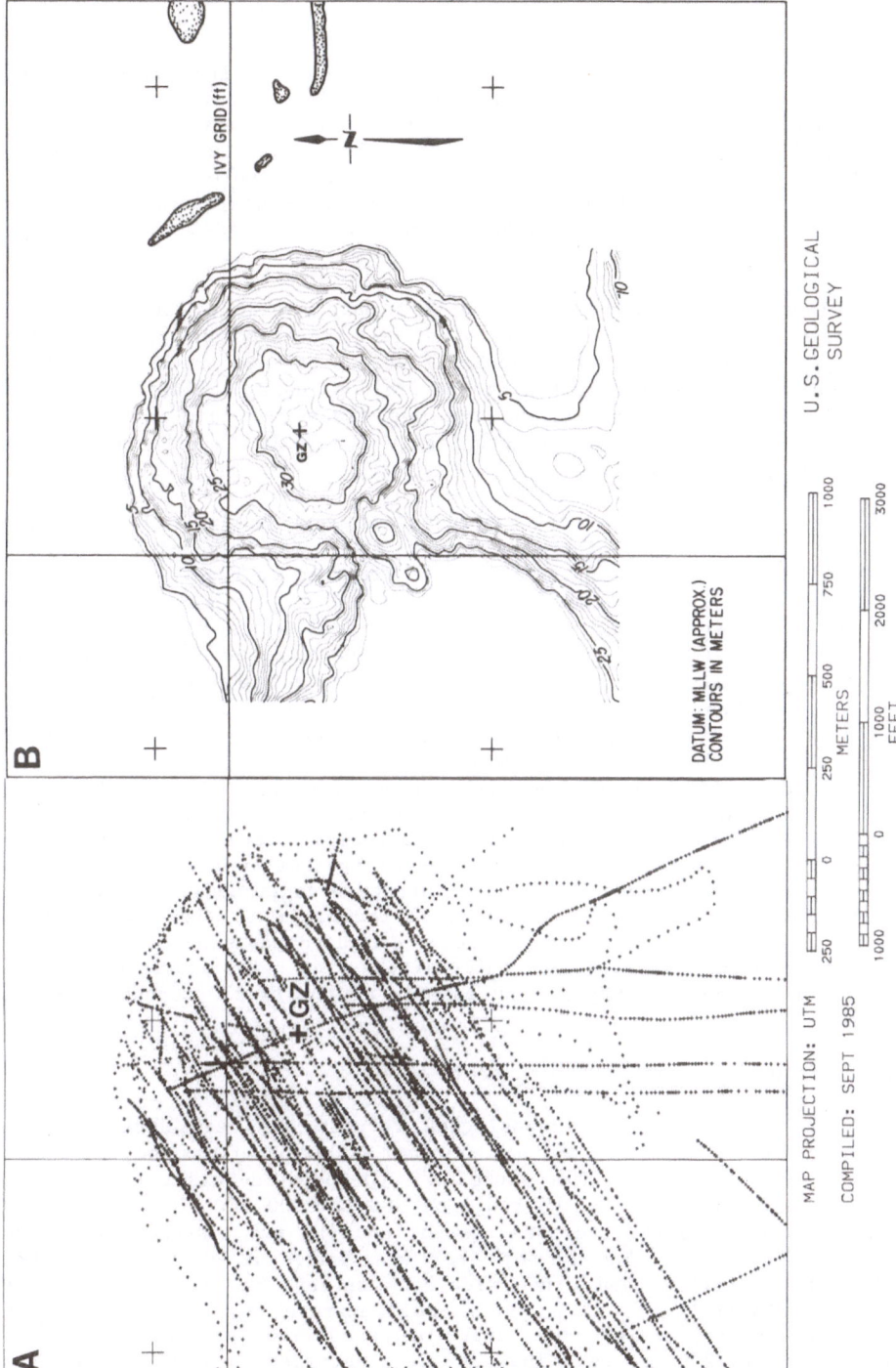

Figure 3. A. Point plot of data used to construct bathymetric map. B. 1-m bathymetric map of KOA crater. GZ is ground zero.

2.3.   Sidescan Sonar.

2.3.1.   Field survey.   The sidescan sonar data were acquired with a Klein
Model 422S-101 AF transducer unit operating at an acoustic frequency of 100
kHz and a nominal resolution of 0.15 m (0.5 ft).   The transducer was mounted
in a 545-kg (1,200-lb) fish towed by the EGABRAG II for all but the shallowest
portions of the sidescan mosaic.   The same system in a lighter fish was towed
behind the MIKE boat in shallow water.   The data were displayed on a Klein
531T wet-paper recorder, and the analog signals were recorded on 1-in. tape
with a Honeywell 14-track tape recorder.

The tracklines were spaced 75 m (246 ft.) apart, with a 100-m (328-ft)
swath of image to each side of the towed fish recorded on paper.   This
provided complete coverage from both look-directions between tracklines.   As a
result, a mosaic could be constructed in the field by using a consistent look-
direction, simulating illumination of the sea floor from one side.   These
mosaics were photographed and made immediately available to submersible and
scuba crews during phase two.

2.3.2.   Field corrections.   The real time acquisition of position data allowed
the Klein sidescan to display slant-range corrected and orthogonally justified
images.   Both corrected data and raw data were recorded on tape.

2.3.3.   Laboratory processing.   In the laboratory at Woods Hole, the analog
sidescan tapes were digitized on a MASSCOMP MC-500 computer.   The digitizing
interval yielded cross-track pixel dimensions of about 0.075 m (0.25 ft).
Because the scan lines were gathered at 200-ms intervals and the typical ship
speed was 4 kts (2 m/s), the along-track pixel dimension is about 0.4 m (1.3
ft).   Digital image enhancement techniques were applied to the data by the
USGS in Flagstaff, AZ.   These techniques removed noise and normalized and
improved overall contrast in the raw digitized data.   A precise slant-range
correction then mapped pixels into an orthogonally corrected image.   The
digital sidescan images were mosaicked by using navigation and bathymetry to
align the images.   An airbrush was used to create a continuous image with the
seams removed on an overlay.   This airbrush technique is the same used to
create lunar mosaics of orbital image data.   An added advantage was that the
enhanced mosaic could incorporate information from two overlapping look
directions in the sidescan images.   An example of the final product is shown
in figure 4.

2.4.   Single-Channel Survey.

2.4.1.   Field work.   Because inconsistent results had been obtained from
surface-towed, high-resolution systems during earlier surveys on Enewetak,
this project used a Huntec Hydrosonde DTS (Deep-Towed-Seismic) profiling
system.   The system operates in the range of 500 to 6000 Hz.   It was deployed
on the same tow-fish as the sidescan system, and the two systems could have
operated simultaneously.   The survey area was small enough to permit the
single-channel profiling and the sidescan survey to be carried out
consecutively.   In that way, each system could operate at its most efficient
depth, without any cross talk.   The Huntec profiles were collected at a 0.5-s
firing rate and a 0.125-s sweep.   The data were recorded on an EPC 3200 paper
recorder and as analog signals on 1/2-in. magnetic tape.   The system achieved
excellent subbottom penetration around OAK crater, one-half of which lies in
lagoonal sediments.   Penetration was poor around KOA crater, which lies in
indurated reef rock.

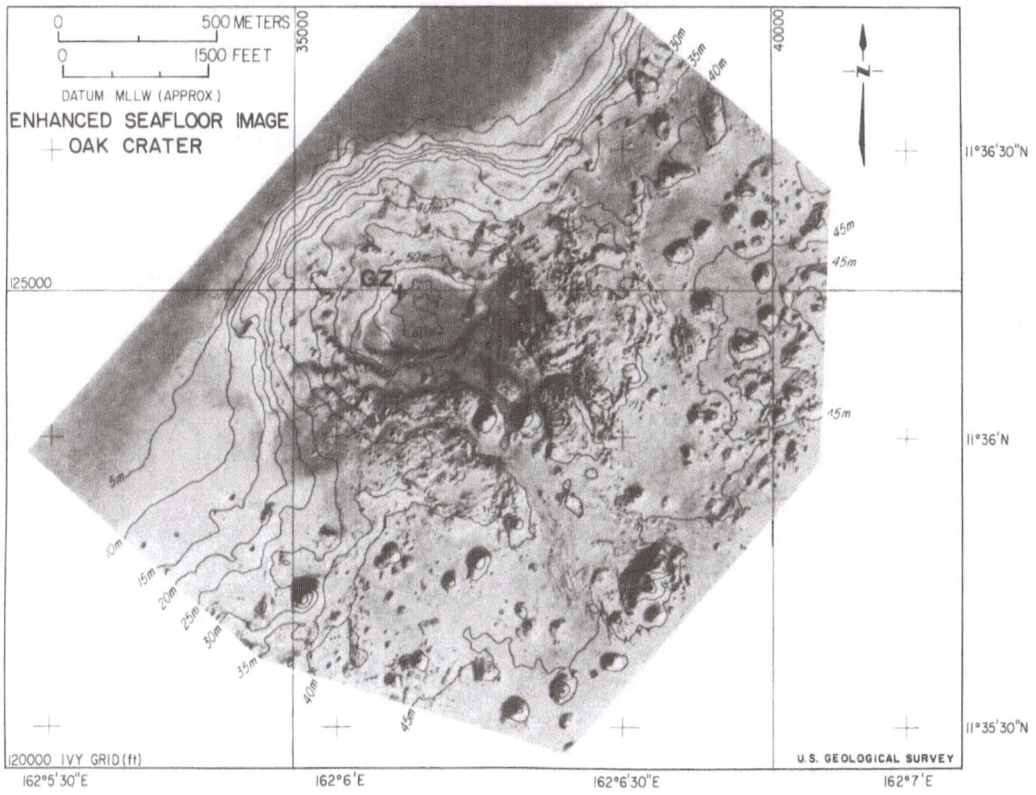

Figure 4.   Airbrush image of OAK crater derived from digital sidescan
             mosaic with bathymetric overlay. Contour interval is 5 m.
             GZ is ground zero.

2.4.2. <u>Field corrections</u>. The data were corrected automatically for fish
depth and heave. The fish-depth correction did not always work properly;
however, we were able to correct for fish depth when we made the position
correction described below.

2.4.3. <u>Laboratory processing</u>. The problem of how to correct the Huntec deep-
towed reflection profiles to a vertical datum and how to correct for fish
position relative to ship's position was solved by integrating the Huntec
profiles with the bathymetric profiles. The digitized bathymetry was printed
on a film overlay at the same vertical and horizontal scale as the Huntec
profiles. By registering the sea-floor reflection found on the Huntec data
with the sea-floor bathymetric profile collected at the same time, the Huntec
data were registered simultaneously to the bathymetric datum and to the
corrected positions described in 2.2.2. Interpretation of the Huntec profiles
was drawn on the film overlays, and the overlays were digitized for computer
processing and plotting of reflector-depth and isopach maps.

3. DISCUSSION AND CONCLUSION

To integrate three different geophysical data sets in phase one of this survey, we established one data set, the bathymetry, with as much accuracy as possible. The other data sets then could be registered to that vertical and horizontal datum.

The bathymetry was corrected in the vertical dimension for transducer depth, tides, and the local velocity of sound in seawater. In the horizontal dimension, the primary correction was from the navigation transponder to the depth transducer. This requires a vector correction that is dependent on the ship's heading (versus the course made good). This correction probably introduces the largest source of possible error in a tightly controlled grid. The simplest solution to the problem is to mount the navigation transponder directly over the depth transducer. This was the solution we used on the MIKE boat and the whaler. It was not possible on EGABRAG II, and a vector correction had to be applied.

To coordinate with phase 2, the submersible and scuba phase, the researchers were provided with large-scale bathymetric maps at 1:3000 and 1:1500. In that way their plots and sample locations also could be corrected and registered to the bathymetric data base.

In phase three of the survey, where the surface-towed multichannel system was the primary profiling system, the multichannel data were positioned directly by vector transformation from the position of the navigation transducer. The bathymetric data collected at the same time had to be repositioned to the depth-transducer in order to be incorporated in the bathymetric data base.

Registering the high-resolution data and dive data to the bathymetry proved to be a quick and effective technique to maintain internal consistency among numerous data sources. In surveys of larger areas, where it is not possible to collect such high-density bathymetric data, awareness of such sources of positioning error may help to modify these techniques or to design others to improve accuracy in integrated marine mapping.

REFERENCES

Folger, D. W. (ed.), 1986. Sea-floor observations and subbottom seismic characteristics of OAK and KOA craters, Enewetak Atoll, Marshall Islands. U.S. Geological Survey Bulletin 1678, Denver.

Woodworth, H. V., 1985. Eniwetok Atoll surveys; memorandum for record. Defense Mapping Agency, San Francisco.

SEA FLOOR MAPPING USING GLORIA DIGITAL TECHNIQUES

James W. Schoonmaker, Jr.
and
Bonnie A. McGregor
521 National Center
U.S. Geological Survey
Reston, Virginia 22092

ABSTRACT

GLORIA imaging sidescan sonar data were collected in 1982 and 1985 over the northern part of the Gulf of Mexico seaward of Texas, Louisiana, and Florida. They were digitally processed using the USGS-developed Mini Image Processing System (MIPS). The radiometrically and geometrically corrected 1982 image strips were film-mosaicked to enhance the interpretation of the sea floor morphology. Digital mosaicking is presently being utilized on all data to produce a marine geology atlas of the Gulf.

An overview of the geologic interpretation and the various technical elements includes descriptions of the amount and types of data gathered; the cruise location; and the MIPS hardware, software, and processing procedure. MIPS processing steps include geometric corrections for water column variations, slant range to ground range projection, differential scale, and ship's velocity changes; and radiometric corrections for shading due to power drop-off, nadir power build-up, speckle noise, and striping. The final step of digital mosaicking the imagery into 2- by 2-degree map quadrangles is discussed.

1. MAPPING PROGRAM

As an initial response to the presidential proclamation establishing the Exclusive Economic Zone (EEZ) over the submerged lands extending 200 nautical miles seaward from the coast of the United States and certain other possessions, the U.S. Geological Survey (USGS) is producing a series of reconnaissance maps of the sea floor of the EEZ area (McGregor and Twitchell, 1985). Data for these maps are being gathered using the GLORIA (Geological Long-range Inclined ASDIC) system, designed and developed by the Institute of Oceanographic Sciences (IOS), United Kingdom. The GLORIA system (as presently configured) can map the sea floor in depths from 150 meters to the deepest parts of the oceans (McGregor and Twichell, 1985). GLORIA sends out an array of beams of acoustic energy to both port and starboard, imaging a swath of approximately 15, 22, or 30 km on each side depending upon the pulse rate, and records the reflected energy. Because the exact swath width achieved varies with water depth as well as the pulse rate, the spacing of adjacent ship tracks is adjusted to allow some overlap of the imagery data to ensure total sea floor coverage.

Navigation of the ship is based on best available data. Loran-C, transit satellite, and GPS are all logged simultaneously when available. During the Gulf of Mexico Survey, Loran-C was used to position the ship. Besides navigation information, water depth and magnetic field data are also logged. A merged file of data with a 2-minute time increment is constructed and includes date, time, latitude, longitude, water depth (corrected and uncorrected for sound velocity), total

M. Kumar and G. A. Maul (directors), Marine Positioning, 259–264.
© 1987 by the Marine Technology Society.

magnetic field, and residual anomaly (McGregor and Schoonmaker, 1986). The digital
sidescan data are recorded separately and merged with the navigation file as part
of the post-cruise processing. A preliminary photomosaic of imagery data is constructed at sea by hand to allow real-time geologic interpretation as the survey
is being conducted.

During early 1982 and summer 1985, most of the continental slope and rise in
the northern Gulf of Mexico seaward of Texas, Louisiana, and Florida was surveyed
using the GLORIA system. Data collection proceeded for 88 days covering about
500,000 square kilometers of seabed.

2. IMAGE PROCESSING

All the digital processing for this GLORIA data was done on the Mini Image
Processing System (MIPS) at the USGS installations in Reston, Va. and Woods Hole,
Mass. MIPS was originally designed as an office research and development image
processing system based on a Digital Equipment Corporation (DEC) PDP 11/23 minicomputer and the RSX-11M operating system. For this processing, MIPS was installed
on a PDP 11/73 minicomputer with the RSX-11M+ operating system and a 470 megabyte
hard disk.

The heart of MIPS is the software. It was an outgrowth of the highly successful
Flagstaff Image Processing System located at the USGS Computer Center in Flagstaff,
Ariz. Consequently, approximately 70 to 80 man-years of image processing software
development were available for MIPS (Chavez, 1984). All MIPS version software,
however, has undergone modifications to upgrade both internal and external documentation and make it more user-friendly.

Many of the programs used to process the data collected by the GLORIA sonar
imaging system were developed by Pat Chavez with the assistance of Jeff Anderson
and others, all with the USGS in Flagstaff, Ariz. They are fully described by
Chavez (1986). The programs include algorithms that are GLORIA-specific and some
that are general in nature. The GLORIA-specific include corrections for slant-
range geometry, water column offset, aspect ratio distortion, variations in ship's
velocity, speckle noise, shading problems, and navigation data corrections. The
nonspecific programs do such tasks as read data from tape to disk, compile histo-
grams, perform low- and high-pass filtering, combine files, and allow viewing of
the imagery.

2.1 Data Preparation The onboard ship navigation data, recorded on magnetic tape,
are the first to be processed. These data are analyzed for excessively large changes in latitude, longitude, or depth, with any exceeding preset conditions being
flagged. The operator then fixes any data gaps or errors, usually with straight
line interpolation between identified good data points. After iterating the navigation data through this process, a file of data with discontinuities removed is
stored for later use during the image processing.

The GLORIA image data, recorded on magnetic tape, are read onto disk on the
MIPS system. The data are compressed from 16 bit to 8 bit logarithmically and
reformatted for compatibility with later programs. The time and date, latitude
and longitude, and bathymetric value are written onto each line of image data.
Straight line interpolation is used to determine values for image lines occurring
between the 2-minute navigation data fixes.

2.2 Geometric Corrections The major geometric problems encountered in GLORIA
data are caused by water column offset, slant-range scale distortion, aspect ratio
distortion, and changes in the ship's velocity. The true nadir pixels are corrected for the offset to the sides caused by the water depth while other pixels in
each across-track record are corrected for the extreme slant-range distortions

caused by the large change in depression angle of the GLORIA system - from approximately 90 degrees at near range to 5-10 degrees at far range. The along-track resolution is determined by the sampling rate and the ship's velocity; variations in either are corrected. When the geometric processing is completed, each pixel represents 50 meters in both directions.

2.3 <u>Radiometric Corrections</u> Radiometric corrections needed include shading corrections for power dropoff from near to far range, a low power problem at near nadir caused by slow buildup of the transmitted signal, speckle noise, and striping noise. The power-dropoff problem, caused by attenuation of the acoustic wave by the water through which it travels, and the low-power-at-near-nadir problem, are cured by a two-pass normalization, resulting in removal of the low frequency horizontal noise pattern in the GLORIA data. Speckle noise is suppressed by applying a small smoothing filter to the image. Striping noise is removed by running both a high pass filter and a low pass filter over the same data and adding the two outputs together.

2.4 <u>Corrected Image</u> After these numerous steps are performed, the result is an image that has been both geometrically and radiometrically corrected. Figure 1 shows the dramatic results, the before and after of image processing.

3. MOSAICKING

The final step in processing GLORIA data is to mosaic the individual images together. Although the digital image processing done to this stage had greatly increased the interpretability of the data, a mosaic is still needed to allow a synoptic interpretation of the entire area. Photomechanical mosaicking was initially used (Warren, 1977) to produce a reconnaissance mosaic (see fig. 2), but a digital method of mosaicking and tone matching GLORIA imagery into desired geographic boundaries and map projection is presently being tested.

4. GEOLOGIC INTERPRETATION

The GLORIA underwater imaging system can image large areas of deep water terrain. In the Gulf of Mexico, features such as submarine channels can be traced meandering for over 100 kilometers across the sea floor (e.g. fig. 2). Through these sinuous, river-like paths, sediment-laden currents are believed to flow, distributing sediments to the ocean floor far from land. Massive debris flows can also be identified covering hundreds of square kilometers of sea floor. Topographic highs and basins formed by the flow of salt mobilized from depth by the weight of overlying sediments can be mapped on the images. The seaward edge of this salt deformation region is marked by an escarpment shown in figure 1b. Image processing and mosaicking make GLORIA a powerful system for evaluating the geology and dynamic processes of the ocean floor.

REFERENCES

Chavez, P.S. Jr., 1984. "U.S. Geological Survey Mini Image Processing System (MIPS)." U.S. Geological Survey Open-File Report 84-880, 12 p.
Chavez, P.S. Jr., 1986. "Processing techniques for digital sonar images from GLORIA." Photogrammetric Engineering and Remote Sensing, Vol. 52, No. 8, pp. 1133-1145.
EEZ-SCAN 84 Scientific Staff, 1986. "Atlas of the Exclusive Economic Zone, western conterminous United States." U.S. Geological Survey Miscellaneous Investigations Series I-1792, Sheet 29, Sonar Mosaic.

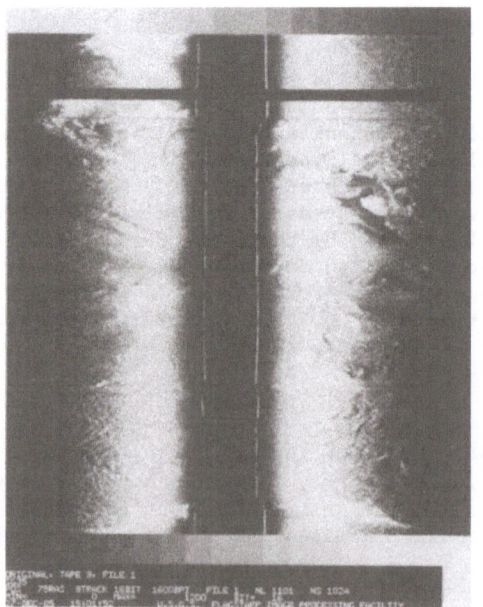

Figure 1.—a. Example of uncorrected GLORIA image data. b. Example of the same data but with full GLORIA processing.

Figure 2.--A photomechanical mosaic sheet from a GLORIA atlas of Pacific Ocean imagery (EEZ-SCAN 84 Scientific Staff, 1986).

McGregor, B.A. and Schoonmaker, J.W. Jr., 1986. "Image processed sidescan sonar
    data: An aid to geologic interpretation." Technical Papers of the ASPRS-ACSM
    Spring Convention, March, Washington, D.C., Vol. 5, pp. 223-232.
McGregor, B.A. and Twichell, D.C., 1985. "U.S. Geological Survey Gulf of Mexico
    GLORIA program." USGS Open-File Report 85-465, 18 p.
Warren, A., 1977. "Mosaicking by photomechanical method." Pan American Institute
    of Geography and History, Quito, Ecuador.

HORIZONTAL DATUM ANOMALIES ON NAUTICAL CHARTS:
A SOLUTION

Oren E. Stembel, Jr.
Nautical Charting Division
Charting and Geodetic Services
National Ocean Service, NOAA
Rockville, Maryland 20852

ABSTRACT

A number of horizontal datums are used for positional control on nautical charts produced by the National Ocean Service (NOS). Inherent to these datums are anomalies which have surfaced over the years that affect the global accuracy of charted information. As more sophisticated electronic positioning systems become available to the mariner, these datum anomalies will become increasingly noticeable. NOS has studied the problem and determined that a new reference datum would be necessary to remove these anomalies. This new datum, the North American Datum of 1983 (NAD 83), has been developed and NOS has decided to adopt NAD 83 as the horizontal datum for the nautical charts produced of the continental waters of the United States, including the Great Lakes, Virgin Islands, Puerto Rico, and Hawaii.

Introduction

Nautical charts are first and foremost navigation tools. They are primarily intended to show routes of safe passage, marine dangers, aids to navigation, and location of port facilities to both large and small craft chart users. Because many charts are published at very large scales, they must be accurate in depicting information used by the mariner to determine a position. A large percentage of the nautical cartographer's time is spent determining the charting control to be used for applying source material to ensure the correct placement of items on the chart. The mariner assumes the chart has an inherent accuracy necessary to promote safe navigation.

At present, NOS uses a number of regional horizontal datums for compiling approximately 970 charts. The vast majority of the charts are compiled on the North American Datum of 1927 (NAD 27).

The NAD 27 has met the needs of the surveying and mapping community for many years, but as surveying technology improved the use of earth centered datums for navigation came into use. Regional differences exist between NAD 27 and earth centered systems. These differences were found to be in the magnitude of tens of meters along the coasts, over 100 meters in Alaska, and even greater in the Hawaiian Islands when compared to other world datums. It was apparent that a new North American Datum was required. The National Geodetic Survey Division (NGSD) has developed a new datum called NAD 83. This datum differs from NAD 27 in that a new reference ellipsoid has been used and it is an earth-centered system. The final adjustment positions on this new datum have recently become available.

Impact on Nautical Charting

In 1984, working with preliminary adjusted figures, the Marine Chart Branch (MCB) of NOS initiated a study to examine the impact of the new datum on its nautical

M. Kumar and G. A. Maul (directors), Marine Positioning, 265–273.

charts. The amount of shift required between the chart's present datum and
NAD 83 was determined. An assessment was made of this shift for each chart to
determine when the chart's projection would require shifting for the chart to be
on NAD 83, or because of chart scale and datum shift, only a revision of the
chart datum note would be necessary.

The study found, overall, the smallest shifts occur in the Great Lakes region
while the greatest shift in both latitude and longitude is in the Aleutian
Islands of Alaska (in excess of 150 meters along both axes). In general, the
range of shift varies from 0 to 25 meters in the Great Lakes, 15 to 50 meters
along the Atlantic coast, 20 to 40 meters along the coast of the Gulf of Mexico,
80 to 100 meters along the Pacific coast, and 100 to 160 meters in Alaska.

Figures 1 and 2 portray graphically the amount of latitudinal and longitudinal
shift in meters between NAD 27 and preliminary NAD 83 values for the coastal
areas of the continental United States, including the Great Lakes.

The shift between datums other than NAD 27 ("orphan" datums) and NAD 83 is
greater in most cases. For example, NGSD reports that for charts of the Hawaiian
Islands the shift between the Old Hawaiian Datum and NAD 83 will be approximately
360 meters in latitude and 285 meters in longitude. The resultant shift between
the Puerto Rico Datum and NAD 83 is about 220 meters.

Once the probable magnitude of datum shift has been determined, it was necessary
to define the criteria that would determine whether the chart projection should
be shifted cartographically. In developing these criterion, the existing line
weights for projections and other charted features were considered. Since the
standard width for a chart projection line is 0.1 millimeters, or 0.004 inch, it
was decided that a projection would be revised whenever the datum shift resulted
in a space between the old and new projections equal in width to one projection
line. This meant that whenever the datum shift at chart scale equaled or
exceeded 0.2 millimeters the projection would be reconstructed; when less than
0.2 millimeters, only a revision to the datum reference note would be required.
The 0.2 millimeter is the centerline-to-centerline dimension between the old and
new projection lines with a 0.1 millimeters space between the lines.

Using this 0.2 millimeter projection shift as criteria, a 1:50,000-scale chart,
for example, would have its projection revised when the datum shift exceeds 10
meters.

Figure 3 shows the relationship between chart scale and datum shift, or the
condition where the user of NAD 83 data would incur a 0.2 mm plotting error if
the existing chart projection was not shifted. Table 1 is a tabulation of the
charts by scale, including controlling insets, which were evaluated against the
0.2 mm projection change criteria. This evaluation identified those charts that
would require only a revision to the existing datum reference note. The
remainder would require a revision of the chart projection. More than 80 percent
of the charts and chart panels are compiled at a scale larger than 1:100,000.
Since most coastal areas have a total shift in either latitude or longitude
greater than 20 meters, it was determined that more than 90 percent of the charts
would require a projection shift to convert to NAD 83.

Once the magnitude of the problem was determined, the constraints to a full-scale
implementation program were examined. These include the State plane coordinate

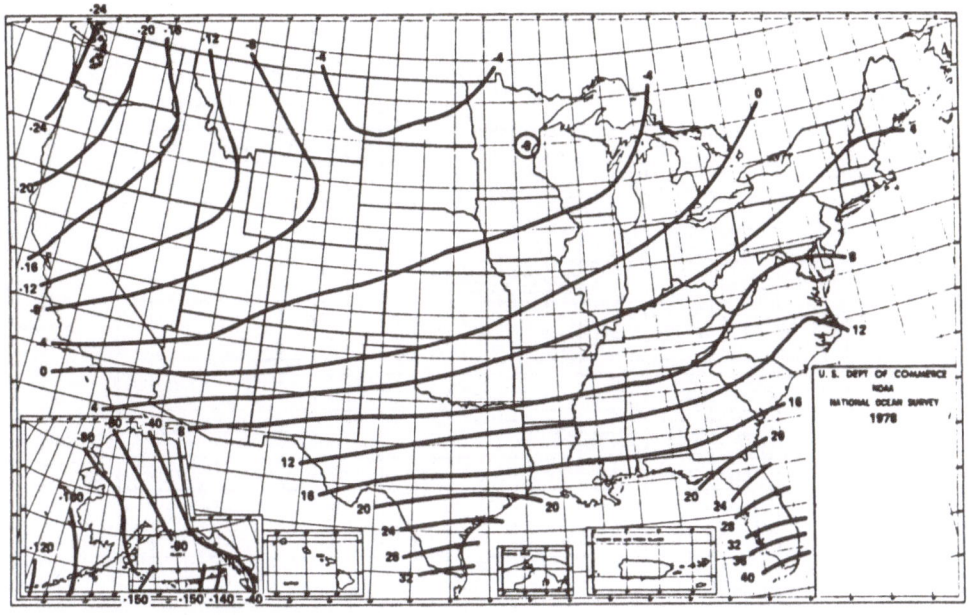

Figure 1.--Expected latitude change from NAD 27 to NAD 83 (in meters).

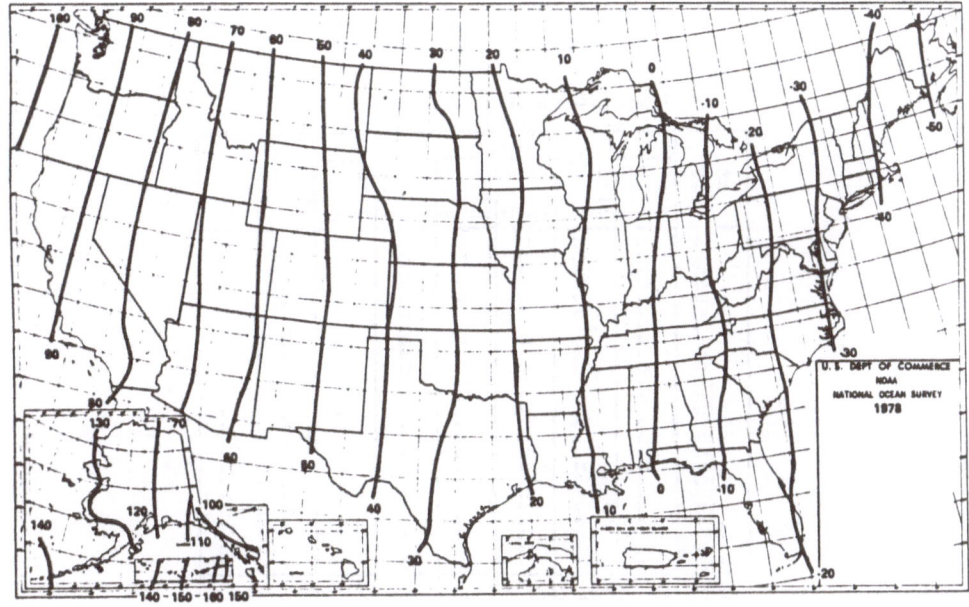

Figure 2.--Expected longitude change from NAD 27 to NAD 83 (in meters).

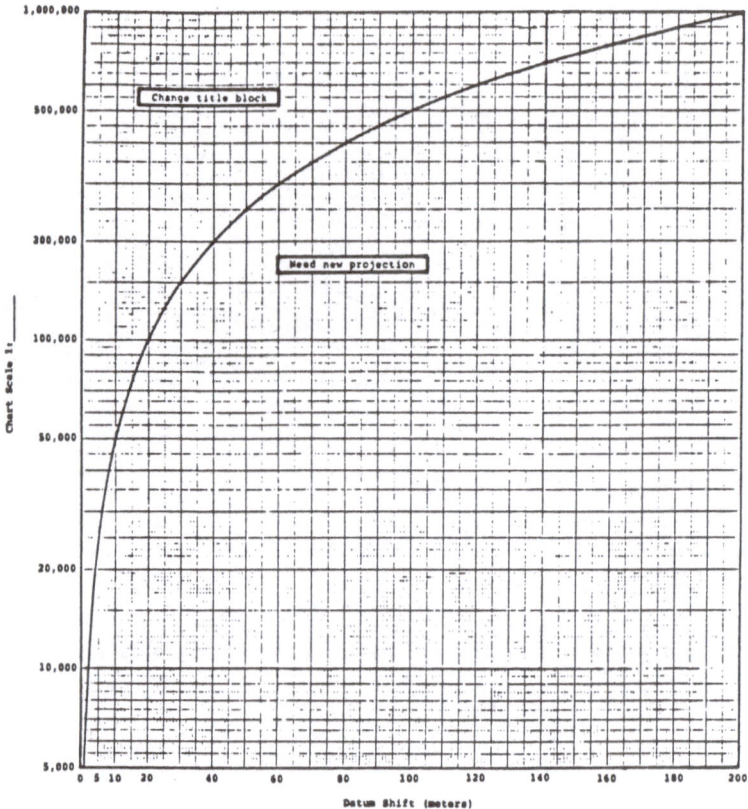

Figure 3.--Plot of projection shift criterion.

Table 1.--NOS charts/insets: summary by scales

Scale (1:K)	Charted area						Total	
	Atlantic Gulf		Pacific		Great Lakes			
	Charts	Insets*	Charts	Insets*	Charts	Insets*	Charts	Insets*
5	4	9	11	24	4	19	19	52
10	44	63	39	84	15	4	98	191
15	14	14	8	13	18	27	40	54
20	64	77	61	73	3	8	128	158
25	7	7	17	16	4	2	28	25
30	3	4	10	9	19	10	32	23
40	126	106	93	66	24	20	243	192
50	6	5	42	40	1	0	49	45
60	1	1	-	-	7	2	8	3
80	89	75	57	34	17	2	163	111
100	5	2	14	7	3	1	22	10
120	-	-	1	0	19	2	20	2
150	4	4	6	4	1	0	11	8
200	2	2	21	16	-	-	23	18
250	1	1	3	3	1	1	5	5
300	2	2	11	8	2	0	15	10
400	11	11	1	1	2	2	14	14
500	1	1	-	-	3	3	4	4
600	2	2	11	10	2	2	15	14
1000	3	3	5	4	-	-	8	7
1500	-	-	4	3	1	1	5	4
2000	1	1	2	2	-	-	3	3
3500	-	-	5	5	-	-	5	5
10000	-	-	1	1	-	-	1	1
TOTAL	390		423		146		959	

*Figures represent the number of charts or insets (panels) in the NOS nautical chart suite for the scale shown. Whenever a chart contained multiple scales, insets or panels, the largest scale was counted.

ticks, electronic positioning lattices, "orphan" datums, personnel resources, cost, timeliness, and lost chart production.

Constraints to be Considered

A.   State Plane Coordinate Grid Ticks

The NOS receives an extensive amount of new data each year which is used to revise existing nautical charts or to compile new charts. Not all of the data received are referenced to the same geodetic projection of the chart, but may be referenced instead to a local or State plane coordinate system. For example, even though most of the charts are compiled on a Mercator projection, the majority of new charting data received from the U.S. Army Corps of Engineers (COE) such as channel improvement depths, breakwater and pier construction, etc., are usually referenced to a local or State plane coordinate system. These data would normally be converted to a projection common to the chart before being applied.

When the COE began using the State plane coordinate systems for referencing or controlling their surveying, dredging, breakwater construction, and other harbor and waterway improvement projects, the NOS cartographers soon recognized the benefit of having State plane coordinate grid ticks on the chart to facilitate application of COE data. NOS has printed these ticks along the borders and within selected charts and insets for almost 40 years. NOS currently publishes 236 charts and 187 insets that contain State plane coordinate grid ticks. Generally, the scale of these charts and/or insets is 1:40,000 or larger.

The implementation of NAD 83 would require every nautical chart and/or inset containing State plane coordinate grid ticks to be revised. Some of the reasons for this are as follows.

1.   The State plane coordinate systems currently shown on charts and insets are mathematically dependent upon NAD 27. NAD 83 is based on a different ellipsoid than NAD 27. This ellipsoid is earth-centered rather than the North American "best fit" position for NAD 27. This would result in a shift of the present State plane coordinate systems when the charts are converted to NAD 83. Whether the shift is noticeable will depend on the magnitude of the shift and the scale of the chart or inset.

2.   The NAD 83-dependent State plane coordinate systems will have grid values published in meters. The present NAD 27-dependent State plane coordinate grid values are published in feet. The impact of the units' change will be evident in the rounding of metric coordinate values assigned to the initial point in each State plane coordinate system. For example, if the longitudinal false easting value for the initial point in a Lambert Conformal Projection Zone was 2,000,000 feet under the NAD 27 system, a value of 609,600 meters would result for the NAD 83 system if a true conversion of units was performed. However, to preserve easy-to-use values for the new metric grid system, a value of 600,000 meters might be selected. This rounding off in selecting values for the initial point in each State plane coordinate system would require every State plane coordinate system charted to be recompiled.

3. Many of the State plane coordinate zones will change numerical values because states have selected new false northings and false eastings that are radically different from the simple feet to meter conversion. For example, most NAD 27 false northings were selected to be 0 feet. Many states are choosing false northings of 100,000 meters, 200,000 meters, or 1,000,000 meters for the initial point in the new NAD 83 State plane coordinate grids.

To assess the impact of and justification for converting the charts and insets to the new State plane coordinate values, three questions had to be addressed: First, the timetable for when the preponderance of NAD 83-dependent State plane coordinate referenced data would begin to arrive from suppliers such as the Corps of Engineers. Second, whether there will be an orderly and timely transition by all COE districts. Third, recognizing that the State plane coordinate grid ticks are shown for the primary use and benefit of NOS cartographers. Therefore, is the cost of correcting the existing charts justified when evaluated against the benefit derived by the cartographers?

Discussions with COE personnel indicated that it would be close to 10 years before they begin extensive implementation of the NAD 83 State Plane coordinate control in surveying and channel maintenance programs. The discussions also revealed that they had come to rely on the state plane grid ticks we had added to our charts and requested that we retain them in the future. This creates a dilemma. The ticks were originally added to aid the NOS cartographers applying source data whose control was based on the grid system. COE primarily provides that information, and is, apparently, now a user of those same grid ticks. However, those grid ticks represent a coordinate system based on NAD 27. To retain those ticks on charts whose horizontal datum is NAD 83 is not good cartographic practice. The final decision is yet to be made on this question.

B. Electronic Positioning Lattices

Because of the great distances between the master and slave transmitting stations comprising a LORAN or OMEGA chain, it is expected that the datum conversion values for the position of each transmission station will vary. What is not known at present is how these variations will affect the hyperbolic lattice for those charts which are located in the outer reaches of the lattice. It is possible that many of the LORAN and OMEGA lattices will have to be recompiled and this might require assigning a higher production priority for these charts. At present, NOS publishes 248 LORAN charts, 22 additional charts containing both OMEGA and LORAN lattices, and 2 charts containing only OMEGA. The same 0.2 millimeter criteria used to justify a projection shift will be employed in assessing the need to shift or replot an electronic lattice.

C. "Orphan" Datums

Charts which are not compiled on NAD 27 are classified as being compiled on "orphan" datums. Of the approximately 135 charts on "orphan" datums, 51 charts of the Great Lakes are currently published on the U.S. Standard Datum. An additional 15 charts are published on astronomic datums (mainly in Alaska and some remote Pacific Islands), 36 charts are on the Old Hawaiian Datum, and 29 charts are on the Puerto Rico Datum. In addition, two charts are compiled on unknown datums, and a recreational chart is published without a datum projection. An assessment will have to be made of each of these charts to determine if NAD 83 datum conversion can be achieved with the information currently available. It

may be necessary to request special field survey support to obtain NAD 83
positions for prominent features on several of these orphan datum charts before
new chart projections can be constructed.

Even though NAD 83 can be extended to the NOS charts of other Pacific Islands,
such as Guam and Samoa, the Director, Charting and Geodetic Services, decided
that these charts will be compiled on World Geodetic System 84 (WGS 84). For
charting purposes, there is virtually no difference between WGS 84 and NAD 83,
and the use of WGS 84 will satisfy the Department of Defense requirement. It is
expected that the Defense Mapping Agency (DMA) will be able to provide the
information necessary to accomplish the datum conversion in these areas.

D.   Personnel Resources

In the past 2 years, the number of production cartographers has steadily
declined. The remaining small staff of cartographers who were not transferred
will have difficulty maintaining these manually produced charts. A correspond-
ingly similar reduction in the number of negative engravers supporting chart
production has occurred also in recent years.

E.   Dollar Costs

To assess the resources required to convert the NOS charts to NAD 83, the
additional workload associated with the conversion of inset or chart panel
projections must be considered. At a cost of $19.98 per staff hour for
cartographic services and $21.50 for reproduction services, the labor cost of
shifting a projection is estimated to be $1,602 per chart. When material costs
of $60 per chart and $30 per panel are included, the total cost of the
implementation is approximately $1,675 for each chart requiring a projection
shift. (An additional cost of $615 per chart and $207 per inset will be required
to revise the State plane coordinate grid ticks.)

When LORAN lattices need revision, even though the plotting is done by
automation, it is estimated that an additional 32 hours of cartographic services
and 35 hours of reproduction services would be required. Material costs would
add an additional $150. In total, an additional $1,540 would be necessary for
each chart requiring a change in the LORAN or OMEGA lattices.

In summary, it would cost $1,507,000 to shift the projections on approximately
900 NOS nautical charts if accomplished by manual cartographic methods, and
$420,000 to revise the electronic positioning lattices on 272 charts if using
manual and automated methods. It is estimated that an additional $145,000 would
be required to publish new State plane coordinate grid ticks on 236 charts and
$39,000 to revise the grids on another 187 insets.

F.   Automated Cartographic Production Capacity

At present, there are two separate chart production groups. The first compiles
and maintains the charts by hand in the traditional method. The vast majority of
charts are maintained this way. The second group, Automated Cartographic
Production Group (ACPG), maintains the charts in a digital data base environment
called the Automated Information System (AIS). The ACPG has been producing
charts in this manner for slightly over a year with some success. Charts
maintained by this group are updated on a workstation where the cartographer can
manipulate data using a keyboard, mouse, and an electronic video display. The

finished product is a series of photographic positives which serve as the color plates of the current chart as it exists in the data base. There is virtually (theoretically, at least) no manual scribing required.

Each production method has its own strengths and weaknesses. In terms of implementing NAD 83, the traditional manual compilation and scribing methods are very time consuming and labor intensive. On the other hand, implementing NAD 83 using the automated chart production method can be accomplished more quickly and inexpensively by either revising the geographic positions in the data base or applying a shift factor to each chart just prior to obtaining the photographic plates. Although relatively cheap and quick, this method has one major drawback. At the present time, there are only five production workstations with no plans to add more in the future. If this system was used to implement NAD 83, the whole program would take a decade or more to accomplish.

G.   Timeliness

The NOS charts are presently compiled using six horizontal datums. The first charts published on NAD 83 would add the seventh. Due to the nature of our production schedule during the period of transition, there would be many cases where the current edition of adjoining or overlapping charts would be based on different datums. During this time, features on overlapping charts would not be in agreement. Besides this, the large positional difference between the present datum and NAD 83 in some areas make a quick conversion desirable. For these reasons, the time it takes to fully implement NAD 83 should be held to a minimum.

H.   Lost Chart Production

Converting 970 charts to a new horizontal datum is going to have a great impact on manual chart production. Given no increase in cartographic or reproduction personnel, the extra time it will take to revise a chart's control grids (projections) is time that would normally be spent compiling the next chart on the production schedule. Each hour spent on converting charts to NAD 83 pushes every following chart back an hour. The study team estimated that implementing NAD 83 will take an extra 24 cartographic and 35 reproduction hours per chart as well as 20 cartographic and 20 reproduction hours per inset. This does not include revisions to electronic positioning lattices or State plane grid ticks. It was estimated this extra workload would decrease manual chart production by about 10 to 15 percent. This could be alleviated somewhat by hiring more cartographic personnel. This is not likely to happen at this time, however, because of the restricted budget we are presently operating under. There will probably be no change in the future.

The implementation of NAD 83 also would impact automated chart production. If the implementation is accomplished by revising the geographic positions in the data base, it may require the shut-down of the entire AIS for several months while the shift software routines are tested, executed, and the shifted data base thoroughly checked. This would result in a large one-time reduction in chart production. After the data base is revised to NAD 83, some extra time will be required to shift new data being entered into the data base to NAD 83.

If NAD 83 is implemented by applying a conversion factor to all chart data after it comes from the data base, there should be little impact on automated chart production.

The Decision to Implement NAD 83

A careful chart-by-chart examination revealed that 93 percent of the nautical charts have a positional difference that exceeds the 0.2 millimeters threshold set for projection revision. In most cases, the shift was far in excess of 0.2 millimeters. For example, a 1:20,000-scale chart along the North Carolina outer banks region has an error of about 1.5 millimeters, in the San Francisco area the error at that scale is 4.5 millimeters, and in the Aleutian Islands of Alaska, the error approaches 11 millimeters.

In the future, ever increasing accuracy from improved electronic positioning systems, especially those utilizing satellite signals, will be available to the chart user. The magnitude of the positional differences between horizontal datums on the nautical charts will be confusing if not potentially hazardous. Therefore, the need to implement NAD 83 outweighs the costs in terms of money, resources, and lost chart production.

The implementation study team determined that with careful planning and the use of a hybrid manual/automated conversion approach the entire suite of nautical charts could be converted in 5 years. This plan specifies that charts with less than 0.2 millimeters projection shift need only have their datum reference note changed. Further, the AIS data base should be revised to NAD 83 positions as soon as the correction values are available from NGSD, and the maximum use be made of the AIS to convert existing charts. The remainder of the charts would be converted using the manual compilation and reproduction methods with careful scheduling to maximize the resources. Moreover, all new and reconstructed charts must use NAD 83 as the horizontal datum.

This datum shift will be invisible to the majority of the chart users. Nautical chart corner limits are usually determined by the importance and extent of the area of interest to the user. Projection lines are added for control after the basic chart limits have been decided on. When a chart neatline closely coincides with a projection line, the projection line is used as the neatline if it does not degrade the usefulness of the chart. The same policy will be used when adjusting projection lines to NAD 83.

REFERENCES

Stembel, O., Monteith, W., 1985. "Implementation of North American Datum of 1983 in the National Ocean Service Nautical Charting Program." NOAA Report NOS 115, CGS 8.

Vincenty, T., 1979. "Determination of North American Datum 1983, Coordinated of Map Corners (second Prediction)." NOAA Technical Memorandum, NOS NGS-16

THE ROLE OF GPS IN NAVAL OCEANOGRAPHY

R. Adm. John R. Seesholtz[1]
Oceanographer of the Navy
Office of the Chief of Naval Operations
Washington, DC 20390-1800

Thank you. It is a pleasure to be here today.

Yesterday and today you have heard a considerable amount about NAVSTAR GPS. Therefore, I am going to give a very brief overview of where we are today with GPS and then discuss how we plan to use GPS in Naval Oceanography.

"Being brief" always reminds me of a speech teacher I once had. She used to tell us, "When you are speaking in public, there are three things you must <u>always</u> remember. Number one: you should stand up straight and tall so that every can see you. Number two: you should speak loud and clear so that every can <u>hear</u> you. And number three: you should be brief so that everyone will <u>appreciate</u> you!" Today (after this nice luncheon) I am going to try to concentrate on the third one!

First of all, let me give a quick reivew of where we are with GPS and how we have already used the system. GPS presently consists of a seven satellite constellation that provides three dimensonal positioning data for the coastal regions of the United States for roughly three to five hours daily. Although this test constellation has too few satellites to provide other than limited service, I am pleased to say we are using it rather regularly.

It was used to relocate the Titanic--along with loran C that was available 24 hours a day--and it was used in the recovery efforts of the Challenger debris.

With this taste of the system, we are now looking forward to the day when we can count on GPS for continuous service. When will this be? Let me take a minute to review the schedule for GPS.

Originally GPS satellites were to be launched exclusively by the space shuttle, achieving worldwide two-dimensional positioning service (based on a 12 satellite constellation) in May 1987, and a three-dimensional positioning service (based on an 18 satellite constellation) in 1988. Because of the Challenger disaster, the GPS constellation has been delayed operationally two years.

As a result of this disaster, it was decided that subsequent GPS launch vehicles would include both the space shuttle and a new medium expandable launch vehicle (MLV) now under development. Given a projected resumption of GPS launches in January 1989, it is anticipated that worldwide two-dimensional capability will be realized in late 1989, and three-dimensional capability in mid-1990.

DOD is currently recommending that the selective availability feature of GPS be implemented when a 12 satellite constellation becomes operational. When or if the National Command Authority approves this recommendation, the standard positioning and universal coordinated time accuracies will be changed to 100 meters and 300 nanoseconds.

---

[1]Guest speaker's presentation at INSMAP 86 luncheon.

M. Kumar and G. A. Maul (directors), Marine Positioning, 275–277.
© 1987 by the Marine Technology Society.

The Biennial Federal Radionavigation Plan (FRP), jointly signed by both the Secretary of Defense and Secretary of Transportation, prescribes policy decisions regarding implementation, operation, and termination of nationally operated radionavigation systems. Since GPS was developed as an all purpose navigation system to satisfy known navigation requirements well into the next century, the federal radionavigation plan calls for the phase-out of military air use of omega and military overseas loran C both in 1992, transit in 1994, and military land-based tacan and vor/dme in 1997. Although use of maritime omega and loran C within the United States' coastal confluence zone will continue into the next century, the FRP indicates that GPS will become the sole source air navigation aid. It is for this reason that FAA has asked that selective availability implementation provide Standard Positioning Service (SPS) positioning accuracy of at least 100 meters.

Now (still remembering number three!), I would like to discuss a few of the ways GPS will be used in naval oceanography. One of our most important tasks is to conduct surveys. With the advent of GPS, naval oceanography will enter a new era in conducting surveys--hydrographic, oceanographic, magnetic and gravity surveys.

I will discuss the role of GPS in hydrographic and geophysical surveying. How do we plan to use GPS in hydrographic surveying? First, some background information. Currently, nearshore large scale hydrographic surveys are dependent on land based radio positioning systems for horizontal control. The establishment and maintenance of the sites for the positioning system transmitters require a significant amount of ship time which could otherwise be spent in conducting the survey. The use of GPS will result in more efficient operations by reducing shipboard and field station navigational logistics time by 25 percent. Moreover, it will save us over $4,000,000 annually in hydrographic surveying when it is fully operational.

We will also use GPS in our new survey system called HYSTAR (the Hydrographic Satellite Tracking and Recording System). HYSTAR consists of a microprocessor interfaced in a GPS receiver, two echo sounders, a helmsman's display, a plotter, a display/keyboard device and a storage device with removable media. Of the 14 units contracted for, several units have been delivered to the ships and are receiving limited use, due to the small number of satellites available. The results are satisfactory, however, and the data collected by these units are being processed in the field using shipboard computers.

The HYSTAR system is capable of collecting all the GPS data necessary for differential processing to permit post-processing to improve positional accuracy. It is anticipated that the survey ships will receive their full complement of HYSTAR units in 1987, although their use will continue to be limited until GPS reaches its full two-dimensional capability.

Another area in which GPS will play a significant role is that of geophysical surveying with our aircraft. A Geophysical Airborne Survey System (GASS) is being procured for installation aboard our project MAGNET P-3 orion airplane. The new GASS will do away with interfaces for navy navigation satellite system receivers and for loran C and omega and in their combine GPS receivers with inertial navigation systems and altimeters to provide very precise three-dimensional positioning and altitude information.

The GASS operator will control the survey (which survey line to fly, at what altitude and at what speed) through keyboard entries at the operator console.

Appropriate guidance information will then be generated and displayed to the pilots. In addition to providing positioning and guidance for the survey, the GASS will also collect and record data from a variety of geophysical sensors including magnetometers, gravity meters and three different types of airborne bathymetric sensors. GASS becomes operational in 1989.

Project MAGNET, worldwide geomagnetic survey operations will be augmented by satellite deployment in the near future. Expendable, low cost probes designed to be launched into low orbit from the space shuttle are presently being developed and will be deployed when the shuttle operations are resumed. The probes will utilize GPS positioning and altitude, thereby eliminating the need for tracking stations, again realizing operational savings of funds.

Beginning in January 1987 we plan to install a Seabeam System with GPS receivers aboard three of our twelve oceanographic ships. The system will provide real-time digital data acquisition of the multibeam data and navigational data provided by the GPS receiver. Real-time contour charting will be produced from these data after they have been corrected for side beam geometry, ray bending and ship's motion, including navigation. There is also a shipboard post-processing capability including computer-aided editing of navigation and sounding data, as well as an in-house capability of data post-processing.

In addition to these four systems in which GPS plays an almost indispensable role, GPS receivers will be installed aboard the rest of our ships and aircraft (a total of 12 ships and three aircraft). On the four deep ocean survey ships with our existing multibeam system, the GPS receivers will reduce reliance on the very expensive inertial systems, reducing costs and increasing accuracy by an order of magnitude. These installations are scheduled to be completed prior to GPS' attaining full two-dimensional worldwide capability.

Tomorrow morning Dr. Chris Mooers, of the Naval Postgraduate School, will chair a workshop on GPS Applications in Oceanography. He will be discussing other aspects of oceanography: The Application of GPS to Acoustic Tomography, Seafloor and Shipboard Measurements, Drifting Buoys, and Remotely Operated Vehicles. I am looking forward to hearing the results of this session.

In concluding, I would like to say that I believe GPS is one of the most significant technological improvements in the history of naval oceanography. We have probably not had anything so valuable since the invention of the echo sounder. GPS will extend accurate survey positioning for our ships from the present limit of 200-400 nm offshore to cover the entire world. It will replace our medium-range survey positioning systems and will eliminate our shore stations. We will be able to conduct our surveys on a worldwide scale, in all weather situations, and with a high degree of accuracy. GPS is a real technical breakthrough, and it heralds a new, very exciting era for all of oceanography-- indeed, for all of us. Thank you.

THE CANADIAN HYDROGRAPHIC SERVICE EXPERIMENT
IN ELECTRONIC CHART DISTRIBUTION
By: G. Morse, IDC and M. Casey, CHS
October, 1986

1.0   Introduction

The Electronic Chart (EC) is a generic name for a new class of bridge management
systems for use on board commercial shipping and recreational vessels. Its' widespread
use will cause fundamental changes in the way in which Hydrographic Offices (HO's)
like the Canadian Hydrographic Service (CHS) manage and distribute its' products since
to be effective it must allow the user to have some manipulative control over the
presentation of the chart data. This can have obvious catastrophic safety and legal
implications unless Hydrographic Offices retain some control over minimum standard
display sets. Since EC technology is still in its' formative stage, it is difficult to predict
in which form information will be displayed. However, it seems certain that given
progress in the various enabling technologies such as global navigation, data storage,
artificial intelligence and very high speed integrated circuitry, some form of bridge
management system which incorporates an electronic chart-like feature will emerge in
significant quantities. This situation gives the motivation for CHS's interest and
experiments in this field. Interested readers are referred to the Proceedings of the Joint
Workshop,"The Electronic Chart", April 19, 1985, held in Dartmouth, Nova Scotia,
Canada, for more information and an extensive list of references on this subject.

2.0   How The Electronic Chart May Affect CHS

The world of hydrography can be broadly broken down into four basic
sub-functions: Survey, Chart Compilation, Construction and Distribution. Each aspect can
be, and often is, considered an end in itself with its' own specialists and special
interests. The wide scale introduction of electronic charts will affect each of these
areas in a different way, but will likely affect the Construction and Distribution
functions the most. The objective of this paper is to describe some of the work being
carried out at CHS to explore various ways in which the Distribution function can be
carried out.

2.1   Issues

Two of the major distribution issues which must be solved by nearly all HO's (and in
particular for CHS) are:

i)   how to convert the extensive graphical chart source data base to a digital form,
      and
ii)   how to provide the digital file that is the EC source to their clients.

The term data base is not exclusive to digital data but refers to any collection of data
in any form. All HO's, for example, have extensive data bases in the form of graphic
documents. CHS, like other HO's, now faces the challenge of providing their data to
internal and external users in a computer digestible form. Computer assisted compilation
tools like the CARIS system employed within CHS for example, require source data from
hydrographic surveys, public works plans, bridge and power line crossings, etc., in
digital form. In addition to this, the advent of the Electronic Chart brings a demand for
the compiled digital EC files as well.

M. Kumar and G. A. Maul (directors), Marine Positioning, 279–284.

For most countries the size of the existing (non-digital) data base is so enormous that the major challenge of the digital age is seen as the conversion of data from graphic to digital. Conversion is seen as the most important enabling step without which, no further progress can be made. While true to some extent, this attitude fosters a mind-set that the conversion of data is the problem to be faced. In fact there are still a host of major challenges to be faced - what to give consumers and how to present it are two questions that are yet to be answered.

For a country like Canada, the challenges are amplified by the country's vast geographic area and the decentralized nature of the CHS. Four, semi-autonomous regions exist within CHS, each containing both survey and compilation capabilities. Chart distribution is centralized and coupled to a network of about 100 chart dealers handles the majority of sales to customers. Thus, the distribution of charts in any form will always be a formidable issue.

For a number of reasons the distribution of EC files should also be centralized. These reasons range from practical issues like costs, how and what to provide, to those concerning legal liability. The latter will require the assurance that the EC files are identical in each region and at each chart dealer - not a trivial task in the case of a dynamic data base.

Chart corrections pose an additional challenge. To maintain the present level of service, the EC files will have to be fully up-to-date, and incorporate all the latest chart corrections at the moment of leaving CHS. Users must also have access to subsequent changes on a convenient medium.

These distribution problems can be summarized as how to link together the widely dispersed CHS regions and the chart dealers into a united network. A network in which the regional offices are free to draw from and contribute to (a 2-way system) a National Chart Data Base, and the dealers are free to withdraw (a 1-way system) the EC files that they require. In this paper we discuss an experimental private satellite network which we feel has the potential to solve this challenging, but often ignored, problem

2.2   The Experimental Solution

This network is presently being assembled with some new features in an attempt to meet the challenge noted above. It will be used to broadcast complete chart files and as well as data which may be used in chart compilation. Charts will be reproduced on a local electrostatic plotter, but since the system is only experimental, these charts will not be used for navigation. Charts and data may also be edited and used in a variety of ways to determine the viability of such a system for nationwide data distribution. Initially, equipment will be installed in Ottawa and two sites near Halifax and linked by an uplink in central Ontario. These sites were chosen for a variety of reasons but, they could be placed almost anywhere.

Most of the equipment is produced by the International Datacasting Company (IDC), which offers several capabilites including:

1. High Speed, 6.3 Megabits per second
2. Low Error Rate, better than 10^{-9}
3. Security through scrambling and receiver authorization
4. Unlimited number of users
5. Nationwide coverage from one satellite transponder

In addition, some new capabilities are being developed.

One of these capabilities is the Print On Demand (POD) station[1]. This includes the interface of a large electrostatic plotter with an IDC receiver so that full-feature charts may be printed to satisfy local demands. Because this system will be able to produce any chart within thirty minutes of a request, storage and manual updating costs can be kept to a minimum.

The plotter used as part of this system is a Versatec model 3436 which produces 3'x4' charts with 400 dot per inch resolution This color output provides the capability to accurately reproduce nautical charts from files broadcast on the satellite. Material costs for reproduction may actually be lower using the plotter than the current printing process. In remote areas, it may be economical to produce the majority of the charts sold on a local plotter.

The major benefits of such a system can only be realized however, if the chart files can be kept continually up to date so that charts made on the plotter are always current. To accomplish this, a return link for the satellite system is under development. When complete, changes to any file may be transferred from one office to another. Two VAX computer systems currently in use at CHS offices are being interfaced to the IDC satellite system so that data may flow over the return link at 56 Kilobits per second. This network is illustrated in Figure 1.

The chart data base for this experimental system will be stored on an Intel 310 computer at the central uplink. This system will continually broadcast the data in order using an IDC technology known as the "Wheel"(Figure 2). When updates arrive via the return link, the data base will be edited immediately, so the next time that file is transmitted, the changes will be effective. The high data rate of 6.3Mb/s permits rapid availability of any file in the system. If one considers the present case of 150 files in the system averaging 1.5 million bytes the time required to cycle through the data base is given by:

> 8 bits per byte x 1,500,000 = 12,000,000 bits
> times 150 files = 1,800,000,000 bits total
> divided by 6.3 Mb/s = 285.7 seconds or
> less than 5 minutes to transmit all the
> charts in the data base.

---

[1]. "LOOKING AHEAD: New Technology for Publishing the Nautical Chart", D.H.Vachon and T.V.Evangelatos, Canadian Hydrographic Service, Ottawa, Ontario

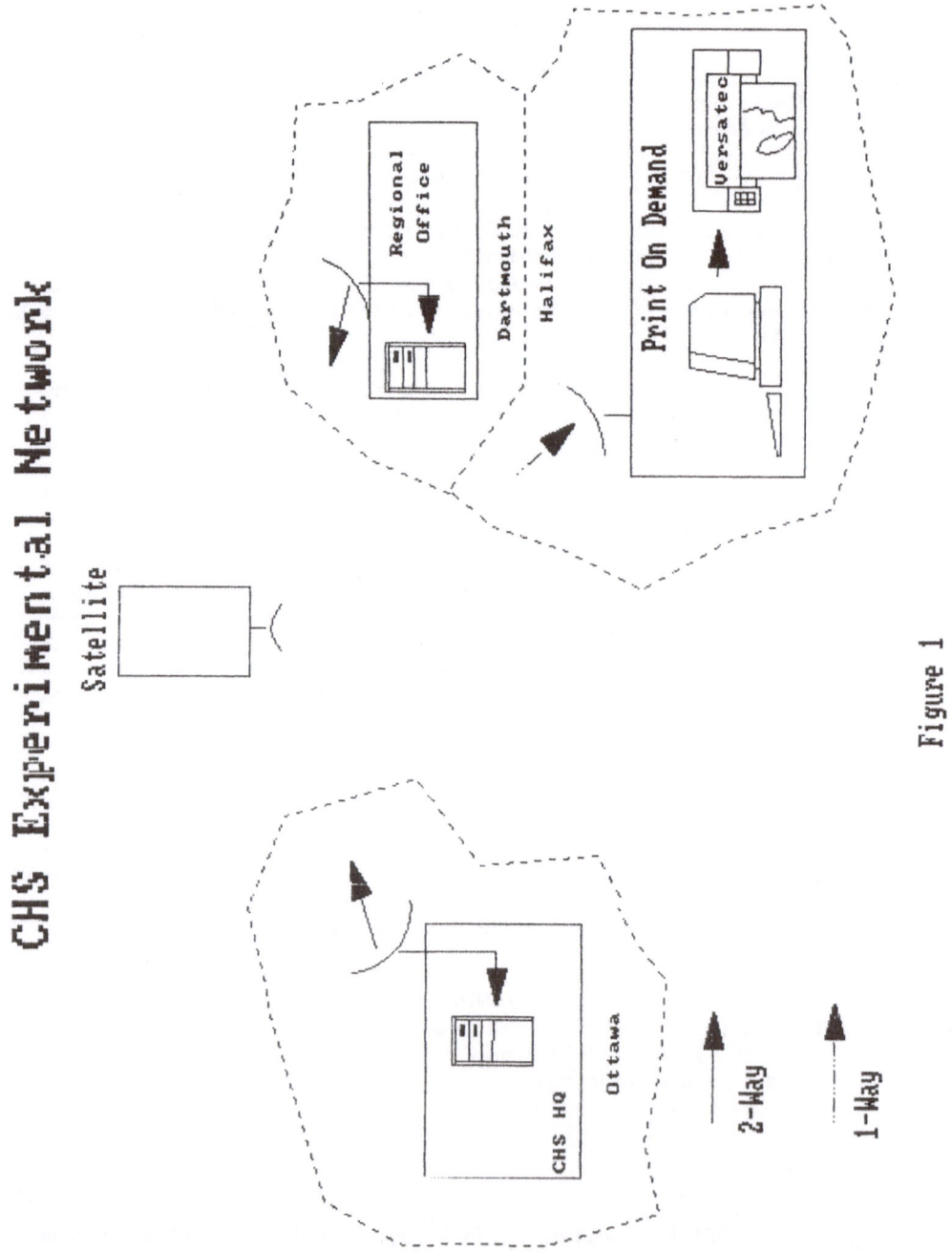

CHS Experimental Network

Figure 1

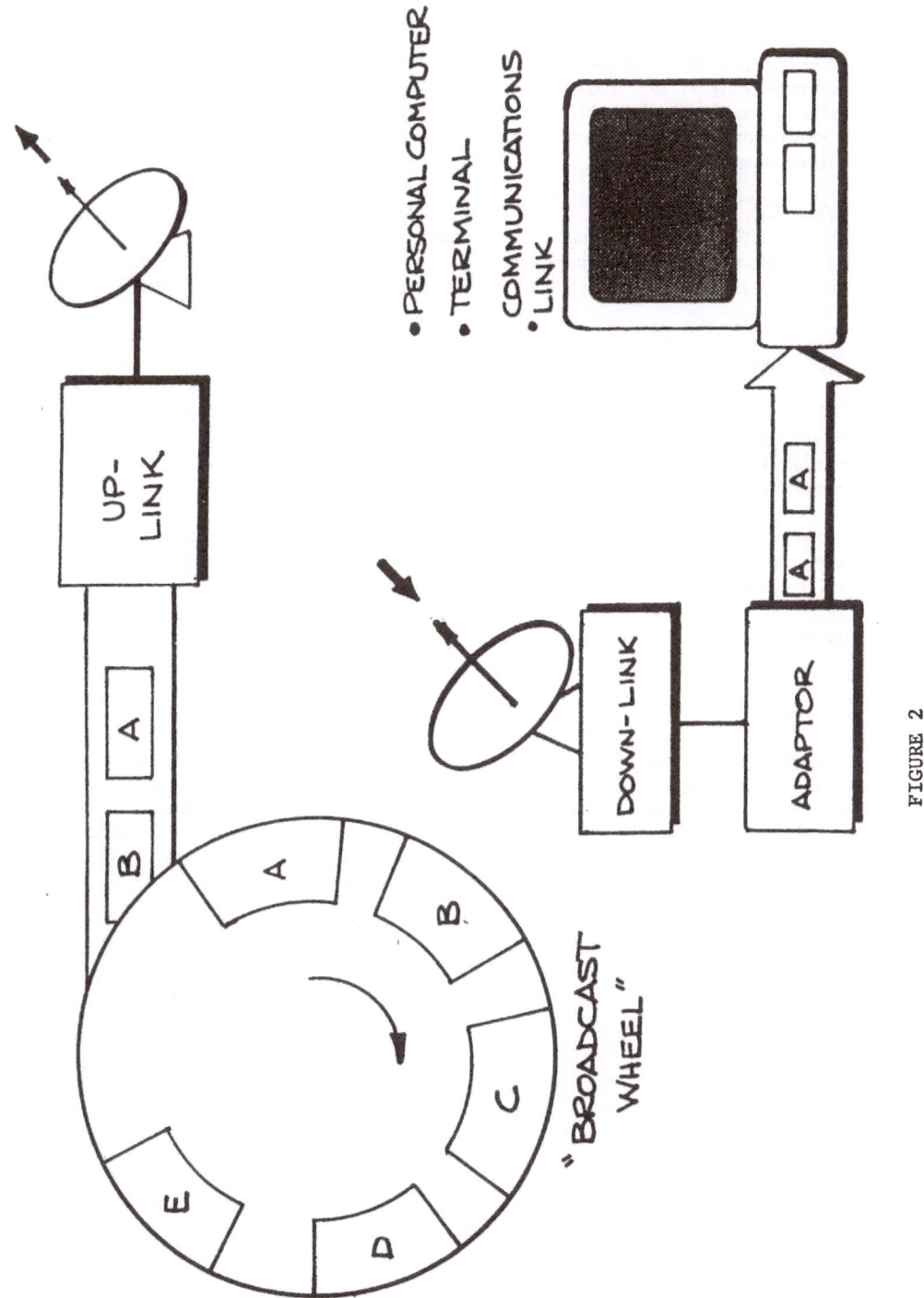

FIGURE 2

Before any updates are made available to users, an approval stage may be required. The security capability noted previously will facilitate this requirement. File changes can be addressed only to the receiver at the central site for quality control prior to being released for distribution. No other receiver will be able to pick up this data until approved and addressed to the Intel system for updating. Data is also scrambled to prevent unauthorized use.

When files are transmitted, they will be available to any suitably equipped receiving station in North America. One of the advantages of a satellite distribution system is the coverage provided and the unlimited number of potential users. Should additional dealers be required they may be quickly set up for reception and use of the system.

2.3   Operation

When assembled, the network will be operated for a period of about two months during which many operational schemes will be employed. The purpose is to judge the feasibility of such an approach and, if judged suitable, to determine the most favorable method of using the system.

Actual charts will be produced on the plotter although they will include a notice that they are not to be used for primary navigation purposes due to the experimental nature of the project. The chart files will be edited with both real and imaginary changes to determine the ability of the network to meet the demands of CHS and the user community.

3.0   Conclusion

The Electronic Chart offers potential benefits to the shipping and recreational boating communities, but, not without problems arising for supporting agencies such as CHS. To deal with these issues only when demand has become great would obviously not result in a considered approach, and hence, an optimum solution. Therefore, work is now underway to determine the feasibility of a new distribution methodology to meet this challenge which seems certain to come in the near future.

THE MAPGEN CARTOGRAPHIC SYSTEM

Gerald I. Evenden
U. S. Geological Survey
Woods Hole, MA 02543

ABSTRACT

MAPGEN is a software system that facilitates production of cartographic displays in the research and production environment. The system generates a set of metagraphic overlays of application-defined geographical information that can be aggregated in any combination for display without reprocessing the original data. An overview of the control files, available cartographic projections, graphic attributes, overlay generator and ancillary support programs, and the device-independent graphic subsystem is presented along with examples of usage. System transportability and associated host hardware and operating system requirements are also addressed.

1. Introduction

The MAPGEN (MAP GENerator) digital cartographic software system provides the basic tools with which to transform digital geographic data into a map, including a device-independent graphic subsystem to control plotting hardware. This system is open-ended so that new functions can be readily added without modification of existing software, thus providing both the dynamics required in the research environment and the system stability required by production usage.

Intended application of MAPGEN falls into two basic categories: production of work-sheet charts and publication quality maps. The work-sheet usage generally demands rapid production with little concern for cosmetic details and, unless the map is used for navigation or other mensural purposes, minimal concern for the finer points of cartography. For publication quality maps, such details become one of the primary criteria as well as does the meaningful and careful selection of the displayed data.

The preplanning of mapping requirements may combine the work-sheet needs of field applications with the publication quality requirements of the end product. MAPGEN facilitates this planning by treating the map as a set of overlays. These overlays, which form a graphics data base, can be quickly aggregated in any combination to make meaningful and appropriate composite maps of the information pertinent to a specific need.

Before continuing, the principle limitations of the MAPGEN system should also be noted. First, it will not produce cartoon graphics common to maps designed for visual impact (e.g., media presentations) because this display emphasis conflicts with cartographic accuracy and introduces considerable additional software and usage complexity. Secondly, MAPGEN is designed for large- (1:10,000) to medium-scale (1:10,000,000) scale maps and situations where the cartographic projection employed is reasonably well behaved over the geographic range selected. In general, this does not pose a problem for most local, regional, or continental mapping.

M. Kumar and G. A. Maul (directors), Marine Positioning, 285–294.
© 1987 by the Marine Technology Society.

2. System considerations

The MAPGEN system is developed for operation with the UNIX operating system and consequently reflects usage and operational characteristics similar to UNIX system utilities. It is currently operational on 68000 microprocessor systems.

The sources for MAPGEN and support graphics currently consist of about 11,000 lines of 'C' language code. Reasonable effort is made to maintain modular and structured programming practices as well as to ensure that code constructs and system function calls are a subset of the proposed 'C' language specifications (American National Standards Institute, 1985). Documentation consists of a user's manual (Evenden and Botbol, 1985), graphics documentation (Evenden, 1984), as well as "READ_ME" type files and the self-documenting 'makefiles' for system installers.

3. Overlays

A frequent method of making maps using computers has been the one-pass approach where the original geographic data have been collated with cartographic parameters and ancillary information such as titles, scale bars, and coastlines and passed to the map generating program as one basic unit. Besides causing a great deal of repetitive computing, the one-pass approach discourages the user from looking at the map-making process as a series of steps and at the data as elements of a graphic data base that are selectable and combinable in a manner appropriate to display requirements. Older manual methods, because of the arduous step-by-step process involved and the necessity of organization to minimize these efforts, can still serve as a model for the modern map maker.

In the manual process, the map maker either prepares a graticule sheet for a coordinate base or secures existing base material, such as a quadrangle map, which will perform as a working base for subsequent drafting of data on transparent overlay sheets. In this process, separate overlay sheets must be prepared for each of the basic data elements because of several factors: the data may be preliminary or not required for the final product; separate printing processes (i.e., color, grey levels, etc.) may be applied to each overlay; and preparation errors requiring rework will only affect one overlay.

In the computer analog of the manual process, the transparent sheets now become metagraphic files -- information describing graphic operations in a generic manner -- which can be readily retrieved and displayed on both graphic terminal devices or output to hardcopy plotters. This process results in important side effects:

θ Creation of an overlay file generally requires skill in graphic system operation, yet output of the graphic requires minimal technical knowledge.

θ Only the overlay file needs to be kept on-line for ready display, and the source data creating the display can be archived. This typically reduces disc storage requirements because the compressed nature of the metagraphic file usually requires considerably less storage space than the original data.

θ Computationally complex aspects of generating the overlay are only performed once.

4. Defining the map

Like the indexing pins or tabs that register the overlays of a manual system, a digital system also requires its "indexing pins" to bind the overlays to the carto-graphic base. In the MAPGEN system, this is achieved through an ancillary defini-tion file that contains all of the cartographic information needed for generating the overlays and subsequent display output.   Before proceeding to specifics of creating this definition file, some aspects of cartographic planning need to be examined.

4.1 Initial planning

This initial stage of determining the cartographic requirements of a project is critical to effective use of the maps produced. As previously emphasized, carto-graphic products are usable by all aspects of project activities and it behooves the ad hoc cartographer to evaluate these overall needs rather than merely addressing the details of a specific use.   Some of the questions that should be asked at this stage are:

θ Are there traditional requirements such as navigation maps for the chart-house?

θ Is there mensurable usage intended for the maps?

θ Does the selected scale require that separate maps be generated and, if so, do these maps need to join or overlap?

θ Do various aspects of the project require different scales or partitioning of the area into subregions?

θ Are the maps required to match existing maps, either as overlays or abutments and, if so, what are the cartographic attributes of the originals?

After the basic cartographic needs and constraints of the project are determined, there are often details which still need to be resolved. The following suggestions may assist the beginning map maker in choosing appropriate cartographic parameters:

Scale - Choose a scale so that the map is of reasonable size for handling.   Unless the map is intended as a wall-hanging, the gross size should not be much more than one meter on an edge. Consider making the map in adjoining panels if predetermined scale requirements cause an excessively large map.

Geographic range - Geographic range should obviously enclose the area of interest, but should also provide sufficient additional area for aesthetic purposes. Select boundary values which will coincide with graticule overlay intervals.

Margins - Provide sufficient margins for potential legend annotation.   The bottom margin area is usually reserved for scale bars, scale fraction, and explanation of the cartographic attributes of the map. Left and/or right margin areas are normally employed for information related to the data plotted.

Map projections - An appropriate map projection is highly dependent upon the geo-graphic location and extent of the region, as well as the intended application of the map. "Map projections used by the Geological Survey" (Snyder, 1984) is an excellent source of additional map-making information. MAPGEN provides about twenty of the projections described by Snyder.

Geoid parameters - Unless involved with specialized applications, these parameters
(major-minor axis radius, eccentricity, etc.) can be ignored. MAPGEN will supply reasonable values automatically.

5. Program mapdef

Given the project mapping requirements, the next step is execution of the MAPGEN
program mapdef to create the map definition file. It should be noted here that generating the map definition file can be a repetitive process since it is usually
desirable to test out the parameters using a display of a graticule overlay with
MAPGEN program grid. This not only demonstrates the viability of selected parameters, but it is also required when performing some of the windowing options. Spending extra time at this stage to ensure the efficacy of the definition file is certainly time well spent!

A small scale map of the continental United States can serve as an example of this
process. Assume that the map's basic requirements, besides the geographic range,
are a scale of 1:20,000,000 and the Alber's Equal Area projection. Figure 1
represents the typical results of a first attempt for this kind of map. For a map
with a more traditional appearance, the geographic window can be trimmed by a rectangular data window based upon representative geographic coordinates. A second
pass with mapdef will then produce the map shown in Figure 2.

6. Map definition file

A summary of the structure of the map definition file will clarify its role in the
MAPGEN system.

6.1 Cartographic projection section

Because of the number of projections available to this system and the computational complexity of many of these transformations, the MAPGEN system employs a
approximation system whose coefficients are determined by two hidden programs: a
general-purpose geographic coordinate projection program proj and the program bivar,
which determines bivariate Tchebychef polynomial coefficients to approximate the
projection. Use of polynomial approximations not only enhances the speed of projection evaluation, it also eliminates having to link a myriad of separate projection
modules with each of the programs requiring geographic projection operations.

The currently employed approximation scheme does have problems when the transformations are sufficiently non-linear, and although this approach has been functional
for most current applications, additional research should be made for more suitable
methods.

6.2 Graphics section

The graphics section of the definition file consists of information typical of all
graphics: scaling, rotation and windowing control. Scaling, however, has a compli-
cating factor in an environment where there are several different plotting devices
and where some devices (typically terminals) will require alteration of the specified scale of the map to fit the display. In the MAPGEN system, it is necessary to
designate one local plotter as the principal plot device (typically a drum or flat
bed plotter of moderate size) and to employ its counts-per-centimeter as the scale

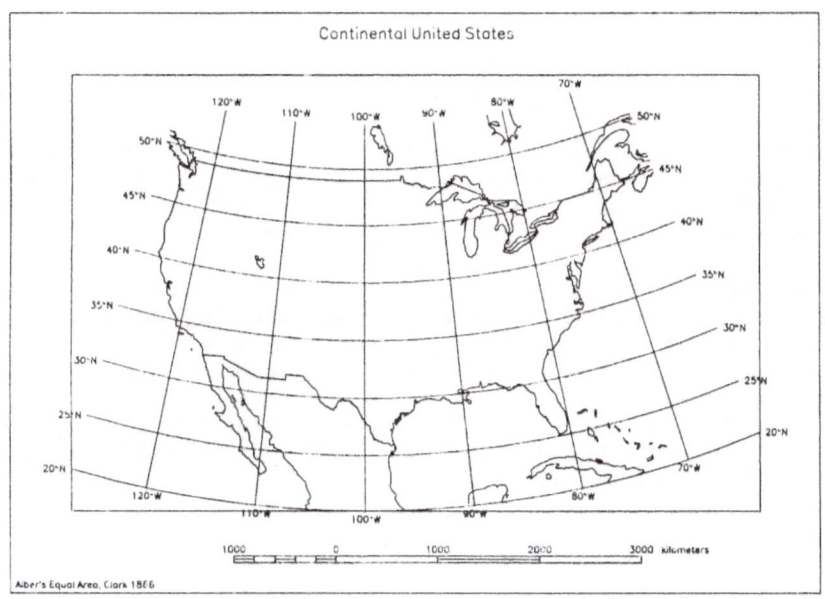

Figure 1.   Preliminary base map of continental United States region.

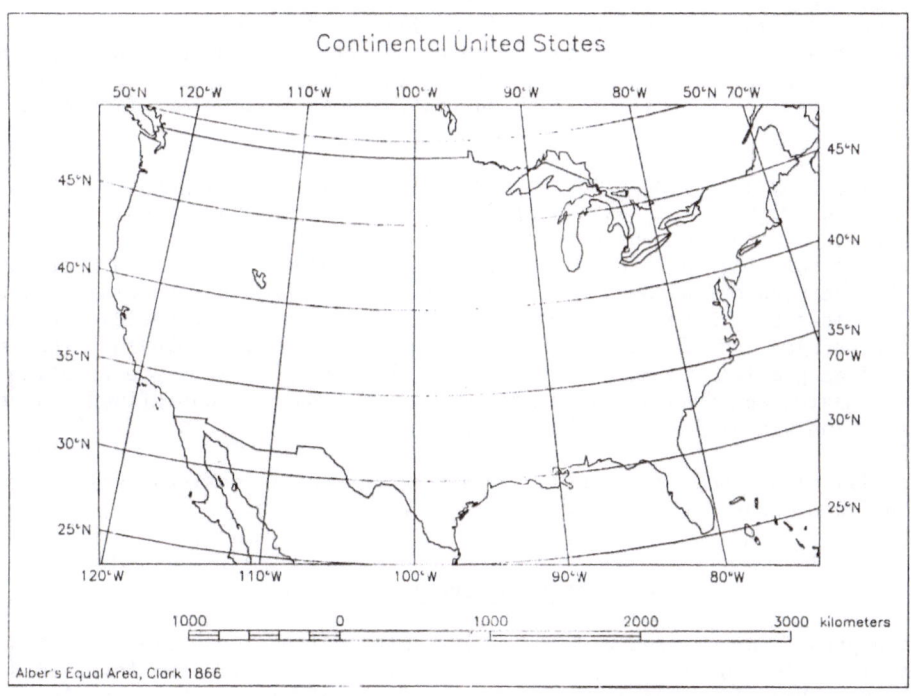

Figure 2.   Second pass base map of continental United States region.

conversion target. When plotting on other devices, the graphics system can automatically rescale output to fit the device.

There are three basic windows in the MAPGEN system: the global window, which limits all graphic operations (legends as well as data); the geographic data window applied prior to projection; and the projected data window. The projected data window is often set (a default operation) to include all possible values resulting from passing the geographic window, but it is sometimes useful in limiting the range for continental-scale conic maps where equatorial fanning out of the projection may create a great deal of useless space as shown in Figure 1.

Executional steps of program mapdef follow the order of the previous discussion: get and process geographic information into approximating polynomial coefficients and get and process windowing and scaling information into graphic control data. Because the projection approximation phase is independent of the graphics phase, it need not be repeated (saving considerable CPU time) when adjusting the graphic parameters. For example, reprocessing the map definition file for creating Figure 2 did not require redetermining the approximating polynomial employed for Figure 1. Mapdef can also be used to display the principal data within a map definition file and generate projection conversions of geographic data for debugging purposes.

7. Grid and legend overlay programs

The use of two MAPGEN overlay-generating support utilities, grid and legend, have already been demonstrated in the previous examples (Figs 1 and 2). The program grid can provide a graticule overlay for the defined map. In the examples shown in Figures 1 and 2, solid-line drafting of parallels and meridians as well as complete marginal annotation of coordinates were selected options. The user has considerable latitude in selecting various options such as the primary and secondary interval for each axis, control of annotation size and font, and solid-line or tick-mark graticule.

The overlay program legend allows the user to do line drafting and character plotting anywhere within the basic plot area defined by the map definition file. The program provides complete control over line and character features and allows ready reference to both the basic plot and data area coordinate windows so that positioning can be carried out without having to remember or compute the actual values of these coordinates. Note that legend only employs centimeter coordinates and does not perform projection of geographic data. In addition, by employing the symbols '>' and '¦' as prefixes to coordinate values, the coordinates become relative to the respective maxima and mean of the window axis. Of course, unprefixed values are relative to the window's origin (lower-left corner).

Legend will also create scale bars for either internally defined units or user-defined units of measure, and will post the scale fraction.

8. Coastline overlay program

The main problem associated with coastline data is the sheer volume of information required to define complex features adequately. For example, the formated data for the World Data Bank II (WDBII) (National Technical Information Service, 1984), which defines coastlines, rivers and international boundaries, requires approximately 110 million bytes. Adequate disc space, deformating, and scanning this information would cause considerable impact on small computer systems. Consequently, preprocessing and data compression need to be performed so that a more reasonable data set may be available for quick processing.

In the MAPGEN system, coastline data are preprocessed into a two file set: (1) a directory file and (2) a line segment data file. The directory file contains an entry for each of the line segments of the data file that describes feature type-code, geographic range, geographic coordinates of the beginning of the segment, number of nodes, and a pointer to the location in the data file. The nodes of the line segments in the data file are stored in incremental form (distance between points). The resolution of the data may be specified so that least significant bits need not be retained. As a result of this conversion, WDBII can be reduced to less than 15 million bytes for a 25-meter-resolution coastline data base: about a 10 to 1 compression. Further reduction in the size of the coastline data base can also be achieved by limiting the geographic region covered.

The preprocessed coastline files, along with ancillary information to control graphic parameters, are input to the overlay program coast. From the user's stand-point, coast is one of the simplest MAPGEN programs to use and only requires that he be aware of which coastline database files are available. The local MAPGEN system manager has the burden of establishing and maintaining appropriate coastline files and, in certain situations, may have to develop procedures to format source data into appropriate form suitable for preprocessor input.

Although the term "coastline" has been employed, this part of the MAPGEN system should be considered for any similar data (i.e., closely spaced points defining line segments), especially those that are frequently used in generating overlays. It offers considerable advantages in speed and storage requirements when compared with using the program lines for overlay output.

9. Line and point overlays

Two MAPGEN programs are currently available for making overlays of geographical data sets: points and lines. The programs are similar in function except that lines can connect the sequence of input points with lines and offers rotational control of character posting data so that they can be plotted either orthogonally or parallel to the direction of the line vectors. In addition to traditional control over the characteristics of the drafted lines and characters, the program provides for flexi-ble specification of the format of the input data.

One problem in implementing programs such as lines and points is the need to develop a reasonably flexible method for deformating data from a wide variety of sources without trying to solve the problem for all cases. In the MAPGEN system, data fields are assumed to be delimited with a tab character (usually alterable to any character) because that option is superior to the fixed-field formats charac-teristic of punched card emulation. When data fields are in this form, it is simple to select the appropriate fields that define the geographic coordinates and posting data as one-number entries. The capability to handle some forms of fixed-field data is provided by allowing additional character offset and length specifications to be appended to the field number. Any data form that cannot be directly processed must, of course, be preprocessed or filtered by system utilities.

A more difficult problem associated with handling input data is the form of the geographic coordinates. Variants seem almost infinite and, as such, cannot be han-dled with any one program without a complex lexical analyzer. As a compromise, MAP-GEN uses DMS (Degrees, Minutes, Seconds) format for all geographic data input

(including control information for other MAPGEN programs) wherein several combinations of degrees, minutes, and seconds can be used. Without resorting to details of
the lexical definition of this format, its flexibility can be demonstrated by the
equivalency of the following set of longitude entries:

```
150 15 22.3w
-150d15'22.3"
-150 15.3716667
150.25614444W
```

Note that because there is no degree mark in the ASCII character set, the letter "d"
is used as a degree subfield suffix. This schema will not handle some packed forms,
like -1501537, that omit blanks and decimal points; pre-filtering the data through a
system editor program will be required in these cases.

Figure 3 shows a typical application of the overlay program <u>line</u> where the track
of a cruise leg is to be plotted together with annotation of the date and time. In
actuality, these data were generated in two overlay passes: the first pass only
drafted the cruise track line and the second pass posted the date and time at every
fourth track coordinate.

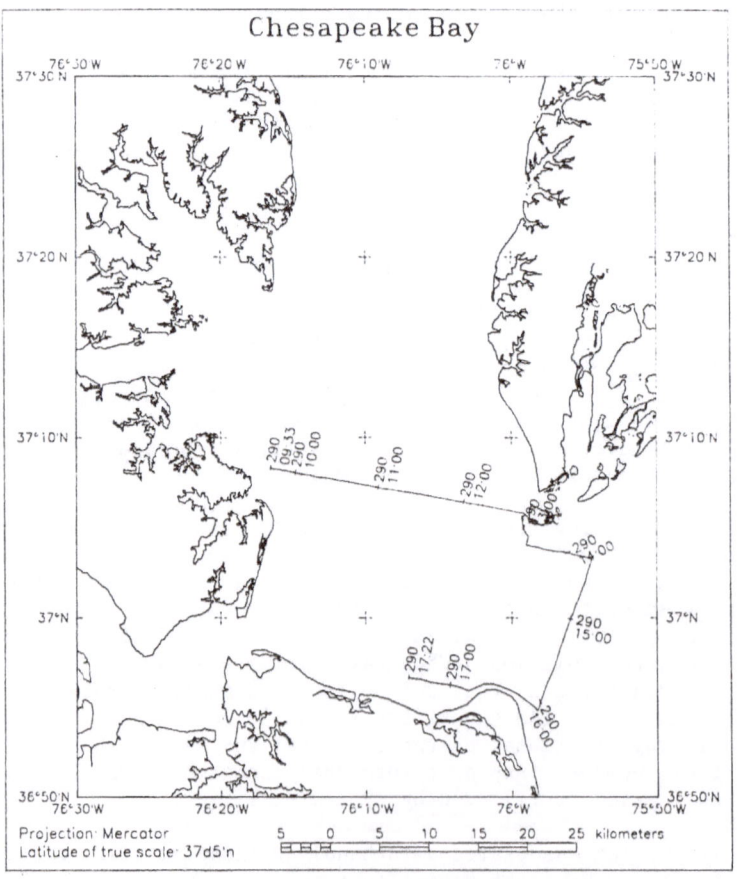

Figure 3. Example map of track line data.

Although points is similar to lines and may seem superfluous, its use is recommended in those cases where discrete point plotting is involved, as in the case of well-site data.

10. Preview program

Selection and concatenation of the overlays in order to create a composite map may be performed by executing the programs plotter or preview. Plotter is generally employed where the overlays are to be drafted on the normal hard-copy device and scaling is not altered. When previewing the composite map on an interactive graphic terminal, it is usually desirable to force the map to fit the screen. This rescaling of the overlay files is performed automatically by preview.

11. Graphics subsystem software

Two developmental criteria of the MAPGEN system were (1) to be as independent of graphic hardware as possible and (2) to avoid using proprietary or commercial graphic software. Failure to meet either of these criteria would inhibit potential transportation of the system. Consequently, a graphics support package, also useful for general graphics applications, is included as part of the MAPGEN system.

In this graphics system, the applications programs use a small library of functions that translate the typical commands associated with vector graphics into a device-independent metagraphic stream which can be saved as a file (e.g., MAPGEN overlay) and/or transferred directly to the graphics driver program plotter for immediate display. The graphics driver program, which must convert the metagraphic stream to appropriate device control codes, is designed to be readily modifiable for most common plotting devices.

Basic services provided by the program plotter are dashed line generation, windowing, Hershey character set (Wolcott and Hilsenrath, 1976), cursor control, and up to 255 logical pens. The logical pens allow for several concurrent graphical operations. The application program may intermix graphic operations by selecting the appropriate pen identifier along with its associated character font, line properties, and other attributes without having to perform local bookkeeping control for the selection operations.

Specific plotting device control is limited to a set of one or more small, one or two page routines which respond to requests for device initialization, pen motion, basic scaling properties, and optional cursor control. In cases where the device cannot perform specific tasks, such as selecting multiple mechanical pens, the requests can be ignored or replaced with substitute operations. Installation of a new plotting-device control routine can typically be made in one or two days depending upon the programmers proficiency and familiarity with the graphics package and the new plotting device. In some cases, these functions have been performed by using manufacturer-supplied support software.

12. Conclusions

The two-fold purpose of this discussion is to present the concept of map making as a process of building the map in layers or overlays and give a short summary of some of the tools to achieve this end. MAPGEN is not, nor will ever be, a complete sys-

tem but is designed instead as a foundation for indefinite expansion and adaptation
to fit the needs of the local environment.

13. Disclaimer

UNIX is a trademark of AT&T Bell Laboratories. Any reference to tradenames within
this paper is for purposes of identification only and does not constitute endorsement by the U. S. Geological Survey.

14. References

American National Standards Institute, 1985, Draft proposal of the C programming
     language of the X3J11 committee: Cardiff, New Jersey, Plum Hall Inc., 157 p.

Evenden, G. I., 1984, Device independent vector graphics: unpublished administrative
     report on file at Woods Hole, MA., 43 p.

Evenden, G. I., and Botbol, J. M., 1985, MAPGEN user's manual: unpublished adminis-
     trative report on file at Woods Hole, MA., 150 p.

National Technical Information Service, 1984, World data bank II: U. S. Dept. of
     Commerce, v. 1, no. PB-271 868/2; v. 2, no. PB-271 870/2; v. 3, no. PB-271
     871/2; v. 4, no. PB-271 872/2; v. 5, no. PB-271 873/2

Snyder, J. P., 1984, Map projections used by the Geological Survey: Geological Sur-
     vey Bull. 1532, 313 p.

Wolcott, M. W., Hilsenrath, J., 1976, Tables of coordinates for Hershey's repertory
     of occidental type fonts and graphic symbols: U. S. Dept. Commerce, NBS Special
     Publication 424, 166 p.

CONTINENTAL MARGIN MAPPING PROJECT
AND
CARTOGRAPHIC MANAGEMENT INFORMATION SYSTEM
FOR THE
U.S. EXCLUSIVE ECONOMIC ZONE*

Edward C. Escowitz
U.S. Geological Survey
Reston, Virginia 22092

The U.S. Geological Survey has established the Continental Margin Mapping
(CONMAP) Project as part of its overall research program focused on the nation's
Exclusive Economic Zone (EEZ). The project was commenced in response to the
need for a systematic means to reference and disseminate geographically oriented
data and information. The primary goal of this project is preparation of a map
series that provides continuous coverage of the EEZ at a scale of 1:1,000,000.
The series consists of 21 map panels prepared by computer using an Albers conic
equal-area projection. All data presented on the maps are stored in digital
data bases which can be easily updated with new information and subsequently
replotted.

A software system for map generation (MAPGEN) is used to processes thematic
data sets for each map panel into registered digital cartographic overlays which
can be combined and plotted on a variety of computer output devices. Maps and
overlays are organized in a hierarchical structure which is equitable to a set
of map cabinets and map drawers; an electronic map library for EEZ cartographic
data. Access to the map library has been designed so that a particular thematic
map can be drawn using a single command. This capability provides the base for
the Cartographic Management Information System (CMIS) which allows the entire
CONMAP library, as well as other geographically oriented data files, to be
referenced and displayed.

---

*To be published as a U.S. Geological Survey Open File Report in spring 1987.

M. Kumar and G. A. Maul (directors), Marine Positioning, 295.
© *1987 by the Marine Technology Society.*

PREDICTIVE LORAN-C POSITIONING IN THE MARINE ENVIRONMENT USING AREA MODELING FOR REAL TIME PROPAGATION CORRECTIONS

by

J. Ralph Johler

JOHLER & ASSOCIATES
16796 W. 74th Pl., Golden, Colorado 80403
(303) 422 1033

ABSTRACT

Loran-C has long been used successfully for general navigation in the marine environment, but modeling for the effects of the land masses whichch inevitably intercept the propagation paths from the transmitters is necessary for accurate positioning. This modeling is of course specific for each area of interest and involves complicated propagation theory to accurately determine the distortion of the wave constant phase surface.

A unique form of area modeling has been developed to remove propagation errors from various radio navigation systems so that accurate positioning can be obtained. The problem has been reduced to a simplied algorithm that can be used on a small computer in almost real time. In this paper propagation errors are depicted for Loran-C near Alaska in the Bering Sea and in an area off the coast of Louisiana in the Gulf of Mexico where comprehensive testing was performed. The effects of various types of land masses, especially those which exhibit topographic features or unusual resistivity characteristics are discussed with the aid of area modeling. Predictions from area modeling are compared with observations at independently located positions. It is concluded that Loran-C can give roughly 10 meter baseline positioning accuracies in the marine environment.

1. INTRODUCTION

The determination of the precision of Loran-C measurements in the natural or man-made noisy environment involves the phenomenon of biases in the TD (Time Difference) measurements. Such biases can arrise from a number of sources such as for example, weak skywave contamination of the leading edge of the pulse or interference from continuous wave signals near the Loran-C frequency. Ground wave propagation distortion of the surfaces of constant phase of the wave from the Loran-C transmitter produce systematic biases in the TD's. In hyperbolic and circular coordinate systems such propagation errors cause geometrical errors independent of the random fluctuations. If propagation errors are removed, biases may still remain from other causes.

M. Kumar and G. A. Maul (directors), Marine Positioning, 297–310.

Systematic propasation errors exist in the marine environment. Such errors are of the order of 1000 to 2000 nanoseconds or roughly 300 to 700 meters which can be an order of magnitude greater than the resolution of the receiver. These errors have a highly nonlinear spatial distribution and are a consequence of the intersection of the propasation paths with land massses between the transmitters and the receivers. This ground wave phenomenon is quite predictable since it is a consequence of the natural laws soverning the propasation of the surfaces of constant phase for the ground wave. Therefore, for accurate positioning in the marine environment in the 10 meter resime, propasation compensation is an absolute necesssity.

Full wave area propasation modeling has been developed to remove the propasation errors from most any ground wave radio navisation system so that accurate positioning can be obtained. Although the physical theory of ground wave propasation over the non-homoseneous, irresular surface of the earth is quite complex, the problem at hand has been reduced to simplified alsorithms that can be used on a small computer in almost real time coordinate conversion. In this paper, two areas are used to demonstrate marine propasation phenomena: (1) effects of irresular ground on propasation in the Berins Sea near Alaska; and (2) demonstration of the removal of propasation effects in real time in the Gulf of Mexico while cruising off the Louisana coast. From these considerations it is concluded that 10 meter absolute baseline accuracy should be our soal in the marine environment.

The subject of hish accuracy radio positioning in the marine environment requires some technical comments to put the discussion into context. Radio transmitters are located in various known and select positions, usually on land, and measurements are made on the sisnal propasated from such transmitters to the observer. The only reliable measurement the observer can make is a phase measurement. Amplitude measurements of the crest of continuous sisnal can usually be made within 3 decibels but such are not very accurate or helpful. On the other hand, phase can be measured to hish accuracy, say 8 to 10 sisnificant fisures. Here is the only measurand for the electromasnetic sisnal and indeed it can be used for accurate positioning. This is true because the sisnal is propasating outward from the transmitter senerally at the speed of lisht, c=299.792458 meters per microsecond. Thus the continuous wave has crests, troushs and zeroes. The zeroes are easy to tas in the propasating wave. Of course, for purposes of navisation the radio sisnal must be organized in some manner so that the time the zero crossing leaves the transmitter and the time it reaches the receiver are known. Also it is nessary to know which zero crossing is being used to resolve cycle ambisuity or resolve the "lanes". Assuming this has been accomplished, and the time of arrival at the receiver is known, the distance, D, is determined from the fact that the propasation time, t = D/c. This is the time for the surface of constant phase to leave the transmitter and arrive at the receiver. Unfortunately this condition does not obtain ever, since the presence of matter, such as the earth and the ionosphere distort the surfaces of constant phase so that the formula: t = D/c does not hold.

   The analytic field, $E(\omega,D)$, which describes mathematically both the amplitude and the phase of the propagated field as a function of frequency and distance must satisfy Maxwell's equations and satisfy boundary conditions near objects that occupy the space in which the wave propagates including' the disturbing influence of the transmitter. The distance ,D, is usually formulated as a geodesic in radio navigation problems. In essence the solution for the analytic field takes on the form:

$$E=|E|exp(i\omega t-i\omega \eta_1 D/c-i\phi_c)$$ (1)

where $exp(i\omega t)$ is the complex time function of a cosine time harmonic wave, $cos(\omega t)$, $\omega =2\pi f$, $i=\sqrt{-1}$, f is the frequency in Hertz, $\eta_1 =1.000338$ and, is the wavefront distortion term resulting from the boundary conditions. Equating the exponential argument to zero (zero crossing of the cycke) one finds for time, t_p :

$$t_p =\eta_1 D/c+t_c$$ (2)

where $t_c =(\phi_c/\omega)(10^6)$ in microseconds and t is the propagation time or the surface of zero cobstant phase arrival time relative to its departure time from the transmitter. The quantity   is quite fundamental to all radio systems and especially those used for navigation. It vanishes only under the most extraordinary circumstances, such as free space of infinite extent. Near the earth's surface it is a highly nonlinear function of geographic position and altitude above the surface. In any case, in the real world in which one uses radio navigation it is always finite and must be taken into account for accuracy. Fortunately ,it is a small number relative to the quantity $\eta_1 D/c$ and hence, cast in the above formulation of the problem can be regarded as a correction or a number with 3 or 4 significant figures that is added to a large number of 8 or 9 significant figures. But this small correction is nevertheless of such extraordinary importance that it makes the difference between a useful high precision positioning device and a useless or mediocre positioning device at best.

   Loran-C uses a hyperbolic coordinate system by forming a set of time differences that register on the receiver console as TD_w, TD_x, TD_y, TD_z, for secondary transmitters W,X,Y,and Z if indeed four such transmitters exist in the chain. Only two are necessary to determine a position. The propagation filter utilized in this analysis can calculate a corresponding set of coordinates: Y_w, Y_x, Y_y, Y_z. This is accomplished as follows:

$$Y_k=(D_k-D_m)\eta_1/c+t_c(D_k)-t_c(D_m)+ED_k$$ (3)

where $D_k=D_k(\phi,\psi)$ and $k=w,x,y,or z$. ED_k is the emission delay constant. Since Y_k and t are functions of D_k, the geodesic distance they are really functions of latitude and longitude, $Y_k=Y_k(\phi,\psi)$ and $t_c=t_c(\phi,\psi)$. The error, dY_k, of observation or prediction is always written as observed minus predicted:

$$dY_k=TD_k-Y_k$$ (4)

where dYk is still in units of microseconds. The conversion of this number to units of meters of error is subtile in hyperbolic space. If Ak is the azimuth to the secondary and Am is the azimuth to the master transmitter, then :

$$dAk=|(Ak-Am)/2|$$

or,

$$dAk=|\pi-|(Ak-Am)/2| \qquad (0 <= dAk <= \pi/2) \qquad (5)$$

and,

$$dDk=0.5dYk(c/\eta) \qquad (6)$$

where dDk is now in units of meters, provided c=299.792458 meters per microsecond. Since one can set a position with only two time differences, say Yy and Yz, then the distance error, dDr, is given by:

$$dDr=[(dYy/\sin(dAy))^2 +(dYz/\sin(Az))^2$$
$$+2(dYy/\sin(dAy))(dYz/\sin(Az))\cos(Tyz)]^{\frac{1}{2}}/\sin(Tyz) \qquad (7)$$

where,

$$Tyz=dAy+dAz$$

or,

$$(0 <= Tyz <=\pi).$$

Equation (7) is deterministic for non-orthogonal lines of position. Under such circumstances there exists a geometrical amplification of error,GAOE, that can be calculated:

$$dDro=[(dYy)^2 + (dYz)^2]^{\frac{1}{2}} \qquad (8)$$

$$GAOE=dDr/dDro \qquad (9)$$

The GAOE is minimized when the lines of position intersect at 90 degrees and both TD's are on the baseline, a situation which never exists. However, the third term in equation (7) may be negative such that the GAOE may be less than unity, (0 < GAOE < 1.).This is a peculiar property of nonorthogonal lines of position.

It is quite possible to operate Loran-C in the circular coordinate system instead of the hyperbolic coordinates. Thus using equation (3) and subscript "c" for circular one can write:

$$Ykc=Dk\eta/c+t_{c}(Dk)+EDk \qquad (10)$$

$$dDkc=(c/\eta_i)(TDkc-Ykc) \tag{11}$$

$$dDrc=[(dYy)^2+(dYz)^2+2(dYy)(dYz)\cos(2Tyz)]^{\frac{1}{2}}/\sin(2Tyz) \tag{12}$$

$$dDroc=[(dYy)^2+(dYz)^2]^{\frac{1}{2}} \tag{13}$$

$$GAEC=dDrc/dDroc \tag{14}$$

In equation(11) TDkc is the observed quantity by analogy to TDk in hyperbolic coordinates. The latitude and longitude,(ϕ,ψ), can be found from equation (3) and an initial estimate and the following differential equation:

$$\begin{bmatrix} dYy \\ dYz \end{bmatrix} = \begin{bmatrix} Ayy & Ayz \\ Azy & Azz \end{bmatrix} \cdot \begin{bmatrix} d\phi \\ d\psi \end{bmatrix} \tag{15}$$

where,

$$Aij = \frac{\eta_i}{c}\left[\frac{\partial Di}{\partial Uj} - \frac{\partial Dm}{\partial Ui}\right] + \frac{\partial t_{ci}}{\partial Uj} - \frac{\partial t_{cm}}{\partial Ui}$$

and,

$$D1,2=Dy,z \ , \ U1,2=\phi,\psi.$$

The quantity t_c has been called the secondary phase correction. It is the crux of the matter of accurate positioning over land or in the marine environment where land masses are involved. While it is a small number compared to the primary propagation time,$\eta_i D/c$, it is sufficiently great as to produce errors that may render a positioning system inferior or even useless. Furthermore, the geometry of the hyperbolic or circular coordinate systems can amplify such error. Finally, it is a function of geographic position and indeed it is this quanity that is displayed in this paper for each area under consideration. Thus,

$$t_c=t_c(\phi,\psi).$$

2. PROPAGATION MODEL

The propagation model is based upon an organized ensemble of data that the author has accrued over a period of many years concerning the effect of the ground, its geologic and soil structure and electrical properties upon the electromagnetic ground wave. An introduction to the techniques employed has been given by Johler (1979) and applications and discussion of the techniques have been given by Johler and Cook (1984,1985). Figure 1 depicts the scheme employed in analyses used in this paper.

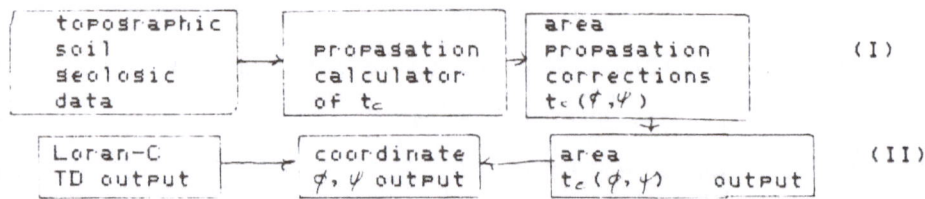

Figure 1.   Illustrating propagation correction calculations (I) w ith
             real time playback (II) for coordinate converter and
             Loran-C sensor.

   In essence, the soil,subsoil or hardpan and the genrral geologic
structure of   the   area   under   consideration are introduced into a
propagation calculator which satisfies the boundry conditions at the
surface of the ground for the ground wave.   The results are reduced
to an area representation of the propagation corrections for $t_c(\phi,\psi)$
for the area of navigation or position determination. This is labeled
Roman I in Figure 1.   The area propagation corrections are made
available in the field (Figure 1., Roman II) with a coordinate
converter operating on the output of a Loran-C sensor.

3.   IRREGULAR, NONHOMOGENEOUS GROUND EFFECTS ON $t_c(\phi,\psi)$.

Figure 2 depicts an area in the Bering Sea that exhibits propagation
characteristics typical of regions in the marine environment that
involve propagation from the transmitters over irregular land masses.
Propagation corrections were calculated for the cross hatched area
and a diagonal navigation lin was scrutinized. The TD's along this
line were calculated because the propagation path from the St.Paul
transmitter intercepted the land mass of St. George Island as the
line reached the south west coener of the area. An apparent phase
anomaly existed which apparently was attributed to this island but
this analysis only indicates a small perturbation due to St. George
island. Rather, the perturbations along the navigation line were
found to be primarily caused by the Alaskan Peninsula topography and
Kodiak Island topography.

   The distortion of the wave front traveling along a radial from the
Narrow Cape transmitter to the area of interest in Figure 2 is
depicted in Figure 3. Here the value of the correction t is traced
along the geodetic from the transmitter toward the area. The various
topographic features are readily correlated with the perturbations in
the value of t . The transmitter is located on land but after a few
kilometers the wave traveling outward encounters Usak Bay where the
phase recovers to a smaller value. Part of this reduction in phase
is also a consequence of the near field of the transmitter. After a
series of perturbations due to a mountain range, the Shelikof strait
is crossed by the wave front and another phase recovery is indicated.
The Alaskan Penninsula in now crossed and another mountain range
causes perturbations in the wavefront.   The   most prominent
perturbation is found near the Aviachek volcano crater. After
crossing mudf lats and moving out to sea the perturbations soon
cease, but the effects on the wavefront propagation time remain.

Figure 2.   Map depicting Bering Sea and Shelikof Strait in relation
            to Narrow Cape and St. Paul transmitters with the area of
            interest indicated as the shaded trapezoid. The dotted
            line across the area indicates the navigation line of
            interest with respect to topographic effects in the Bering
            Sea caused by the Alaskan Penninsula.

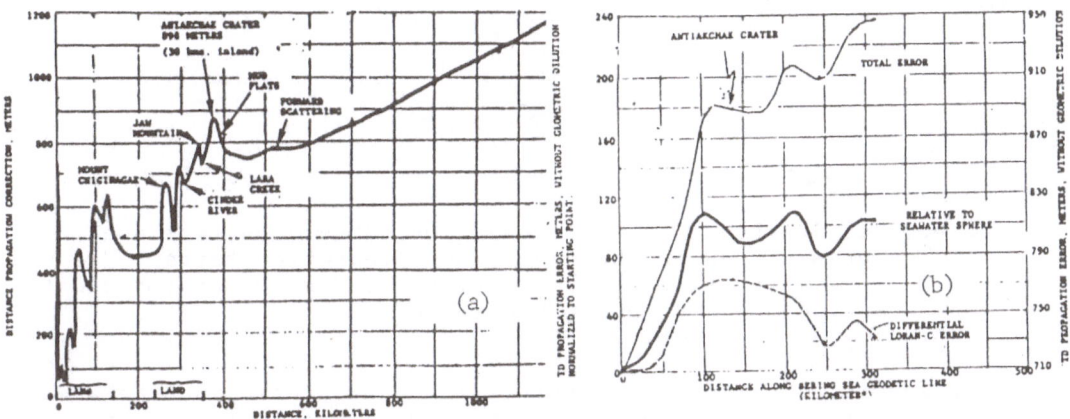

Figure 3.   (a) $t_c(D)$ depicting wavefront traveling from Narrow Cape
            transmitter across Alaska Penninsula to area of interest;
            (b) $t_c(\varphi, \psi)$ along navigation line illustrating far at sea
            effects of propagation path intercepted land mass.

     In Figure 3b the complete effect of the land on the TD measurement
for the Narrow Cape-St. Paul pair of transmitters is depicted as a
propagation error in meters. The total error, the error normalized
to the sea water propagation error, and the error normalized to the
starting point (assuming one had a known position at the starting
point). In all cases depicted the error due to the land could not be
removed without resort to the modeling described.   This is a
consequence of the nonlinear spatial distribution of the error due to
propagtion.

4.   REAL TIME COMPARISON BETWEEN PREDICTIONS AND MEASUREMENTS.

The practical value of the propagation filter was tested in a comprehensive manner in the Gulf of Mexico near Louisana. This required independent positioning at sea and also required continuous playback of position every two seconds while the ship on which the test was conducted was traveling about 8 knots. A second, more accurate radio system known as ARGO was therefore used in the test. Other redundant positioning was also obtained with Autotape and GPS (Global Positioning System). Positioning was also obtained with the Navcube which employs Kalman filter estimates of position based upon several sources of position.

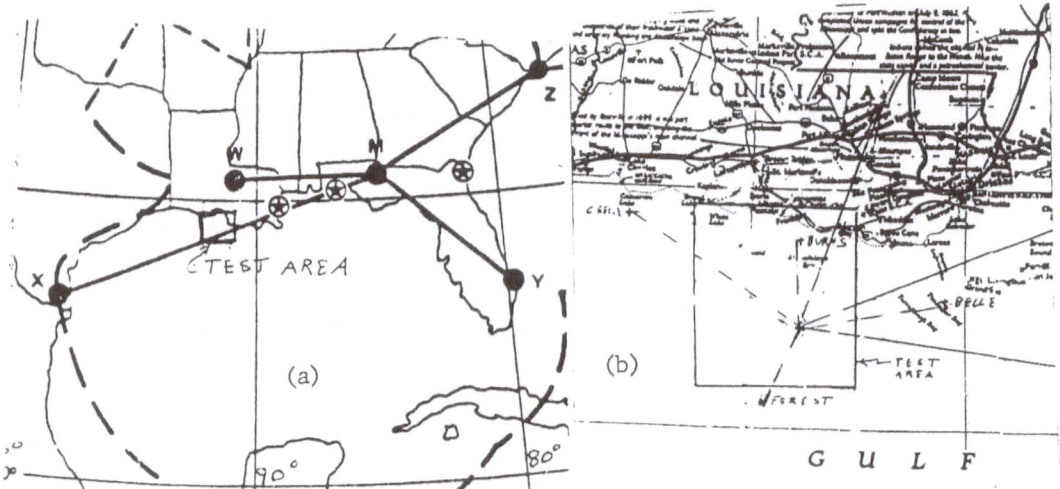

Figure 4.   (a) Southeast Loran-C chain (GRI 7880) and test area for Loran-C used in the real time propagation tests.
(b) Closer view of test area showing ARGO transmitters: Forest, Creole, Burns and Belle.

Figure 4 (a) shows the U.S.Southeast Loran-C chain and the area in which the test was conducted. Only two TD's were employed using the Master station at Malone, Florida, the W-Secondary at Granseville, Louisiana and the Y-Secondary at Jupiter, Florida. Figure 4 (b) gives a closer view of the area with the ARGO stations indicated: Forest, Creole, Burns and Belle. Also lines indicating the direction to Granseville, Malone and Jupiter Loran-C transmitters from a point of interest in the area are shown.

The configuration of ARGO and Loran-C depicted in Figure 4 in essence comprises a precision marine positioning system. Thus, propagation corrected Loran-C can be used to resolve the lane ambiguity of the ARGO. At a frequency of 1750 kHz the ARGO lane width is 85.7 meters. Predictive Loran-C can give 10 meter accuracy. Once the ARGO lane is identified the ARGO can pull the position accuracy to one meter.

As a byproduct of the test, the scheme shown in Figure 4 represents a concept for high accuracy positioning in the marine environment. Both Loran-C and ARGO were propagation corrected in the test area shown. It is especially important to correct ARGO when any land is involved in the propagation paths to the area under consideration, Johler et al,1985. Also, the height of the antennas becomes important for ARGO. All these effects were taken into account in the modeking for the test.

Although the land areas involved in the propagation of the Loran-C and ARGO signals to the area are comparatively flat, some topography exists. More important is the modeling of the geology,the soil and freshwater and saltwater areas of the ground throughout the area and as distant as the transmitters. The results of the propagation/modeling computations were compacted into the propagation filter before the test was begun, and the playback was used during the test on a HP9920 computer.

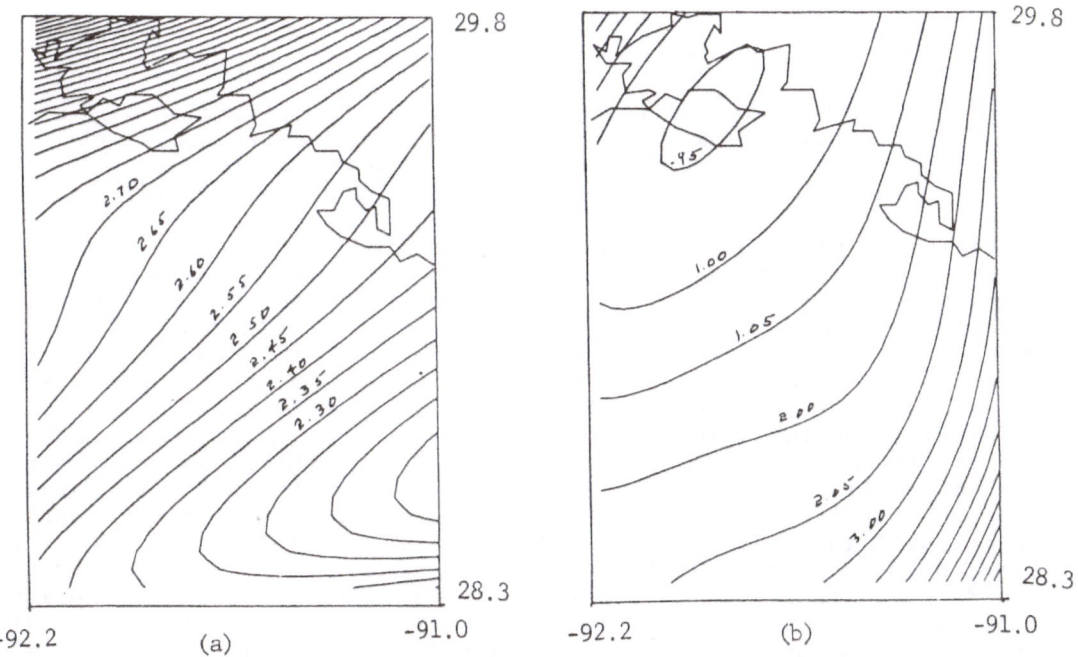

Figure 5.   (a) Playback of surface of constant phase distortion of propagated wave from Malone Master transmitter, $t_c(\phi,\psi)$. (b) Playback of surface of constant phase distortion of propagated wave from Grangeville W-Secondary transmitter, $t_c(\phi,\psi)$. Contour interval is 0.05 microseconds.

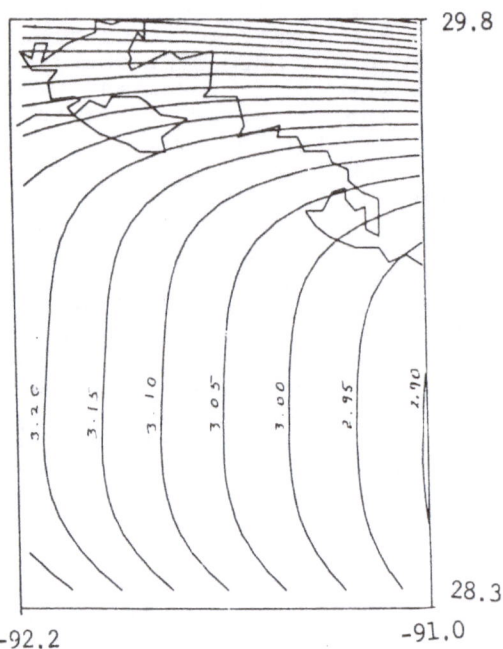

Figure 6.   Playback of surface of constant phase distortion of
            propasated wave from Jupiter Y-Secondary transmitter,
            t_c (ϕ,ψ). Contour interval is 0.05 microseconds.

    Figures 5 and 6 display output from the propasation playback
filter in sufficient density to draw contour lines of the
seosraphical (ϕ,ψ) distortion of the constant phase surface of the
propasated wave from the Master Malone Loran-C transmitter, the
W-Secondary Granseville transmitter and' the Y-Secondary Jupiter
transmitter.   The contour intervals are 0.05 microseconds or 5
nanoseconds. The presence of the land masses of concern in the area
are also indicates. In seneral, the distortion, t_c(ϕ,ψ) is sreatest
when the amount of land involves is sreatest.   There is of course
some distortion even over a propasation path that is enirely
seawater.   In fact, this has been referenced as the ."secondary phase
correction" in Loran-C jarson.   The effects of land masses were then
noted and described as an "additional secondary phase correction".   I
resard this approach as a conceptually incorrect physical notion.
Thus, the effect of the land is at least the arsument of the
convolution intesral of a complex sea water field with the analytic
field over land.   In seneral it is even more involved.   In essence
and quite to the contrary the quantity t_c (ϕ,ψ) alone can describe
the   measured   surfaces   of   constant   phase   propasated   from   the
transmitter.

    Equations (1 throuh 15) were prosramed on the HP9920 computer   to
work in conjunction with the propasation filter playback.   The HP9920
was   installed   aboard   a   survey ship of John E. Chance & Associates,
Inc., at Intracoastal City, Louisiana.

Data from the Loran-C receiver was sampled at intervals of two seconds and navigation information such as latitude and longitude were displayed on the computer screen. The full wave solution provided by the playback filter corrected distances and coordinates automatically in real time through the HP9920 computer both at the dock and while the ship was in motion in the Gulf of Mexico test area. During the period August 6,7 and 8 1985, approximately 150 000 Loran-C data points were collected by the expedition at known positions. The complete analysis of these data is beyond the scope of this paper. Herein, a sample of the data will be presented to enhance the discussion of Loran-C technology.

TABLE 1.

Loran-C Standard Deviation of Position Error For Various of Time Windows of Travel in Test Area.

8/6/85,8/7/85;
Julian Dates 219,220

Time Window Hours	Loran-C TD Residual,Meters	
	sDw	sDy
---------	---	----
2300-2400	3.4	11.8
0100-0200	4.1	13.7
0200-0300	3.6	12.7
0300-0400	4.8	13.6
0400-0500	5.2	15.7
0600-0748	6.8	6.4
0748-1000	7.8	6.3
1000-1200	5.2	6.1
1200-1330	2.4	1.4
1330-1500	2.7	2.3
1500-1600	2.4	4.0
1600-1800	2.7	6.9
1800-2000	3.7	5.9
2000-2151	2.3	3.0

Table 1 comprises 20.85 hours of operation in the test area or approximately 37530 data points.

TABLE 2.

Specific Point Analysis of Absolute Error Using Dock Emmission Delay Determination.

Point No.	dDw	dDy	dDro	GAOE	dDroc	GAEC
---	----	-----	----	----	-----	----
1	5.0	-4.9	7.1	2.7	14.1	0.9
2	11.8	-8.8	14.7	2.3	29.4	0.9
3	10.9	2.9	11.3	5.2	22.6	1.0
4	-5.4	8.4	10.0	3.9	20.1	0.9
5	6.8	3.4	7.6	6.1	15.2	1.1
6	-12.4	-1.9	12.6	4.3	25.1	1.0
7	1.2	-11.0	11.1	5.5	22.2	1.0
8	11.2	-0.9	11.2	3.6	22.4	1.0
9	-6.0	-19.1	20.0	7.0	40.1	1.1
10	-0.1	0.0	0.1	4.3	0.2	0.9

sDro=11.66 meters aGAOE=3.02
EDw=12809.137   EDy=45201.522

TABLE 3.
Same as Table 2 with Propagation Filter Turned Off.

1	178.2	94.7	218.3	5.9	436.6	1.1
2	184.9	90.8	218.3	5.8	436.6	1.1
3	189.8	108.2	175.3	6.1	350.6	1.1
4	139.9	87.2	183.2	6.3	366.4	1.1
5	152.1	82.1	149.4	6.1	298.8	1.1
6	153.1	86.9	150.1	5.7	300.1	1.1
7	166.7	77.8	163.6	5.4	322.3	1.1
8	156.5	77.8	170.9	6.0	341.8	1.0
9	139.4	59.6	174.6	5.8	349.3	1.1
10	-1.6	-0.3	151.4	11.5	302.9	1.1

sDro=175.03 meters aGAOE=1.96
EDw=12806.710   EDy=45201.537

Table 1 summarises the standard deviations of the position error expressed as standard deviation sDw for the Granseville Secondary-W and Malone Master-M time difference error and sDy for the Jupiter Secondary-Y and Malone-M time difference errors, expresced in meters of the same units as dDk,k=w,y, equation (6). Thus, the residual error is the observed minus the predicted time difference multiplied by the speed of light in air and divided by two:i.e. the baseline error of a hyperbolic coordinate system. The equivalent circular coordinate system error would be twice this value, equation (11). Variable biases in the average error ranged from 0.1 to 13. in hyperbolic coordinates or 0.2 to 26 in in circular coordinates. This depended upon time of day and was greater at night, especially on the Jupiter measurements, dDy. Thus, while all of the ground wave propagation error had been removed, some biases existed based on the noise level from various sources suvh as skywaves, atmospherics, other interference etc. Widely used and relatively inexpensive Loran-C receivers were employed so that at this level of precision the quality could only be improved with a more advanced receiver design. Throughout the test, it was quite possible to resolve the lane ambiguity (85.7 meters) of the ARGO navigation system. It was found to be quite possible to correct ARGO lane count during this test and hence position was know in the one meter regime . The accuracy of ARGO in the presence of land is discused in some detail in a previous paper, Johler,Cook and Sampson (1985). Also, the NAVCUBE used in conjunction with the line of sight positioning equipment known as AUTOTAPE gave continuous position over long periods during the test.

Table 2 presents an analysis of some specific points in the test area. In particular, Point No. 10 is a measurement at the dock at Intracoastal City, Louisiana,ϕ =29.7829332,ψ =-92.1552627,near the top right corner of the area depicted in Figure 4 (b). Measurements were made here on 8/6/85 at 20:45 hrs before departure of the ship. A first order survey marker, N.A. Datum, was found on the dock and the coordinates were obtained. From this and two other survey points in N.A. Datum, 1927 it was possible to accomplish with the propagation filter absolute prediction of positioning coordinates in contrast with differential positioning. Points No. 8 and 9 were observed and recorded at a point in the test area known as 129 Platform,ϕ =28.9596292, ψ =-91.6245267, a location at a distance of 104.797 Km at an azimuth of 150.490 degrees, Figure 4, from Point NO. 10 at Intracoastal City. Point No. 9 was measured with an Accufix Loran-C receiver and data was sent to the ship by radio. Measurements were made at 22:12 hrs on 8/7/85. Point No. 8 was also observed and recorded on Platform 129 during the visit of the expedition ship at 21:06 hrs on 8/7/85 using an Arnav Loran-C receiver that was taken aboard 129 Platform. It is interesting to note the discrepancy of 9 meters for dDro and 18 meters for dDroc between the two receivers. This is consistant with the argument given above that the accuracy is limited to the performance of the receivers in the noise environment when the propagation effects have been removed.

Point No. 6 and Point No. 7 of Table 2 were recorded at ℓ=28.8547500, ψ=-91.4834522, according to the Navcube/Autotape or ℓ=28.85505208, ψ=-91.4835841, according to the ARGO, an azimuth a distance of 121.829 Km from the dock, at an azimuth of 154.480 degrees. For both the Navcube/Autotape one finds close to the same discrepancy with Loran-C. This gives us further confidence in using ARGO or Navcube for a standard to check Loran-C while in motion. It should be emphasized that each of these points are instantaneous among many taken at two second intervals. These points were recorded on 8/8/85 at 10:07 hrs.

Point No. 4 and Point No. 5 of Table 2 were recorded at ℓ=28.9598889, ψ=-91.6255417, according to the Navcube/Autotape or ℓ=28.960041, ψ=-91.6248092, according to the ARGO, a distance of 104.771 Km from the dock, at an azimuth of 150.473 degrees. These points were recorded on 8/7/85 at 22:12 hrs. Also Point No. 3 was recorded at ϕ=28.6215139, ψ=-91.6649694, according to Navcube/Autotape at a distance of 137.272 Km from the dock at an azimuth of 158.603 degrees. Finally, Points Nos. 1 and 2 of Table 2 were recorded at ℓ=28.6983056, ψ=-91.6237722 according to Navcube/Autotape or ℓ=28.6984286, ψ=-91.6238736 according to ARGO at a distance of 130.848 Km from the dock, at an azimuth of 156.666 degrees. It is noted that the standard deviation of the baseline error, sDro=11.66 meters. The average geometric effect for hyperbolic coordinates for these data is, aGAOE=3.02. A byproduct of the calculation are the emission delay EDk, k=w, y.

The removal of the propagation corrections from the problem reduced the measurements to the Loran-C system and receiver errors in the presence of noise given in Table 2. It is a simple matter to turn off the propagation filter and recalculate Table 2. This is shown in Table 3. In effect Table 3 demonstrates the very great errors that develope if the Loran-C is operated in a "differential mode" relative to the dock without propagation corrections. It is apparent that Loran-C becomes useless for positioning to great accuracy without the propagation filter.

5. CONCLUSIONS.

The accuracy goal for Loran-C should be at least 10 baseline meters in the marine environment. or it should be limited by the receiver performance in the noisy environment. Ground wave propagation can be completely removed with the aid of propagation corrections for each area of interest. Full wave propagation corrections have been demonstrated in this paper in real time using a propagation filter on a small computer on board a moving ship.

The results of this analysis suggest a precision radio survey system using Loran-C to resolve the ARGO lane count and using the ARGO system to set meter accuracy. Thus, the ARGO acts as the precision vernier measurement for Loran-C.

6. REFERENCES.

Johler, J.R.,1979,"Loran-C Pulse Transient Propagation." DOT Report
        No.CG-D-52-79, U.S. Assession No.AD AO 77551, U.S. Coast Guard,
        Washington, D.C.
Johler, J.R. and Cook, A.R.,1984,"Accurate Position Determination
        in the Bering Sea Using Loran-C" Proc.13th Annual Symp.,
        Wild Goose Assoc., P.O. Box 556, Bedford, Mass, pp 125-154.
Johler, J.R. and Cook, A.R.,1985,"Accurate Overland Radio Positioning
        in Southern California With the ARGO System Medium Frequency
        Ground Wave", Proc. Natl. Technical Meeting, January, 1985, The
        Institute of Navigation, 815 15th St,N.W., Suite 832,
        Washington, D.C., pp 167-176.

7. GLOSSERY

The following navigation or special equipment was mentioned in the text:
ARGO:A positioning/navigation system manufactured by Cubic Precision,
Tullahoma, Tennessee.
NAVCUBE: Same.
AUTOTAPE: Line of sight positioning, same manufacturer.
ACCUFIX: Loran-C receiver manufactured by Mesapulse, Bedford,Mass..
ARNAV: Loran-C receiver manufactured by Arnav Systems Inc.,Portland,
Oregon.
HP9920: Computer manufactured by Hewlett Packard, Palo Alto,California.

8. ACKNOWLEDGEMENTS.

The success of the expedition to th Gulf of Mexico was in large measure a consequence of the gracious hospitality of John E. Chance & Associates,Inc. of Lafayette, Louisiana, who supplied a survey ship and ARGO equipment, Loran-C equipment and personnel to operate the ship and the navigation equipment. Information quoted from the Gulf of Mexico tests of August 6,7 and 8 1985 were included with the permission of Cubic Precision of Tullahoma, Tennessee. These tests included both ARGO and Loran-C systems with propagation corrections for the ground wave in real time at sea. The outstanding work of the many participants from Cubic Precision led to the development of the excellent data base cited. Also, special mention of a penetrating data analysis performed by Monty Griffin after the test and his able contributions before the test is in order. Finally, the author wishes to cite the untiring and courageous efforts on this and related projects of his vice president, colleague and friend the late Alan Ryan Cook who passed away December 13,1985.

The RAFOS Navigation System
H. Thomas Rossby

Graduate School of Oceanography
University of Rhode Island
Kingston, RI 02881

ABSTRACT

The RAFOS (SOFAR spelled backwards) navigation system uses the deep sound channel to insonify large areas with precisely timed acoustic signals for position determination. The idea is not new, but it is only in the last few years that it has been put to practical use. Three moored SOFAR sound sources, one off Cape Hatteras, one near Bermuda and one on top of a seamount, send three signals/day to provide ~ 5 km tracking accuracy between 75°W and 55°W. A complete system description is in press and will appear later this year (Rossby et al., 1986).

Through repeated launches of constant density floats in the center of the Gulf Stream, we are studying the dynamics of how water is transported, dispersed and re-entrained by the meandering current (Rossby et al., 1985).

REFERENCES

Rossby, T., A.B. Bower and P.T. Shaw, 1985. "Particle Pathways in the Gulf Stream." Bull. Am. Met. Soc., Vol. 66, No. 9, pp. 1106-1110.
Rossby, T., D. Dorson and J. Fontaine, 1986. "The RAFOS System." J. Atmos. & Oceanic Tech., Vol. 3, No. 4 (Dec.) (in press).

M. Kumar and G. A. Maul (directors), Marine Positioning, 311.
© 1987 by the Marine Technology Society.

INTEGRATING MARINE NAVIGATIONAL SYSTEMS IN POST CRUISE PROCESSING

Peter Buhl
Lamont-Doherty Geological Observatory
of Columbia University
Palisades, New York   10964-0190
and
Paul Manning

ABSTRACT

The R/V ROBERT D. CONRAD uses a mix of systems for mid-ocean navigation during geological and geophysical research cruises: GPS, Transit Satellite, Rho-Rho Loran-C, Two-Axis Doppler Speed Log, and Gyro Compass. These systems have complementary error-spectra characteristics. GPS has high long period (>10 minute) accuracy, but can show unrealistic ship velocities when consecutive one-minute readings are used. The infrequent transit satellite fixes have only long period accuracy. Loran-C has high medium period accuracy (< hours). The speed log and Gyro show high short period (< hour) accuracy, but currents and wind introduce long period errors.

The information from all systems is combined to achieve the best possible navigation over all time scales. A group of transit fixes establishes average Loran-C station ranges. Short period GPS and Loran-C errors are diminished by smoothing difference navigation formed by subtracting speed log/Gyro dead reckoning positions from GPS and Loran-C navigation points. Comparison of raw gravity and the Eotvos correction provide a check on this procedure.

1. INTRODUCTION

The R/V ROBERT D. CONRAD, operated by Lamont-Doherty Geological Observatory, is engaged in worldwide geological and geophysical research. Its underway instrumentation includes a Proton precision magnetometer, single and multichannel seismic systems using both watergun and airgun sound source arrays, a 3.5 kHz shallow penetration profiling system, and a gravimeter. In addition, a SEABEAM swath bathymetric mapping system is operated by URI. Station work uses piston and other types of sediment cores, rock dredges, and heat flow instruments. Each of these systems require varying degrees of navigational accuracy. To core particular sedimentary layers we need to return to locations selected from the seismic system records. Reconnaissance seismic surveys have less stringent navigational specifications. The high accuracy demanded by the SEABEAM system is discussed elsewhere by Robert Tyce.

2. GRAVIMETER NAVIGATIONAL REQUIREMENTS

A new Bell Aerospace BGM-3 gravimeter was installed on CONRAD in February, 1984. (Bell and Watts, 1986, Geophysics, pp. 1480-1493). This instrument and its Graf-Askania GSS-2 predecessor present a slightly different navigational problem. The gravity values of interest to geophysics are anomaly values; the difference between that on a theoretical spheroid and actual measurements. Theoretical gravity is a function of latitude and has a maximum gradient of approximately 0.8 milligals/kilometer at 45° latitude. A larger potential source of error is the east-west component of ship's speed, the Eotvos effect,

313

M. Kumar and G. A. Maul (directors), Marine Positioning, 313-316.
© *1987 by the Marine Technology Society.*

approximately 14 milligal/meter/second (approximately 7 Mgal/knot). Bell and Watts show the accuracies required to address various geological problems. Errors of less than a milligal are highly desirable. The instruments easily achieve this. It is navigational uncertainty that imposes the limits on interpretation. A one milligal error can be caused by 0.7° heading error at 10 knots on a north-south course, a 0.14 knot speed error on an east-west course or a 1.2 kilometer north-south positional error. Clearly ship's velocity is the major source in Marine gravity inaccuracy.

As vertical accelerations of the ship cannot be distinguished from changes in gravity, gravimeters are highly damped. The filter time constant is on the order of 100 seconds. The ship's velocity should be subject to the same low pass filter before the Eotvos correction is calculated and applied. Ship's velocity, therefore, needs to be accurate at periods greater than one minute.

3. NAVIGATION SYSTEM ERRORS

Navigation systems on CONRAD include: Magnavox T-set GPS, Magnavox 1107R transit satellite, Internav LC-408 range-range Loran-C, Furuno CI-30, two-axis Doppler speed log, Sperry MK-27 Gyro, and HP 5065A Rubidium frequency standard. Each of these has varying inherent accuracy over different time periods. Transit satellite fixes every 100 minutes clearly tell us nothing about the ship's motion between these points. Individual fixes have uncertainties on the order of 10 to 100 meters, but in the aggregate have absolute accuracy.

Range-range Loran-C can have high accuracy (approximately 10 meters) in the 1 to 10 hour range. The individual range measurements need calibration by, for instance, comparison with 10 or more transit satellite fixes. We have used Loran-C signals to calibrate our Rubidium frequency standard in Bermuda, St. Johns, New Foundland, and L'Orient, France. These studies show that Rubidium frequency offset can be determined to 0.1 microsecond/day (300 meters/day) and that Loran-C long range propagation can be stable to this same accuracy. Exceptions occur at dawn and dusk when $\pm$ 0.5 microsecond excursions were noticed in Bermuda from the inland Seneca, New York and Caribou, Maine stations.

The current GPS system has high absolute position accuracy (approximately 10 meters), but our experience is that the independent velocity measurements can have 1 knot errors. To be accurate to 0.1 knot, velocity from successive one minute positions requires approximately 3 meter accuracy which is often not obtainable.

The combined Gyro/Doppler speed log is somewhat difficult to evaluate. Individual speed and heading readings appear to vary by $\pm$ 0.5 knot when plotted as east and north speed components. We vectorially add the individual three second readings to produce one minute averages. Minute to minute variation is reduced to $\pm$ 0.1 knot. We obtained approximately 16 hours of GPS during a 1-1/2 day transit across the northwest Australian continental shelf in March, 1986. The speed log was in bottom track and speed was approximately 11 knots with a two hour period at 7 knots. A plot of GPS minus dead reckoning (DR) position shows maximum departure rates of 0.1 knot with long periods with much lower rates. This is rather remarkable since the log readings are given in 0.1 knots and the Gyro repeater is a 1/6 degree stepper. Winds were light with a two foot swell at this time. The change in ships speed from 11 to 7 knots is not discernible on the position difference plot.

Most of the 28 day N.W. Australia cruise was in deep water with the speed log in water track. Further checks of GPS-DR difference navigation showed no slope changes during course or speed changes. This cruise was devoted to multichannel seismics, so speeds hovered around 5 knots. Slopes exceeded 1 knot at times, indicating the presence of currents of this magnitude. In the latter part of the cruise, as the weather deteriorated, (20 knot winds) there were larger excursions in the DR speed components, but these were matched in the GPS derived speeds, indicating that these ship accelerations were real.

The conclusion is that the Gyro/speed log combination has an accuracy approaching 0.1 knot in each component, absolute in bottom track and relative in water track.

4. NAVIGATION SMOOTHING

The speed log/Gyro appear to have high relative accuracy in the one minute to one hour range. Averages over one minute are needed to reduce the jitter in individual readings. The variability of ocean currents on time scales greater than one hour is apparent from GPS-DR difference plots. However, they did not appear to have significant shorter period components off N.W. Australia during March and April, 1986.

Velocities from successive one minute GPS positions appear to have jitter that varies from 0.2 knots to several knots. In this intercomparison of speed systems we must make some assumptions. The prime one is that ships have considerable inertia, and absent a change in driving force, the velocity remains constant. Thus, short period jitter is not real.

Operations that average positions before and after the target point can only be used while the ship's velocity is constant. That is, they cannot be used near or around ship speed or heading changes. The basic ship's maneuvers are, however, recorded by the Gyro/speed log system. We propose to average or smooth the GPS-DR difference which has these maneuvers removed. The smoothed difference is then added back to the DR positions to produce the final navigation.

For this operation to be successful both the speed log and Gyro need to be accurately calibrated. A speed log error will manifest itself as a current directed along the ship's axis. This current changes abruptly during course changes. The smoothing operation assumes only slowly varying currents of periods longer than the operator length. A small Gyro error shows as a thwart ship current which also changes abruptly during maneuvers.

Simple smoothing operators reduce random noise with a power proportional to length. It is assumed that the position error spectrum decreases with increasing period, as the operator can only suppress error with periods less than its length. The operator length must be less than the current periods. The noise must have this short period characteristic. Occasional spikes in the position will not be removed by smoothing, but will simply bias the smoothed position away from the true position.

An independent check on ship's east-west velocity component was propose by Bell and Watts. The Eotvos corrected gravity and the Eotvos correction itself

should be uncorrelated. That is true gravity changes should not be related to
ships velocity changes. Their method provides a test of the proposed
smoothing method and would allow proper selection of operator characteristics.

5. CONCLUSIONS

The speed log/Gyro combination can be used to improve GPS or Loran-C
positions. A dead reckoning track from the speed log/Gyro is subtracted from
the GPS/Loran-C positions. The difference is smoothed and added back to the
DR track to produce final navigation. This is beneficial due to the superior
short period accuracy of the speed log/Gyro. The efficiency of this technique
can be tested in calculating the Eotvos gravity correction.

A NEW SOUND VELOCITY MEASUREMENT SYSTEM

Marie C. McIntyre and Dwight E. Boegeman
Scripps Institution of Oceanography
Marine Physical Lab, A-005
La Jolla, California 92093

ABSTRACT

At present, ranging with acoustic signals is the primary means for determining positions on the sea floor. Converting travel time measurements into distances requires accurate knowledge of the effective sound velocity along the propagation path. In fact, marine positioning accuracy is presently most limited by the uncertainty in determinations of the ocean's sound velocity field. The best sea-going sound velocity meters have accuracies of a part in 10^4, which translates into position uncertainties of 100 centimeters in 10 kilometers. A recently developed sound velocity meter improves this accuracy by a factor of ten, using a time-averaged, phase comparison process. A description of this sound velocity meter and a discussion of its calibration and accuracy are presented here. Ultimately, knowledge of sound velocity in the ocean is limited by the variability of the ocean's temperature and salinity structure. Therefore, the requirement is to measure sound velocity accurately on time and length scales appropriate for the position measurement being made, and a brief discussion of statistical methods of measurement in the ocean in order to adequately model the spatial coherence and temporal variability of sound velocity follows.

1. INTRODUCTION

In order to locate a point on the sea floor, a marine positioning system is proposed in which a network consisting of at least three transponders is installed on the sea floor (see figure 1). By ranging on the transponders from a sea surface platform near the surface-projected geometric center of the transponder network, the transponders become sea floor reference points precisely located relative to the platform which, in turn, may be located in an absolute frame of reference by conventional positioning techniques such as GPS, laser ranging, etc. Alternatively, the relative positions of the transponders to each other may be determined by ranging on each from a deeply towed vehicle. In either case, the acoustic travel times must be converted to distances using the effective sound velocity along the propagation path. The effects of increasing sound absorption with increasing frequency (Fisher and Simmons, 1977) as well as typical noise levels (Urick, 1975) over ten kilometer ranges, are that travel times can be determined to within ten microseconds, given reasonable output power (Spiesberger and Worcester, 1981).

A newly developed precision transponder is capable of measuring travel times to within ten microseconds or less (Spiess, 1980). This indicates that distances can be measured to a centimeter over kilometer ranges, once sound velocity is known to a part in 10^5. Sound velocity may be determined directly with a sound velocimeter or calculated from conductivity, temperature and pressure measurements. In order to calculate sound velocity to 1 in 10^5, temperature must be measured to about a millidegree Centigrade, pressure to < 1 decibar and salinity to better than .01 ppt. Although these levels of measurement accuracy are achievable with off the shelf instruments, the equations relating conductivity, temperature and pressure to sound velocity are only good to 1/3 of a part in 10^4 (Lovett, 1978). Commercially available sea-going sound velocimeters, however, are only good to 1 part in 10^4 (Mackenzie, 1973). Thus we have recognized and attempted to meet the need to develop a new instrument capable of measuring sound velocity to at least a part in 10^5.

2. THE INSTRUMENT

2.1. <u>Transducer Probe.</u> The transducer probe used with our system is a single transducer, folded path probe originally built by Nusonics, Inc., (see figure 2). The pressure case rates to 1000 bars and is built out of Hastelloy

M. Kumar and G. A. Maul (directors), Marine Positioning, 317–326.
© 1987 by the Marine Technology Society.

Figure 2.   Sound Velocimeter Probe and Electronics

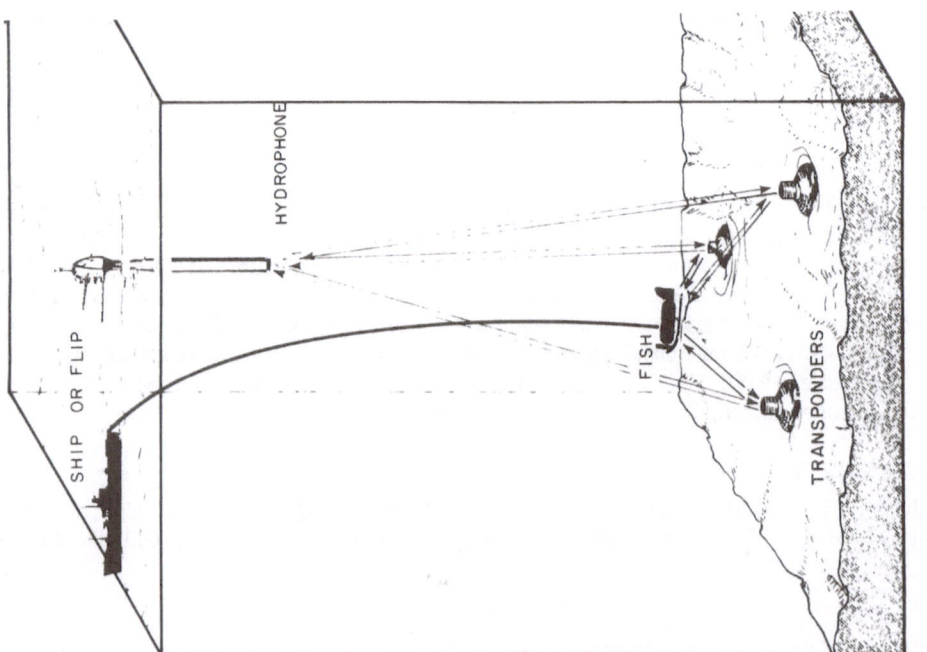

Figure 1.   Sea Floor Reference Points

C, a metal alloy whose coefficients of thermal expansion are small and well known. The transducer crystal is resonant near four Megahertz and the total path length from crystal to reflector and back is about eight and a half centimeters. A baffle has been added surrounding the transducer face to limit multiple echos.

2.2. Circuitry Logic. A recent literature survey indicated that sing-around type velocimeters seem to be capable of the greatest accuracy, (Greenspan and Tschiegg, 1957; Wilson, 1959; Carnvale et al, 1968; Millero and Kubinski, 1975; Chen and Millero, 1976). These velocimeters receive a pulse train, triggering the transmission of the next pulse train and so on. The pulse repetition frequency measured is then related to sound velocity. Their resolution is thus limited by wavelength or frequency. Our newly designed sound velocimeter is an extension of the sing-around concept, using a time-averaged, phase comparison process to more accurately clock the round trip travel time.

It operates near four Megahertz, repeating a four-part timing cycle: transmit, wait, receive, and wait again. During transmit, 129 cycles from a continuously running voltage controlled oscillator (VCO) are sent to the transducer (see figure 3). A count down, which starts on the second cycle of the transmit pulse train, continues during the wait period and ends in the first part of the receive period, when an OK time window one cycle wide is generated. The received echo is expected within this window. If the echo arrives in the receive period before or after this window, the voltage sent to the VCO is changed to increase or decrease its frequency, bringing the leading edge of the echo inside the OK time window.

Once this is accomplished, a phase lock loop takes over and fine-tunes the frequency of the VCO, matching the phase of the VCO output with 32 cycles of the echo pulse train. This phase matching technique permits round trip travel time to be clocked with an accuracy greater than first cycle identification only. During the wait again period the further reflected echos die down and much later another pulse train is transmitted. This is also an improvement over conventional sing-around techniques which retransmit as soon as the first reflected echo is received. Further reflected echos then interfere with the first reflected echo of the next transmitted pulse, resulting in an apparent phase shift, delaying recognition of the real echo.

The countdown between transmit and the OK time window can be changed manually on a thumbwheel (see figure 4), yielding a range of VCO operating frequencies which result in correct phase lock. Since the countdown represents the integral number of cycles between the second cycle of transmit and the second cycle of receive (the first cycle being lost in the noise), a different operating frequency is attained for each thumbwheel setting. The count divided by the resulting frequency, however, yields travel time, which should be a constant independent of thumbwheel setting or frequency.

3. CALIBRATION

3.1. Discussion. To determine sound velocity from the operating frequency of our sound velocimeter, there are two unknowns which must be modeled: path length and electronic delays. The equation relating frequency to sound velocity is:

$$\frac{count}{freq} = traveltime = \frac{L(1 + \alpha T)}{C} + B \tag{3.1}$$

where L is the path length, α is the thermal coefficient of expansion, C is sound velocity and B is the electronic delay, which may also be modeled as a function of temperature and frequency. Least squares fits of L with temperature and B with temperature and frequency, yield estimates of L, α, and all time delay coefficients.

The path length and thermal coefficients of expansion are specified by the probe's manufacturer, Nusonics Inc., but are also measured indirectly during the calibration. Changing the path length by a precisely known amount and measuring the resulting change in travel time, results in an independent estimate of path length:

$$\frac{L}{t - B} = \frac{\delta L}{\delta t} \tag{3.2}$$

where L is the path length, t is the travel time and B is the electronic delay. This is done using gage blocks whose widths are known to 0.02 μ meters.

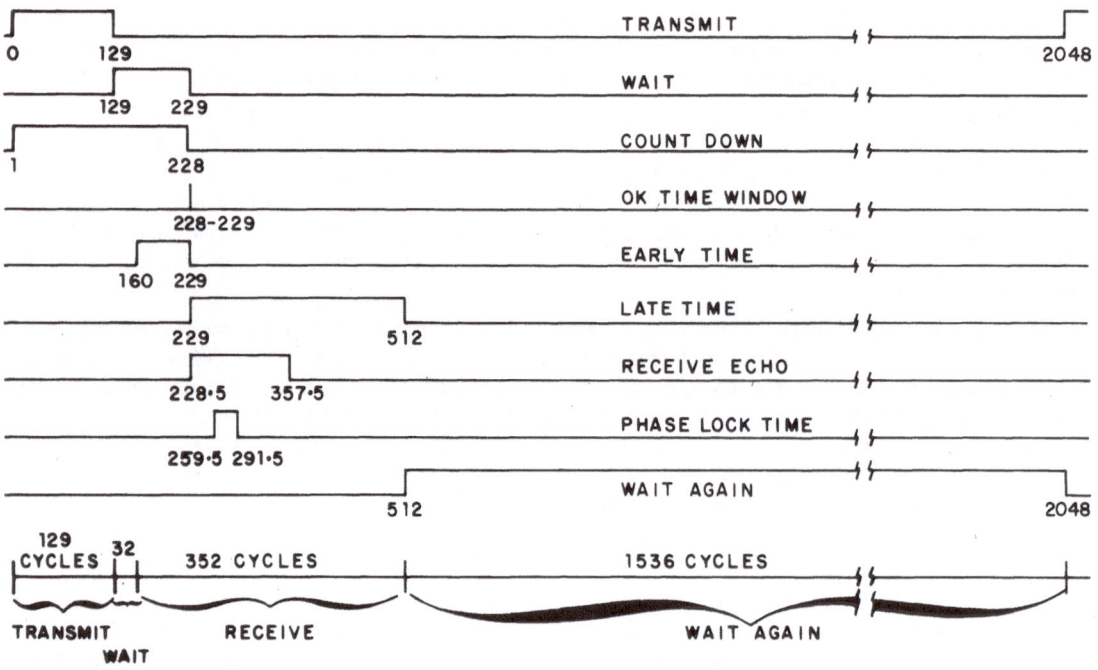

Figure 3. Timing Diagram

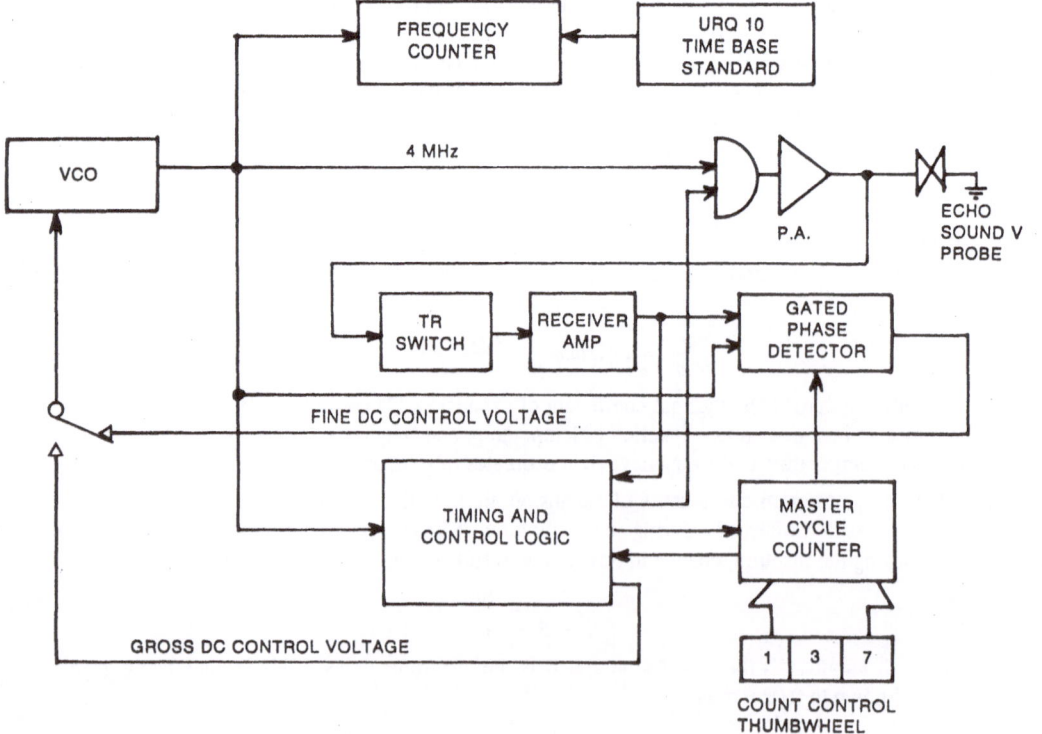

Figure 4. Block Diagram

3.2. Set-up. The sound velocimeter was calibrated inside a temperature controlled water bath in a chill room. The room was cooled to 4°C and remained between 4°C and 6°C for the duration of the experiment. The sound velocimeter itself resides in a specially constructed plastic tank, filled with 1 M Ω pure water. This is suspended in a stirred outer water bath which in turn is surrounded by glycol circulating continuously through refrigeration coils. There are rheostat controlled heater blades in both the glycol and the outer water bath.

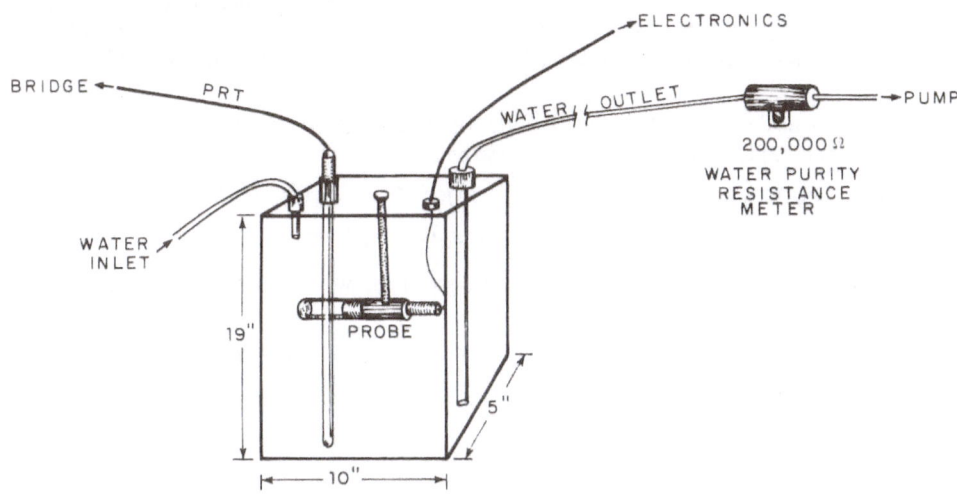

Figure 5. Calibration Tank

The tank is air-tight and water-tight. The probe is suspended about 9 inches deep by a rod, bolted to the cover, (see figure 5). The platinum resistance thermometer (prt) used to measure temperature was placed perpendicularly alongside the sound path with the axis of the sound path at the mid-point of the prt. Water can be drawn off the bottom and fresh 1 M Ω water added at the surface as needed. Nitrogen blankets the top of the tank to keep the water from absorbing carbon dioxide, raising the impurity level. The temperature of the pure water is measured on a Mueller bridge outside in a lab, to 0.001 °C. The sound velocimeter electronics and the frequency counter are also in a lab outside the chill room.

3.3. Procedure. First, all the refrigeration units were turned on cooling the bath to 0°C, the glycol to -4.45°C, and the chill room to 4.2°C. Then, the heater blades were turned on in the glycol and outer water bath to reach a starting temperature of nearly 30°C. After the bath temperature had stabilized and the sound velocimeter frequency was changing by only 1 part in 10^6 or so, temperature and frequency measurements were taken. A minimum of ten sets of simultaneous prt bridge readings and frequency measurements were taken at each temperature for a mid-range thumbwheel setting. In addition, a minimum of six sets of data were taken for ten to twenty different thumbwheel settings at each temperature. The prt bridge measurements were taken with current flowing alternately in a normal and reverse sense.

Data were taken from 28°C to 2°C, decreasing by about 3° increments. Additional measurements were made with the shorter path length at four different temperatures: 28°C, 20°C, 10°C, and 0°C. Before and after measurements at each temperature the water purity was monitored by drawing water off the bottom and pumping it through a 200,000 Ω resistance meter. Throughout the experiment a nitrogen blanket was kept over the pure water and the water level above the probe was monitored in order to subtract out any possible effects of water pressure on sound velocity and path length.

3.4. Results. It proved difficult to hold the temperature in the bath exactly constant, so most of the data were taken as the temperature increased or decreased over a small range. A least squares fit of bridge resistance to operating frequency (which also increases or decreases slowly in response to the temperature drift) smoothes the data and from this fit the averaged resistance and frequency are calculated (see figure 6). The double rows of data in figure 6 are the result of an approximately constant offset between normal and reverse bridge readings and the average between them is the correct bridge resistance for that temperature, in the 1968 temperature scale. The mean square error of fits to all the normal readings and reverse readings separately was about 0.0001 ohms. This translates to a temperature uncertainty of about 0.001°C or better than a part in 10^5 in sound velocity.

Sound velocity was calculated from these temperature data to less than 3 parts in 10^6 using the equation of Del Grosso and Mader (1972) for pure water at atmospheric pressure.

$$C = 1402.3876 + 5.037111 \; T \; - \; 5.808522 \times 10^{-2} \; T^2 + 3.341988 \times 10^{-4} \; T^3 \; -$$
$$1.478004 \times 10^{-6} \; T^4 + 3.14643 \times 10^{-9} \; T^5. \tag{3.3}$$

A correction was then made to sound velocity for the weight of the water above the probe, using only the highest order terms from Chen and Millero (1976):

$$C_o = C_p - (0.153563 + 6.8982 \times 10^{-4} \; T - 8.1788 \times 10^{-6} \; T^2)P \tag{3.4}$$

where P is in bars. This correction was barely significant at the part in 10^5 level, as can be seen in the data table below.

Table 1. Resistance, Temperature and Sound Velocity Data

Resistance	Temperature	Sound Velocity	C, corrected
28.49574	28.7430	1506.1705	1506.1671
28.50088	28.7939	1506.2922	1506.2888
28.16209	25.4442	1497.8673	1497.8639
27.65408	20.4278	1483.6488	1483.6454
27.30755	17.0104	1472.7894	1472.7861
26.95488	13.5361	1460.7084	1460.7051
26.67964	10.8272	1450.5203	1450.5170
26.26816	6.7815	1433.9765	1433.9733
25.84978	2.6731	1415.4434	1415.4402

The effects of water pressure on path length, due to linear compression and bowing (Mackenzie,1971), are negligible in this case but will be significant at higher pressures in the ocean. The path length does need to be corrected for thermal expansion, however. The probe's manufacturer lists the round trip path length as 8.6108 cm at 0°C and gives the thermal coefficient of expansion as 15×10^{-6} per degree C. The gage block used to shorten the path length was exactly 5mm wide at 20°C and has a thermal coefficient of expansion of 8.5×10^{-6} per degree C. The resultant path lengths at each temperature are given in Table 2 below.

The travel times listed in Table 2 include some electronic delay time. To calculate this delay, the theoretical travel time, length divided by sound velocity, is subtracted from the actual travel time listed here. These delays are listed in the last column of Table 2.

Table 2. Path Length and Time Delay Data

Count	Frequency, MHz	Travel Time, μsec	Length, cm	Delay, μsec
220	3.823629	57.536962	8.614513	0.3420299
220	3.823889	57.533050	8.614519	0.3426958
220	3.803155	57.846708	8.611409	0.3375712
220	3.767857	58.388627	8.613439	0.3327128
220	3.740929	58.808921	8.612997	0.3279440
220	3.710862	59.285416	8.612548	0.3238364
220	3.784913	59.710751	8.612198	0.3374517
226	3.742440	60.388409	8.611676	0.3337622
230	3.758015	61.202523	8.611145	0.3653008

Before modeling these delays as functions of frequency and temperature, they were normalized to the same count value, 220. This was done by running a least squares fit on the set of data taken over a range of thumbwheel settings at 28°C, (see figure 7). This accounts for electronic delays which are a known fraction of a cycle and accumulate additively. Plotting the rest of the data on the same graph revealed an additional constant delay of unknown origin in all of the data taken with a thumbwheel count of 230. This constant delay was also removed.

These scaled delays were then modeled as functions of frequency and temperature. A nonzero quadratic thermal coefficient of expansion, β, would look like an electronic delay that was a function of T^2/C:

$$\frac{count}{frequency} = \frac{L(1 + \alpha T)}{C} + \frac{\delta T^2}{C} \tag{3.5}$$

where $\beta = \delta/L$. A least squares fit of scaled delay to T^2/C, however, had an rms misfit of 0.0015 μsec which is equivalent to an error in sound velocity of 4 parts in 10^4.

Phase shifts due to the transducer operating over a range of frequencies near resonance would look like an electronic delay which was a function of the product of frequency and temperature. A least squares fit of scaled delays to frequency times temperature is shown in figure 8, and models the electronic delays quite well, with an rms misfit of only 0.00048 μsec, or less than a part in 10^5 of sound velocity. Further fits of the remaining delay with frequency and temperature individually don't improve the fit.

Thus the best equation relating operating frequency of our velocimeter to sound velocity is given by:

$$C_c = \frac{L(1 + \alpha T)}{\dfrac{count}{frequency} - B} \tag{3.6}$$

where $\alpha = 15 \times 10^{-6}$ per degree C, L = 8.6108 cm and B is calculated from B = -0.3175984 + 0.2847424 $\times 10^{-2}$ count + 0.3026129 $\times 10^{-3}$ freq. $\times$ Temp.

An independent estimate of the path length may be obtained using the data taken over shorter path lengths. Equation 3.2 was used to calculate total path length at each temperature and the results are shown in Table 3.

Table 3. Indep. Est. of Path Length using Gage Blocks

Temperature	Length, cm	Length at 0°C
28.6912	8.61987	8.61503
20.2372	8.63893	8.64157
10.6522	8.64295	8.63632
-0.0408	8.61503	8.61617

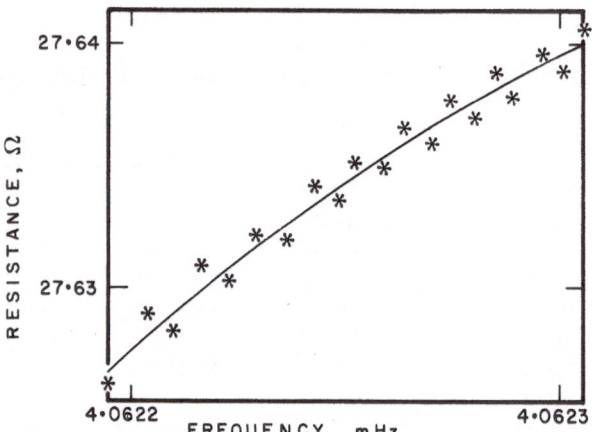

Figure 6. Least Squares Fit of Resistance and Frequency

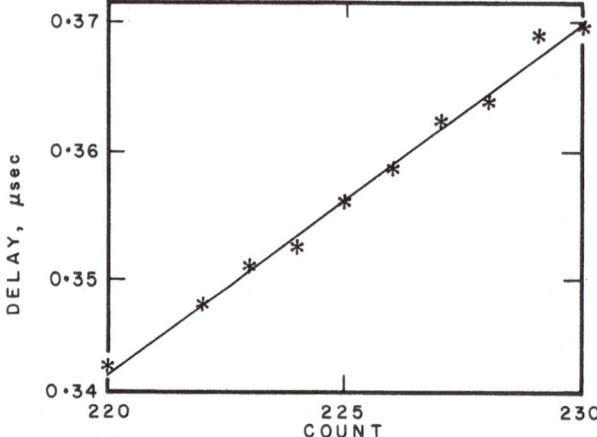

Figure 7. Least Squares Fit of Delay and Count

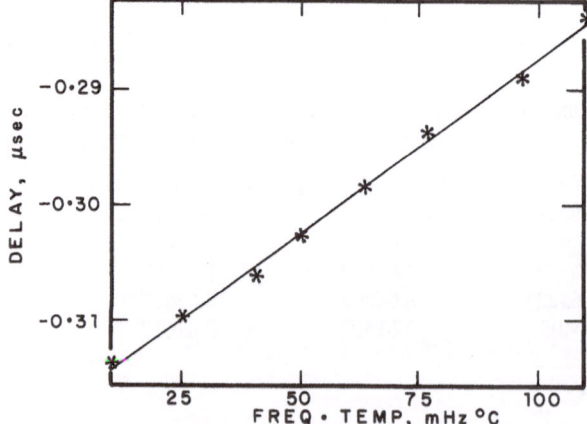

Figure 8. Least Squares Fit of Delay and Frequency × Temperature

In order to know the change in path length exactly, the gage block must be attached to the reflector on the probe with a gap less than 0.02 μmeters. This was obviously not achieved with the measurements at 20°C and 10°C, but may have been achieved with the measurements at 28°C and 0°C, in which case the best estimate for path length would be 8.6156 cm +/- 0.0006. More work needs to be done before a real estimate of path length can be measured, however.

4. SUMMARY OF ACCURACY

The standard error of estimate, s, for equation 3.7 relating sound speed to operating frequency is given by:

$$s = \left[\frac{(C_c - C_o)^2}{p - q} \right]^{\frac{1}{2}} \tag{4.1}$$

where C_c is the sound speed calculated using the velocimeter, C_o is the sound speed in pure water, p is the number of data points and q is the number of independent coefficients or one plus the order of the independent variable. The resulting standard error of estimate for equation 3.6 is 0.013 m/sec, which is slightly better than a part in 10^5 of sound velocity. Thus we have met the need to develop an instrument capable of measuring sound velocity to at least a part in 10^5. This velocimeter could now be used in saline solutions to measure sound velocity where empirical equations aren't accurate enough.

Some improvements could be made in accounting for time delays, however. More work needs to be done on modeling time delays using data at different thumbwheel settings. Furthermore, it should be possible to isolate more physical causes of time delays. For example, the probe shape itself causes guided wave-mode propagation, resulting in diffraction. A rough modeling of this effect predicts 12.5 degrees of phase shift (a 0.01 μsec delay), which varies with water temperature. Finally, some of the time delays may be removed by thermally isolating the electronics and using higher frequency components.

5. FUTURE DEVELOPMENTS

Ultimately, this instrument will be used in the ocean to model sound speed both spatially and temporally. First, however, it must be calibrated at higher pressures in the lab. Next, the dispersion relation between sound speed measured at 4 MHz and ranges measured near 10 kHz must be determined.

The dispersion of sound waves in the ocean is a result of the relation between the real and imaginary parts of the index of refraction of sea water. The imaginary part relates to the attenuation of sound waves with frequency as well as chemical relaxation processes (Fisher and Simmons, 1977). The real part relates to the dispersion of sound velocity with frequency, and can be predicted from the imaginary part by the Kramers-Kronig relation. There is an extensive literature on this dispersion relation (Ginzberg, 1955; Mangulis, 1963; O'Donnell, Jaynes and Miller, 1978 and 1981; Weaver and Pao, 1980). In our case the dispersion between 4 MHz and 10 kHz is likely to be from 0.017 m/sec in the deep ocean to 0.035 m/sec in warmer, shallow water and will be a function of water temperature, salinity and depth. This is clearly at or above the part in 10^5 level at which we hope to measure sound velocity and will need to be determined empirically.

Finally, in order to make use of our ability to accurately measure sound velocity, the length and time scales over which it changes must be modeled, using the autocorrelation coefficients of the index of refraction in space and in time (Chernov, 1960; Flatte, 1979). From these, correlation times and distances can be estimated, within which the mean square phase fluctuations (and therefore transit time fluctuations) remain below the part in 10^5 level. Events likely to affect the usefulness of sound velocity measurements include turbulent microstructure, internal waves, and tides. The first two are mostly limited to the upper half of the water column and the last can be adequately measured and subtracted out of the data. The characteristics of water in the upper half of the water column change by about a part in 10^3 over time scales too short to model, whereas the deeper ocean is probably stable to a part in 10^5 or so. Thus the stability of the medium is an important limitation to our ability to usefully model sound velocity over travel paths in the ocean and it is really environment, not engineering which ultimately limits achievable accuracies in range measurement.

REFERENCES

Carnvale, A., Bowen, P., Basileo, M., and Sprenke, J., 1968. "Absolute Sound- Velocity Measurement in Distilled Water." J.Acoust. Soc. Am., Vol. 44(4), pp.1098-1101.

Chen, C., and Millero, F.J., 1976. "Reevaluation of Wilson's Sound-Speed Measurements for Pure Water." J.Acoust. Soc. Am., Vol. 60(6), pp. 1270-1273.

Chernov, L.A., 1960. "Wave Propagation in a Random Medium." McGraw-Hill, New York.

Del Grosso, V.A., and Mader, C.W., 1972. "Speed of Sound in Sea-Water Samples." J.Acoust. Soc. Am., Vol. 52(3), pp. 961-974.

Fisher, F.H., and Simmons, V.P., 1977. "Sound Absorption in Sea Water." J. Acoust.Soc. Am., Vol. 62(3), pp. 558-564.

Flatte, S.M., 1979. "Sound Transmission through a Fluctuating Ocean." Cambridge University Press, Cambridge.

Ginzberg, V.L., 1955. "Concerning the General Relationship between Absorption and Dispersion of Sound Waves." Sov. Phys. Acoust., Vol. 1, pp. 32-41.

Greenspan, M., and Tschiegg, C.E., 1957. "Sing-Around Ultrasonic Velocimeter for Liquids." J.Acoust. Soc. Am., Vol. 28(11), pp. 897-901.

Lovett, J.R., 1978. "Merged Seawater Sound-Speed Equations." J.Acoust. Soc. Am., Vol.63(6), pp. 1713-1718.

Mackenzie, K.V., 1971. "A Decade of Experience with Velocimeters." J. Acoust. Soc. Am., Vol. 50(5), pp. 1321-1333.

Mackenzie, K.V., 1973. "Calibrations of Oceanographic Research Velocimeters." J. Acoust. Soc. Am., Vol. 53(3), pp. 869-875.

Mangulis, V., 1964. "Kramers-Kronig or Dispersion Relations in Acoustics." J. Acoust. Soc. Am., Vol. 36, pp. 221-222.

Millero, F.J., and Kubinski, T., 1975. "Speed of Sound in Seawater as a Function of Temperature and Salinity at 1 Atm." J.Acoust. Soc. Am., Vol. 57(2), pp.312-319.

O'Donnell, M., Jaynes, E.T., and Miller, J.G., 1978. "General Relationships between Ultrasonic Attenuation and Dispersion." J.Acoust. Soc. Am., Vol.63(6), pp. 1935-1937.

O'Donnell, M., Jaynes, E.T., and Miller, J.G., 1981. "Kramers-Kronig Relation-ship between Ultrasonic Attenuation and Phase Velocity." J. Acoust. Soc. Am., Vol.69(3), pp. 696-701.

Spiesberger, L., and Worcester, P.F., 1981. "Fluctuations of Resolved Acoustic Multipaths at Long Range in the Ocean." J. Acoust. Soc. Am., Vol. 70(2), pp. 565-576.

Spiess, F.N., 1980. "Acoustic Techniques for Marine Geodesy." Marine Geodesy, Vol.4(1), pp. 13-27.

Urick, R.J., 1975. "Principles of Underwater Sound." McGraw-Hill, New York.

Weaver, R.L., and Pao, Y., 1981. "Dispersion Relations for Linear Wave Propagation in Homogeneous and Inhomogeneous Media." J. Math. Phys., Vol. 22(9), pp.1909-1918.

Wilson, W.D., 1959. "Speed of Sound in Distilled Water as a Function of Temperature and Pressure." J. Acoust. Soc. Am., Vol. 31(8), pp. 1067-1072.

EVOLUTION OF POSITION-LOCATING SYSTEM ON DATA BUOYS

Phillip J. Kies and Dennis W. Mahar
National Data Buoy Center
NSTL, Mississippi

ABSTRACT

The National Data Buoy Center (NDBC) operates and maintains a fleet of environmental data buoys on the Atlantic and Pacific Oceans, the Bering Sea, the Gulf of
Alaska, the Gulf of Mexico, and the Great Lakes. This fleet of moored buoys,
numbering approximately 50 as of this writing, occasionally suffers a failed
mooring, at which time the buoy "gets underway." In this event, up-to-date,
accurate positions are required to facilitate tracking the buoy by a recovery
vessel. In addition, daily position reports from all deployed buoys are required
to ensure the buoys are "on station" and have not moved from their reported
positions, which could create a hazard to navigation.

For the past decade, NDBC has used Loran C and Service Argos Platform Transmit
Terminals (PTT's) for deep-ocean buoy positions. These systems are not without
their share of problems when used in remote, unattended situations. NDBC is now
looking to the future and the possible use of GPS or GEOSTAR to provide position
and tracking information. This paper takes an in-depth look at NDBC's past,
present, and future involvement with various electronic navigation/positioning
systems, including hardware and software.

1. INTRODUCTION AND BACKGROUND

The present-day system of buoy platforms operated by NDBC has evolved from the
original National Data Buoy Development Project begun in 1967. The design goal
of this project was to develop a large buoy platform capable of being moored at
various deep-ocean sites. The US Coast Guard (USCG) was tasked with this project
because of their long-standing experience with buoys of all types, and because
they had the in-house resources to support the project. The USCG was at that
time actively engaged in the design and development of a Large Navigational
Buoy (LNB), which could be used to replace the aging and expensive fleet of manned
lightships near major US ports. Since many of the engineering design goals for
both buoy projects were very similar, it seemed to be the most cost-effective
approach. The data buoy would house all of the power, signal, control, and radio
equipment needed to support the hourly acquisition and transmission of meteorological and oceanographic sensor data from a remote ocean site. The meteorological
data obtained from a system of these buoys would be invaluable in providing
information on long-term weather patterns, and in forecasting hurricanes and
other potentially dangerous storm systems approaching the continental US.

In 1970, the data buoy project was transferred to the National Oceanic and
Atmospheric Administration (NOAA) in the Department of Commerce, and relocated to
the NASA Mississippi Test Facility near the Mississippi Gulf Coast. A nucleus of
Coast Guard officers from the project were retained to assist in the transition and
continuing buoy development. Today, the project has evolved into the National
Data Buoy Center as an organizational element of the National Weather Service,
and the NASA facility is now titled the National Space Technology Laboratories.

M. Kumar and G. A. Maul (directors), Marine Positioning, 327–336.
© 1987 by the Marine Technology Society.

Under the auspices of an interservice agreement, the Coast Guard continues to
provide a minimum of 15 personnel to NDBC on a full-time basis and provides
ships, aircraft, and small boat support on an as-required basis. All buoy
deployments, service visits, and retrievals are carried out from Coast Guard
vessels.

The first experimental deep-ocean moored buoy was deployed in 1971, and since
that time the program has continually expanded. In 1974, the deep-ocean network
was supplemented by the Continental Shelf (Conshelf) moored buoys, a second network
of buoy stations deployed within 100 nautical miles of the US East and West
Coasts. Figure 1 is a photograph of a 12-meter discus buoy in a typical moored
configuration. Figure 2 shows the operational buoy network in 1975, consisting
of 14 moored buoys. By 1980, this number had more than doubled, and, currently,
there are over 50 moored buoys in the NDBC operational network, supporting
several domestic and international programs. Figure 3 shows the current buoy
network with Loran C coverage indicated.

In the years 1971-1980, there were major evolutionary developments in the data
acquisition and transmission equipment. The first electronic payloads comprised
older technology of discrete component systems, with a 100-watt HF transmitter to
telemeter the data to shoreside. These systems required tremendous amounts of
power to operate, making them very expensive to use in a continuous mode. In the
mid-70's, the National Environmental Satellite Service (NESS) was created and
established the Geostationary Operational Environmental Satellite (GOES) system.
The GOES system consists of two satellites in stationary orbits over the equator,
one (GOES EAST) at $75^{\circ}W$ longitude, and the other (GOES WEST) at $135^{\circ}W$ longitude.
The satellite visibility of this system is excellent, with full coverage on the
surface of the earth over the entire North American continent.

The second-generation electronic systems were hybrids that mixed discrete
component technology with TTL devices and featured dual HF/UHF telemetry capability
to test the ability of the GOES system to handle the data flow from the platforms.
Figure 4 shows a simplified data flow diagram for both HF and UHF telemetry of
the synoptic messages from a data buoy. After successful testing of the UHF
telemetry link through GOES, NDBC began conversion of the buoy electronics to
operate only on UHF. The changeover resulted in a considerable reduction of
power consumption while increasing the successful error-free data rate. Late in
1979 the last HF equipment was removed, and the shoreside HF receiving station
was closed. Rapid technological progress in the integrated circuit industry
resulted in the availability of sophisticated CMOS microprocessors and VLSI chips
for the design of the current generation of electronic data acquisition packages
to be used by NDBC.

2. BUOY LOCATION SYSTEMS

Early in the moored buoy program, NDBC realized the importance of developing a
reliable buoy-locating-/-positioning system. When the inevitable occurred, and
one of the buoys came loose from its mooring, it was essential to be able to
recover the buoy without requiring an intensive (and expensive) search, and
before the buoy was permanently lost or became a navigational hazard. Beginning
in the early 1970's and progressing through 1980, several studies were conducted
by, or for, NDBC to determine what type of buoy-locating-/-reporting system best
suited NDBC program objectives. Many methods of position location were discarded
for such disadvantages as one or more of the following: limited range,

Figure 1. Photo of a 12-Meter Discus
Buoy in a Typical Moored
Configuration

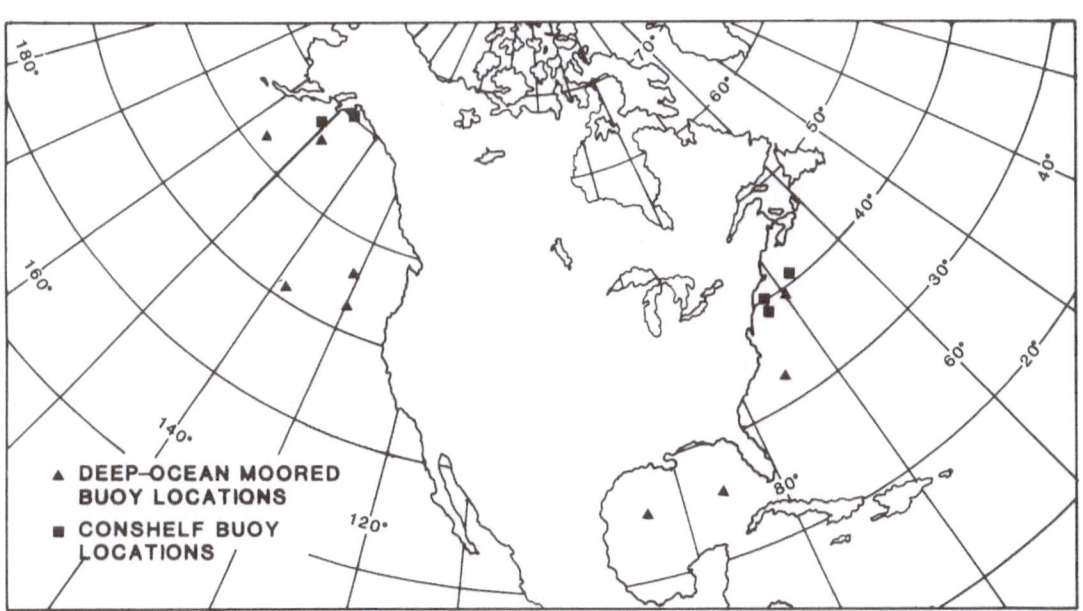

Figure 2. 1975 Operational Buoy Network (14 Buoys)

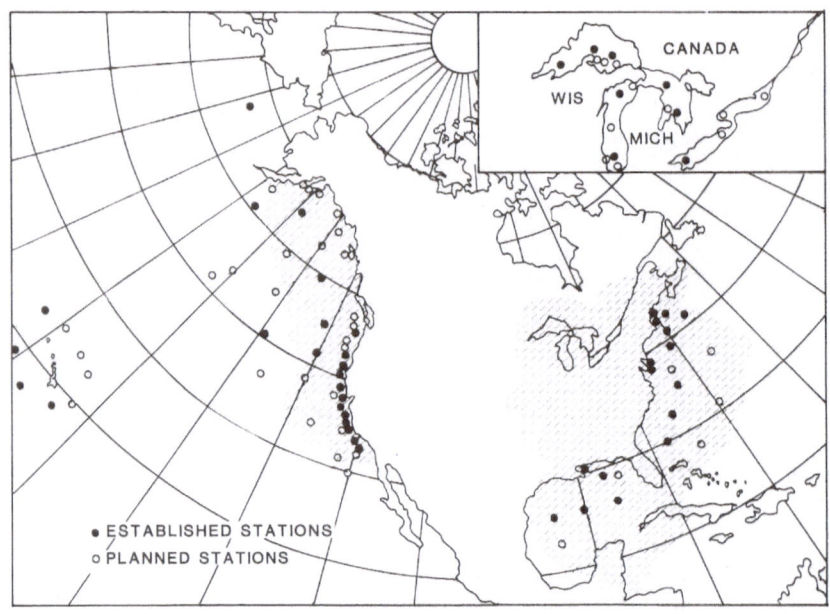

Figure 3.   Current Buoy Network With Loran C Coverage Indicated by Shaded Area

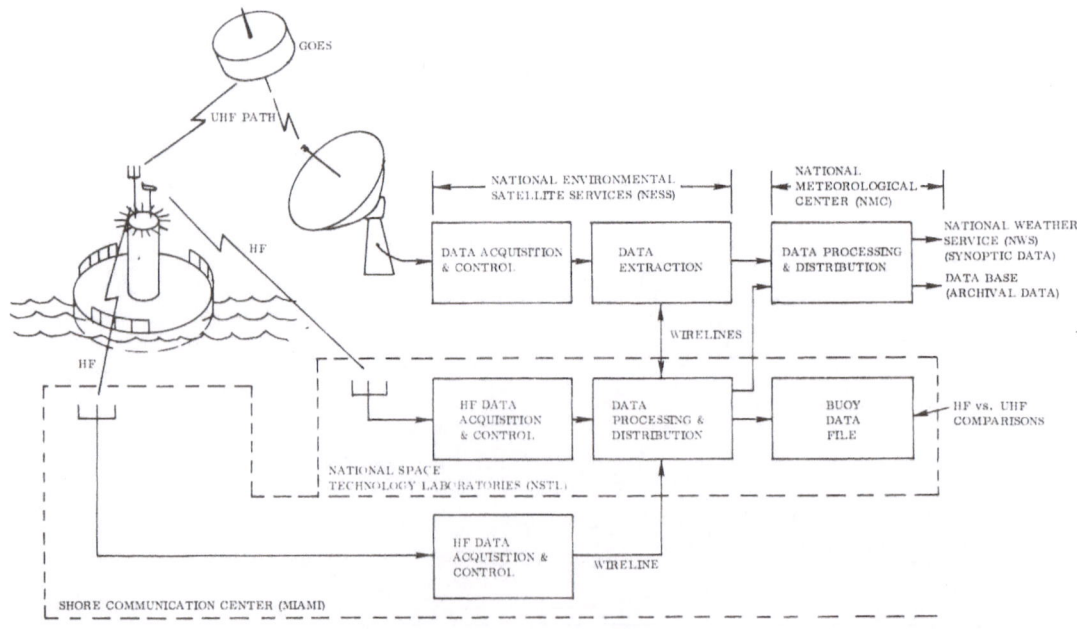

Figure 4.   Simplified Data Flow Diagram for Both HF and UHF Telemetry of
            Synoptic Messages From a Data Buoy

poor accuracy, unavailable in the required coverage areas, scheduled phase-out, or high cost. This group included Loran A, celestial observations, inertial and radar navigation systems, DECCA, RAYDIST, HF direction finding, CONSOL/CONSOLAN, OMEGA, transmit, GOES satellite ranging, IRLS, and GPS. Loran C was looked at favorably but was ultimately rejected because of the expense and poor performance of the receiving equipment available during that period. The Random Access Measurement System (RAMS) was finally selected as the most suitable position-locating system for NDBC's moored, deep-ocean buoys.

3.   RANDOM ACCESS MEASUREMENT SYSTEM

The principle of operation in the RAMS is based on traditional Doppler signal shift measurement techniques. In some other Doppler systems, such as TRANSIT, the satellite contains the transmitter, and the user determines his position with onboard receiving and computing equipment. The RAMS reverses the transmit-receive relationship; the RAMS equipment on the buoy transmits a signal to the satellite. The RAMS message is transmitted at a UHF frequency ($\approx$401 MHz), for approximately 1 second every 6 seconds. Each message contains platform identification information and a carrier signal, and may also contain sensor data, if desired. When a satellite passes overhead, it receives and detects the RAMS message from the platform, records and stores the information for later transmission to a ground control station. Since the platform transmits several times during a satellite pass, there will be several successive RAMS messages stored for each platform operating in the path of the orbit. Carrier frequency changes detected in the successive messages are later used to provide Doppler position information when the data are processed. The position reporting accuracies of the RAMS does (over time) vary slightly, but is normally within 3 to 5 kilometers (CEP). Random access is achieved by not having any time or frequency synchronization in the transmission mode. The platform operates completely independent of any other device, and the noncoherence in time and frequency allows the satellite to accommodate a large number of nearly simultaneous signals from several platforms in the same general geographical area.

NASA launched the NIMBUS-F satellite in 1975 with the RAMS capability. In the NIMBUS system, the platform data accumulated and stored by the satellite were transmitted to a ground control station during each orbit. This station relayed the raw data to Goddard Space Flight Center, Maryland, over high-speed data links. At Goddard, all raw data and position information were processed and distributed. The NIMBUS system demonstrated the effectiveness of a RAMS system with its simplicity and reliability. The biggest advantage to NDBC was that all the parts were already in place. NASA had the satellites, ground station, and Goddard Space Flight Center. The only component of the system to be added was the low-cost buoy transmitter. The platform transmitters used by NDBC on this system were Buoy Transmit Terminals (BTT's). The disadvantages of RAMS were few in the NIMBUS system. The major drawbacks were that the data were not available in real time and that the number of satellites available limited the number of fixes available each day to 2 to 4.

When the NIMBUS performance declined in 1977-1978, a new series of RAMS satellites, called the TIROS-N series, was made operational. With the TIROS system in place, NDBC obtained and used TIROS Drift Detectors (TDD's) as the platform transmitter. TDD's were also used extensively in the (expendable) drifting buoy program in the mid-1970's. Technically, the TIROS system is nearly identical to NIMBUS, except that it is an international cooperative venture with the US providing the satellites and France doing the data processing at the Service Argos facility in Toulouse, France. The big change is that France charges a user fee for each platform. Having to pay for this service caused some discontent with the system, and when the costs rose significantly, renewed interest was generated in finding an alternative system. That interest created the opportunity to re-evaluate Loran C as a useable system on data buoys.

4.   LORAN C

In 1979, the result of Service Argos user fees (approximately $3.5 K/year/ platform) was to initiate a search for an alternative position-reporting system. Technological advances in the late 1970's in microprocessor-based systems allowed several marine electronics manufacturers to market low-cost receivers that automatically lock on and track Loran C signals. These receivers were in the $1500 to $2500 price range. Currently, the price of Loran receivers is in the $800 to $1000 price range. A project began in 1979 to install a Loran C system on several test buoys and evaluate the results. The design goal was to develop a system that would take advantage of the best features of the Loran system, (e.g., signals available to users continuously, position fixing of one-quarter nautical mile or less, coverage at 95 percent of moored buoy locations, excellent repeatability (95% 2 DRMS), and cost-free to all users). The project milestones included setting operational requirements, selecting and functionally testing the receiver, developing the hardware and software interface, and field testing of the package.

Operational Requirements of NDBC Loran Reporting System: NDBC's operational requirements for a Loran Reporting System were as follow:

a. Require no extensive redesign of host payload electronics
b. Operate in an unattended mode with power cycling
c. Use a battery power supply of 12 to 28 VDC at 5.0 AH days or less
d. Accomplish automatic acquisition of a master and one pair of
     secondaries from a selected Loran C rate within the 1:3 SNR coverage area
e. Provide stable operation from 0°C to +50°C
f. Provide Loran position data to the host payload during the normal
     synoptic data acquisition period each hour

Receiver Selection: Selection of a suitable automatic Loran C receiver could have been a very difficult task because there were approximately two dozen models on the market at the time. Fortunately, the task was simplified because another agency, the USCG, had just concluded an extensive, year-long test and evaluation program to select a good Loran C receiver for their search and rescue boats. NDBC reviewed the results of the CG project and selected the Sitex-Koden Model 757 as the best Loran C receiver available for use on moored buoys. Several were bought for the development program in 1980. Initially, the sets were independently tested for form, fit, and function to check out the range of features and become familiar with the performance. This was necessary to determine the best method to interface the sets to buoy electronics systems. The Sitex receivers lived up to the excellent ratings earned in the Coast Guard test program.

Receiver/Buoy Electronics Interface: To be useful on moored buoys as a position-locating/-reporting device, the Loran sets had to be interfaced to a host electronics data acquisition system or buoy payload. Because the payload had been designed to acquire analog sensor signals, the receiver had to appear as an

analog device. Circuit cards were designed that would "latch" onto the Loran time
differences (TD's) being displayed in digital form and translate them in a
BCD-to-analog convertor, providing the correct analog output to the payload when
sampled. The interface card also contained power and control circuitry to allow
the host payload system to completely control the Loran set during a normal
meteorological data acquisition period. The interface card was designed to be
mounted piggyback on an existing receiver circuit card in the set, and all
interconnections were made through an unused connector available on the back of
the receiver, resulting in a neat, compact package. Since Loran receivers are
normally designed to be used by an operator on a vessel using front panel controls,
some software interfacing was needed in both the Loran set and the host payload.
The Loran programming was modified by the vendor so that it would only acquire
and track a master on a specified pair of secondaries on a predetermined Loran
rate (GRI) when initialized. Preselection of the rates/secondaries prior to
deployment of a buoy could then be accomplished by simply changing a set of
PROMS. In this manner, the inventory of Loran sets could be preprogrammed for
various geographical locations, and no field setup was required during actual
deployment of a buoy. The buoy payload software required a simple modification
to control the timing of the length of the signal sampling period, both to
conserve power and to prevent data from being taken before a full signal lock was
attained.

In 1984, NDBC placed a newer generation electronics payload into use that
contained a modem and serial data port. As a standard feature, the Model 757 had
a serial data output port that was normally intended for autopilot control on a
vessel, so it became a simple matter to replace the modem with a simple I/O
interface whenever this system was used on a buoy. As with the previous payloads,
a simple software change enabled the new system to control the Loran set and
sample the data during normal hourly data acquisitions.

Coordinate Conversion of Loran C TD's: The installation of Loran C receivers
(systems), coupled to buoy electronic payloads, provided TD's in near-real time.
However, NDBC did not have a coordinate conversion system to obtain geographical
coordinates. Initially, due to time and resource constraints, NDBC decided to
use a modified form of coordinate conversion (Rogoff and Winkler, 1977). This
method requires a "calibration point" near or at a buoy location. The calibration
point data includes; Loran stations (master - 2 secondaries) geographical locations, calibration point (Cal Point) geographical location, and the observed or
predicted TD's for the Cal Point. Once these data are established in a read
file, conversion from TD's to latitude/longitude, or the inverse, is easily
obtained.

As one can imagine, the system required considerable attention as each buoy
location required its own data bank of Cal Point information. With the expansion
of buoy stations nearing 40 in number, the "care and feeding" of the data base
became intolerable. Murphy's Law took over everytime a new station was added, and
it became almost a full-time job to ensure buoy positions were correct.

In 1984, NDBC decided to fully automate the coordinate conversion system.
Although Rogoff's method had served NDBC well for some 5 years, it was apparent
a system that would be totally interactive, using a common data base and including
all Loran C chains presently in operation, was the only way to proceed. A version
of the Coast Guard's EEE-10 coordinate conversion program (Gaszley, 1986) was
modified to run on NDBC's Data General MV/8000 super-minicomputers. The program

requires only the Loran chain Group Repetition Interval (GRI) and station (triad, e.g., x, y) identification in its initialization for a buoy location. All chain constants are stored in a common data base which facilitates a worldwide model capable of coordinate conversion for any Loran C chain without prior information (except for the Loran chain GRI and secondary stations selected). Figure 5 is a typical plot of a buoy station depicting Loran and TIROS position information over a period of several days. Normally, a daily computer listing of buoy locations is provided to data analysts and the program manager. If a buoy has exceeded its watch circle and/or has experienced a mooring failure, a graphical presentation will be produced in a matter of minutes for recovery planning.

This program, together with a program presently under development to determine actual anchor location based on Loran position, water depth, buoy type, wind speed and direction, and mooring length, are the backbone of the NDBC Moored Buoy Position Location System.

One area still under development is a data base to establish an Additional Secondary Phase Correction (ASF) table for all buoy locations. The ASF value is required to correct the conversion program from a model using all salt-water propagation paths to include empirical or DMA-corrected data. The ASF adjustment provides a "predicted" location versus a "computed" location (i.e., the predicted location is derived taking into account any anomalous propagation paths, which, in turn, provides the "best," absolute, geographical position). The use of a portable GPS receiver will greatly enhance measurement of the ASF values and will be instituted during 1987.

Operational Results: During the years 1980 through 1983, over 50 of these Loran systems have been bought and placed into service, averaging 30 operating systems a year during that period. The Loran set itself has been extremely reliable, with a mean time between failures exceeding 2.5 years. The data rate and positional accuracy has also been excellent, averaging close to 9,000 position reports per buoy year, with reportable and absolute accuracies exceeding the Coast Guard's published values for any particular area. When a buoy does go adrift, the search vessel can generally home in on it because real-time updates of the Loran C TD's are made available from NDBC on an hourly basis. The systems are now beginning to age and become unsupportable since the model 757 has been out of production for nearly three years. NDBC is now looking toward a future generation of buoy position-reporting systems.

5.   FUTURE PROGRAM PLANS

NDBC plans to continue to use Loran C as the primary moored buoy position-reporting system through the next 5 to 10 years, since it will be available as a marine position-locating system at least that long. NDBC presently has new Loran receiver specifications in process for the procurement of 50 to 75 new sets in fiscal year 1987 as replacements for the existing sets. Besides the normal performance specifications for Loran C receivers, there are mission-oriented requirements included for such things as normal operation at +12 VDC, substantially reduced power consumption, NMEA-183, standard interface, automatic operation with host payloads (including signal selection and power control), and small physical size. NDBC also plans to obtain a portable Global Position System (GPS) receiver in fiscal year 1987 for use as a "ground truth" measurement device that will provide a means of establishing a secondary phase correction (ASF). This in turn will be used to establish "absolute" geographical coordinates. The TDD equipment

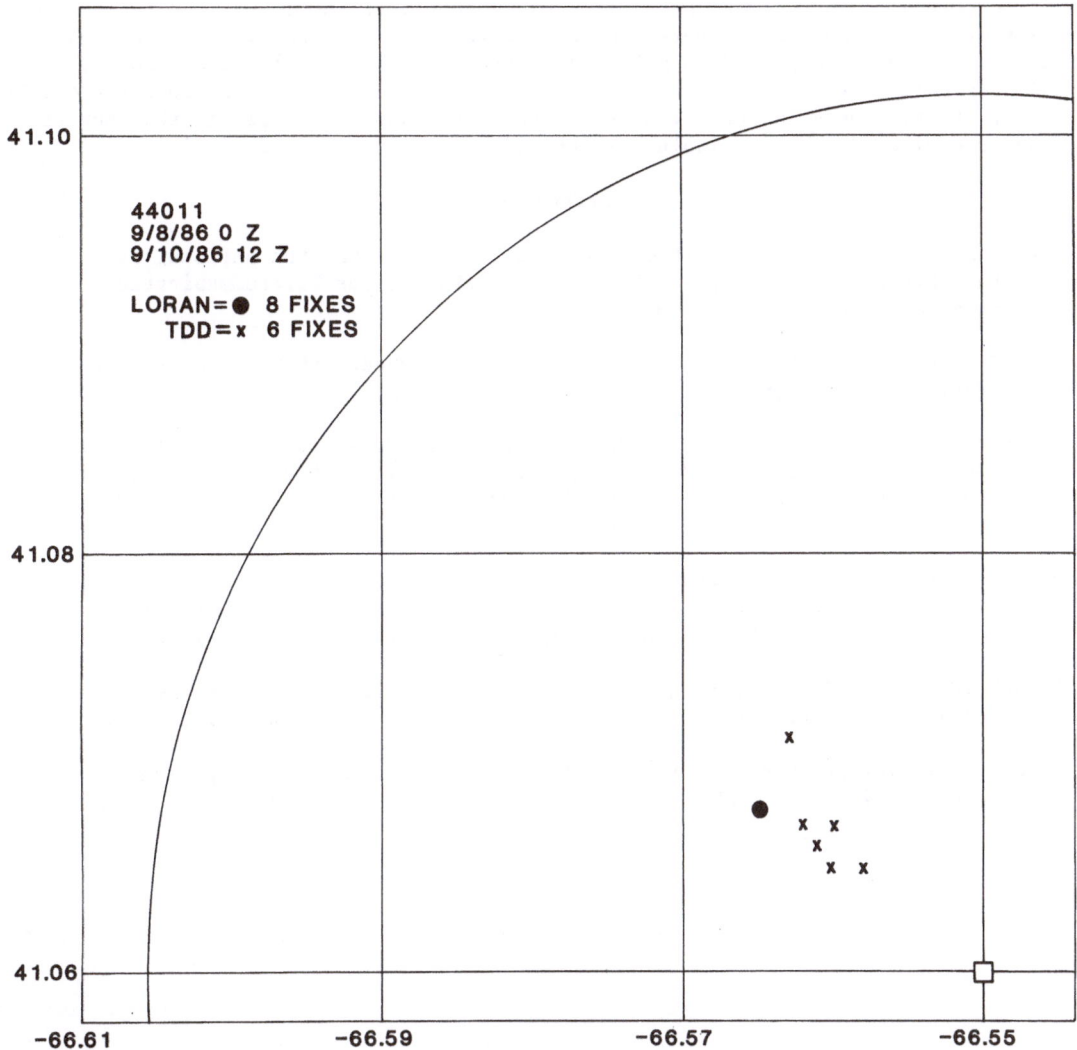

Figure 5. Typical Plot of a Buoy Station Depicting Loran and TIROS Position
           Information Over a Period of Several Days

will continue to be used on the smaller expendable drifting buoys, and to provide position reports on moored buoys located outside 1:3 SNR Loran C coverage areas. The GPS program is expected to get back on track in the next 18 to 24 months, and, hopefully, GPS receiving equipment prices will eventually drop to a reasonable level. Present predictions are that suitable low-cost receivers (in the $800 to $1200 range) should be available for moored buoy use in the mid-1990's.

REFERENCES

Beery, Wesley M., 1971. "Position Location Techniques for the National Data
    Buoy Systems." Paper presented at the Institute for Telecommunication
    Sciences, Boulder, Colorado.

Gaszley, LCDR Robert L., 1986. "Untitled." Unpublished. USCG Document, EEE-10
    (revised) CG Headquarters, Washington, DC.

Jespersen, James., Clements, Al., Davis, Dick., and Weiss, Marc., April 1980.
    "Study of the Feasibility of Using GOES for Position Location." Study
    prepared by the National Bureau of Standards for National Oceanic and
    Atmospheric Administration under Contract No. 31 USC 686.

Mahar, D. W., May 1980. "Loran-C Position Reporting System." Study published
    by the U.S. Department of Commerce, National Oceanic and Atmospheric
    Administration, Rockville, Maryland, for the National Data Buoy Center.

Plusch, CDR S.P., 21 February 1980. "Selection of a Loran-C receiver for CG
    Small Boats." USCG Memo from CDR S. P. Plusch to File, Washington, DC.

Rogoff, Mortimer and Winkler, Peter. 1977. "Loran Calculator Navigation."
    Report CG-74, 516-B/WDC.

SEAFLOOR BENCHMARK POSITIONING SYSTEM EXPERIMENT
INTERIM REPORT

Lester Q. Spielvogel
SEACO. INC.
146 Hekili Street
Kailua, HI   96734
and
Narendra K.Saxena
Civil Engineering Dept.
University of Hawaii
Honolulu, HI 96822
and
Steven P. Tucker
Naval Postgraduate School
Monterey, CA 93940

ABSTRACT

During May 1985, the Seafloor Benchmark Positioning System (SBPS) Experiment was conducted about 35 km off Monterey California. In this experiment, the position of the R/V ACANIA was determined by Global Positioning System receivers. Concurrently, acoustic positioning, miniranger positioning and other data sets were collected.

Discussed here are the details of the survey cruise and the accuracy of the acoustic positioning data. Some of the details of the 1986, Phase 2, cruise are discussed.

1. INTRODUCTION

The aim of the Seafloor Benchmark program is to position physical markers on the ocean floor at precise geodetic locations. The experiment that I will report on was performed in two phases and was meant to collect and process data to determine how precise the current state of the art is and to predict the accuracy achievable in the next decade. The experiment was conceived by Dr. Narendra Saxena and incorporated the skills, funding and equipment from a great variety of sources. I will mainly be reporting on one aspect of Benchmark, acoustic navigation and its errors. An understanding of the methods, sources of errors, hardware and software is necessary before incorporating GPS devices for the precise placement of markers on the ocean floor.

The basic problem is one of communications with navigational satellites for absolute position from a ship of uncertain motion on a varying sea surface and combining this with communicating with acoustic transponders on the ocean floor through a medium of varying properties in time and space where single measurements take place over intervals of several seconds, so that when you finally know where you are, you're not there anymore.

M. Kumar and G. A. Maul (directors), Marine Positioning, 337–347.
© 1987 by the Marine Technology Society.

Near real time processing is needed to interpret ship position in a matrix of measurements so as to infer absolute position of the acoustic transponders, which are temporarily fixed in their remote position close to the ocean floor.

The experiment was conceived with several projects in mind. The interest of SEACO is its BENCHMARK project. BENCHMARK is the combined hardware, software, and service needed to provide absolute sea floor positioning with sub-meter accuracy. While we are not in the position of delivering absolute sub-meter accuracy in the present, we feel, and have felt for several years, that this accuracy is a suitable goal for the near future for some oceanic regions. This demonstration has shown that relative underwater sub-meter accuracy is attainable in near real time. We hope that it may convert some remaining skeptics as to the feasibility of that same accuracy in absolute positioning. The uses for this accuracy will be to deploy precisely located markers to locate scientific and military sites and possibly to indicate locations of national and economic boundaries. Others associated with this experiment are interested in results similar to those sought by SEACO.

2 BENCHMARK EXPERIMENTS & ERROR SOURCES

The experiment was meant to develop criteria and to evaluate methods for at-sea geodetic positioning for the project sponsors. The experiment was performed in two phases, Phase 1 in 1985 and Phase 2 in 1986. Phase 2 was more ambitious, covering an area about 4 times that of phase 1 and employing more instrumentation.

The numerical results presented in this report are entirely from the Phase 1 data, which is still in the process of being analyzed.

The acoustic data was collected to compare methodology and computer algorithms for high accuracy positioning. The data is meant to show that sub-meter relative positioning is the state of the art in real time. Because the acoustic and GPS data were collected simultaneously, the accuracy of the relative acoustic positioning also provides a means for modeling the relative accuracy of sequences of GPS fixes in a region. The acoustic data was reduced by several schemes and the results compared. Methods of data selection and processing were compared. While the BENCHMARK results were not obtained in real time, the methods used can certainly be implemented in near real time.

Phase 1 took place in late May 1985. The location was situated approximately 30 KM off the Monterey-Carmel, California Coast (Fig. 1). The depth was 1330 to 1400 meters and gently sloping to the northwest. The bottom was featureless, although the nearby bathymetry had the potential of producing tidal currents which might have had a deleterious effects on our assumptions of steady, very slowly varying sound speed profiles with no horizontal variations. As a result of comparing different data sets, the variation of the sound speed produced identifiable errors, which were interpreted in terms of sound speed and tidal phase. There appeared to be no changes of sound speed great enough to produce errors that would seriously

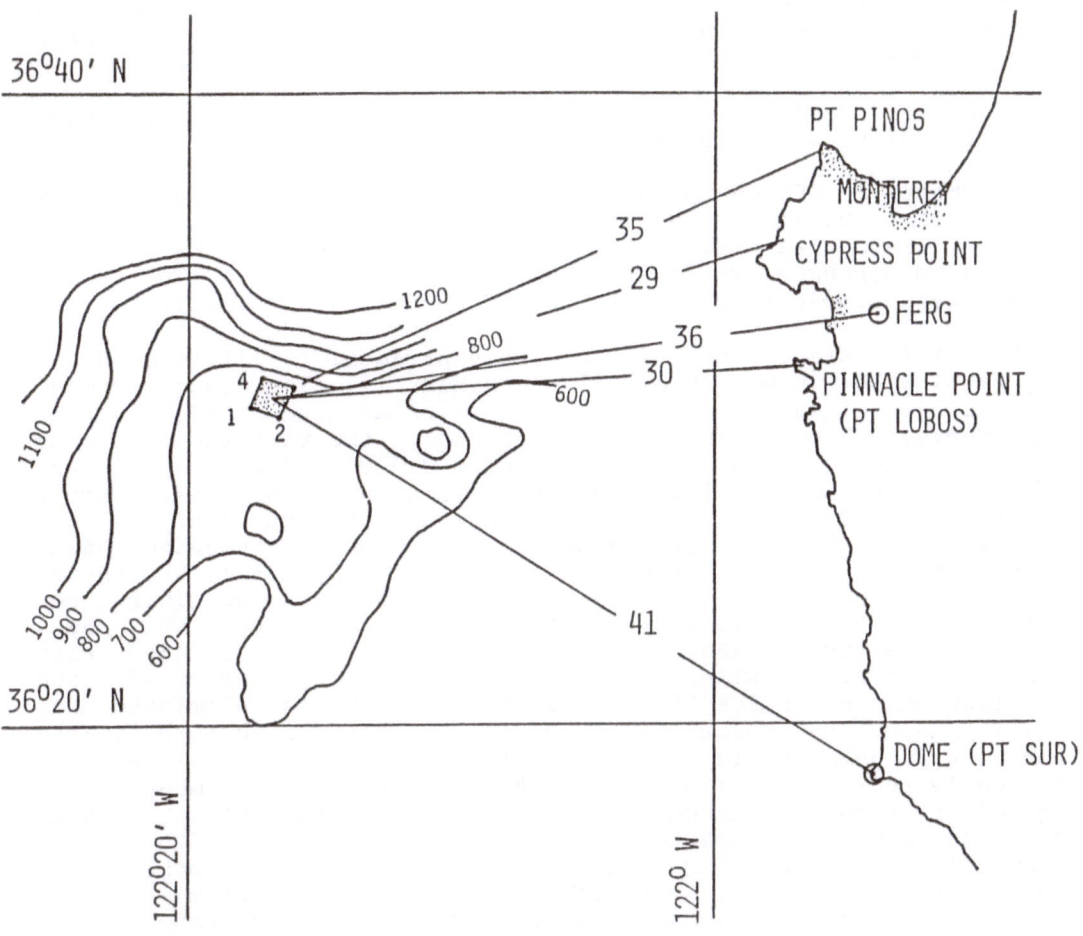

FIG. 1, BENCHMARK SITE. Phase 1, May, 1985, experiment. Distances are given in kilometers and depths in Fathoms. Miniranger antennae are at FERG and DOME.

degrade the desired and expected accuracy. We expect to verify our simple model of tidal effects and sound speed effects when the GPS differential measurements are fully reduced. We hope to see similar effects in the Phase 2 data reduction. The size and depth of the main part of the Phase 1 array is shown in Fig. 2. Phase 2 data was collected in the same region but covered a greater area.

The principle reference navigation equipment for Phase 1 was a miniranger with line of site antennae at Pt. Sur (DOME) and above Carmel (FERG). Data was collected at 15 to 16 second intervals. LORAN C ranges were also recorded. The accuracy expected in the experiment site was expected to be within 5 meters for the miniranger. This error includes a 2 to 3 meter error from surveying in the antenna.

GPS devices aboard supplied were Magnovox Phase II-B, Texas Instruments 4100, and Interstate (Astrolabe II).

The Phase 2 GPS equipment was more extensive. It included 4 devices all operating in the differential mode: TI 4100 (with GESAR software), Motorolla Eagle, Trimble 4000S and Magnavox MANPAC. We used the miniranger FALCON for real time navigation. The shore stations were almost identical to those of Phase 1 (Fig. 1).

STD's at the start and finish of the experiment (a little over two days apart) were taken with a Neil Brown Mark III. The acoustic sound speed was determined from a Wilson's Equation as used by the National Ocean Survey. From the processed STD data, the variation in sound speed over the duration of the experiment did contribute some sub-meter errors. From these four casts we estimated a " Minimax, " error of from 10 to 25 cm in our processed baseline measurements, attributable to slow variations in the sound speed. Processed data indicated this should be between 30 and 100 cm. Realistic hypotheses imply the extreme low end of this range is to be expected. This error was big enough to resolve and was systematically removed. This was a correctable error between data sets and should not be confused with the short term errors within a data set. A greater number of STDs and the placement of a sea-floor pressure gage would eliminate any ambiguity here. The pressure gage should be incorporated into similar experiments, data could be acoustically telemetered to the surface and used in real time calibrations. For deep ocean experiments, many deep STD's require increased ship time on the site. In some locations, many of these measurements may be replaced with medium depth STD's and a sea floor sound velocity meter. The errors due to changing sound velocity profiles within a data set (from 2 to 4 hours), are an order of magnitude less provided an accurate CTD is taken within a short interval of the acoustic data.

Anomalous errors are eliminated with the data editing procedures if they are beyond realistic preset limits, smaller magnitude errors are minimized through over-sampling and averaging in data processing. We concluded that monotonic small sound speed changes and medium tidal excursions must be modeled in the data reduction or they may very well bias the results giving unacceptable error magnitudes when sub-meter accuracy is demanded. In a low noise environment we may

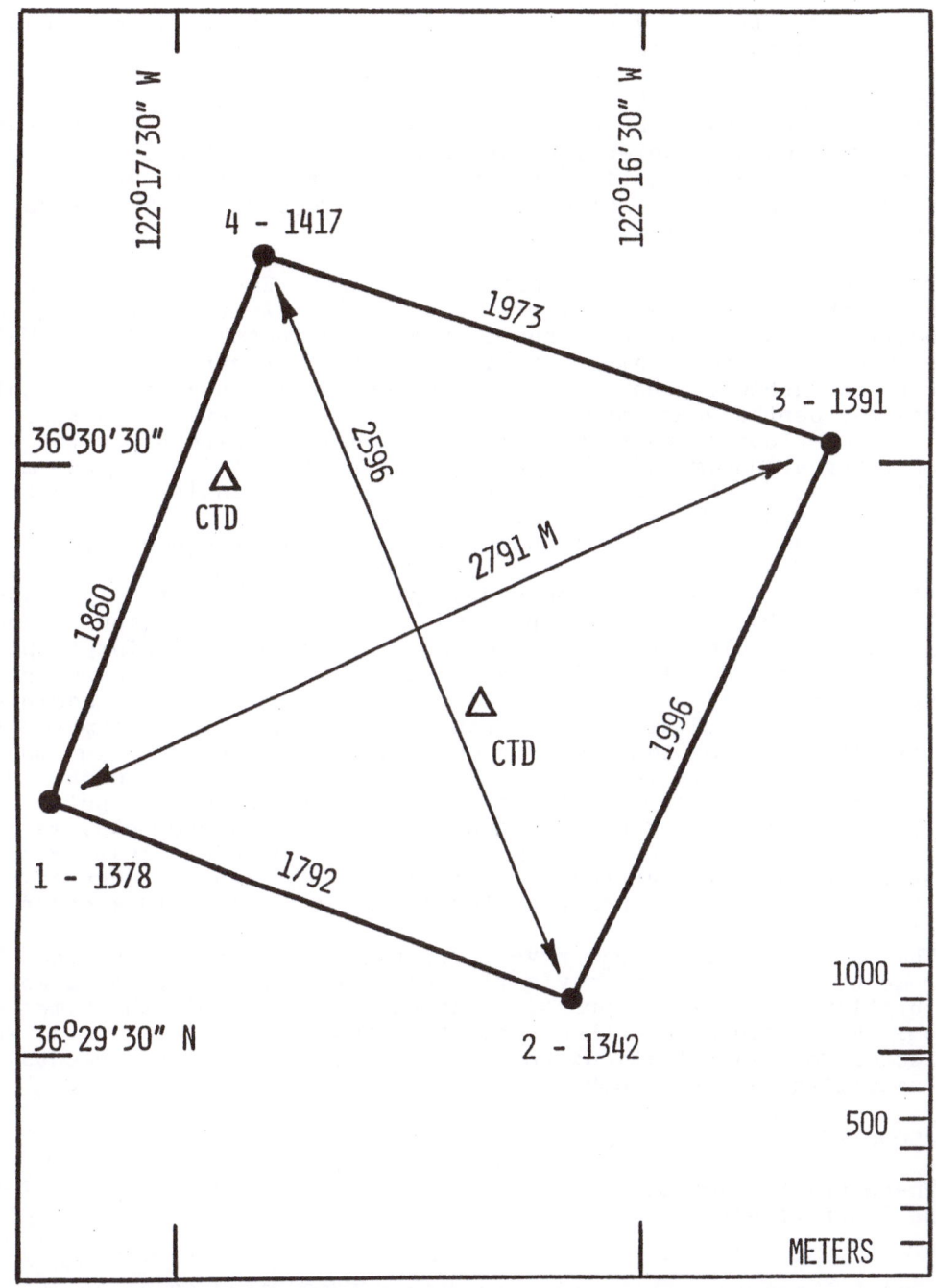

FIG. 2, PARTIAL BENCHMARK ARRAY. Phase 1, May, 1985, experiment. All distances are in meters. CTD represents the positions of full depth CTD profiles. The Phase 2 experiment in Aug, 1986, covered this and adjoining areas.

extract tidal data in a multitiered site calibration. We have found
that we may extract a 1 parameter family of sound velocity changes
from data set to data set.

Data was sampled at 5 to 10 meter intervals on the CTD casts. The
harmonic mean speed to the transponders was computed. A 14 point
subset was used to determine ray bending corrections. After the
corrections were made, it was determined that a single HMS could be
used at each transponder site for Phase 1. This will not be true, in
general, at deeper sites.

The acoustic gear was supplied by OCEANO. Each transponder is
mounted on a 9 meter tether (Fig. 3). A taut mooring was
accomplished by using three buoyancy spheres instead of one. While
this minimized the watch circles of the transponders in the
background current, the size of the watch circles becomes a problem
at greater depths where the transponders must be elevated further off
the ocean floor so as to be able to communicate directly with each
other (sing-around mode). In Phase 2, we incorporated a current meter
in the center of the array, to provide an indirect means of
determining the transponder attitudes (horizontal displacements in
the background current). In both phases, a low frequency (8 to 16
KHz) Long Base Line System was used. In the final Phase 1
configuration, we used 4 double transponders in a relatively square
array (Fig. 2). They were operated in both the conventional manner
and also as relay transponders for determining point-to-point
distances. An additional Sonotech transponder was deployed near one
corner of the array. The published accuracy for the OCEANO equipment
is quoted as +/-15 cm for the transponder delay. We estimate the
overall accuracy for ideal measurements as less than +/-30 cm modulo
lost wavelengths (10 to 13 cm). Precise transponder technique uses
the leading edge of the first cycle in the acoustic pulse for
ranging. To maintain the precise nature of our measurements, we must
also measure the length of the entire pulse to ensure that we do,
indeed, satisfy the criterion of measuring the leading edge of the
first cycle. The pulse interrogation period was set at 10 seconds.

The acoustic gear was never properly interfaced with the GPS
equipment for real time navigation and for near real time absolute
calibration. This was a practical impossibility within the time and
fiscal constraints of this project. Simultaneous data sets were
collected for all situations when data was available (instruments up
and satellites up) for both phases.

3 ACOUSTIC DATA

There are 4 major data sets that were used for our analysis of the
Phase 1 experiment.

```
1) E-W Tracks            76 x 10 Points 5/19 21:33:57-00:42:17
2) N-S Tracks            82 x 10 Points 5/19 17:07:34-20:57:20
3) E-W M. Kuo Tracks     86 x 10 Points 5/20 17:18:23-19:62:06
4) N-S M. Kuo Tracks     78 x 10 Points 5/20 22:10:10-00:31:03
```

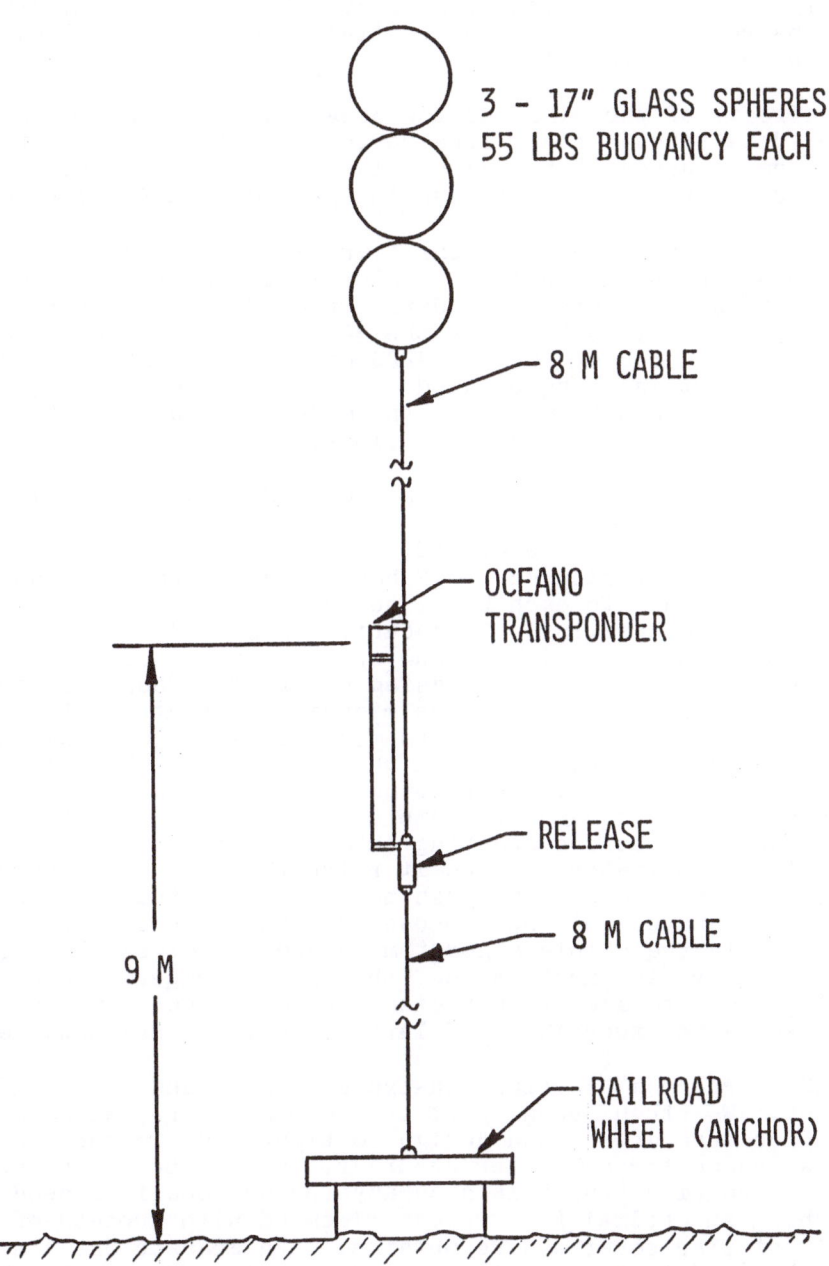

FIG.3, TRANSPONDER CONFIGURATION.

The positions of the data stations is shown in Fig. 4. Each data station consists of 10 consecutive pulse responses while sailing at a relatively constant heading at a relatively constant and slow speed. These constraints are needed to attain the most accurate range correction for ship velocity.

The effects of the wind and the ocean swell was different in the separate data sets. This allows us to determine the effects of sea state on the accuracy of the data. The ship motions (pitch and roll) were recorded. These have not been integrated into the data as yet.

Fig. 5 shows an overlay of the four data sets. While not a pretty slide, this shows the ability to shuffle the stations from the four data sets to study, with real data, the effect of ship's track on data accuracy. Any subset of the 322 stations may be used as a simulated track. Subsets of the 3220 data points may be used also if a denser data set is to be studied. It must be remembered, however, that the closer the data points, the more they must be corrected for dynamic errors to make valid conclusions.

4 ANALYSIS OF AND RESULTS FROM THE ACOUSTIC DATA

For his master's thesis at NPS, Mike Kuo analyzed the data by a baseline method attributed to Fubara and Mourad. This required sailing orthogonal to each baseline at several locations near the bisector of the baseline. Consecutive pulses on each track were plotted to find the minimum sum of the responses from the end transducers. This gives the crossing point. The four crossing points were analyzed further to find the midpoint of that baseline and its resultant length. This was performed with all the baselines. The results, which are spelled out in the thesis, and which I have taken the liberty of adding to, is that no appreciable ship time is saved with this method. A very accurate guess is needed for the positions of the transducers before starting data collection. The method fails as a precision indicator, since it relys too much on a sparse set of data. The data needs to be much more accurate than other methods because of the nature of the processing. The main advantage in using this method is the simple algorithm needed to solve the equations. The simplicity is moot since the data loggers necessary for collecting the data are of sufficient speed for handling the data and the data reduction techniques of larger, conventional data sets.

OCEANO performed a fine analysis at sea and then after the experiment. We then used part of their software with a careful editing procedure of our own design to tighten up on the accuracy of the data needed for sub-meter accuracy. We included subroutines of our choice to fine tune this accuracy for our specific needs. After establishing an optimal fit, we experimented with subsets of the data to see if similar results were obtained with sparser data sets. The results obtained from the data subsets were in general agreement with each other. The baselines measured from this method of from 2 to 3 km agreed within 25 cm with each other. The actual data reduction method that we used works in two stages.

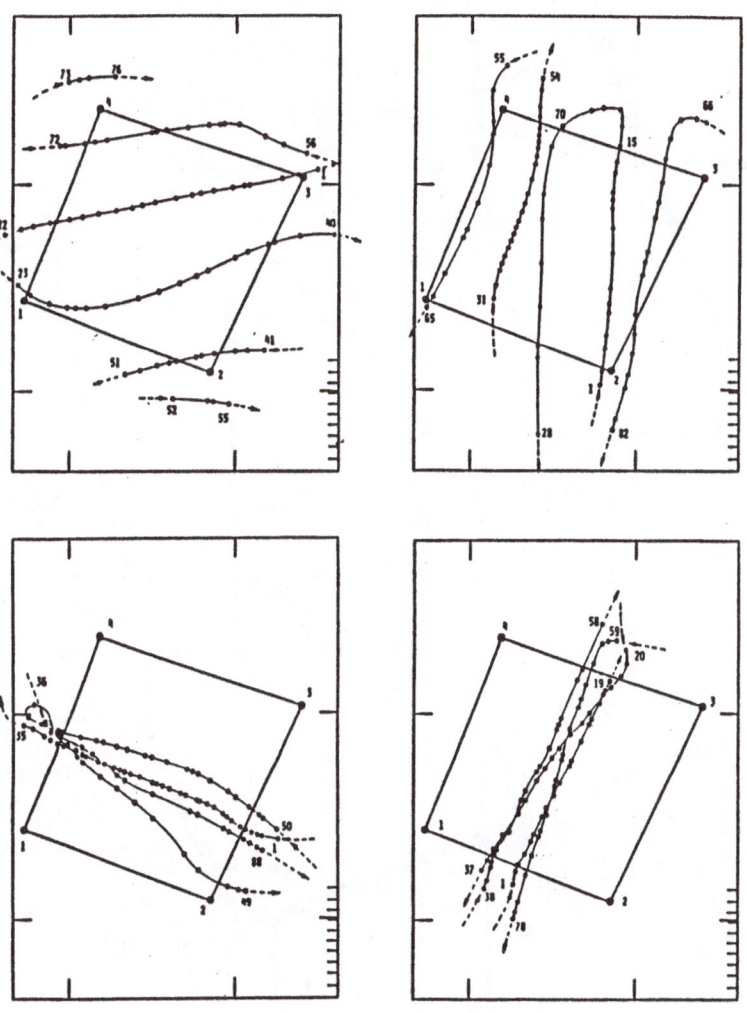

FIG.4, BENCHMARK SHIP TRACKS. Phase 1, May, 1985, experiment. Each data set contains approximately 80 stations or 800 interrogations and responses.

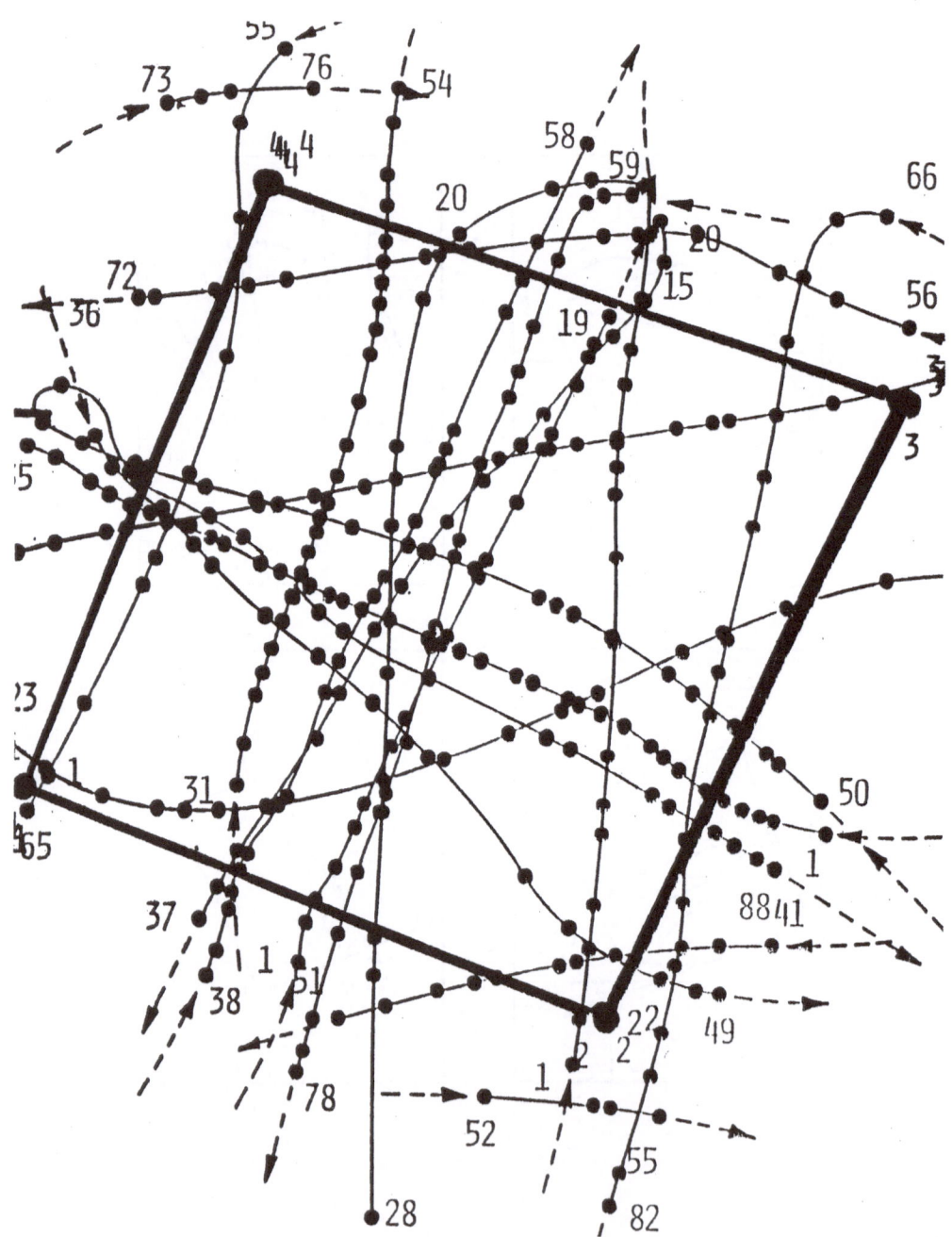

FIG. 5, OVERLAY OF FOUR DATA SETS. Synthetic ship tracks can be constructed easily from the composite data set to evalute the accuracy and logistics of different methods of data collecting.

The first stage reduces each station of ten pulse responses to a station of one representative pulse response. After correcting for the sound speed and raybending, the radial velocity to each transponder is determined from a regression on the remaining data and the leading non-trivial order correction is made. The most representative corrected data point from the velocity reductions is selected.

The second stage reduces the remaining representative data set or an edited data set by an algorithm attributed to Fletcher and Powell using quadratic convergence factors.

The software is designed for the HP 9836. The software runs on other HP series 200 machines with some limitations, in particular, screen oriented editing sometimes runs amuck. This can be handled in other ways however.

We stopped at this stage in our software. It is at this point that ship-motion errors can be corrected for. It is at this stage that we may compute any meaningful tidal motions from the data set. The data would normally be recalibrated to determine the most accurate baseline calibration.

A data set of 80 stations can be quickly calibrated in 15 minutes. With an interactive set of rules, this can stretch out to 60 minutes with apparently better results.

Comparing independent data sets, we found the baseline accuracy to be : +/- 25 cm. without correcting for ship motions. The normal error for the navigation data is 1.25 cm. (0.75 cm. using 90% of the data). This, also, is without correcting for ship motions.

CALIBRATION OF A RANGE TO RANGE LORAN CHAIN
WITH THE USE OF GLOBAL POSITIONING SYSTEM[*]

Michel LE GOUIC
Service Hydrographique et Océanographique
de la Marine
B.P. 426
29275 Brest Cédex
FRANCE

ABSTRACT

A new radionavigation LORAN C chain has been set up in France in 1985. This chain works with two transmitting stations : a master located in Lessay (Normandy) and a slave located in Soustons (Landes) near the Spanish-French border. Therefore this chain can only provide a range to range fix to an user equiped with an appropriate LORAN receiver and an accurate clock.

French Hydrographic Office (SHOM) carried out in 1985 and 1986 two calibration campaigns of this LORAN chain in the eastern part of North Atlantic Ocean. GPS provided the references of position, and of time with a time and frequency monitor linked with an atomic Caesium clock.

Results of these campaigns are discussed here from two points of view : the adequacy of the equipment installed inboard to realize the purposed calibration and the modelling of propagation conditions of the French LORAN chain transmissions.

---

[*]Paper was submitted for inclusion in proceedings but was not presented at symposium.

M. Kumar and G. A. Maul (directors), Marine Positioning, 349–355.

A new LORAN C chain has been set up in 1985 in France. This chain works with a Group Repetition Interval (GRI) of 99 600 μs and consists of two transmitters : the master located at LESSAY in the northern part of France and the slave at SOUSTONS near the French-Spanish border. A control station is installed in BREST. The coverage of the chain extends offshore of Europe up to the middle of the North Atlantic Ocean where the user achieves a range to range fix (fig. 1).

Because of the large offshore coverage of the French LORAN chain and of its particular type of lines of positions (which requires a very precise reference time) GPS has been the master piece of the equipment used by the French Hydrographic Office (SHOM) during two calibration campaigns carried out in 1985 and 1986.

This paper describes the equipment, the methods and the main results of these calibration campaigns.

1. THE GOALS OF THE CALIBRATION CAMPAIGN OF THE FRENCH LORAN C CHAIN

The time of transmission (TOT) of a LORAN C pulse depends most certainly of the electrical properties of the earth's surface over which the signals propagate, but also on the terrain variations of the surface and of the refractive index of the atmosphere (and its rate of change with the altitude) at the surface. Theoretical and practical conditions of the propagation of LF waves have been discussed by a lot of authors (Johler, Horovitz, Doherty, Millington, Sammadar..) and their results can be briefly summarized as follow.

1.1. If n is the refractive index of the atmosphere and Co the velocity of the light in vacuum, the velocity of the LORAN pulses in the air is Co/n.

1.2.The LORAN C signal generally propagates over seawater (ground wave) and a first corrective term is introduced taking in account the complex impedance of the water : this term is usually called secondary factor (SF)*.

As the electrical sea water properties are rather constant one can compute the SF effect with for example the SALT model of the RTCM (Radio Technical Commission for Marine Services) which gives the correction in microseconds expressed in terms of distance (d in meters) for standard seawater conditions (conductivity #5 mhos per meter, dielectric permeability #80 and relative permeability #1) :$\Delta t = - 0.4146 + 40\ 173/d + 2.1\ 645\ 10^{-6} \times d$

The effects of the changes of seawater conductivity are weak and generally lesser than 0.05μs and can be easily derived from simple graphs.

1.3. When the LORAN signal propagates over land the formulation of the LF propagation is much complicated. The computation of the correction induced by a land path (called Additional Secondary Factor or ASF) requires the input of the surface impedance and terrain elevation.

The surface impedance is determined by the nature of the soil, its moisture content, its geological structure, its temperature, the depth of wave penetration... . Some of these factors present temporal fluctuations with the seasons or the weather conditions. Therefore it is very difficult to determine at a given time and a given place the true impedance of the path on which the LORAN pulse propagates in order to derive the ASF corrections especially if they concern a wide coverage area.

(*) Certain authors include in this secondary factor the additional secondary factor described in § 1.3.

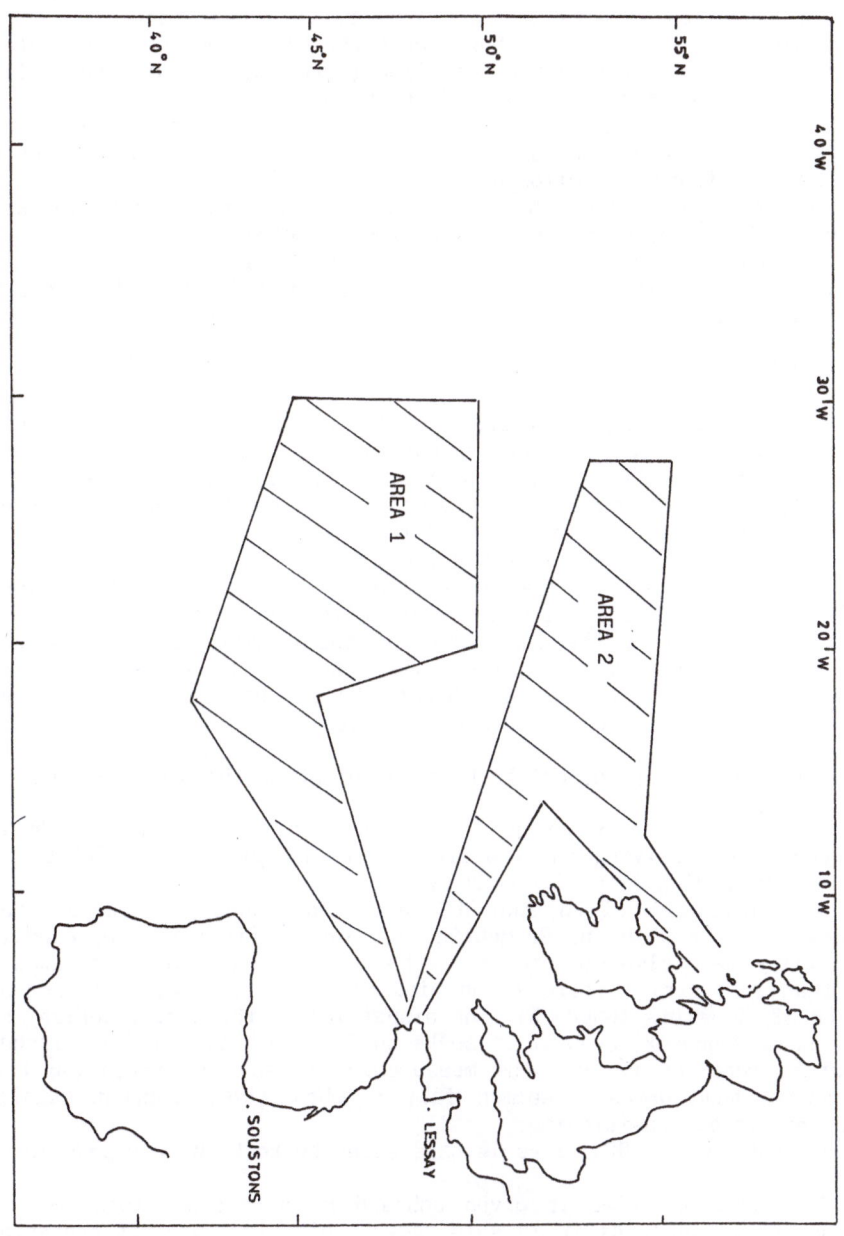

FIGURE 1

RANGE IN SUMMER (mean)

Terrain elevation effects on a LF groundwave propagation are very smoothed with the distance to the point where the perturbation has been generated. If the user is located far away from the coastline (more than 100 Nautical Milles) these terrain elevation disturbances can be neglected.

1.4. By taking in account the analysis presented above, the calibration campaigns of the SHOM aimed to meet following points :
- to verify the adequacy of the SALT model for the propagation over seawater of the French LORAN signals, in regard with local environment conditions,
- to estimate the value of ASF conditions in the coverage area,
- to ascertain that the user who applies correctly SF and ASF corrections will get an accuracy better than ±.3µs in LORAN C and ± 100 m or better in position.

2. EQUIPMENT

2.1. Calibrating a radiopositioning system implies the use of a reference system which can meet the accuracy requirements and provide the range needed. In the case of the French LORAN C chain a third condition was imposed : to know the absolute time during a month with a precision at least as good as the equivalent in seconds of the location accuracy requirement.

The only system able to answer the requirements in terms of location and range was GPS. The requirement of time precision made us use an atomic clock : but as the relative drift of an atomic clock installed in a ship is about 10^{-12} (with fluctuations induced by the variations of the geomagnetic field when the ship moves) it was necessary to control the rate of the clock with an external reference. One more time GPS was choosen to be this reference, because of the high stability of the time provided by its satellites.

2.2. The sketch plan of the calibration equipment is shown on figure 2.

2.2.1. The position is given by one of the two GPS receivers : MAGNAVOX RPU3 (with Pcode and receiving the two frequencies L1 and L2) and SERCEL TR5S (with C/Acode and receiving only L1 frequency).

The intercomparison of the two GPS receivers showed an agreement better than 15 meters in x,y,z when the Geometric Dilution of Precision was good (GDOP≤ 5). The GPS fixes were also compared near the coastline with Syledis fixes (accuracy better than 3 meters) : there is no bias on a GPS fix statistics and a standard deviation of 15 meters (GDOP < 5). The dispersion of the Sercel series is the same as the one of Magnavox, even if it works on C/A code with only one frequency : in fact Sercel receiver removes the measurement noises by processing the carrier phase rather than using a Kalman filter, what gives a great stability to a sequence of dynamic positioning.

The accuracy of the GPS fixes is thus assessed to be better than 15 meters.

2.2.2. The reference time is given onboard by a Caesium clock manufactured by Oscilloquartz Society. As it is said above, this clock was not protected against the geomagnetic field variations, and it was necessary to control its drift. For that, a time and frequency monitor TRIMBLE 5000A was linked to the Caesium clock. The TF monitor tracks a selected GPS satellite and derives from the knowledge of the accurate position and the time of the given satellite, the shift between GPS satellite and derives from the knowledge of the accurate position and the time of the given satellite, the shift between GPS time and at the time of an external clock connected with it.

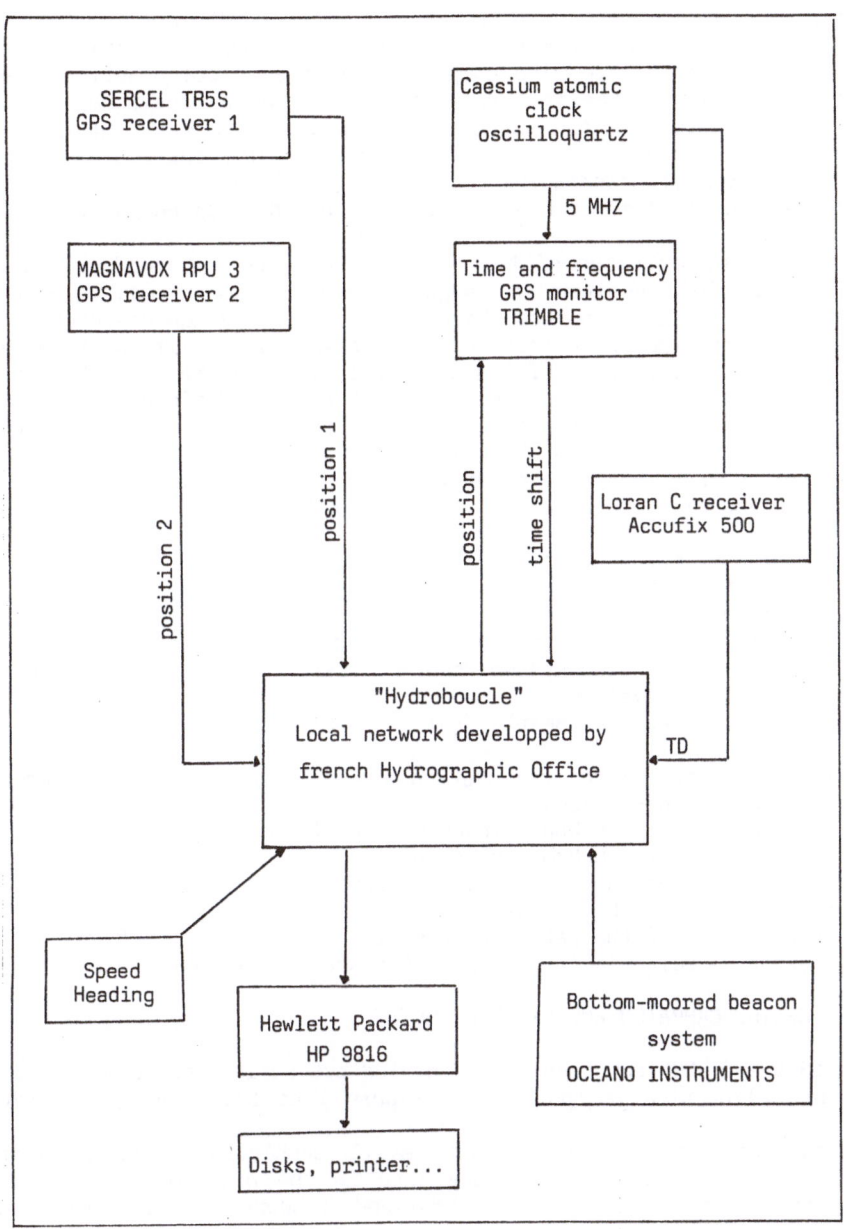

FIGURE 2

The principle of the control of the atomic clock's rate was the following. At the control station of the LORAN chain in Brest and aboard, two Trimble receivers track the same GPS satellite. The comparison of the shifts observed between the given satellite and the time in Brest control station on one hand, and observed between the same satellite and the time on board on the other hand, provides a corrective synchronization term to the aboard time. This correction was recorded for a later use when processing the LORAN data (§ 2.2.3).

The accuracy of the time on board is thus assessed to be better than 100 ns.

2.2.3 Times of transmissions of the LORAN pulses were indirectly measured on a MEGAPULSE Acufix 500 receiver (a range to range receiver was damaged at the beginning of the two campaigns and the MEGAPULSE spare device was used).

Acufix 500 is an hyperbolic receiver which displays two Time Differences (TD). On one TD set, the difference of Times of arrival (TOA) from LESSAY (Master) and SOUSTONS (slave) was displaid, on the second TD set the difference of TOA between LESSAY and an emission generated on board by a LORAN simulator syncronized at the beginning of the campaign with LESSAY emissions. TOTs are then easily deduced from the displaid TD's (with the error induced by the drift of the aboard clock).

The TOT's measurement is assessed to be better than 0.1µs.

2.2.4. The other potential error sources are the following :
- The synchronization of the chain is better than 50 ns in absolute time or 30 ns in relative time between stations,
- The GRI and the Emission Delays are controlled at each emission,
- WGS72 is the only geodetic system used and there are no residual errors due to transformation between systems,
- The antennas were installed nearly from eachother.

2.2.5. The total error of the calibration system is therefore reckoned as being (in terms of standard deviation)
$\sigma = \sqrt{\sigma_S^2 + \sigma_T^2 + \sigma_L^2 + \sigma_{GPS}^2} = 0.08\mu s$ (independance of the error sources)
where σ_S is the error of synchronization of the chain $\sigma_S = 0.03\mu s$
          σ_T is the error on the reference time aboard $\sigma_T = 0.05\mu s$
          σ_L is the error of the measured TOTs $\sigma_L = 0.03\mu s$
     σ_{GPS} is the error of the reference position $\sigma_{GPS} = 0.04\mu s$
This total error is suitable in regard of the accuracy requirements.

3.   COURSE OF THE CAMPAIGN AND FIRST RESULTS

3.1. Both 1985 and 1986 campaigns were carried out during the summertime. In 1985 the area of calibration is spot on the figure 1 as the area 1, in 1986 as the area 2.

GPS satellite configuration is not still completed : GPS 2D fixes were available during two periods in a day (one in the nightime, the other in the daytime) each of about 2 hours in 1985 and 3 hours in 1986. Therefore only discontinuous calibration stations were possible (in particular evolution of the LORAN pulses during sunset and sunrise has been observed only during two days when using bottom-moored beacons). However the propagation conditions are steady enough in the calibration area to bear out the extrapolation between two separate measurement stations.

3.2. The first results are conformable to the theoretical predictions :
- The range of the groundwave is about 1 100 Nautical Miles in the day and 900 in the night,

- The correlations signal to noise ratio versus distance and measurement dispersion versus distance are strong and can be successfully used to estimate quickly the likely accuracy. The correlation coefficients are the same either on a seawater path or on a mixed land/sea path,
- The salt model of RTCM is well fit to determine the SF corrections,
- The assessment of the ASF corrections has been leaded in statistics terms : computation of the correction in several azimuths from the LORAN stations and determination of a probable velocity of LORAN waves over land (the different types of ground paths have been separated : Ireland, United Kingdom, Brittany, Spain). The results are partial and other calibrations campaigns in various weather conditions (especially in other seasons) are needed before to draw up a complete ASF model.

4. CONCLUSION

Intensive use of GPS was necessary to carry out these 2 campaigns of calibration of the French LORAN C chain. Even if the constellation of GPS satellites is not still completed, GPS is already a very efficient tool in offshore activities as well for precise positioning as accurate time control.

LORAN C remains the most widely used long range radionavigation system : propagation conditions of LF waves are much perturbated by the environment conditions. Therefore it is necessary to develop methods of prediction of ground wave propagation time anomalies in the LORAN C signal transmissions.

GPS and LORAN C can today be usefully associated in offshore positioning : GPS provides the input data necessary to predict LORAN corrections and LORAN is used in the periods of no reception of GPS satellites.

CURRENT ISSUES IN EXPRESSING
UNCERTAINTY OF POSITION IN GPS NAVIGATION

S. Mertikas[1]

C. Field[2]

D.E. Wells[1]

[1] Department of Surveying Engineering, University of New Brunswick, P.O. Box 4400, Fredericton, N.B., Canada E3B 5A3

[2] Department of Mathematics, Statistics and Computing Science, Dalhousie University, Halifax, N.S., Canada B3H 4H8

ABSTRACT

This paper is an investigation for reliable and efficient scale estimators, which will summarize one-dimensional variability of position fixes in GPS navigation. It also includes several statistical techniques used for simple and direct revelation of the variability and shape of the distribution of position fixes.

Strong and weak points of the presently used scale measures are discussed and some cautionary remarks concerning their applications are made. A comparative study of several scale estimators applied to the Global Positioning System (GPS) is presented as well.

1. INTRODUCTION

Quite often, one is required to calculate some statistical summary —either graphical or numerical— which will give an idea of how widely the values are spread on either side of an arbitrary central value (usually called a location estimate). A statistical summary or collective of this kind is called a *measure of scale* or *measure of width*. Clearly, a certain measure of scale will represent what we intuitively call the variability of a given sequence of values. The smaller the variability, that is the closer the individual values to the central value (e.g., mean, median, etc.) the smaller the scale and vice versa.

In this paper, measures of scale will be used as accuracy measures of one-dimensional data, rather than used for standard errors in the subsequent analysis of interval estimation. Various scale measures have been proposed as standard procedures of communicating accuracy of navigation systems. Undoubtedly, estimating a measure of scale of a sample is essential if only for the following reasons:
1. It condenses information of variability. Although it fails to describe detail, a single scale estimate can reveal much of the variability information contained in the sample.
2. It permits comparative accuracy studies among data sets. In other words, it can be used as a statistical device to examine whether or not a certain data set is more spread out than another.
3. It enables us to use it as a quality measure to prevent overbuilding (Monahan, 1986).
4. It can be used to construct confidence intervals for location estimates, i.e., position fixes.

It has to be admitted that the doctrine of accuracy estimation is quite vague, not very well founded, and more complicated than location estimation. This is, firstly, because it is not very clear what should be taken as a "natural" measure of accuracy and, secondly, because we do not really know what an accuracy measure should estimate. Accordingly, any choice of accuracy measure is rather arbitrary. The reader is referred to Lax (1985, section 2) for discussion and references.

In the past, most of the formal procedures concerning accuracy measures have concentrated on parametric models, whose inferences primarily rely on distributional models. Let us consider an example. One scale estimate which has been deeply entrenched in our statistical practice is the standard deviation. It has been extensively used as a reasonable and simple choice for expressing uncertainty. In principle, when some distributional assumptions are satisfied, it has very attractive properties. For example, when our position errors happen to follow the Gaussian distribution, it represents the most

M. Kumar and G. A. Maul (directors), Marine Positioning, 357–370.
© 1987 by the Marine Technology Society.

efficient scale estimator (i.e., 100% efficiency). Nonetheless, we have to acknowledge the existence of some problems for this particular scale estimate.

1. It breaks down even when only a single gross error is present in our data, and the value of the estimator is virtually controlled by the large error. Indeed, such catastrophic behaviour of the estimator is avoided primarily because in practical situations common sense generally saves the standard deviation from inflation by rejecting all the flagrant gross errors (usually called outliers). However, a huge gross error can easily escape our attention in large and complicated data sets, such as those of the GPS navigation system. Furthermore, rejection of large values entails the danger of discarding useful information. Indeed, most of the information associated with scale resides in the tails of the distribution and not in the centre where all the "legitimate" observations are located. It remains true that rejection of data is a very controversial issue and certainly involves some problems. For example, it is nearly impossible to recognize and separate "proper" observations from "improper" ones in large multivariate data and in areas close to the flanks of the distribution (these outliers are called "dubious" outliers) (Barnett and Lewis, 1978). Additionally, we may falsely retain a "bad" observation and discard a "good" one, since most of the rejection techniques are susceptible to statistical errors (Huber, 1981). Masking effects and clustering constitute some extra problems (Barnett and Lewis, 1978).

2. Although the standard deviation does perform excellently under the Gaussian distribution, its performance deteriorates when the Gaussian assumption is relaxed. This has been demonstrated by a striking example due to Tukey (1960). He considered the mixtures of two normal distributions of the form:

$$F(x) = (1 - \varepsilon)\Phi(x) + \varepsilon\,\Phi(x/3) \tag{1}$$

where $\Phi(x)$ is the standard normal cumulative distribution and ε is a small number ranging from 0 to 0.10 (0% to 10.0%). Each observation can be considered as being generated by the above "chance mechanism" with probability $(1-\varepsilon)$ to be a "good" one and ε to be a "bad" one. Next he compared the relative asymptotic efficiency of the standard deviation:

$$S_D = \sqrt{\Sigma(x_i - \bar{x})^2) / (n - 1)} \; ; \quad \bar{x} = \text{sample mean, } n = \text{sample size} \tag{2}$$

relative to the mean absolute deviation:

$$S_{AD} = (\Sigma|x_i - \bar{x}|) / n \;\;, \tag{3}$$

using data from the above contaminated Gaussian distribution. Although the mean absolute deviation is 88% efficient, compared to the standard deviation at the strict Gaussian distribution (Fisher, 1920), it requires only $\varepsilon = 0.18\%$!! (e.g., just 2 observations in 1000) to make the mean absolute deviation more efficient than the standard deviation. This shows that the mean absolute deviation is preferable to the standard deviation for some contaminated distributional models; though neither of these two is here being recommended as the "best" estimate. It should also be stressed that it is almost impossible to distinguish observations coming from a purely Gaussian distribution ($\varepsilon = 0.0\%$) and observations coming from the contaminated model when $\varepsilon = 1.0\%$.

We do not wish to conclude from these arguments that all the parametric procedures are poor choices in stating navigational system performance. In many situations, they may still be an ideal choice. We do wish to point out, however, that when we are concerned with safety combined with good efficiency in the potential presence of GPS gross errors, parametric techniques may not be the "ideal" choice. It is precisely for the above reasons that we hesitate to evaluate uncertainty using accuracy measures whose performance strictly depends on a distributional model, such as the standard deviation.

Some might argue that we are overemphasizing minor nuances, but this is not true. The following explanation is necessary to substantiate our stance. In the first place, normality is not guaranteed regarding the distribution of the GPS position errors. As has been shown elsewhere (Kalafus and Chin, 1986; Mertikas et al., 1986), even if meticulous rejection procedures have taken place, the remainder of the data does not necessarily follow the magic bell-shaped curve of Gauss. Therefore, even using some of the classical parametric procedures, best efficiency is not promised unless the distributional assumption is strictly satisfied. In the second place, safety is a major issue in navigation and should be insured at all cost. Accordingly, we should look for accuracy measures which are not affected by a relatively small percentage of outlying values, no matter how extreme they are. Indeed, there is evidence that a small fraction of outliers is even present in cases where good and stable geometry exists in the GPS navigation system (see Figure 1). In the third place, in our highly computerized age it is to be expected that large amounts of position data will be statistically evaluated without human interference. Evidently, we wish to develop accuracy measures for fully automatic data processing, since it becomes impossible to examine the quality of data in detail when masses of them are fed into computers daily.

But how shall we treat the uncertainty problem in navigation? It is our opinion that we should aim at developing procedures that will describe the behaviour of position errors not only based on strict parametric models but also in the neighbourhoods (i.e., approximations) of such models. We should seek out estimating procedures which are reliable over a wide range of potentially true distributional models with marginal loss of efficiency when the ideal model of Gauss is present.

The main intent of this paper is to evaluate and compare the reliability of several scale estimators as applied to the GPS position data over a variety of potential situations. It is noteworthy that here the term *navigation* is used in the sense that only four range measurements are being made simultaneously to four satellites resulting in a unique position determination of the user (e.g., three Cartesian coordinates X_u, Y_u, Z_u and reciever bias B_u).

2. GENERAL DESCRIPTION OF PERFORMANCE

Range measurements to four GPS satellites have been collected using a Texas Instruments TI-4100 GPS receiver at a stationary location. Our sample consists of about 1200 position fixes made in May, 1984, in Ottawa, Ontario. Figure 1 displays the variation of only one position component ($X_u(t)$) from its median value (med(X_u)) with respect to time. Figure 2 shows the Dilution of Precision factor for the corresponding observational period.

Tables 1, 2, and 3 are a set of summary values called *letter-value* displays (Tukey, 1977), which are used to reveal the general behaviour of our data. Letter-value displays are a collection of observations extracted systematically from the sample and are so called since they are customarily labelled with single letters, such as M, F, E, D, C, B, A, and so on. These letter tags are shown in the first column of the tables. Note that the first label (letter N) refers to the number of values and is located in the upper left corner. The second column represents the corresponding depths of the letter values. By depth, we mean the least relative position or standing of a single observation in the ordered sample relative to the extreme values (minimum, maximum). For example, the minimum and maximum have depth 1 (one), while the median has depth [(n+1)/2], where n is the sample size. A letter-value display is formed in an inverted U-like enclosure. The middle top value designates the median (M), the right side (fourth column) represents the upper letter values while the left side (third column) represents the lower ones. Note that we have attached the following tags to letter depths: F (for fourth), E (for eights), and then in reverse alphabetic order D, C, B, A, Z, Y, X. On balance, letter values are observations from the ordered samples extracted in such a way that the fraction of data remaining beyond each of them is a power of 1/2. For example, for the median M, the fraction is (1/2 = 50%); for the fourths (F) ($(1/2)^2 = 25\%$) [halves the data between the median and the extremes]; for the eighths (E) ($(1/2)^3 = 12.5\%$) [halves the data between the fourths and the extremes], and so on.

In addition to depths and letter values, the display includes two extra columns designated "midrange" and "spread." Midrange provides information with respect to symmetry (or equivalently skewness) of

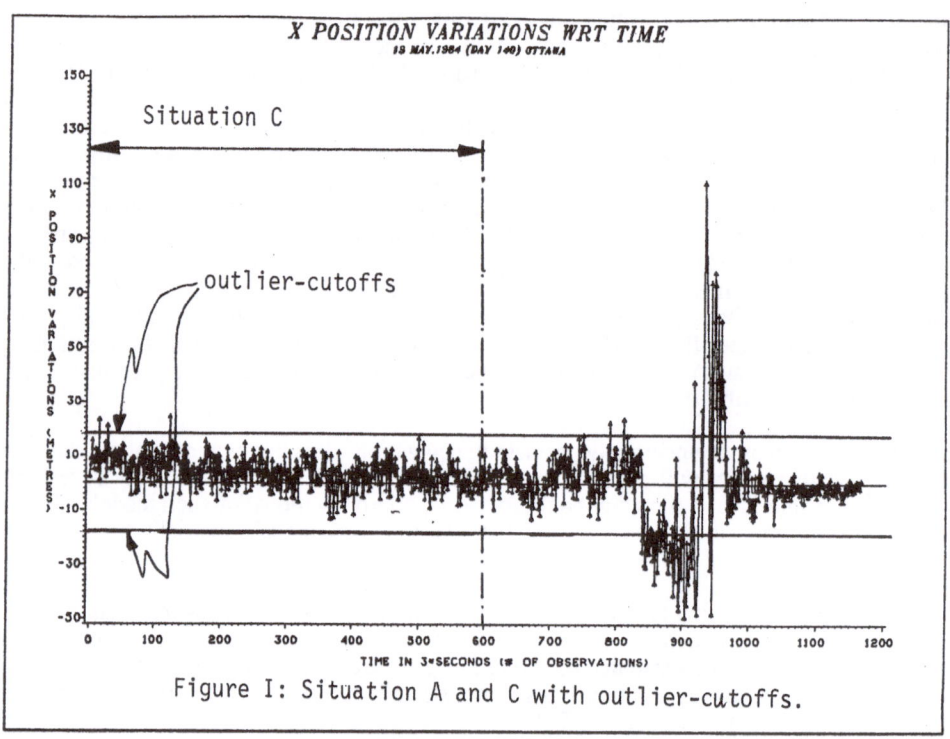

Figure I: Situation A and C with outlier-cutoffs.

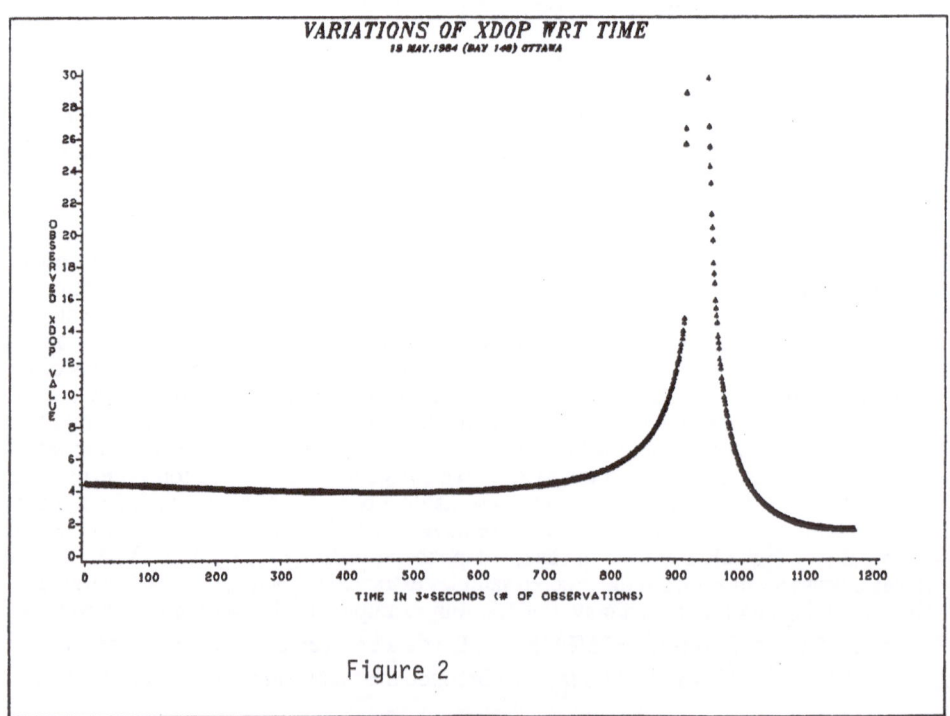

Figure 2

Table 1:Letter-value displays for X-deviates

	DEPTH	LOWER	UPPER	MIDRANGE	SPREAD
N=	1167				
M	584.0	0.000		0.000	
F	292.5	-4.382	4.950	0.289	9.342
E	146.5	-8.068	8.288	0.110	16.356
D	73.5	-18.268	11.039	-3.614	29.308
C	37.0	-29.191	14.796	-7.197	43.987
B	19.0	-42.871	35.750	-3.561	78.621
A	10.0	-51.098	60.732	4.817	111.830
Z	5.5	-75.070	129.985	27.457	205.055
Y	3.0	-139.102	268.582	64.740	407.684
X	2.0	-354.131	365.350	5.609	719.481
	1	-7187.574	1168.039	-3009.768	8355.609

Table 2:Letter-value displays for Y-deviates

	DEPTH	LOWER	UPPER	MIDRANGE	SPREAD
N=	1167				
M	584.0	0.000		0.000	
F	292.5	-6.018	4.627	-0.695	10.645
E	146.5	-16.515	7.889	-4.313	24.404
D	73.5	-34.873	12.731	-11.071	47.604
C	37.0	-53.474	36.961	-8.257	90.435
B	19.0	-85.392	94.207	4.407	179.599
A	10.0	-113.084	182.587	34.751	295.671
Z	5.5	-186.336	338.217	75.940	524.554
Y	3.0	-343.969	701.758	178.895	1045.727
X	2.0	-891.098	947.551	28.227	1838.649
	1	-18493.633	3024.160	-7734.734	21517.789

Table 3:Letter-value displays for Z-deviates

	DEPTH	LOWER	UPPER	MIDRANGE	SPREAD
N=	1167				
M	584.0	0.000		0.000	
F	292.5	-7.276	7.534	0.129	14.810
E	146.5	-21.563	19.466	-1.048	41.029
D	73.5	-41.166	61.242	10.038	102.408
C	37.0	-95.949	107.320	5.685	203.269
B	19.0	-236.823	176.348	-30.238	413.171
A	10.0	-430.437	243.460	-93.488	673.897
Z	5.5	-785.927	406.936	-189.496	1192.863
Y	3.0	-1606.519	770.833	-417.843	2377.352
X	2.0	-2168.086	2008.427	-79.829	4176.512
	1	-6889.910	41995.738	17552.914	48885.648

our data, and is defined as the sample average of the upper and lower letter value. Examining Table 1, we can see that the midrange remains almost unchanged (its value is almost zero) up to the E letter-value inclusive. This indicates that about 75% of our GPS X-deviations are symmetrically located around the median M. Moreover, spread is the difference between two letter values and normally gives an indication on how spread our values are. For example, from Table 1 we can infer that about 75% of our data have a spread of 16.356 metres (E spread = 16.356 m) or that 87.5% of them have a spread of 29.308 m (D spread = 29.308 m).

3. COMPARATIVE SCALE PERFORMANCE

The performance of a family of ten scale estimators was investigated for sample sizes n=20 and n=100. Table 4 contains a description of these estimators (see Inglewicz (1982) and Lax (1975) for further details). The variance of the natural logarithm of the scale estimator, $Var[ln(S)]$, was used as a measure of performance. This choice of a logarithmic expression is justified for the following two reasons. Firstly, the logarithmic transformation often normalizes samples of non-negative numbers, such as scale estimates, and secondly, the $Var[ln(S)]$ is not affected by multiplicative factors since:

$$Var[ln(kS)] = Var[ln(k) + ln(S)] = Var[ln(S)],$$

where k can be any constant. Efficiency of the estimators was calculated relative to the smallest computed variance for the given family of estimators. That is:

$$(\text{efficiency})_j = \frac{\text{smallest computed variance}}{\text{actual variance}} = \frac{\min_{\text{all } i}\{Var[ln(S_i)]\}}{Var[ln(S_j)]}. \tag{4}$$

At this juncture, it should be mentioned that this investigation relates to the GPS X-axis deviates only. However, very similar results were obtained using the other two components (i.e., Y and Z). By and large, we will search for scale estimators that yield a high efficiency over a range of three potential situations. Such estimators are commonly called robust estimators (Huber, 1981; Hamplel et al., 1986). With this in mind, we choose three "situations" within our GPS data. Situation A refers to the entire sample of raw deviates as illustrated in Figure 1. Take heed of the fact that these data contain a satellite outage as is clearly depicted in the dilution of precision diagram (Figure 2). Situation B is composed of data from the previous situation A after all the identified outlying values have been removed. It is, thus, useful to establish some "outlier-cutoff" technique. A rule of thumb that works very well for reasonably large data sets (e.g., when n > 100) is as follows. We define as fourth-spread (d_F) the difference between the upper fourth (F_u) and the lower fourth (F_l). For example, in our case F_l = -4.382 m, F_u = 4.960 m, and therefore $d_F = F_u - F_l$ = 4.960 m - (-4.382 m) = 9.342 m. Potential outliers will be considered —and in our case they are removed from the data— any values that are outside the following outlier-cutoff interval:

$$[F_l - 1.5 \, d_F, \, F_u + 1.5 \, d_F]. \tag{5}$$

Figure 5 exhibits situation B, which is constructed by discarding all outliers from the original set. Note that some outliers could also be identified even when the satellite geometry is strong and stable. Finally, situation C corresponds to the stable and strong geometrical condition and encompasses all observations contained within the first and the 600th one inclusive.

As a next step, small (n=20) and moderate (n=100) samples were randomly drawn from each particular situation. This drawing process was repeated 200 times and each time estimates for each scale estimator were computed. Incidentally, the random selection procedure was performed by a uniform pseudo-random generator. At any rate, this sampling with replacement method is a widely applicable tool for assessing variability of complicated estimators and is called the ***bootstrap method*** (Efron, 1979). The rationale lies in using the available samples (i.e., data from situation A, B, and C) as models of population; then from them draw new samples (n=20 and n=100) to evaluate variability.

TABLE 4

No.	Scale (width) Estimator	Ancillary Estimates for Location (T) and Scale (S_o)	(Standardized) Deviations from Location Estimate	General Description/Algorithm
1	Standard deviation (S_D)	$T=\overline{X}$=(sample mean) $S_o=1$	$U_i=X_i-T$	$S_D=\sqrt{\dfrac{\sum\limits_{i}^{n}U_i^2}{n-1}}$ X_i = sample value n = sample size
2	Trimmed standard deviation (S_T)	$T=T_{2,\alpha}$=two-sided trimmed mean α=trimming proportion	$U_i=X_i-T_{2,\alpha}$	$S_T= T_{1,\alpha}[U_1^2,U_2^2,\ldots,U_n^2]$ where $T_{1,\alpha}$=one-sided trimmed mean (α=20%) $T_{2,\alpha}$=two-sided trimmed mean (α=10%) $T_{2,\alpha}=\dfrac{1}{n(1-2\alpha)}\{(1-r)[X_{(K+1)}+X_{(n-K)}]+\sum\limits_{k+2}^{n-K+1}X_{(i)}\}$ $r=\alpha\cdot n-[\alpha\cdot n]=\alpha\cdot n-K; K=[\alpha n]$
3	Mean absolute deviation (S_{AD})	$T=\text{median}\{X_i\}$	$U_i=X_i-T$	$S_{AD}=\dfrac{1}{n}\sum\limits_{i=1}^{n}|U_i|$; X_i=sample value n= sample size
4	Normalized median absolute deviation (S_{MAD})	$T=\text{median}\{X_i\}$	$U_i=X_i-T$	$S_{MAD}=\dfrac{\text{median}\{|X_j-T|\}}{0.6745}$
5	Normalized fourth-spread (or pseudo-sigma) (S_F)	$T=\text{median}\{X_i\}$	$U=F^{-1}(\frac{3}{4})-F^{-1}(\frac{1}{4})$	$S_F=\dfrac{F^{-1}(\frac{3}{4})-F^{-1}(\frac{1}{4})}{2\Phi^{-1}(\frac{3}{4})}=\dfrac{F^{-1}(\frac{3}{4})-F^{-1}(\frac{1}{4})}{2*0.6745}$ where $F^{-1}(\frac{3}{4})$ and $F^{-1}(\frac{1}{4})$ are the sample 75th and 25th quantities $\Phi^{-1}(\cdot)$=standard Gaussian quantile
6	Gaussian skip(S_G)	T_{BI}=biweight estimate of location (see Inglewicz(1982))	$U_{(i)}=\dfrac{X_{(i)}-T_{BI}}{E[Z_{(i)}]}$	$S_G=[T_{BI}(U_1^k,U_2^k,\ldots,U_n^k)]^{1/k}$;$k=2$ where $E[Z_{(i)}]$=expected value of the ith ordered Gaussian random variable.
7	Huber's scale estimate (M-estimate)(S_H) (robust)	$T=\text{median}\{X_i\}$ $S_o=MAD$	$U_i=\dfrac{X_i-T}{S_o}$	Solve for S using the following eqn $\dfrac{1}{n-1}\sum\limits_{i=1}^{n}\psi^2(U_i)=E[\psi^2(Z)\,|\,Z{\sim}N(0,1)]$ where $\psi(U)=\begin{cases}-b \text{ when } u<-b\\ u \text{ when } |u|\le b\\ +b \text{ when } u>b\end{cases}$;$b=1.7$ The above eqn is usually solved iteratively (e.g., use Newton-Raphson technique).
8	Biweight A-estimator (S_B)(robust)	$T=\text{median}\{X_i\}$ $S_o=MAD$ c=Tuning constant	$U_i=\dfrac{X_i-T}{c\cdot S_o}$	$S_B=\dfrac{n}{\sqrt{n-1}}\dfrac{[\sum\limits_{|U_i|<1}(X_i-T)^2(1-U_i^2)^4]^{1/2}}{|\sum\limits_{|U_i|<1}(1-U_i^2)(1-5U_i^2)|}$ $c=8.5$=tuning constant;n=sample size
9	Wave A-estimate (Andews' sine (S_W) (robust)	$T=\text{median}\{X_i\}$ $S_o=MAD;c=2.1$	$U_i=\dfrac{X_i-T}{2.1.S_o}$	$S_W=\dfrac{2.1\cdot n\cdot S_o}{\sqrt{n-1}}\dfrac{[\sum\limits_{|U_i|<\pi}\sin^2(U_i)]^{1/2}}{|\sum\limits_{|U_i|<\pi}\cos(U_i)|}$ $\pi=3.14159\ldots$
10	Modified wave A-estimator(S_{MW})	$T=\text{median}\{X_i\}$ $S_o=MAD;c=2.1$	$U_i=\dfrac{X_i-T}{2.1.S_o}$	$S_W=\dfrac{2.1\cdot n\cdot S_o}{\sqrt{n-1}}\tan^{-1}\left[\dfrac{[\sum\limits_{|U_i|<\pi}\sin^2(U_i)]^{1/2}}{|\sum\limits_{|U_i|<\pi}\cos(U_i)|}\right]$

TABLE 5: Situation A when n=20

	Estimator	Minimum Value	Low Adjacent Value	Lower Fourth F_L	Median	Upper Fourth F_u	High Adjacent Value	Maximum Value
1	S_D	4.126 m	4.126 m	7.889 m	11.433 m	16.673 m	27.172 m	1607.646 m
2	S_T	3.162	3.162	5.090	5.975	7.139	10.059	22.293
3	S_{AD}	3.262	3.262	5.694	7.062	9.383	14.544	366.217
4	S_{MAD}	2.253	2.253	5.423	6.510	7.814	11.319	11.756
5	S_F	2.583	2.583	5.569	6.660	8.058	11.484	19.751
6	S_G	3.313	3.313	5.956	7.024	7.811	10.522	12.777
7	S_H	3.309	3.309	5.504	6.372	7.138	9.552	13.932
8	S_B	3.441	3.441	5.878	6.732	7.800	10.506	14.037
9	S_W	3.514	3.514	5.973	6.810	8.100	11.135	14.378
10	S_{MW}	3.512	3.412	5.969	6.807	8.094	11.120	14.360

TABLE 6: Situation B when n=20

	Estimator	Minimum Value	Low Adjacent Value	Lower Fourth F_L	Median	Upper Fourth F_u	High Adjacent Value	Maximum Value
1	S_D	3.604	3.604	5.263	5.824	6.496	8.036	8.036
2	S_T	2.769	2.769	4.100	4.653	5.325	7.010	7.010
3	S_{AD}	2.861	2.861	3.961	4.510	5.080	6.750	6.852
4	S_{MAD}	2.838	2.838	4.606	5.471	6.733	9.685	9.685
5	S_F	2.804	2.804	4.728	5.842	6.923	9.739	11.118
6	S_G	3.237	3.237	4.894	6.023	6.688	9.221	10.440
7	S_H	2.701	2.701	4.485	5.171	5.906	7.755	8.312
8	S_B	3.290	3.290	5.024	5.762	6.486	8.303	8.953
9	S_W	3.499	3.499	5.023	5.730	6.446	8.221	8.803
10	S_{MW}	3.496	3.496	5.021	5.728	6.443	8.217	8.796

TABLE 7: Situation C when n=20

	Estimator	Minimum Value	Low Adjacent Value	Lower Fourth F_L	Median	Upper Fourth F_u	High Adjacent Value	Maximum Value
1	S_D	2.694	2.764	4.725	5.324	6.042	7.943	8.098
2	S_T	2.107	2.107	3.670	4.194	4.908	6.636	6.958
3	S_{AD}	2.124	2.124	3.629	4.096	4.674	6.170	6.412
4	S_{MAD}	2.110	2.110	4.148	5.044	6.145	8.924	10.695
5	S_F	2.027	2.027	4.193	5.316	6.386	8.742	10.757
6	S_G	2.567	2.567	4.547	5.317	6.244	8.566	9.062
7	S_H	2.100	2.100	4.058	4.684	5.525	7.282	7.859
8	S_B	2.657	2.657	4.470	5.200	5.913	7.821	8.083
9	S_W	2.666	2.666	4.501	5.182	5.892	7.786	8.038
10	S_{MW}	2.664	2.664	4.499	5.179	5.888	7.780	8.035

TABLE 8:   Situation A when n=100

	Estimator	Minimum Value	Low Adjacent Value	Lower Fourth F_L	Median	Upper Fourth F_U	High Adjacent Value	Maximum Value
1	S_D	7.229 m	7.229 m	12.324 m	17.337 m	38.588 m	52.977 m	729.660 m
2	S_T	4.537	4.537	5.511	5.926	6.603	8.150	8.950
3	S_{AD}	5.163	5.163	7.372	8.572	11.730	18.079	92.655
4	S_{MAD}	4.431	5.001	6.434	6.981	7.554	8.835	10.034
5	S_F	4.482	5.001	6.433	7.003	7.612	9.214	10.179
6	S_G	5.552	5.552	6.564	7.079	7.664	8.918	9.880
7	S_H	5.058	5.058	5.935	6.327	6.879	8.074	8.496
8	S_B	5.691	5.691	6.566	7.011	7.748	9.180	9.899
9	S_W	5.836	5.836	6.784	7.351	8.145	10.145	10.367
10	S_{MW}	5.836	5.836	6.783	7.350	8.144	10.143	10.366

TABLE 9:   Situation B when n=100

		Minimum Value	Low Adjacent Value	Lower Fourth F_L	Median	Upper Fourth F_U	High Adjacent Value	Maximum Value
1	S_D	4.692	4.893	5.621	5.944	6.225	6.907	6.907
2	S_T	3.797	3.797	4.459	4.748	5.034	5.656	5.646
3	S_{AD}	3.812	3.812	4.471	4.743	5.025	5.673	5.673
4	S_{MAD}	4.270	4.270	5.444	5.985	6.430	7.838	8.371
5	S_F	4.214	4.214	5.427	6.037	6.570	8.196	8.360
6	S_G	4.534	4.534	5.630	6.067	6.431	7.453	7.751
7	S_H	4.263	4.263	4.986	5.325	5.646	6.375	6.375
8	S_B	4.813	4.813	5.700	6.028	6.380	7.155	7.155
9	S_W	4.767	4.767	5.659	5.981	6.309	7.109	7.109
10	S_{MW}	4.767	4.767	5.658	5.981	6.309	7.108	7.108

TABLE 10:   Situation C when n=100

		Minimum Value	Low Adjacent Value	Lower Fourth F_L	Median	Upper Fourth F_U	High Adjacent Value	Maximum Value
1	S_D	4.275	4.440	5.223	5.497	5.749	6.467	6.645
2	S_T	3.415	3.433	4.123	4.359	4.594	5.159	5.159
3	S_{AD}	3.357	3.559	4.120	4.342	4.556	5.114	5.235
4	S_{MAD}	4.163	4.163	5.040	5.447	5.863	6.983	7.290
5	S_F	4.040	4.040	5.138	5.490	5.977	7.228	7.488
6	S_G	4.138	4.138	5.117	5.464	5.779	6.650	6.866
7	S_H	3.849	3.849	4.604	4.896	5.175	5.784	5.784
8	S_B	4.320	4.379	5.214	5.516	5.792	6.513	6.513
9	S_W	4.296	4.360	5.191	5.488	5.754	6.500	6.500
10	S_{MW}	4.296	4.360	5.190	5.488	5.754	6.499	6.499

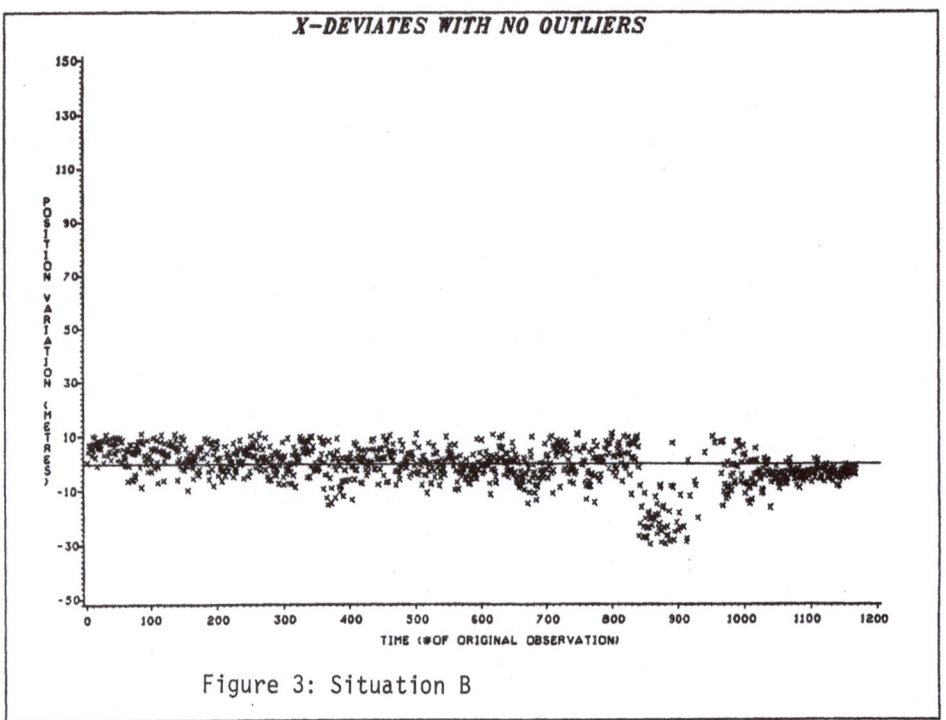

Figure 3: Situation B

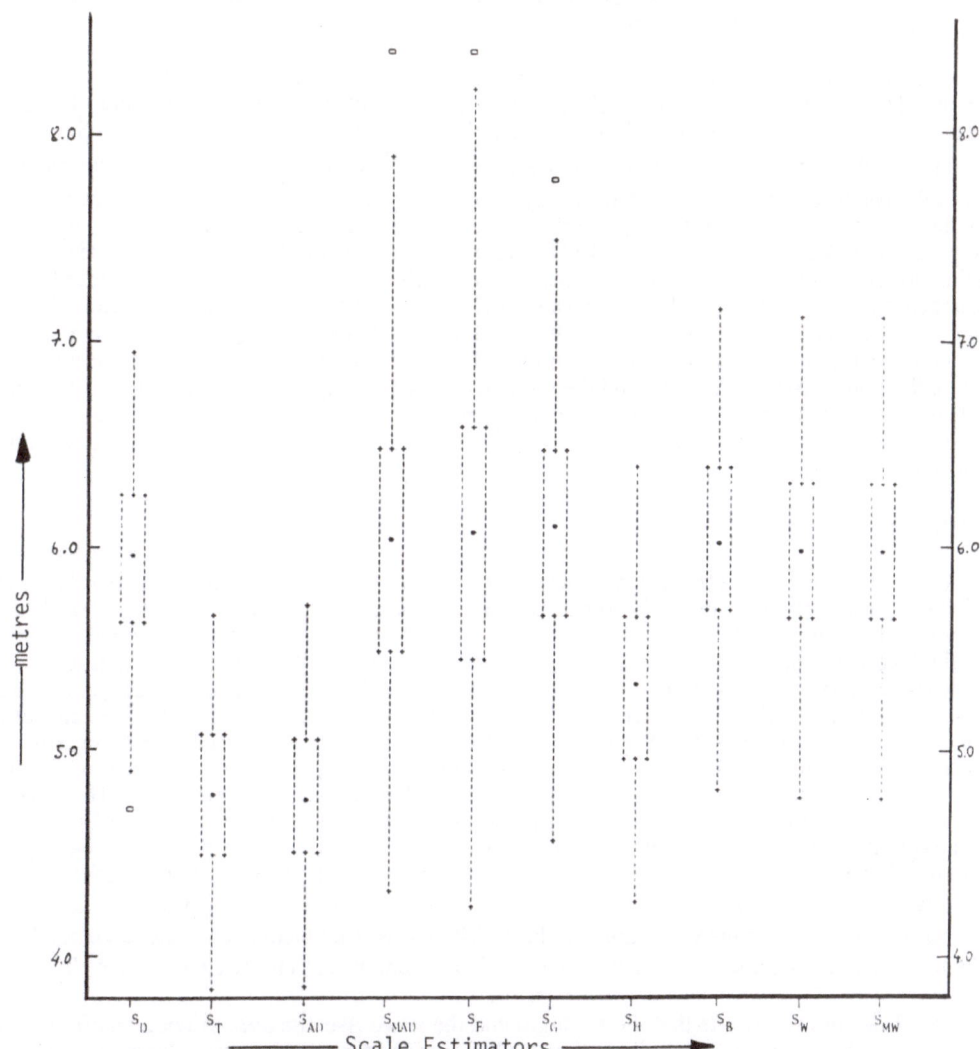

Figure 4: Box-plots for situation B, when n=100

At all events, 200 draws from each situation (i.e., A, B, and C) will provide a good aproximation of the sampling distribution of each scale estimator. At this point is should be mentioned that the choice of three such different situations was determined by what we would expect to arise in practice when position data are treated by practitioners.

Tables 5 to 10 highlight seven important summary values of the sampling distribution of the estimators. Here the term "adjacent value" applies to that value closest to each outlier-cutoff which is not an outlier. A simple examination of these three tables shows that the standard deviation S_D and the mean absolute deviation S_{AD} bring out an erratic behaviour in situation A. On the other hand, the values of the robust scale estimators change very little even when substantial changes exist in a small part of the data. Figure 4 presents some simple and convenient schematic plots for cross-comparisons. These plots are called ***box-plots*** (Tukey, 1977) and correspond to situation B for sample size n=100. Each thin box in the middle is customarily extended vertically from lower fourth to upper fourth, while a cross inside the box pinpoints the median's position. The vertical lines outside the box show the extent of the data except for outlying observations; which appear individually beyond the end of these lines and are marked with O. The general intent of the box-plot is to indicate a location estimate (median), the middle half of the data, and the range with detailed exposition of the stray values. Note that when careful outlier rejection techniques have taken place, such as that of situation B, the three last robust estimators (i.e., S_B, S_W, S_{MW}) not only estimate the same quantity as the standard deviation, but also exhibit the same variability.

4.   DISCUSSION

Naturally, an estimator is efficient with respect to certain distributional situations, if it has a low variation (i.e., minimum $Var[ln(S)]$) in comparison with all other scale estimators; bias is another issue but very hard to appraise (Mallows, 1977; Hamplel et al., 1986). In addition to the above condition, our final intention has been to select those scale estimators which exhibit the smallest variability (largest efficiency) across situations (e.g., data with outliers, without outliers, with good conditions). Tables 11 and 12 present the variability, in terms of $Var[ln(S)]$, the efficiencies and the trieffciencies of the scale estimators described in Table 2. Incidentally, triefficiency of an estimator is the minimum of its efficiencies for the three different distributional situations (i.e., A, B, and C). Inspection of those tables reveals that only three estimators performed very well across the three situations. The biweight A-estimator attained the maximum triefficiency (82.80% for n=20 and 79.62% for n=100), followed by the modified wave A-estimator (79.70% for n=20 and 69.9% for n=100) and the wave A-estimator. Although the standard deviation S_D performed extremely well in both B and C situations, it failed completely, as anticipated, in situation A. The trimmed standard deviation S_T protected its estimate against outliers, but its efficiency was not very high. The normalized median absolute deviation S_{MAD} and the fourth-spread S_F, however resistant, they could not achieve high triefficiencies as well.

In general, we may conclude that the biweight and the wave A-estimators provide an efficient and reliable method for estimating one-dimensional GPS positional accuracy (note the results are quite comparable with Lax (1985). These robust estimators are designed to work even when a small part of the data is interspersed by "bad" ones, such as outages, stray values, etc., and they are suitable for automatic processing. Moreover, they avoid the conceptual difficulty of identifying "legitimate" data and provide the same results as the parametric procedures when "improper" observations have been meticulously removed from the data.

ACKNOWLEDGEMENTS

We are grateful for discussions with and comments provided by Dr. M. Tingley from the Department of Mathematics and Statistics of the University of New Brunswick. The first author expresses his appreciation to the *Alexander S. Onassis* Public Benefit Foundation for financial support. The GPS data were made available to us by the Geodetic Survey of Canada. This work was supported by the Natural Sciences and Enginering Research Council of Canada and particularly by a Strategic Grant entitled *Marine Geodesy Applications* held by Dr. D.E. Wells.

TABLE 11: Variances, efficiencies and triefficiencies when n=20

Esti-mator	Situation A		Situation B		Situation C		Triefficiency (%)
	Var [ℓn(S)]	Efficiency (%)	Var [ℓn(S)]	Efficiency (%)	Var [ℓn(S)]	Efficiency (%)	
S_D	0.91781	5.59	0.02591	100.00	0.03923	100.00	5.59
S_T	0.09092	56.44	0.03198	81.02	0.04598	85.33	56.44
S_{AD}	0.48785	10.52	0.02874	90.14	0.04163	94.25	10.52
S_{MAD}	0.07644	67.13	0.07323	35.38	0.09657	40.63	35.38
S_F	0.08379	61.24	0.06849	37.83	0.09434	41.59	37.83
S_G	0.05682	90.31	0.04877	53.12	0.06074	64.60	53.12
S_H	0.05132	100.00	0.03687	70.26	0.05308	73.91	70.26
S_B	0.06042	84.94	0.03129	82.80	0.04736	82.85	82.80
S_W	0.06442	79.66	0.02979	86.95	0.04487	87.44	79.66
S_{MW}	0.06438	79.70	0.02980	86.94	0.04489	87.40	79.70

TABLE 12: Variances, efficiences and triefficiencies when n=100

Esti-mator	Var [ℓn(S)]	Efficiency (%)	Var [ℓn(S)]	Efficiency (%)	Var [ℓn(S)]	Efficiency (%)	Triefficiency (%)
S_D	1.61348	0.64	0.00502	100.00	0.00651	92.72	0.64
S_T	0.01998	51.73	0.00679	74.02	0.00658	91.79	51.73
S_{AD}	0.53768	1.92	0.00630	79.80	0.00610	98.91	1.92
S_{MAD}	0.01690	61.16	0.01529	32.86	0.01227	49.21	32.86
S_F	0.01656	62.43	0.01540	32.63	0.01312	46.00	32.63
S_G	0.01287	80.34	0.00966	52.01	0.00791	76.34	52.01
S_H	0.01034	100.00	0.00740	67.86	0.00672	89.86	67.86
S_B	0.01298	79.62	0.00588	85.46	0.00604	99.91	79.62
S_W	0.01479	69.90	0.00563	89.23	0.00604	100.00	69.90
S_{MW}	0.01479	69.90	0.00563	89.23	0.00604	100.00	69.90

REFERENCES

Efron, B. (1979). "Bootstrap methods: Another look at the jacknife." *Annals of Statistics*, Vol. 7, pp. 1-26.

Fisher, R.A. (1920). "A mathematical examination of the methods of determining the accuracy of an observation by the mean error and the mean square error." *Monthly Not. Roy. Astron. Soc.*, Vol. 80, pp. 758-770.

Hamplel, F.R., E.M. Ronchett, R.J. Rousseeuw and W.A. Stahel (1986). *Robust Statistics*. John Wiley, New York.

Huber, P.J. (1981). *Robust Statistics*. John Wiley and Sons, New York.

Iglewicz, B. (1982). "Robust scale estimates." In *Understanding Robust and Exploratory Data Analysis*, eds. D.C. Hoaglin, F. Mosteller, J.W. Tukey, John Wiley, New York.

Kalafus, R.M. and G.Y. Chin (1986). "Measures of accuracy in the NAVSTAR/GPS: 2 DRMS vs CEP." Presented at the Third Annual Technical Meeting of the Institute of Navigation, Long Beach, Ca, 21-23 January.

Lax, D.A. (1975). "An interim report of the Monte Carlo study of robust estimators of width." Technical Report No. 93, Series 2, Department of Statistics, Princeton University.

Lax, D.A. (1985). "Robust estimators of scale: Finite-sample performance in long-tailed symmetric distributions." *Journal of the American Statistical Association*, Vol. 80, No. 391, Theory and Methods, pp. 736-741.

Mallows, C.L. (1979). "Robust methods—Some examples of their use." *American Statistician*, Vol. 33, pp. 179-184.

Mertikas, S., C. Field and D.E. Wells (1986). "GPS accuracy measures: Tests and proposals." *Proceedings of the Fourth International Geodetic Symposium on Satellite Positioning*, University of Texas at Austin, Austin, TX, 28 April - 2 May.

Monahan, D. (1986). Personal communication. Canadian Hydrographic Service, Ottawa, Canada.

Tukey, J.W. (1960). "A survey of sampling from contaminated distributions." In *Contributions to Probability and Statistics: Essays in Honor of Harold Hotelling*, ed. I. Olkin et al., Chapter 39, Stanford University Press, Stanford.

Tukey, J.W. (1977). *Exploratory Data Analysis*. Addison-Wesley Publishing Company, Reading, MA.

COMPUTING THE HEADING EFFECT
AT M.F. AND L.F. POSITION FIXING

C.D. de Jong
and
J.C. de Munck
Section Mathematical and Physical Geodesy
Thijsseweg 11
2629 JA Delft
The Netherlands

ABSTRACT

On medium and low frequency position fixing the combination antenna-ship may cause heading dependent additive errors to the measured distances or pseudo-distances.

In the hyperbolic mode it is not possible to measure the additive corrections directly because a measurement is essentially the result of the values for a pair of fixed stations. Such a problem can also arise in case of range-range measurements if other unknowns have to be solved.

The authors propose to develop the corrections in a Fourier series similar to the classical method of circle calibration for a theodolite. The method is explained with two examples.

1. INTRODUCTION

One of the errors in position fixing on medium frequency and on low frequency is the heading dependent phase error caused by the ship-antenna combination.

The signal causes electric currents in the ship, which radiate to the receiving antenna. The result is a phase error or time delay in the effective antenna signal depending on the heading of the ship relative to the direction of the fixed station. This error being a function of the relative heading γ, must have a period of 2π.

The heading effect can be seen when a controled course with varying heading is sailed. The "Meetkundige Dienst Rijkswaterstaat" did such an experiment with a dredging vessel on HiFix6 in 1980 (Meetkundige Dienst, 1981), of which figure 1 gives an illustration. In this figure differences between HiFix 6 readings in one

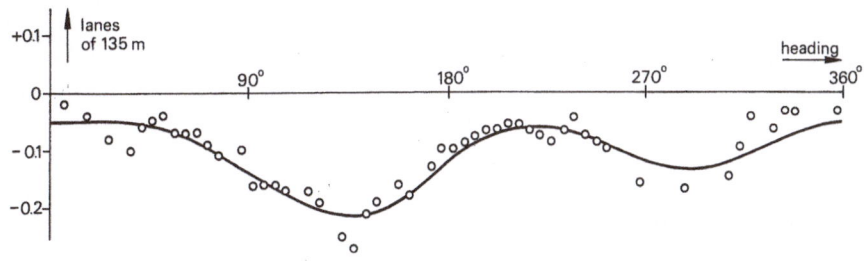

Figure 1. An example of the heading effect (after "Rijkswaterstaat")

M. Kumar and G. A. Maul (directors), Marine Positioning, 371–375.

hyperbolic pattern (lane width 135 m) and Motorola fixes are shown as a function of the heading. In this case variations of up to 200 m occur.

A curve like the one of the figure 1, however, can only be used in a very limited region because it depends on the geometry of the local position lines:
Each hyperbolic fix is the result of two or more pseudo distance measurements, each subject to its own phase error.

The problem can be compared with the calibration of a theodolite circle using the classical "Heuvelink-method" (Bakker, 1970). Here the error in the angle is the sum of the errors in the two directions. Analogously to this Heuvelink method, the phase error (or distance error) for the heading effect is developed in a Fourier series with a period of 2π radians.

If the wave length of the radio waves is comparable to the dimensions of the vessel, or larger, we may expect not more than a few Fourier terms (say up to the fourth harmonic). A large heading effect may be expected if the length of the ship is about a half to one wave length, and also if the ship with antenna is largely asymmetric. In the sections 2 and 3 the method will be explained in some more detail.

2. GENERAL THEORY OF THE METHOD

Radio positioning by ranges, pseudo-ranges or range differences may be considered as a combination of range measurements, where some additional portion of the range may be unknown, but satisfying some conditions of invariability.

In figure 2 the situation is shown for a ship measuring a (pseudo-)range to a fixed station k. A rectangular coordinate system (E,N) is introduced. If desired

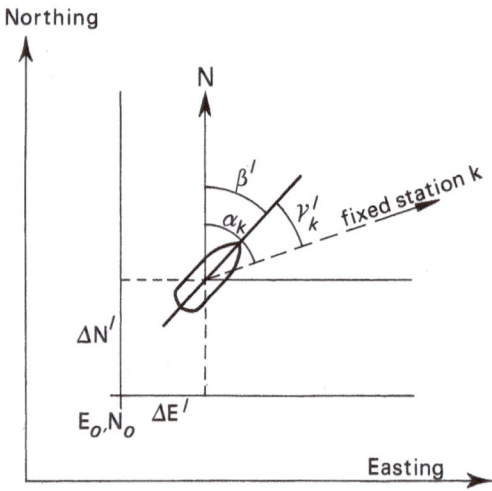

Figure 2.   The general situation

so, the problem can be linearized by expanding the range measurements between the ship in position 1 and the fixed station k into a Taylor series for some point (E_o, N_o) in the vicinity of the ship.

So in first order we find for measurements on one frequency:

$$S_k^1 = S_{k_o}^1 - \sin \alpha_k \, \Delta E^1 - \cos \alpha_k \, N^1 + P_k^T(\gamma_k^1) \tag{1}$$

or

$$\Delta S_k^1 = -\sin \alpha_k \, \Delta E^1 - \cos \alpha_k \, \Delta N^1 + P_k^1(\gamma_k^1) \tag{2}$$

where:

S_k^1 = (measured) range from the ship at position 1 to station k;

$S_{k_o}^1$ = distance from the approximate position, point (E_o, N_o), to the station k;

α_k = bearing of the fixed station. If the distance sailed during the experiment is small compared to S_k^1, α_k may be considered as independent of the position 1;

$\Delta E^1, \Delta N^1$ = corrections to E_o and N_o for position 1 (see fig. 2);

$P_k^1(\gamma_k^1)$ = unknown additional part of the range for position 1 for station k;

γ_k^1 = $\alpha_k - \beta^1$ = the angle between the ship and the direction to the fixed station k;

β^1 = bearing of the ship in position 1.

The periodic function $P_k(\gamma_k^1)$ is now developed as a Fourier series:

$$P^1(\gamma_k^1) = a_o^1 + \sum_{n=j}^{N} (a_n \cos n\gamma_k^1 + b_n \sin n\gamma_k^1) + P_r \tag{3}$$

where a_n and b_n can be considered as constants, except if the different stations k work on different frequencies like Decca Navigator. The coefficient a_o^1 is an additional constant that may depend on the measurement 1. P_r is a rest term containing the terms of the order n N and will be considered as noise. Substituting (3) into (2) and neglecting P_r yields:

$$\Delta S_k^1 = -\sin \alpha_k \, \Delta E^1 - \cos \alpha_k \, \Delta N^1 + a_o^1 + \sum_{n=1}^{N} (a_n \cos n\gamma_k^1 + b_n \sin n\gamma_k^1) \tag{4}$$

3. SPECIFIC CASES

Several cases may be distinguished. E.g. the ship positions may be known, absolutely or relatively, by external observations or they have to be determined simultaneously with the position system under trial. Sometimes, only range differences can be observed for hyperbolic systems.

The theory and calculations become somewhat more complicated if the different fixed stations k operate on different frequencies.

For a more detailed treatment see (De Jong and De Munck, 1986). Here we will confine ourselves to two examples.

3.1. Pseudo-ranges on one frequency; relative ship positions known. The ship sails a small route in which all headings occur, for instance a small circle. At discrete moments (positions 1) the compass heading is recorded, pseudo-ranges are measured relatively to some unknown point (ΔE_o, ΔN_o) with a second system.

The unknowns, ΔE_o, ΔN_o, the additional terms a_o^1 and the Fourier coefficients a_n and b_n ($n=1,2,\ldots,N$), may be solved by a least squares adjustment from the observation equations that can be found from equation (4) by stating:

$$\left.\begin{aligned}\Delta E = \Delta_o E + \Delta_r E^1 \\ \Delta N = \Delta_o N + \Delta_r N^1\end{aligned}\right\} \tag{5}$$

where $\Delta_p^1 E$ and $\Delta_r^1 N$ are the known relative positions of the ship.

The (linear) observation equations can be written as:

$$\Delta X_k^1 = -\sin \alpha_k \Delta_o E - \cos \alpha_k \Delta_o N + a_o^1 + \sum_{n=1}^{N} (a_n \cos n\gamma_k^1 + b_n \sin n\gamma_k^1) \tag{6}$$

where ΔX_k^1 are considered as the observations and defined as

$$\Delta X_k^1 = \Delta S_k^1 + \Delta_r E^1 \sin \alpha_k + \Delta_r N^1 \cos \alpha_k \tag{7}$$

in which ΔS_k^1 are the observed pseudo-ranges.

3.2. Range differences (hyperbolic) on one frequency; relative ship positions known.
In this case the difference in ranges is measured in a hyperbolic pattern, or in more than one pattern. If such a pattern is produced by two fixed stations s and t, we find the observation equations by taking the difference of equation (4) for k=s and k=t and applying equation (5):

$$\left.\begin{aligned}\Delta S_t^1 - \Delta S_s^1 + (\sin \alpha_t - \sin \alpha_s)\Delta_r E^1 + (\cos \alpha_t - \cos \alpha_s) \Delta_r N^1 = \\ \\ = -(\sin \alpha_t - \sin \alpha_s)\Delta_o E + (\cos \alpha_t - \cos \alpha_s)\Delta_o N + P_t^1(\gamma_t^1) - P_s^1(\gamma_s^1)\end{aligned}\right\} \tag{8}$$

The left hand side, which could be called X_{st}^1, may again be considered as the observation, in which ($\Delta S_t^1 - S_s^1$) is the observed range difference aboard the ship in the pattern st minus the calculated range difference from ($E_o N_o$) to s and t. The right hand side of (8) contains the unknowns: the shifts $\Delta_o E, \Delta_o N$ and in $P_t^1(\Delta_t^1)$ and $P_s^1(\Delta_s^1)$ the (required) Fourier coefficients a_n and b_n (the constant a_o^1 disappears in the difference).

4. APPLICATIONS

If a significant heading effect is observed or expected our theory provides the possibility to calculate the corrections which have to be applied to each of the (pseudo-) ranges. So for a hyperbolic observation in one pattern, (a combination of) two corrections have to be applied, namely one correction for each fixed station.

During our "Navgrav" campaign some experiments were made about the heading effect, from which Salzmann has computed the above mentioned corrections. He will report about these results at this symposium. (Salzmann e.a., 1986).

Bakker, G., 1970. "The adjustment of primary direction measurements with special
    reference to circle testing methods" Netherlands Geodetic Commission,
    Publication on geodesy, new series, volume 3, number 2, Delft.
De Jong, C.D. and De Munck, J.C., 1986? "Determination of phase errors of an
    antenna-ship system for position fixing with M.F. and L.F. radio waves" to be
    published in Hydrographic Journal.
Meetkundige Dienst, 1981. "Onderzoek naar het funktioneren van plaatsbepaling m.b.v.
    HiFix6 op een sleephopperzuiger in en nabij de Eurogeul". Meetkundige Dienst
    Rijkswaterstaat, Delft.
Salzmann, M.A., Husti, G.J., Strang van Hees, G.L., Teunissen, P.J.G., Seeber, G.
and Heimberg, F., 1986. A comprehensive combined NAVigation and GRAVimetry
    experiment on the North Sea". Paper presented at the INSMAP International
    Symposium on Marine Positioning, Reston, USA.

THE USE OF CALIBRATED AND MONITORED LORAN AS A MEANS FOR PRECISION VESSEL LOCATION

Mortimer Rogoff
President and Chief Executive
The DIGITAL DIRECTIONS CO., INC.
P.O. BOX 9957, Friendship Station
Washington, DC 20016

ABSTRACT

A method of using Loran C by means of a prior calibration and stationary monitors is described which results in much higher accuracy than usually obtained from conventional use of this system of radionavigation. Accuracies of the order of 5 yards are being experienced in areas where favorable Loran conditions exist, and the results almost never are worse than ten to twenty yards. The method of calibration and monitoring is discussed and reference is made to the equations employed for the calibration process.

1. Introduction

Loran C continues to be one of the most important and useful types of long range radio navigation used in many places around the world. Its most notable use is in the United States, where it is employed along all of the coastlines, waterways and lakes, and far at sea; along the nation's airways and airspace; and increasingly by land vehicles. Not surprisingly, its long use - more than 40 years (including the earlier form of Loran A) - has tempted electronic designers and navigators to obtain increasing levels of accuracy. The use of ground wave transmission of pulsed signals at 100 kilohertz produces a degree of stability and predictability which, if left alone, can result in accuracies of the order of 500 yards. By understanding the causes of the inaccuracies that give rise to errors in computed position it is possible to formulate a system capable of far better performance.

The subject of this paper is the description of a method yielding accuracies that are as good as 5 yards when circumstances are favorable, and typically 10 to 15 yards under almost all conditions. Two approaches are combined: the use of a pre-calibration that compensates for most propagation errors, and the continuing use of a fixed monitor that corrects for temporal shifts.

Once accuracies of this level are achieved it becomes reasonable and useful to show vehicle or vessel position on a carefully constructed electronic map or chart. Thus, a navigation system which allows the user to locate himself, and to plan routes or courses to reach any destination becomes a practical possibility. The Loran system then becomes more than just a means to compute accurate positions; it becomes an input sensor to a tactical and strategic navigation display system that has greater utility than that of the Loran itself.

M. Kumar and G. A. Maul (directors), Marine Positioning, 377–384.

One of the valuable features made possible with this degree of accuracy is the result obtained when a marine radar map is combined with an electronic chart. These two can be combined only if the radar map can be accurately placed and oriented on the electronic chart. One way to achieve this is to use an external system, such as Loran, to locate the center of the radar map on the electronic chart. A compass is used to rotate the radar map so that its angular orientation is the same as the chart, permitting alignment of the two maps. The combined use of radar and electronic chart is another example of the increased utility available from an accurate Loran system.

This paper describes the method of achieving the required calibrations and the use of a monitor receiver to stabilize the computed results in the face of seasonal and atmospheric disturbances.

Much of the practical methodology of achieving these results is the subject of an issued United States Patent (Number 4,590,569), entitled "Navigation System Including An Integrated Electronic Chart Display". The methods and instrumentation utilized to operate such a system are disclosed in that patent, and are the subject of its claims. The author of this paper was one of the inventors of that patent.

2. Description of the Method

2.1. Background - The Need for Calibration and Monitoring

The need for accurate positioning in marine applications arises most often as vessels are navigated close to shore. Unfortunately, the proximity to the land-water interface is also the place where Loran accuracy is degraded in unpredictable amounts. If the earth were a perfect sphere, covered only by salt water the accuracy of Loran would be predictable to a very large degree. Since it is not, we have to contend with a number of sources of inaccuracy, especially near or on land.

The first departure from theoretical behavior lies in the fact that there is a difference in the velocity of propagation of the radio wave when comparing the free space condition to that of a salt-water surface. This distortion can be treated by mathematical shifts in the theoretical model; it is distance sensitive, and can be corrected by equations that apply to the actual separation between shore transmitter and vessel receiver. The second form of this distortion, the Additional Secondary Phase factor (ASF) is far less predictable, since it depends on the conductivity constants of the complex land-water (differential) paths between master station, secondary stations and the vessel location. It is clearly a factor that changes with location, and one that cannot be predicted unless one knows the fine structure of conductivity values of all three radio paths for their entire length.

The second fact is that the earth is not spherical. All computations made with radio waves that traverse the earth's surface must give recognition to the non-spherical shape.

Finally, there are changes in the wave propagation factors which complicate the Loran accuracy problem, since the propagation constants are not stable over either short or long time intervals. Changes in climate, which affect both temperature and ground moisture (conductivity) are experienced with weather changes and with seasonal influences. Thus, even though a calibration is made at a particular locality, reducing errors in Loran down to the level of a few yards, this precision will not be maintained for more than a few hours, and certainly not for more than a few weeks. The figure of 500 yards stated earlier is an amount that is somewhat larger than the expected result of all of the error-producing influences at work in any one locality over the twelve months of all seasonal variations. Thus, the Loran charts or coordinate converters that use the same correction-data built in to the charts, are useful in those situations where the quarter nautical mile accuracy will suffice. Notice that there are no Loran charts prepared for harbors, nor is any correctional data supplied for such places. The reason is clear enough: 500 yard precision is likely to result in Loran-assisted groundings in many places in the restricted waters of a harbor!

2.2 The Approach

The practical and effective method for compensating for the effect of both of these distorting effects is to calibrate the area to be used for precise Loran positioning. Calibration in this instance involves measuring the actual Loran time differences at a place, and simultaneously measuring the latitude and longitude of that place. To achieve 5 yard levels of accuracy, the Loran coordinates need be measured to a precision of about 5 nanoseconds, and the geodetic position needs to be determined to a precision better than 5 yards. (Of course, the latter must be obtained by some method other than Loran; a typically practical method of achieving the required accuracy is the use of horizontal sextant angles measured between objects of known location).

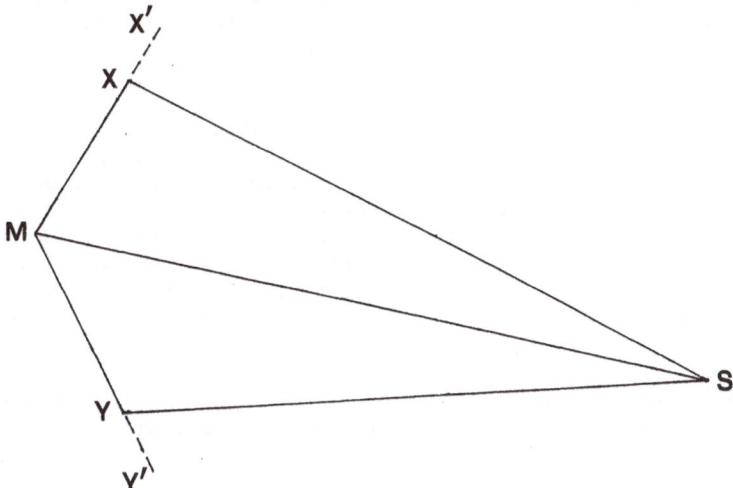

Figure 1.

Reference to Figure 1. will disclose the method used to create the calibration at any place. The master Loran station is located at M, and the two secondaries are, respectively, at X and Y; the vessel is located at position S.

The calibration procedure consists of measuring the Loran time differences at S, due to secondaries X and Y relative to the Master at M, and, simultaneously, the latitude and longitude of S.

The mathematical procedure involved in the calibration has the effect of shifting the location of X and Y along their respective baselines. These shifts produce path-length differences such that the observed Loran values at S would be produced by waves originating from transmitters at the shifted locations X' and Y'. The changed values of baseline length (the lengths B1 and B2 in the figure) are then used in position determination computations. When modified this way they are referred to as virtual baseline lengths.

At each place where a calibration measurement is made, the net effect is to reduce errors to a theoretical value of zero. This results from the definition of the calibration process. If the path lengths used for computation have been adjusted to yield the observed time differences, then using the shifted geometry would yield the measured position when the observed time differences are used in the computation. Since area coverage is desired, it is only necessary to calibrate throughout the area with sufficient density of measurement locations to make the errors at intermediate points small enough to be acceptable.

The equations used to convert Loran time differences to geodetic position assume that the earth is spherical. While this introduces some error, it is small compared to the errors that arise when the vessel is not actually at a calibration point. In effect, this omission places an outside limit on the distance that the vessel can travel from a calibration point without exceeding a maximum allowable error in computed position. This would present a problem if the propagation errors were constant over a large area. Since they are not, it is the propagation effect that dominates the error budget, and the lack of compensation for the true shape of the earth can be ignored.

The actual number of calibration points required in any area depends upon the Loran characteristics of the area. It is not possible to easily predict just how many calibrations need be made to keep errors down to some level; measurements need to be made to reveal the size of the errors and their rate of change with respect to location before any prediction can be made. In the author's experience, maintaining error below levels of approximately five yards in many harbors along the East Coast of the United States requires calibration at places spaced by approximately 1,000 to 2,000 yards.

The details of the equations employed in the computation of position are contained in the United States Patent mentioned in section 1. of this paper.

While the calibration process is underway, one or more Loran receivers are installed on shore to track and record the Loran time differences obtained at fixed locations. Their purpose is to measure the shifts in time difference that occur during the entire interval of the calibration. These shifts from the value at some arbitrary "reference time" are used as offsets to the observed Loran time differences at the calibration points. This procedure is necessary because the Loran signals are subject to continuous perturbations in propagation constants, especially noticeable when one is making measurements at the level of five nanoseconds. Weather and climate changes induce variations in earth conductivity constants. A weather front sweeping across the signal paths during the time of calibration (which can take many days) will superimpose a pattern of drifts on the observed Loran measurements which has to be removed from the data. These shifts can be removed by noting the changes in Loran time differences observed at monitor receivers in the area of the calibration. By using these shifts - measured at the time that each calibration data point is observed - as offsetting corrections to that data point, the required correction is made. The corrected values are used to compute the shifted or 'virtual' baselines that are involved in the computation of position.

Many measurements have been made to determine the areal extent over which a single monitor can accurately reflect changes. Again, it is not easy to predict the size of an area that will be capable of reducing the shifts to any specified amount; measurements should be made to verify the coverage limits. In a typical harbor situation it is been found that a single monitor is effective over a radius of approximately 25 miles; but this figure has to be verified in every case.

Once the calibration procedure is completed, the monitors play a continuing role in system operation. A reference value of monitor reading is established as the value observed at the reference time during the calibration procedure. When the system is in use, the monitors broadcast the shifts from the reference values as they occur (usually at about six-minute intervals); these shifts become the offsets that are used on each vessel in the area when Loran positions are computed aboard those vessels.

Whenever more than one monitor can be installed within the calibrated area this should be done, since the performance of the overall system is critically dependent on the accuracy of the monitor receiver. A system of "voting" can be established where at least two monitors have to yield the same shift values before being broadcast.

2.3 The Implementation

Once the calibration process is completed, a table of virtual baseline lengths is stored in the on-board computer to be used in the computation of Loran position. Since the vessel's position is usually at some place near rather than at the actual calibrated position, a process of interpolation is used to produce a value of virtual baseline that is appropriate for the actual location. Since the interpolated value depends upon position, the actual value used is arrived at by a process of iteration. This procedure is built in to the algorithms from which the on-board programs are derived.

A typical harbor might have as many as 50 calibration constants for each of the two or three secondaries involved. The program selects the constants that apply to the three nearest calibration points and interpolates between these three values.

Monitor shifts are sent to the vessel by any one of two or three methods. These range from a voice radio broadcast to an automatic transmission at periodic intervals that might be as often as every few minutes.

Another method to obtain the corrections is to use the vessel's own Loran receiver as a monitor. If the vessel is at a known location, and if the reference value of Loran time differences is known at that place, then the Loran receiver can be used temporarily as a monitor to obtain the shifts that have taken place.

The process of computing position is continuous, employing the corrected Loran time differences and utilizing the virtual baseline lengths obtained by interpolating within the table of stored constants. Smoothing the results to minimize the effects of noise is built in to the programs, with typical time constants of ten seconds employed to keep position fluctuations to an acceptable low level. In a typical situation, the Loran position is updated at one or two second intervals, with the ten second time constant in place to smooth noise fluctuations to an RMS value of approximately three to five yards.

3. Results Obtained

3.1 The Accuracy of the Method

The methods described in this paper have been in use since 1981. During this five year period, a number of harbors have been calibrated, including Boston, Providence, New York, Baltimore, and portions of Norfolk, Miami, and Tampa. Overall results depend upon the Loran situation in each of these places. For example, the results obtained in the Northeast United States – from Boston to Baltimore – have been uniformly good, at levels that are consistently of the order of five yard accuracies. When each measurement of position is preceded by a monitor reading, this can be improved to even lower levels, e.g. two yards. Miami is the poorest performer in this group because of the poor gradients and crossing angles of the Loran geometry. Tampa is intermediate, with results consistently within ten to twenty yards available.

If enough smoothing is available even the effects of bad geometry can be avoided. The propagational shifts can be observed with time constants at the monitor receivers of the order of ten minutes. If the vessel receiver can use long smoothing intervals, up to few minutes in length, then significant improvement in precision can be observed. This might be the case in seismic applications, or in any situation where unmoving or slowly moving vehicles are in use.

3.2 Initializing by Radar

There is an extremely powerful combination available to the user when accurate Loran positioning is used in conjunction with radar. If a display is constructed from the superposition of a radar map on an electronic chart, and if a match exists between the map and chart, then it is evident that the center of the radar map has been accurately located on the electronic chart. This can be used in either of two ways: if vessel position is known the two charts can be matched; if the map and chart are matched, the vessel's position can be determined. In the first case, the fact that vessel position is known (and also assuming that vessel heading is known) means that the radar map when combined with the electronic chart will have its center at the vessel's exact location on the chart. In that case all of the shore contours of the radar will match those of the electronic chart, and the radar returns from fixed lights and other structures will fall at the charted locations of those objects.   In other words, the radar map will make a match to the chart without any action on the part of the user, or by means of any computer program.

If vessel position is not known, and if the radar map is combined with the electronic chart, the two will not be aligned, nor will their shore contours or other details be matched. If manual adjustments (left-right and up-down) are made to the radar map as seen on the electronic chart, and if these adjustments result in an overall match of the contours, then the center of the radar map marks the location of the vessel. (As before, this assumes that angular alignment is made via ship's compass).

This characteristic of the matched charts can be used in a variety of ways. It can be the basis for finding position. The location of the center of the radar map can be located on the electronic chart by means of a movable cursor once the match is made; the cursor coordinates can be transformed into latitude and longitude by means of a program running in the computer.   This latter quantity can be used in a Loran calibration routine and a calibration point can be established at that place.   Similar measurements made at other places in a harbor would result in calibration of an entire area.

In this instance, radar and the electronic chart have replaced horizontal sextant angles, or some other form of radiodetermination in the calibration process.

The accuracy of positioning the vessel by this matching technique is of the order of 10 to 20 yards; hence the Loran calibration made this way will result in subsequent Loran positioning (with monitors running and feeding data to the vessel) of the same order of accuracy.

4. Conclusions

This paper has described a method of Loran calibration and subsequent monitoring that is the basis for a high-precision differential Loran system. The method has been in use for the past five years in the VIEWNAV® system and has produced consistently accurate results in a number of different harbors. The method appears to extract a practical level of usable accuracy from standard Loran C transmissions, far beyond that obtained when Government charts are employed, and in areas not covered by those charts. Where chart accuracy is nominally stated as 500 yards, using the calibrated differential system described in this paper can yield accuracy of 5 yards.

Limits to attaining this level of performance arise from situations where the Loran geometry is sufficiently bad to produce excessively noisy results. When noise fluctuations are big enough to mask the underlying accuracy, and when filtering or smoothing cannot be extended to reduce this noise then attainable accuracies are reduced.

This type of Loran has been successfully used in a navigation system that combines radar mapping with electronic charts, and where the vessel is shown in actual position on the chart. The resulting display is unique, constituting a fundamental improvement over previous forms of radar presentation.

VIEWNAV is the Registered Trade Name of Navigation Sciences, Inc., Bethesda, MD.

ACCURACY OF MULTIPLE LINES OF POSITION FOR HYDROGRAPHY

Lieutenant Samuel P. De Bow, Jr., NOAA Corps
National Ocean Survey, N/MO11
National Oceanic and Atmospheric Administration
Rockville, Maryland 20852

Professor Narendra Saxena
Department of Civil Engineering
University of Hawaii at Manoa
Honolulu, Hawaii 96822

Captain Glen R. Schaefer, NOAA Corps
Mapping, Charting, and Geodesy Program
Naval Postgraduate School, Code 68Sc
Monterey, California 93943

ABSTRACT

Multiple lines of position (MLOP) have long been acclaimed by
some hydrographers as giving more accurate positions than the
traditional two lines of position. The best method to achieve this
increase in positional accuracy has been debated for an equally
long period of time. This paper does not discuss the virtues of
the various methods but simply displays the results of one MLOP
method utilizing data acquired with typical field procedures and
time-tested methods. The results show that a significantly more
accurate position can be obtained if the MLOP method of position
computation is utilized. Data were acquired with relatively
inexperienced personnel and available (old) equipment. For some
projects, a change in the positioning method--rather than
purchasing new expensive state-of-the-art equipment--may be the
best way to achieve an increase in the accuracy of positioning
data.

1. INTRODUCTION

Hydrographic surveys often have been conducted using two lines
of position (LOPs) which barely provide sufficient positional
accuracy for the survey. Because surveyors seldom have funds
readily available to purchase new, more accurate, state-of-the-art
equipment, surveys may even be conducted at less than desired
accuracies. However, frequently surveyors do have available
"spare" equipment which when used in conjunction with the normal
equipment may substantially increase the positional accuracy of the
hydrographic survey project. Therefore, a scenario which utilizes
additional similar spare equipment is described.

Most conventional hydrographic survey data include position
fixes derived from only two LOPs. Each of the LOPs being the

385

result of an observed range, or an observed azimuth, from a shore
station with a known geodetic position. Producing more than two
LOPs is usually possible with little additional cost. The accuracy
obtained by having multiple lines of position (MLOP) can be
significant.

2. DATE ACQUISITION

Previous works by Kaplan (1980) and Silva (1982) both concluded
an increase in accuracy when using the MLOP positioning method in
lieu of two LOPs for positioning. However, the method was applied
to theoretical data only. Therefore, utilizing available (old)
equipment, students at the Naval Postgraduate School (NPS)
conducted position fixing for a hydrographic survey using various
combinations of ranges and azimuths to position the vessel. Shore
stations occupied by the electronic ranging equipment and the
theodolites were located to Third-order, Class I horizontal control
specifications (FGCC, 1984). Various combinations of the range and
azimuth measuring equipment were used.

Range data were acquired using a Del Norte Trisponder system
loaned to NPS by the National Ocean Service. The Del Norte system
is a short-range positioning system transmitting at about 9 GHz
with a pulse repetition interval which can be selected from 304 to
998 microseconds. Thirty-two consecutive valid time-difference
measurements are processed to provide each range output (Ingham,
1984). The least reading was 1 meter.

Azimuth data were computed based on observational data acquired
by students using Wild T-2 theodolites (De Bow, 1986).
Observations were based on geodetic control. The least reading
was 1 minute of arc.

A hybrid positioning method of one or more ranges and one or
more azimuths was also used. Data acquisition for each was as
previously mentioned.

3. DATA ANALYSIS

A model was constructed to determine the least squares position
for each position fix from the MLOP. The precision of each MLOP
position fix was compared to the precision of a position fix
obtained by only two LOPs to determine if a major difference did
exist between the two methods.

To determine if position fixes computed from MLOP were more
accurate than position fixes determined by only two LOPs, eight
LOPs were acquired for each position fix. Four of the LOPs were
ranges and four were azimuths.

For any given position fix to be computed using only range data,
only one combination of four ranges can provide the position as
only four ranges were acquired. If only three ranges are used to
compute the position fix and four ranges (identified as ranges

A, B, C, and D) were acquired, then any one of four possible different combinations of ranges may be used to compute the position fix (i.e.; ranges A, B, and C; A, B, and D; A, C, and D; or B, C, and D). The same situations occurs if azimuths are used in lieu of ranges. The above MLOP position fixing requires good placement of the shore stations to provide reasonably good intersection angles of the various LOPs in the work area. To increase the probability of good intersection angles of the LOPs when ranges and azimuths are used for the same position fix, the azimuth shore station was required to be at (or very near) a range shore station.

The following assumptions were made concerning the overall accuracy determination of each position fix:

a. All LOPs were independent and uncorrelated.
b. All data were free of blunders and systematic errors.
c. The standard error of range LOPs was 3 meters at all distances from the shore transmitter.
d. The standard error of azimuth LOPs was 4 meters at all distances from the theodolite.
e. The measured ranges were referred to the same point of time during interrogation by the DDMU time-deskew function.
f. Measurements by the azimuth observers were all of equal weight.

4. RANGE MEASUREMENT METHODS

An analysis was made of the acquired range data by the number of LOPs. The mean value of σ_p from each position fixing method is illustrated in table I.

For MLOP position fixes, the mean σ_p value for each day of data and the minimum and maximum σ_p were computed (table I). For the 3 days in which range measurements were acquired, an average σ_p of 1.3 meters was attained with the four LOPs and 1.5 meters with the three LOPs. The geometric accuracy of the MLOP position fixes exceeded the standard error of the Del Norte system by at least 1.5 meters. The slight variation in the mean σ_p's for the three LOPs was due to the geometrical variation of the ranging net from each shore station.

Analysis of the two-range-LOPs data shows a direct relationship between the angle of intersection of the LOPs and σ_p, as would be expected. A wide variation in the computed σ_p's exists due to the varying angle of intersections. Since the purpose of the field work was to determine the position of the vessel using MLOP, the angle of intersection of each individual pair of LOPs was ignored. Consequently, some large σ_p values have occurred for very small intersection angles; e.g., 237 meters for 1.6° (table I). During normal survey operations this situation would not be accepted and alternate plans would be made to obtain coverage in that area. Herein lies a benefit of using the MLOP position fixing method since the position of the vessel can be determined from selected

TABLE I

POSITIONAL ACCURACIES FOR RANGE LOPS

LOPs	DAY	σ_P MIN (m)	σ_P MAX (m)	σ_P MEAN (m)	ANGLE OF INTERSECTION (2 LOPs)
4 R	332	0.3	2.8	1.2	
4 R	333	0.4	2.9	1.4	
4 R	334	0.5	2.4	1.3	
3 R	332	0.1	3.6	1.1	*
		0.04	4.5	1.7	*
3 R	333	0.04	5.7	1.3	*
		0.01	3.7	2.0	*
3 R	334	0.1	2.4	0.8	*
		0.4	4.5	2.2	*
2 R	332	4.8	7.9	6.0	147°
		7.7	236.7	29.2	2°
2 R	333	4.4	8.1	5.7	28°
		4.4	578.2	12.3	1°

* NOTE: For positions which could be computed by more
than one combination of LOPs (e.g., 4 combinations of
3-range LOP fixes exist when 4 ranges have been
measured), the upper line for each day shows the
combination of LOPs giving the minimum mean σ_P's
and the lower line for each day shows the combination
of LOPs giving the maximum mean σ_P's.

shore stations which yield the best geometric configuration. As
expected, the required accuracy of at least 5 meters was not met
when two range LOPs with a standard error of 3 meters were used
(table I).

Overall, the capability of acquiring MLOP automatically from
measured ranges was successful with mean accuracies of 1.2 to 1.6
meters observed depending on the number of stations and the
geometric combinations used.

5. AZIMUTH MEASUREMENT METHODS

The use of an angle measuring instrument (e.g., a Wild T-2
theodolite) for positional control of sounding lines is a laborious
task. Continuous tracking of the vessel is difficult because the
horizontal tangent screw has a finite range. Observers can be
tracking the vessel with the tangent screw only to reach the end of
the drive mechanism moments before the time of the position fix.
Thus, the problem of tracking a vessel whose aspect is changing
rapidly with respect to a nearby observer is magnified. Older
model theodolites have an inverted image of the target which may
confuse the observer since the vessel seems to be moving in the
opposite direction. Also the reading of the observed direction is
through a secondary scope. Therefore, an observer must stop
tracking the vessel, align the horizontal plates, read the
direction and record the value. The lack of digital telemetering
capabilities requires the observed value to be relayed to the
survey vessel by voice radio. Obviously, there exists a potential
for many errors, particularly if position fixes are required at
frequent (less than 30-second) intervals.

Using a nonautomated system (e.g., a T-2) restricts the position
fixing of the vessel to well spaced intervals. Intermediate
sounding positions along the sounding line must be interpolated via
dead reckoning. In most cases, the actual track of the survey
vessel does not exactly correspond to the plotted survey line.

Although the limited manual azimuth determination methods seem
to be too complicated and error prone for hydrographic position
fixing, it is still one of the most exact position fixing methods
available for a large-scale survey. Experienced observers have
little difficulty in acquiring accurate data.

The azimuth-MLOP position fix computed from three or four
azimuths resulted in the most accurate method investigated even
though a 4-meter standard error was applied to all azimuth LOPs.
Examination of the tabulated σ_p's (table II) shows a mean value of
0.7 meter on day 332 and 1.0 meter on day 333 when the four LOPs
were used. The mean σ_p's from three LOPs ranged from 0.7 meter to
1.3 meters for the 2 days of data. The accuracy capabilities of
the MLOP position fixing method are evident.

The two-azimuth-LOPs condition suffers from the same degradation
in accuracy as the two-range-LOPs case. The error in position is a

TABLE II

POSITIONAL ACCURACIES FOR AZIMUTH LOPS

LOPs	DAY	σ_P MIN (m)	σ_P MAX (m)	σ_P MEAN (m)	ANGLE OF INTERSECTION (2 LOPs)
4 Az	332	0.3	1.7	0.7	
4 Az	333	0.1	2.9	1.0	
3 Az	332	0.00	2.5	0.7	*
		0.01	9.5	1.2	*
3 Az	333	0.00	4.9	0.8	*
		0.05	4.3	1.3	*
2 Az	332	5.7	8.9	6.3	41°
		6.3	423.6	26.7	1°
2 Az	333	5.7	12.9	6.4	26°
		5.7	84.2	8.6	4°

* NOTE: For positions which could be computed by more than one combination of LOPs (e.g., 4 combinations of 3-azimuth LOP fixes exist when 4 azimuths have been observed), the upper line for each day shows the combination of LOPs giving the minimum mean σ_P's and the lower line for each day shows the combination of LOPs giving the maximum mean σ_P's.

function of both the distance of the vessel from the shore station and the angle of intersection of the two LOPs. Analysis of all of the σ_p's obtained shows a mean value of 6.3 meters for all data acquired (table II). Seemingly a large number, the probable cause is the initial assumption that each azimuth LOP was in error by 4 meters at all ranges. A more thorough investigation of the angular error of dynamic azimuth measurements could be made which would likely reduce the size of the assumed azimuth error.

6. HYBRID METHODS

Numerous combinations of control configurations exist when ranges and azimuths are mixed together. Because of the poor geometry which often exists when mixing numerous range and azimuth LOPs, using more than two range LOPs and one azimuth LOP is of little value (table III). Even for the position fix computed from two range LOPs and one azimuth LOP to be of significant value, the electronic ranging equipment for one of the ranges and the theodolite should be located at the same shore station.

7. APPLICATION

Short-range microwave positioning systems require periodic static base-line calibrations as well as daily system checks to detect calibration drifts and/or component failures. Both of these operations take time. A distinct advantage of using the MLOP position fixing method for position control in hydrography is that redundant data are acquired for each position fix. Having more than two LOPs for each position fix permits timely detection of positioning system ambiguities. Thus, the need to halt survey operations for daily system check is eliminated. Additionally, if a large σ_p value is found during, or after, field operations the residuals for each LOP can be examined to determine which range or azimuth values are in error. Once the error producing system is determined, it can be replaced or data from the erroneous LOP eliminated from the data prior to processing. The advantages of using the MLOP position fixing method may realize a financial return to the hydrographer when a single LOP is faulty, for the operation most likely will be able to continue with the faulty LOP system removed.

8. CONCLUSIONS

The MLOP positioning method has the potential of substantially increasing the positional accuracy of a hydrographic survey over the traditional two-LOPs positioning method for little additional cost in time and required resources. The increase may be sufficient to allow use of existing equipment to meet accuracy requirements for the now more common large-scale surveys.

The MLOP positioning method also allows the data processing and analysis portion of a survey to be more complete and exact by increasing the ability to identify and correct for erroneous positioning data.

TABLE III

POSITIONAL ACCURACIES FOR HYBRID LOPS

LOPs	DAY	σ_P MIN (m)	σ_P MAX (m)	σ_P MEAN (m)	ANGLE OF INTERSECTION (2 LOPs)
3 R & 1 Az	332	0.1	11.8	1.1	*
		1.2	6.3	3.0	*
3 R & 1 Az	333	0.1	6.7	1.7	*
		1.2	12.5	5.5	*
2 R & 2 Az	332	0.3	2.2	0.8	*
		1.2	5.3	2.1	*
2 R & 2 Az	333	0.2	3.5	1.4	*
		0.2	5.0	2.4	*
2 R & 1 Az	332	0.01	5.1	0.8	*
		0.3	4.9	1.9	*
2 R & 1 Az	333	0.02	4.2	1.1	*
		0.1	7.7	2.6	*
1 R & 1 Az	332	5.0	5.0	5.0	90°
1 R & 1 Az	333	5.0	5.0	5.0	90°

* NOTE: For positions which could be computed by more than one combination of LOPs (e.g., 4 combinations of 3-range LOP fixes exist when 4 ranges have been measured), the upper line for each day shows the combination of LOPs giving the minimum mean σ_P's and the lower line for each day shows the combination of LOPs giving the maximum mean σ_P's.

REFERENCES

De Bow, S. P. Jr., <u>Application of Multiple Lines of Position for Hydrography</u>, M.S. Thesis, Naval Postgraduate School, Monterey, California, 1986.

Federal Geodetic Control Committee (FGCC), <u>Standards and Specifications for Geodetic Control Networks</u>, Rockville, Maryland, 1984.

Kaplan, A., <u>Error Analysis of Hydrographic Positioning and the Application of Least Squares</u>, M.S. Thesis, Naval Postgraduate School, Monterey, California, 1980.

Silva, F. C., <u>Calculation of Hydrographic Position Data by Least Squares Adjustment</u>, M.S. Thesis, Naval Postgraduate School, Monterey, California, 1982.

MONUMENTING MARINE CONTROL*
(*Invited Paper*)

Narendra Saxena
Department of Civil Engineering
University of Hawaii at Manoa
Honolulu, HI 96822

ABSTRACT

Although the idea of monumenting marine control was first mentioned in 1958, the real work started in 1965. With President Reagan's Proclamation of the Exclusive Economic Zone on 10 March 1983, resource assessment of the ocean, including fisheries, oil, gas and minerals, marine control networks will play an important role in boundary demarcation, law enforcement, scientific and industrial activities. New developments in acoustic instrumentation have led to more accurate establishment of marine benchmarks. Accuracies of better than +/-25cm over a distance of 2600m in water depths of 1400m have been obtained.

---

*Presentation given at symposium; paper to be published in 1987.

M. Kumar and G. A. Maul (directors), Marine Positioning, 395.

NAVIGATION FOR SURVEYS OF TRANS-PACIFIC FIBER-OPTIC CABLES

Alexander Shor
Dale Chayes

Lamont-Doherty Geological Observatory of Columbia University
Palisades, NY 10964

ABSTRACT

Cable route surveys were carried out between Oahu and Luzon using the NECOR multi-narrow beam echosounding system on *R/V Robert D. Conrad* for AT&T Communications, Inc. Navigation systems included CA-code GPS (Magnavox T-set), 2-chain Range-Range LORAN C (Internav LC-408), Transit satellite (Magnavox 1107), 2-axis doppler speed log (Furuno CI-30) and gyro heading (Sperry MK-27 gyro). GPS positions include clock-aided (HP-5065A Rubidium standard) 2-satellite fixes in addition to fixes derived from 3, 4 and 5 satellites.

Real-time positions during surveying at 9 to 11 kts were provided by multiple-sensor displays each minute. Up to 10 simultaneous LORAN C ranges were plotted. GPS, Transit satellite, DR., and positions calculated from one R-R LORAN pair were plotted on a single 4-pen plotter. Real-time LORAN was calibrated to 3-satellite GPS at least twice daily to reduce Rb clock drift errors.

Shipboard processing of navigation data provided smoothed plots for all surveys prior to conclusion of the cruise, and demonstrated the viability of "post-processing yesterday's data today" at sea. Rb-clock drift corrections were applied, using transit satellite fixes for reference, and expanded-scale plots of multiple-sensor positions compared. "Splicing" of 20- or 30-minute R-R LORAN and GPS fixes (and Transit fixes in gaps) formed a continuous fix file; intermediate 1-minute positions were calculated using fixed-end speed/gyro dead reckoning using routines developed by the Seabeam group at University of Rhode Island.

Multibeam bathymetry provides estimates of relative positioning error at many points along the route, and indicates that post-time navigation errors exceed 300 meters only during rare extended periods (>4 hours) during which neither GPS nor R-R LORAN were available. Clock drift, GPS and Transit satellite precision were examined during dockside periods in Oahu, Guam and Hong Kong.

As the Seabeam bathymetric swath width normally exceeded 3000 meters, cable deployment within "acceptable" topography the along the pre-planned or modified route is extremely likely.

INTRODUCTION

Lamont-Doherty Geological Observatory received a contract in 1985 from AT&T Communications, Inc., to conduct bathymetric surveys along the proposed fiber-optic submarine telephone cable routes (see figure 1) TPC-3 (segments D1 and D3), and GP-2. Cable route TPC-3 connects the islands of Oahu and Guam, and route GP-2 extends from Guam to Luzon. Surveys were carried out during the period August to October, 1985, using the Research Vessel *Robert D. Conrad* cruises RC2610 (Honolulu to Apra Harbor, Guam) and RC2611 (Guam to Hong Kong via Baler Bay, Luzon). *Conrad* is

397

M. Kumar and G. A. Maul (directors), Marine Positioning, 397–405.
© *1987 by the Marine Technology Society.*

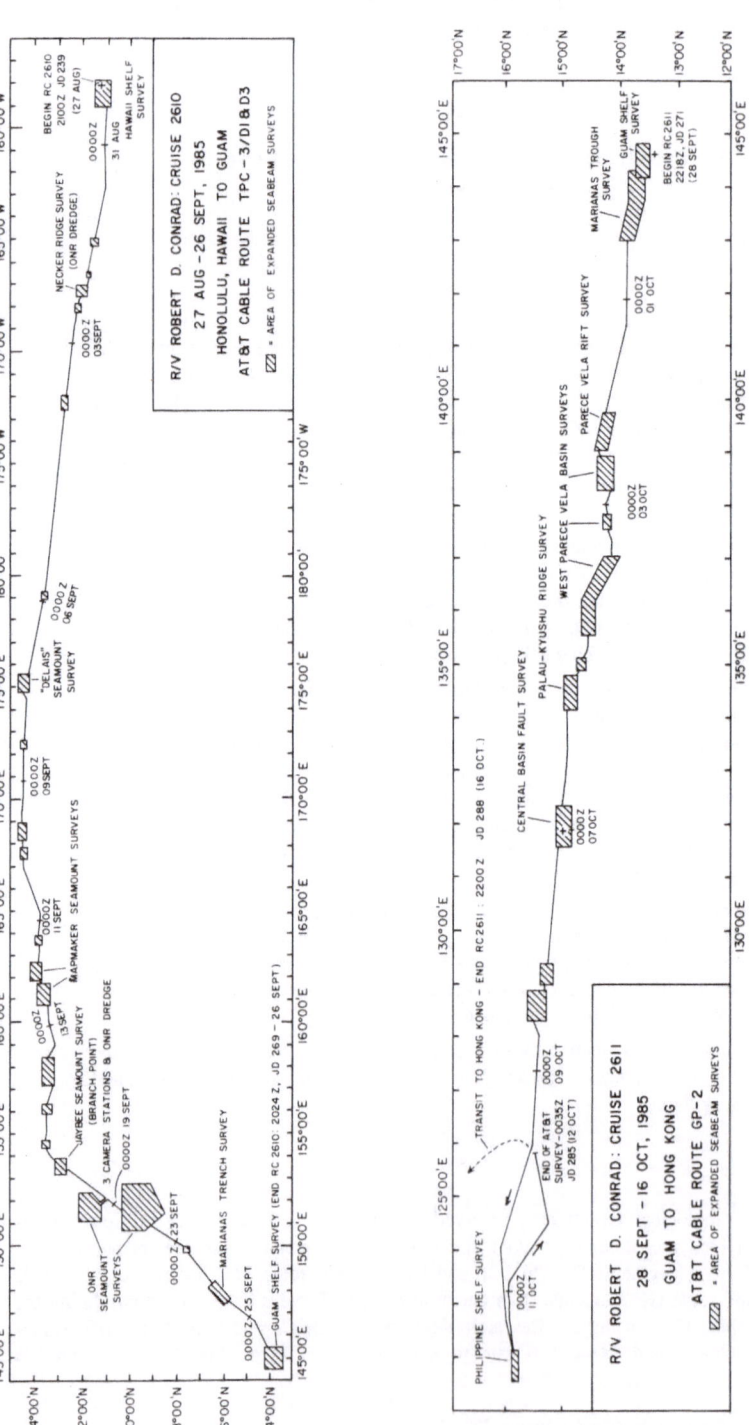

Figure 1. *R/V Robert D. Conrad* cruise track from Honolulu, Hawaii to Apra, Guam and Apra, Guam, to Hong Kong *via* Baler Bay, Luzon.

equipped with a hull-mounted multi-narrow beam echosounding system ("Seabeam"), which is operated by the NECOR Seabeam Operations Group at the Graduate School of Oceanography, University of Rhode Island, under the direction of Dr. Robert Tyce. Figure 2 shows the configuration of the real-time Seabeam system as installed on the R/V Conrad. Surveys were carried out under the direction of Alexander Shor.

Submarine cable route bathymetric surveying has as its principal objective to sufficiently determine seafloor roughness in order to allow determination of necessary extra cable over that required to directly connect the two ends of the cable route. In the case of fiber-optic cable deployment, an additional concern is to determine as accurately as possible seafloor gradients in regions of rough topography, as the strength characteristics of fiber optic cables are not sufficiently well tested in the submarine environment to accurately assess their potential for damage. During the course of surveying and data analysis, seafloor gradients of 1:5 were considered sufficiently steep to require expanding surveys to define possible alternate routing.

Cable route surveying requires navigation sufficiently accurate to prevent deployment of cables over unsuitable or unsurveyed topography. This objective required the ability to reconstruct positions to achieve an absolute navigational accuracy of ±500 meters. It was our goal during surveying to maintain vessel position within 500 meters of the proposed cable route while surveying at 9 to 11 knots.

In order to achieve this level of precision along more than 9,300 route kilometers between Oahu and Luzon, it was necessary to improve both the navigation receiver suite and the real-time display system on the vessel. Techniques used to determine position and to produce smooth plots of occupied positions, as well as methods used to evaluate positional accuracy, are described in the following sections.

NAVIGATION ACCURACY: OBJECTIVES

Specific objectives for route navigation were as follows:

1. Maintain ship position within 500 meters of the designated survey route.

2. Reconstruct positions, using best available navigation, such that bathymetric contours have an uncertainty of less than 500 meters in position along the entire cable route.

In addition to these requirements, which relate to subsequent deployment of submarine cables, a third objective was added solely for purposes of geophysical research. This requirement is to routinely determine vessel velocity, (magnitude and direction), to within ±0.1 knots between 1-minute fixes, in order to accurately determine Eotvos corrections for computation of the earth's gravity field from the raw gravity measured by a BGM-3 gravimeter. This requirement necessitates relative position precision between 1-minute fixes of ±3 meters.

INSTRUMENTATION

On the Conrad, navigation and underway geophysical parameters are routinely logged on a Nova-4 computer. The data logger is shown in figure 3. Navigation instrumentation used during these surveys included the following:

1. MAGNAVOX T-SET GLOBAL POSITIONING SYSTEM RECEIVER.
    Two Magnavox T-set GPS receivers were operated during the course of the surveys. These instruments, S/N 101 and 108, were among the first T-sets produced by Magnavox. They utilize the CA-code GPS positioning signal, which is estimated to have absolute positioning accuracy of ±35 meters horizontal under optimum operating conditions. Positions are computed relative to the WGS-72 reference geoid. Lamont- Doherty has been using T-set receivers on Conrad
    since September, 1984. Immediately prior to the start of cruise RC2610, both sets were upgraded to allow "clock-aided" navigation using two satellites and a frequency standard. Additional modifications were made during the port call in Guam to correct problems detected during the first leg. During routine surveying, time and position from the GPS receiver were logged on tape

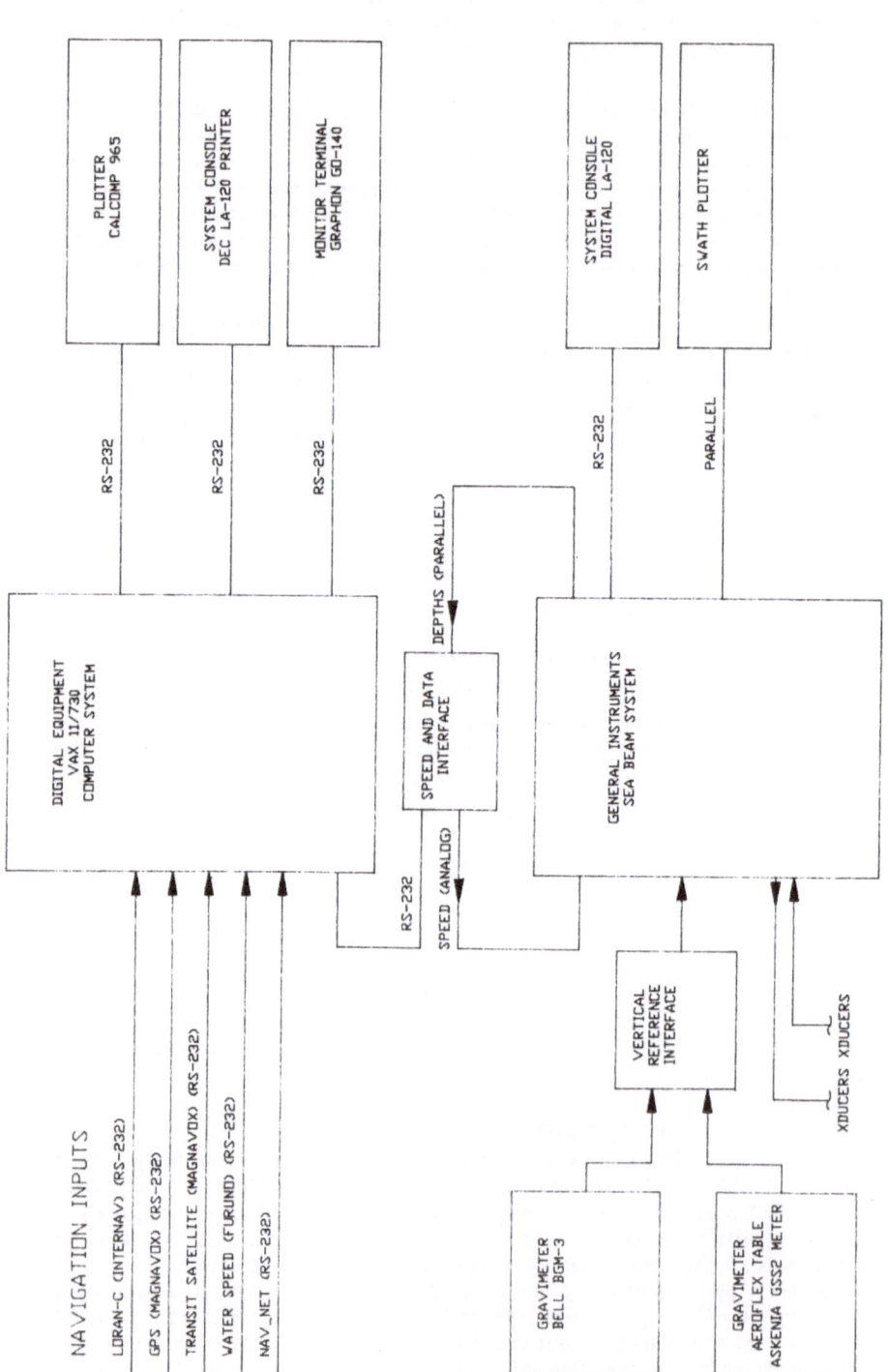

Figure 2. *R/V Robert D. Conrad* real-time Sebeam system block diagram.

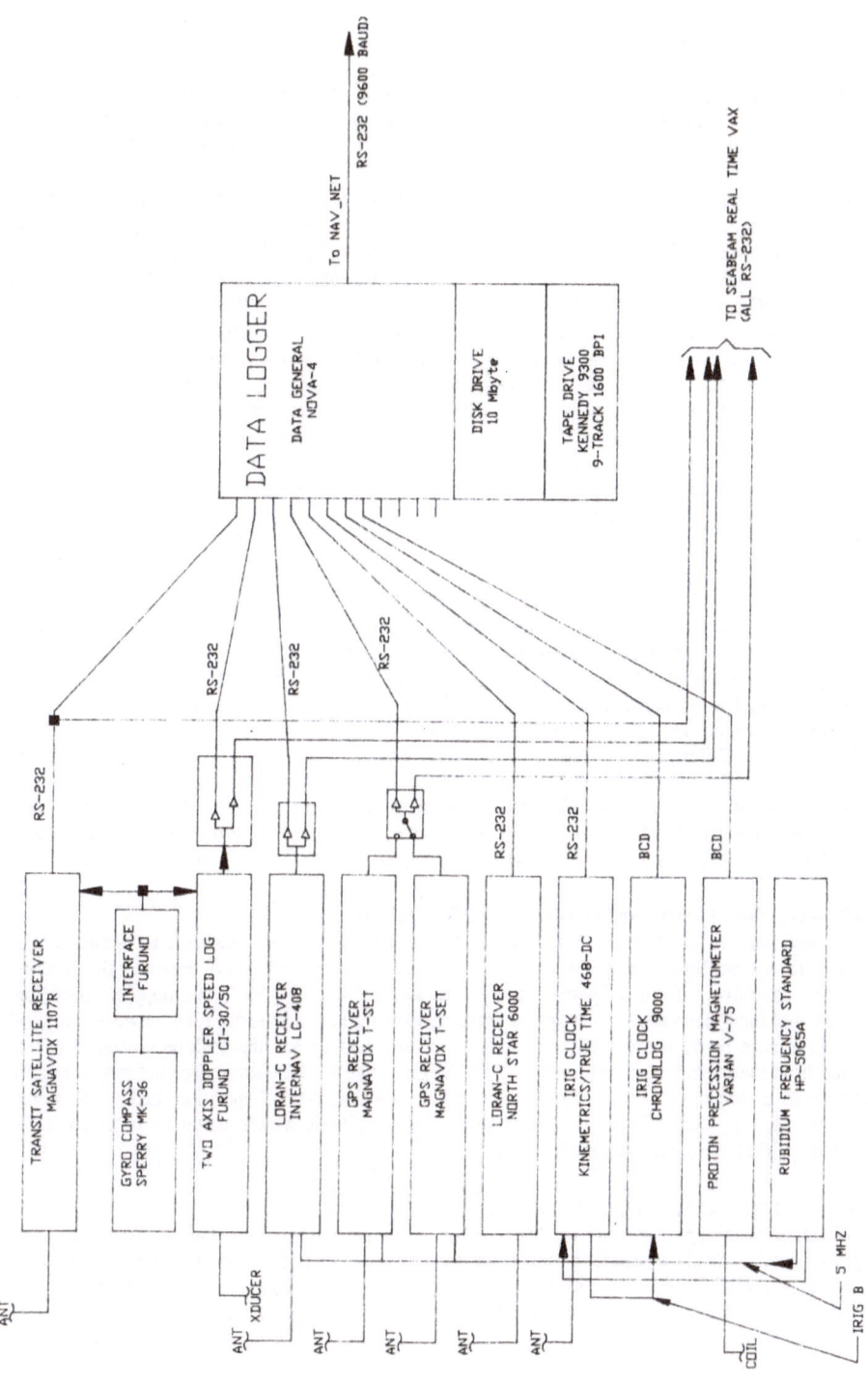

Figure 3. *R/V Robert D. Conrad* geophysical data logger block diagram.

once every 30 seconds by the data logger. In addition to navigation information, an output message containing information concerning GPS satellite constellation geometry and position quality was also logged with each fix.

2. INTERNAV LC-408 LORAN RECEIVER.

LORAN C positioning during the course of the surveys utilized the Internav LC-408 receiver, which has the capability of determining positions in either hyperbolic or range-range mode, and which allows simultaneous acquisition of range/time delay data from two LORAN C station networks. Receiver output includes range and time delay in microseconds from each of up to 10 LORAN C stations as well as time/date information and a position calculated from a single pair of either ranges or time delays. Positions may be calculated from a station pair in a single chain or cross-chain, and calibrations may be entered to both time delay and ranges. Operation of the LC-408 receiver in Range-Range mode requires input of a 5 megahertz reference signal from a frequency standard, which in the present case was a Hewlett Packard Model 5065A Rubidium Frequency Oscillator. Under optimum operating conditions the HP-5065A generates a frequency accurate to approximately one part in 10^{11}, or approximately 1 microsecond per day. During the present surveys the drift rate of the frequency oscillator used was determined to vary between 2 and 3 microseconds per day based on dockside calibrations using multiple LORAN C stations as a frequency reference. Drift rates of this order correspond to 600 to 900 meters per day of cumulative error in each range, or 800 to 1300 meters of error in calculated position per day under optimum station geometry. In order to maintain accurate positions using Range-Range mode navigation computed ranges were generated routinely from GPS fixes based on 3 or more satellites to all Loran stations. Twice daily we recalibrated ranges in the LC-408 for the pair of LORAN stations being used for positioning.

3. HEWLETT PACKARD HP-5065A RUBIDIUM FREQUENCY STANDARD.

A single HP-5065A was used as a frequency reference for two Magnavox T-set GPS receivers and one Internav LC-408 Loran receiver. The 5 megahertz output was buffered to provide sufficient drive for all of these devices. Frequency error throughout the course of the survey cruise resulted in an apparent drift rates of 2.2 microseconds per day in Honolulu (27 August), 2.95 microseconds per day in Guam (28 September), and 3.4 microseconds per day in Hong Kong (13 October). Calibrations performed at sea, using computed ranges from Transit satellites compared to raw LORAN ranges, produced similar results, although dockside data is inherently more precise. Linear regressions of ranges errors computed from 1 to 5 days of Transit satellite fixes were used to calculate drift rates for navigation post- processing; drift corrections were applied to daily LORAN range files prior to calculation of daily fix files in the post-processing routine.

4. MAGNAVOX MX-1107 TRANSIT SATELLITE RECEIVER.

A Magnavox model MX-1107 Transit dual channel satellite receiver was used to calculate fixes from Transit satellites during the course of the surveys. Gyro and speed inputs to the receiver come, respectively, from a Sperry MK-27 gyro and a Furuno CI-30 2-axis doppler speed log. Dead reckoning positions are calculated between fixes by the 1107 using time between fixes and the difference between observed position at the most recent fix and the position computed at that time from the integrated velocity vector from the preceding fix. However, since the receiver is not equipped to accept the cross- track component of velocity, the DR. position, and hence subsequent apparent "current", are less accurate than the positions which would be obtained if the speed log data were used in complete form.

5. FURUNO CI-30 2-AXIS DOPPLER SPEED LOG.

Vessel speed through the water is obtained from a Furuno CI-30 2-axis doppler speed log. Ship velocity data, calculated as along- and cross-track components relative to ship's heading from the ship's gyro, are logged every 3 seconds by the "geophysical data logger". The CI-30 uses a reference depth of 200 meters in the water column (in deep water) for calculation of vessel velocity in two axes; in water shallower than 800 meters the seabed is used as the reference point for velocity computation. Integrated velocity along track provides dead reckoning positions which, when calibrated to fixed navigation points of appropriate spacing, produce a much smoother navigation file than even 1-minute GPS fixes.

6. SPERRY MK-27 GYRO.
Gyro heading is available to the system through an interface provided by the Furuno doppler speed log.

REAL-TIME NAVIGATION
In lieu of any attempt to integrate navigation sensors to produce a single "correct" position, we chose to closely monitor three navigation sources during the survey, relying on experience (and hard copies of position history) to select the best position for real-time navigation. The real-time navigation display system was based on four Compaq-286 microcomputers linked via Ethernet into a single multi-task system shown in figure 4. In order to provide the scientific watchstander and the bridge with the most information and the least confusing presentation, we opted for the following displays:

1   Printouts in the science lab listing, once per minute, a seven line message containing the most recent "geophysical data logger" message in the data buffer. The message as printed included:

   1.a   Time from the Chronolog and True-Time clocks.

   1.b   A three line message from the LORAN receiver including Lat./Long. from the pair of ranges selected, plus the time/date of the most recent LORAN fix and the ranges of all (up to 10) stations on the two chains.

   1.c   A two line message from the GPS receiver, with time/date of most recent fix, plus Lat./Long., altitude, speed and heading, and a series of quality parameters.

   1.d   A one line message from the Transit satellite receiver with time/date of most recent DR. position plus Lat./Long.

   1.e   In addition, Transit satellite fixes were printed with time/date and quality parameters when received. At each fix there is an automatic calculation and print out of ranges to all of the loran stations. This listing, although bulky and wasteful, provided a valuable record of sensor performance which was commonly referred to when one or another of the navigation systems exhibited symptoms of error.

2.   Plots of LORAN range and signal level against time of day. Each of the 10 LORAN range values and the correlative signal level was plotted once per minute on a single plot using a 4-pen HP-7475 plotter. This allowed one to quickly identify which station was at fault if problems such as 10 microsecond jumps occurred, as well as whether the correction should be positive or negative. It also provided a monitoring device to identify stations with low signal levels, and to recognize periods when the set would cease tracking the appropriate slave station and shift to tracking the master.

3.   Plots of Lat./Long at 24"=1 degree longitude, with GPS, LORAN, Transit satellite fixes and DR. positions from the 1107 receiver all plotted on a single sheet. This navigation plot was updated once per minute. It allowed a quick evaluation of the quality of LORAN calibrations, ensuring that the LORAN set was in good agreement with GPS during the limited periods of 3+ satellite GPS coverage. It was also useful for keeping track of the offset between DR. positions and the other sensors.

4.   Screen display of Lat./Long derived from GPS, LORAN and Transit satellite DR., as well as the difference between the various sensor pairs. This display was updated once per minute, and unlike

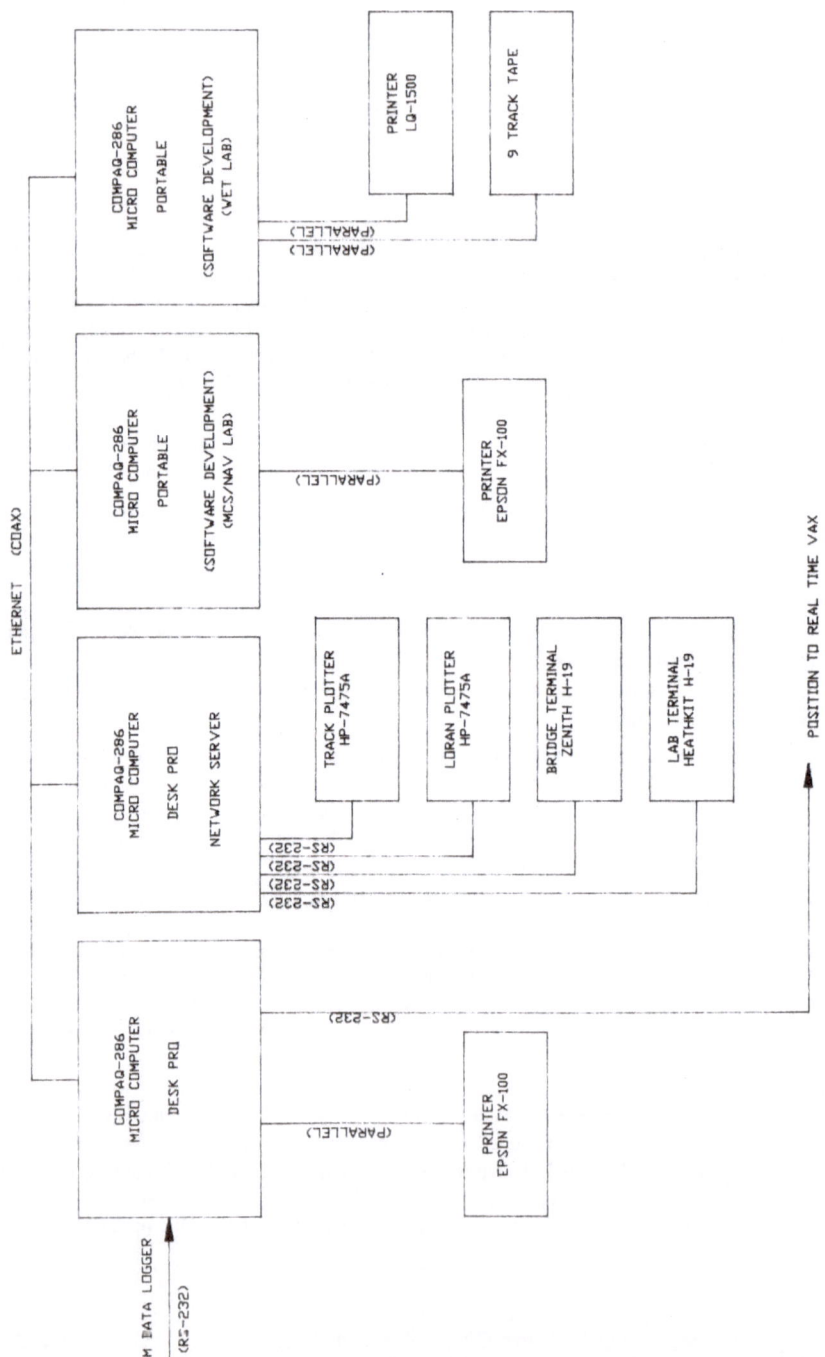

Figure 4. Real-time navigation and geophysical data display system developed for the Trans-Pacific cable route surveys.

the other three displays, this information was displayed both on the bridge and in the science lab. In addition, a "memo" space was provided which allowed the science lab watchstander to type messages from the terminal for display on the bridge. Information such as rise and set times of the GPS satellites, updated calibrations, or performance problems with any sensors were normally transmitted to the bridge in this manner.

5.   In addition to the computer-based navigation display, the ship's officers maintained underway plots on routing sheets prepared in advance of the cruise. These routing sheets, at 1:150000 scale, Universal Transverse Mercator projection, showed the proposed cable route as well as various LORAN range and time delay lines. They were prepared by AT&T Communications, Inc.

CONCLUSIONS

We have shown that it is reasonable (and possible) to post-process "yesterday's data today" with existing technology closely coupled with appropriate human interaction. The techniques described here enabled us to meet A.T.&T.'s requirements for the cable route survey with a minimum of land-based post-processing.

DETAILED STRUCTURAL MAPPING OF AN OCEAN RIFT, USING ACOUSTIC-TRANSPONDER POSITIONING: PROBLEMS OF NET MAINTENANCE (1981-86) AND NEW EXPERIMENTS POSSIBLE WITH LONG-LIFE TRANSPONDERS*

William R. Normark
Janet L. Morton
U.S. Gelogical Survey
Menlo Park, CA 94025

ABSTRACT

Since September 1981, the U.S. Geological Survey has continuously maintained an acoustic-transponder net on the axis of the Juan de Fuca Ridge off west-central Oregon. This transponder net has provided position control for deep-tow vehicles, samplers, and manned submersibles as part of a long-term program to study the structural, volcanic and hydrothermal processes along the 2,200m-deep ridge crest. Field operations in the area each fall utilize a net of four or five limited-lifespan transponders (both recoverable and expendable units); between operations, a skeletal net of one to three transponders is left. In addition to the difficulty of guaranteeing ship time each year for refurbishing the transponder net, servicing the net includes other problems: (1) time lost to deployment, net calibration, and recovery each year, (2) net rotation ambiguity resulting when only one transponder survives to the next field season; and (3) difficulties in reoccupation of sample localities using separate net calibration for each successive field season. We have anchored benchmark floats at several sampling sites to aid in future site reoccupation and to determine absolute accuracy between successive net calibrations. Long-life (5+ years) transponders, capable of measuring round-trip travel times to 10 μs (Spiess and Boegeman, Eos., v. 65, p. 851, 1984), would allow or facilitate several new experiments, including: (1) geodetic determination of rifting rate, (2) monitoring of changes in mineral and faunal assemblages at hydrothermal vents, and (3) detection of sea-floor deformation.

---

*Presentation given at symposium, but paper not available for publication.

M. Kumar and G. A. Maul (directors), Marine Positioning, 407.

HIGH ACCURACY MARINE POSITIONING FOR
DREDGING SURVEYS AT THE PORT OF HAMBURG

By: Dieter Mackenthun,
Richard Muller

Krupp Atlas Elektronik
1453 Pinewood Street
Rahway, NJ 07065

ABSTRACT

The selection of a positional control system for high accuracy applications will usually rely heavily upon actual user experience. However, independent evaluation of a system's capability carried out by recognized authoritative bodies or organizations under carefully controlled conditions can often be of value and interest to potential users.

This paper describes recent surveys with the ATLAS POLARFIX system carried out by the Port of Hamburg.

INTRODUCTION

High accuracy positional control is essential for such applications as sounding and dredging port approaches, rivers and harbor environs. These requirements are usually dictated by a need for improvement of channels, turning basins or anchorages, or alternatively, reclamation of land. Other typical needs include the installation of new navigational aids networks, upgrading of fendering systems and the underpinning of wharves.

The Hamburg Port Authority is among the first to have adopted a digital hydrographic survey sweeping configuration incorporating a position fixing capability designed to meet continuous positional accuracies of plus or minus two meters or better under all conditions. The system also boasts virtually no downtime due to constraints of geometry.

SURVEY VESSEL

The sweeping system is installed on the Port Authority's own survey vessel, MS "DEEPENSCHRIEWER II", whose translated name means "Depth Recorder". Built in 1971, she has an overall length of 27.5 meters, a beam of 8.6 meters and an average draught of 1.5 meters.

409

M. Kumar and G. A. Maul (directors), Marine Positioning, 409–414.
© 1987 by the Marine Technology Society.

SURVEY VESSEL (CONTINUED)

The vessel was subsequently equipped with a swathe system using hydraulic booms of 15 meters on each side. In these, and also in the ship's hull, a total of 37 transducers are installed, sited at one meter intervals. The complete arrangement thus provides a gapless swathe of some 37 meters.

Fig. 1: The Hamburg Port Authority's survey vessel "DEEPENSCHRIEWER II" performs a survey sweep. The vessel is equipped with 37 transducers enabling it to survey 37-meter swathes.

More recent refinements to the Port's total survey capabilities have included the introduction of Krupp Atlas Elektronik's laser range-azimuth position fixing system, POLARFIX. This system, which incorporates automatic tracking facilities and is able to meet required positional tolerances in order to correlate bottom depth contour data to accuracies of better than plus or minus two meters.

At the same time, a new-generation survey processor has been added for precise track-line navigation and plotting as well as annotating distance and position fix marks on both swathe (bottom) recorder and A-0 size track plotter for depth/position correlation.

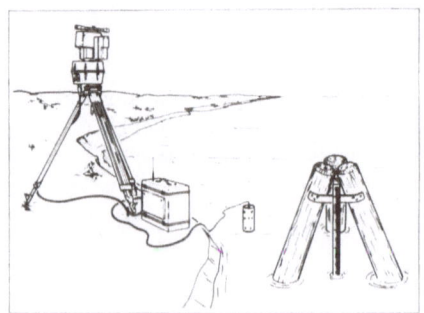

Fig. 2: POLARFIX shore station with level gauge.

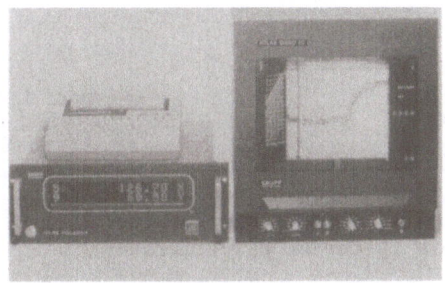

Fig 3: POLARFIX ship station with DESO 20.

Fig 4: Krupp Atlas Elektronik's POLARFIX Laser Range-Azimuth Positioning.

AREA

Aggregate survey and surveillance area assigned for coverage by "DEEPENSCHRIEWER's" swathe system is of the order of 100 square kilometers, of which some 50 kilometers comprises a stretch of the River Elbe. In all, around 50-60 percent of the area is fully surveyed twice yearly, while the more critical harbor basin approaches are appraised up to four times annually. Average survey vessel speed is approximately four knots.

Preparatory dredging operations necessitate on-line production of a bottom contour chart of the Port for subsequent off-line processing of the isobathic charts. From those charts pre-dredge versions are compiled with the annotation of required dredge depths.

AREA (CONTINUED)

Charts are at a scale of 1:1000 or 1:2500 and are projected on to
the track plot of the A-0 size plotter for correlation of the different swathes. They show required minimum depths in isobathic pockets
near quays, whose stability can be threatened by thrust propellers of
container vessels.

OPERATIONS

Annual total dredge volume is approximately 1.2 million cubic meters,
with a maximum volume dictated by the available dumping ground for material.

Given that such a large area is comprehensively surveyed several
times a year for shoal detection, there is an understandable premium
on optimum operational efficiency. This is met firstly by swathing
in preference to cross-profiling, thereby reducing actual survey time
and averting costly neglect of a sector which might otherwise have to
be resurveyed at a later stage. At the same time, the approach also
provides a much higher resolution of the bottom structure. Overall
efficiency is further enhanced by accurate positioning to avoid any
time-consuming overlap of survey runs.

With the recent commissioning of a POLARFIX system and an automated
ATLAS SUSY 30 survey processor for navigation control, the Port is now
able to more rapidly survey the total area under its control.

Considerations which originally led to the adoption of POLARFIX
were largely governed by difficulties previously encountered with
radio positioning systems. All were susceptible to unwanted reflec-
tions in a largely built-up survey area, either by electro-magnetic
or acoustic disturbances.

Moreover, relatively high logistical costs associated with continuous sitting and re-sitting of remote stations in order to secure
adequate survey coverage added to these problems. Surveying a chart
length of 2.8 kilometers, for example, demanded a minimum of four
shore stations for the realization of an acceptable intersection angle
cut.

It was found that no radio system could register a required positional accuracy of plus or minus two meters over more than half the total
survey area.

OPERATIONS (CONTINUED)

The POLARFIX has an operating range of approximately five to eight
kilometers and incorporates an automatic tracking function. The single
shore-based station is linked to a ship-borne receiver and prism assembly. Thus the laser-based POLARFIX system was able to overcome the
previously experienced difficulties while meeting uniform high-order
accuracy requirements.

REFINEMENTS

An important refinement to the system, one which has already made a
significant contribution to improving survey efficiency at Hamburg with
its tidal hub of around 3.4 meters, has been the recent introduction of
the Krupp Atlas PLM 1170 level meter for continuous monitoring of fluctuating water levels.

Consisting of a pressure gauge in the water connected by cable to
the shore station, data is then automatically relayed to POLARFIX's
shipboard telemetry receiver at pre-selectable intervals. The information in turn supplies the swathe system with a true depth data
to the medium tide datum. Accuracy of the level gauge is of the
order of plus or minus one centimeter.

According to port authorities, the introduction of POLARFIX has
ensured that survey runs can now be made as "straight swathes", or
longitudinal profiles, with covered strips matching each other gap-
lessly. With fixes updated continuously every 0.5 seconds, positional
errors are minimized, while between updating intervals, a predictive
filter provides extrapolated values up to five times per half second.

Depth errors are similarly minimized, since the maintenance of
straight survey lines ensures that swathe transducer booms neither
dip nor rise in the water. Instead, they remain horizontal at plus
or minus five centimeters flutter at the tip of the booms.

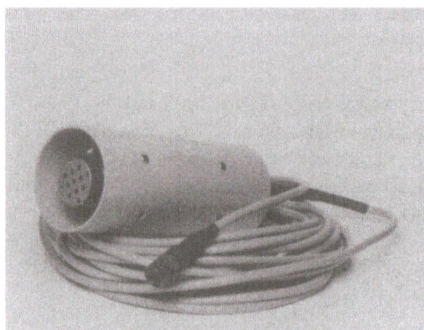

Fig. 5: The PLM 1170 meter for con-
tinous monitoring of fluctuating water
levels.

ADDITIONAL INFORMATION

-1- Hamburg does not stand alone as other port areas in Europe are
     now being surveyed with POLARFIX systems. In North America the
     U.S. Army Corps of Engineers, Canadian Coast Guard, Public Works
     Canada and Canadian Hydrographic Service are among successful
     users.

     While initial operations were primarily in the public sector,
     private companies are now counted among POLARFIX users.

-2- The manufacturer's literature for POLARFIX states positional
     accuracy of the equipment, tracking a moving target, as plus
     or minus 0.5m, plus or minus 0.1m per km of measured range.
     The specification for range is stated as 5000m approximate
     with a capability of approximately 1.5 times visibility in
     fog.

     Reports from the user community indicate performance equal to
     or better than the nominal specifications. Additionally, tests
     performed specifically to evaluate POLARFIX performance rein-
     force these users' reports. In one evaluation Johns Hopkins
     Applied Physics Laboratory was contracted by the United States
     Naval Sea Systems Command to perform system tests. At a nomin-
     al range of 1180m the combined mean error was .78m and the
     combined standard deviation plus or minus .248m. In 66 trial
     runs speeds ranged from approximately 1.75 to 8.75 knots.

-3- Further tests at the NATO FORACS Sounding and Calibration Test
     Range in Stavanger, Norway showed an accuracy of RMS 0.6m at
     a range of 4000. Long range tests showed 10 km as accessible
     in clear weather conditions.

POSITION CHANGES IN MARINE AREAS

Heinz G. Henneberg
Escuela de Geodesia, Fac. de Ing.
Universidad del Zulia
MARACAIBO - VENEZUELA

ABSTRACT

The position changes in marine areas are divided in relatively sudden and slow movements and deformations of the earth's crust and surface. There are to be considered different aspects: Neotectonics and crustal movements, mid ocean ridges, trenches and subductions, deposits and erosions, volcanic activities and lava flow, size of islands, upheavels and subsidences in extraction areas, construction of dykes for flood control and high precision control of large platforms in offshore locations including inclinations of these platforms and oscillations caused by man activities, winds and waves. Interest of industries and private and official institutions in these problems exists. Measurement methods are classical and modern, ranging up to satellite applications through highly developed techniques as VLBI, GPS and SLR. The application of the different methods depends on their precision and accuracy, and the range and significance of the position changes to be measured.

1. INTRODUCTION

The impact of VLBI, GPS, Laser Ranging, and other space techniques on geodesy brings a new, revolutional orientation in professional and scientific activities in positioning on the earth surface in land and marine areas in general and especially in position changes because of the quick and high precision results of measurements. This quick and high precision response enables us to detect and qualify quick position changes immediately after their occurance and enables us to detect very slow movements and deformations of the earth surface. Further, these mentioned space techniques are going into fulfilling the goal to cover the earth with a high precision geodetic network tied to a three-dimensional, universal, inertial coordinate system. In marine areas, all these mentioned items are of extraordinary significance: the unique coordinate system relates all stations and measurements to each other and permits the exact determination of positional coordinates, e.g.: a) The position and area changes in coastal areas can relatively be related to a former position or absolutely to the reference frame.
b) Islands which formerly could only be positioned by astronomical methods, and size changes which could only be detected with higher accuracy through a local network, are now going to be included in the high precision global network, and changes will be detected relatively and at the same time absolutely.
This gives us the possibility to include crustal movement and deformation studies in the Marine Geodesy Field and brings into the marine areas, when needed, the same precision and accuracy as applied to areas on land. That means that in future there will be no difference between measurement methods, results, precision and accuracy in areas on land and on sea. Thus each point on the surface of the earth can be reached, measured, positioned in one common coordinate system, and changes can be related to former positions very quickly

415

M. Kumar and G. A. Maul (directors), Marine Positioning, 415–432.
© *1987 by the Marine Technology Society.*

and with high precision.

For the discussion of position changes in marine areas, there are several important facts to be considered:

The earth's crust is a skin of rock less dense than the underlying man-
tle. We have two kinds of crust, continental and oceanic. The continental
crust is very much older than the oceanic. Parts of the continental crust
are older than 3.500 million years, and very large areas older than 1.500
million years, whereas the oceanic crust is nowhere older than 200 million
years. Common to all oceans are the ridges which form a nearly continuous
submarine mountain range over 40.000 km in length, appearing above sea level
in some places like Iceland, the Azores, and Tristan da Cunha. The evidence
of "sea floor spreading" indicates that the Atlantic , Indian , Arctic and
Antarctic Oceans did not exist 180 million years ago. Sea floor spreading is
a mechanism by which oceanic crust is created and in some parts also destroyed. The spreading rates of new crust are varying in the order of some cm
per year on both sides of a ridge. Oceanic ridges and trenches are the sites
of high volcanic and seismic activities associated with spreading movements
of the litosperic plates. In plate tectonics there are considered seven major plates: Eurasian, African, Indo-Australian, Pacific, North American,
South American, and Antarctic. In addition, there are still some more minor
plates as for example the Nazca,Cocos,Carribbean plates, and others. Further,
island arcs, small ocean basins (Japan Sea), inland seas (Black Sea), transform faults, subduction zones, collision zones and so on form a complex picture of the earth's outer shell. Everything is moving and deforming.

Hurricanes, tsunamis, volcanic eruptions, glaciation, sediments, sea level changes, water household and other influences change coastal areas, crea
te and destroy islands, amplify and reduce areas in marine environments. So
me of these positional changes take place in slow continuous movements and
deformations, others in quick, short periods and sometimes destructive events.
Man made structures as dams modify the contours of coasts, platformes are ins_
talled for industrial purposes (oil, mining, injection plants, etc.).

Dredging creates new water channels and artificial islands made from
dredging deposits etc.

Position changing is always present and actual, slow and sudden, natural
and artificial. The earth is presenting an ever changing face.

Position changes in marine areas are fascinating and sometimes dramatic.

The detection and complete knowledge of these changes is a goal we are
here to discuss, and to give our contribution for science and for the profession.

2. EXAMPLES OF POSITION CHANGES IN MARINE AREAS

In the following there are exhibited some typical examples according to
the former mentioned changes in marine environments. It is naturally understood
that no intention is made to show a complete panorama of all existing types
of movements and deformations. The author thinks that the outlined examples
of multidisciplinary aspects might be helpful for the orientation and development of the Marine Geodesy Field, its science and profession. The examples
include: a) The changes in sea level, b) The blow up of the Atlantic Ocean
Floor (1957-1963), c) Krakatau and other volcanoes, d) Frederic in Alabama,
e) Isla de Aves, f) Earthquake in Alaska, g) Recent crustal movements, Neotectonics, Deformation, h)Marine abrasion and dunes in coastal areas, i) Muds

lides and sedimentation, k) Glaciation and ice movement, l) Maracaibo Lake
entrance, m)Subsidences, vertical displacements, dykes, structures, n)Others

3. THE CHANGES IN SEALEVEL

During the Terciary, sea level was about 80 to 100 m higher as today and
about 25.000 years ago during the glaciation apr. 60 m below the actual state. If we consider the ice household of the earth as a time depending pe-
riodical physical phenomenon, then as a consequence, sea level is always and
instantaneously changing. Further, changes in the actual sea level are due
too to a lot of other components e.g.: evaporation, winter and summer time,
rains, winds, tides, temperatures, salinity, density, currents, air pressure,
hurricanes, storms, earthquakes, tectonic and eustatic movements, rivers,
dredgings, man made structures, dams, harbours, instrumental effects and
others. It is a fact that the correct quantifying of all influences in level
changes is not easy to perform, because of the difficulty to determine what
we call "Mean Sea Level". When we express: "this station has a certain height
above sea level", we do not refer to the mean sea level, we refer normally to
one or more local, or in some cases to national or regional tidal stations.If
we could combine all existing tidal stations of the world through a high precision vertical network then we may come to an expression of a mean sea level
value along the coasts. That still left open the mean value of sea level of the
enormous ocean areas. The recent application of satellite techniques through
altimeter measurements are closing the gap. The GEOS 3 and Seasat measurements
have given results of the mean ocean surface with accuracies of 5-10 cm, which
approximate highly the Geoid. The expected Topex altimeter is announced to give accuracies of about 5 cm and GPS results have given discrete ellipsoidal
heights of stations in the order of cm uncertainties. The combination of space
techniques with local levelling networks produce in each station the knowledge
about the differences between the physical and mathematical reference surfaces.
Changes in sea level, which in some special interpretation, are considered the
up or down movement of the water or the down or up movement of the coast need
a clear physical definition. In fig.1 is seen a typical sequence of sea level
changes at Galveston, Texas, interpreted mainly as subsidence in the Texas
Gulf coast area (J. Small 1973).

4. THE BLOW UP OF THE ATLANTIC OCEAN FLOOR (1957-1963)

In the nearly 6 year period between 1957 and 1963, practically at the same
moment in the geological time scale, three most dramatic events took place
along the Mid Atlantic Ridge: the sudden eruption of the Capelinhos volcano
and the creation of a new island in the Azores region in 1957-58, the eruption
of Tristan da Cunha island volcano the 10th of October 1961, the first
one in historical times which added 86 acres of new land to the island,and the
birth of Surtsey, November 15, 1963, and two islands more off the south coast
of Iceland. These 3 gigantic explosions were the most remarkable geological
events on the Mid Atlantic Ridge in this century. In all mentioned sites a lot
of additional volcanic activities took place in the years shortly after and
between the main events. During the eruption of the Capelinhos were to be distinguished 3 stages, characterized by 2 initial and one final stage. The 2 ini
tial stages (27.8.1957- 5.11.1957 and 6.11.1957- 12.5.1958) produced two new
islands, one submerged again and the second extended until touching the Faial

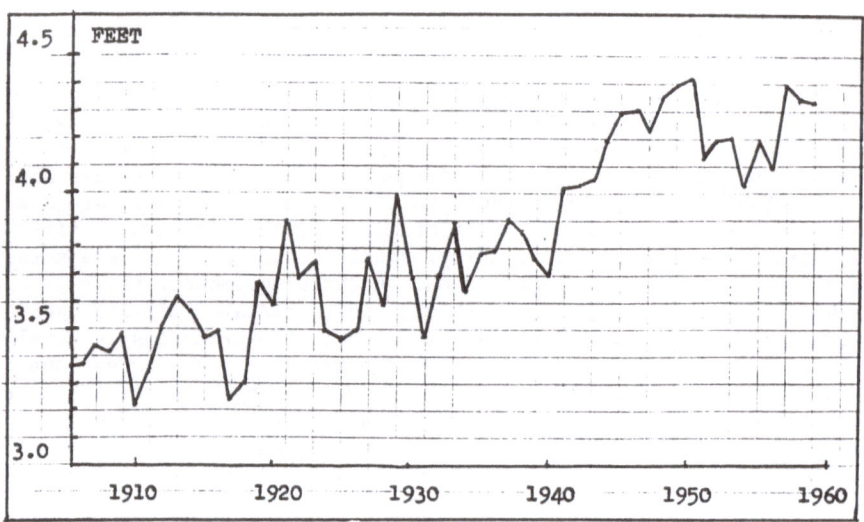

Fig.1   Sea Level Changes At Galveston (after J.Small,1973)

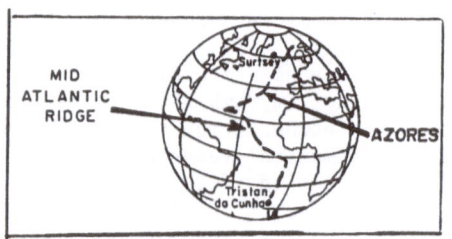

Fig.2   Mid Atlantic Ridge

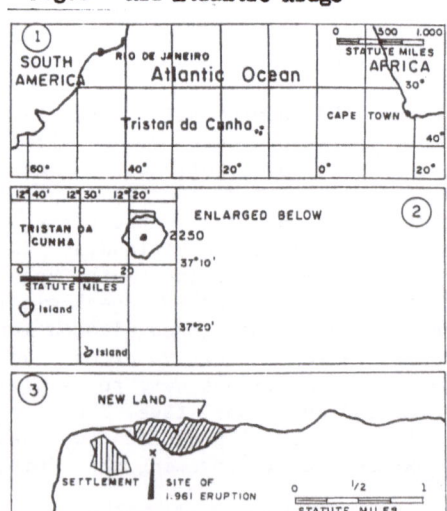

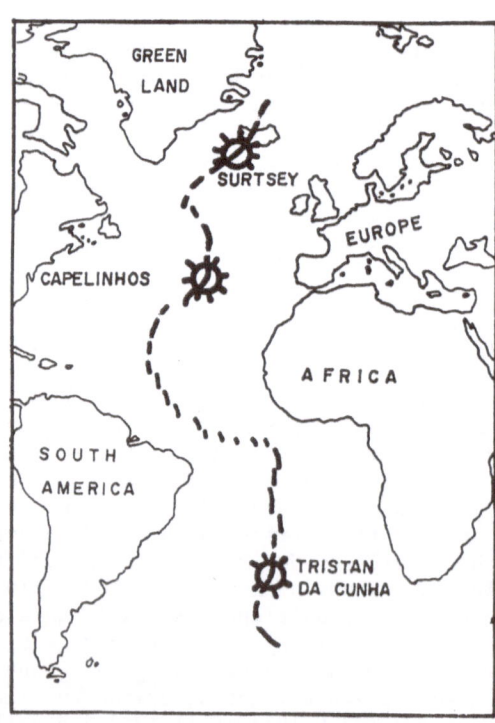

Fig.4   Details of Tristan da Cunha (after NGS,Cart.Div.,1973)

Fig.3   Location of the three Explosions on the Mid Atlantic Ridge

island. The final stage (12.5.58-25.10.58) gave the permanent existence
of the new island and produced a lava effusion of about 85 x $10^6 m^3$. The
Tristan da Cunha eruption in Oct.1961 had as consequence the total eva
cuation of the island population. The central peak of the volcano has a
height of 6760 ft and the island rises 18.000 feet from the sea floor.
The volcano still smoldered a year and a half after the eruption. Quite
at the same time of this eruption in the South Atlantic, in Iceland,the
Askja volcano leaped up in fiery outburst (Oct.26th 1961).The lava flow
covered 11 km^2. The Surtsey birth was a spectacular event with the buil
ding up of 3 islands off the south coast of Iceland during the years
1963-67. Two of these islands lasted less than a year, although both
reached a height of about 70 m and a maximum area of 15 and 28 ha. The
permanence of Surtsey was ensured when the lava outpouring started in
April 1964. Until then the eruption had been wholly explosive. Surtsey
reached a max.height of 178 m. In June 1964 the area of Surtsey was
1.37 Km^2, in 1966 2.4 km^2 and when the eruption ended it was 2.8 km^2.
Rather unexpectedly, an eruption started on Hekla on May 5 th, 1970 and
lasted 2 months. Hekla is a volcanic ridge in Iceland. The lava flow of
this eruption covers 18.5 km^2. The 23 rd of January 1973 another big
eruption took place on Heimaey, the main island in the Vestmannaeyjar
group of Iceland. A nearly 2 km long fissure opened with lava fountains
along its entire length. A month later there was a cone of about 200 m
high and the lava output increased the island's area by about 1.7 km^2.
In figures 2 to 6 we see the locations of the 3 main eruptions along
the Mid Atlantic Ridge with details, including the case of Heimaey with
its size change.

5. KRAKATAU AND OTHER WOLCANOES

The Krakatau islands are located between Sumatra and Java (fig.7).
The famous Krakatau volcano repeatedly has arisen from the sea, exploded
and collapsed. In that area two tectonic plates, the Indo-Australian and
the Eurasian are colliding, forming the gigantic Java trench and a large
subduction zone, which create a fire ring of earthquakes and volcanic
activities and which is responsible too for the upward thrust of the Himala-
ya mountain chain. Indonesia thus ranks as a nation of active volcanoes
and quite permanent seismic activities. Krakatau volcano (fig.8) is one
of the most famous ones, due to its violent and destructive behaviour in
the past. Completely located in the marine environment the changing of si
ze and the disappearing and creation of the land masses in the past can be
seen in fig.8 (after NGS Cartogr. Div. 1985), Fig 8.1 shows ancient Kraka
tau according to former references, 8.2 is a picture of the first collapse in A.D.416 and 8.3 gives the situation before 1680 after the regrowth.
In 8.4 we see the Krakatau islands in their new extension in 1883 before
the big explosion and in 8.5 after the cataclysm and collapse. The cinder
cone, located between the 3 remaining islands (fig.8.6), came up from the
sea floor in 1927. The building of this island, called Anak Krakatau, las
ted nearly two years and the volcano has now a height of about 200m. The
last eruption of Anak Krakatau took place on Nov. 18,1981. The circles
with broken lines in the figures are: the pre caldera fault - circle in
8.2 to 8.4 and outer circle in 8.5 and 8.6-, and the approximate 1883
caldera fault - inner circle in 8.5 and 8.6 -.

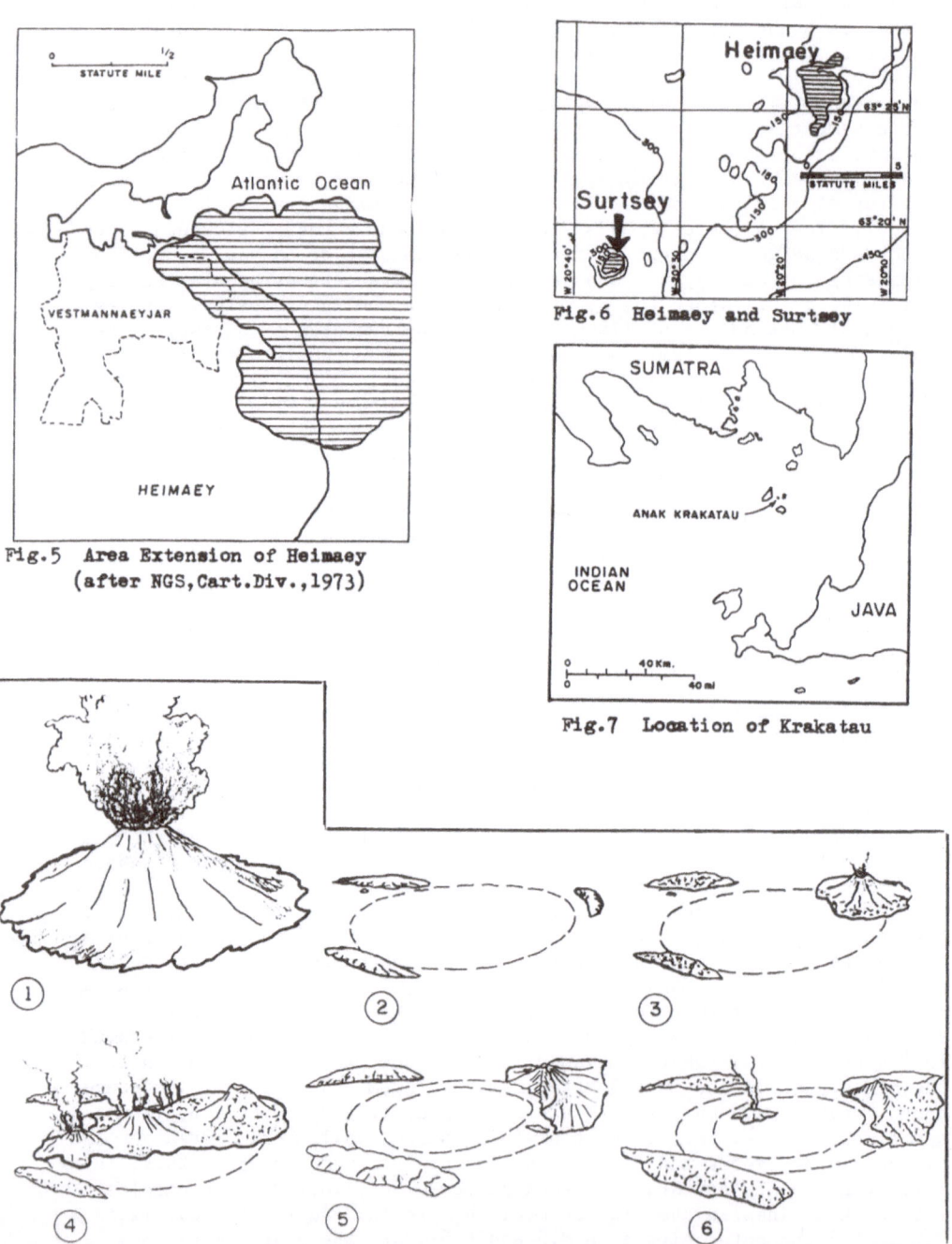

Fig.5 Area Extension of Heimaey
(after NGS,Cart.Div.,1973)

Fig.6 Heimaey and Surtsey

Fig.7 Location of Krakatau

Fig.8 Size Changes of Krakatau Volcano(after NGS,Cart.Div.1985)

From other volcanic avtive parts of the world 2 other volcanoes are
typical for their size changes in marine areas: the Teneguia volcano
(fig. 9) and the Sakurajima (fig.10). The Teneguia belongs to the Canary
islands. The eruption in 1971 amplified the land area of about 29.000 m^2
forming a new coastal platform. The main part of the additional land is
seen in the figure with cross-lines.The SaKurajima is active again since
1972. The lava flows of the former eruptions, 1914 and 1946 are distinguished in the figure. The 1914 activity united Sakurajima, which former
ly was an island, to the peninsula of Ohsumi.

6. FREDERIC IN ALABAMA

In fig.11, A we see the beachfront in Gulf Shores, Alabama in 1978.
When hurricane Frederic came ashore the waves raged over the doorsteps
of the houses. The storm tide as high as 15 feet covered the peninsula.
The condominium was partly destroyed (fig.11,B), houses in other places
disappeared completely, the beach was washed out and offset about 25 m.

7. ISLA DE AVES

Aves island (Isla de Aves) belongs to Venezuela and is the only out-
crop of Aves Ridge above sea level. Aves Ridge is a north-south oriented
submarine mountain range. Aves island is situated in the path of a signi
ficant number of hurricanes (fig.12). The highest part of 3.72 m a.s.l.
consists of a storm terrace. Previous informations on Aves island since
1647 show that the island had progressively been reduced in area until
1968. The historical changes in elevation and in size are probably due
to erosional and depositional processes.Storms and hurricanes play a roll
in it. Long term components could be related to subsidences and crustal
movement of the Caribbean plate. Through figures 13 and 14 we can estimate the changes between 1954 and 1983 (after Schubert, 1984 and P. Herrera,
1971).

8. EARTHQUAKE IN ALASKA

March 27,1964 was the fatal date of one of the biggest earthquakes in
North America. The main shock was located in the north of Prince William
Sound, south-east from Anchorage. 30 foot waves were reported near Kodiak
and in 6 hours the first wave front had reached Hawaii. In Seward (fig.15)
all dock facilities vanished, swept away by the Tsunamies. The shore line
one day after the disaster was displaced 80 and more meters. In the impact
of the Tsunami wave at Valdez, the pier with stevedores and onlookers vanished forever. There the earthquake knocked out the water and sewage systems. Weeks after the quake Kodiak's foundations appeared to have permanently slumped some five feet. In Anchorage fissures opened and streets broke
vertically down to 3 m and more, houses were offset horizontally by 10 and
more meters and the rupture of the earth's crust showed surface cracks of
several meters width. Fig 16 shows apparent horizontal displacements follo
wing the earthquake (after Ch.Whitten, 1973).

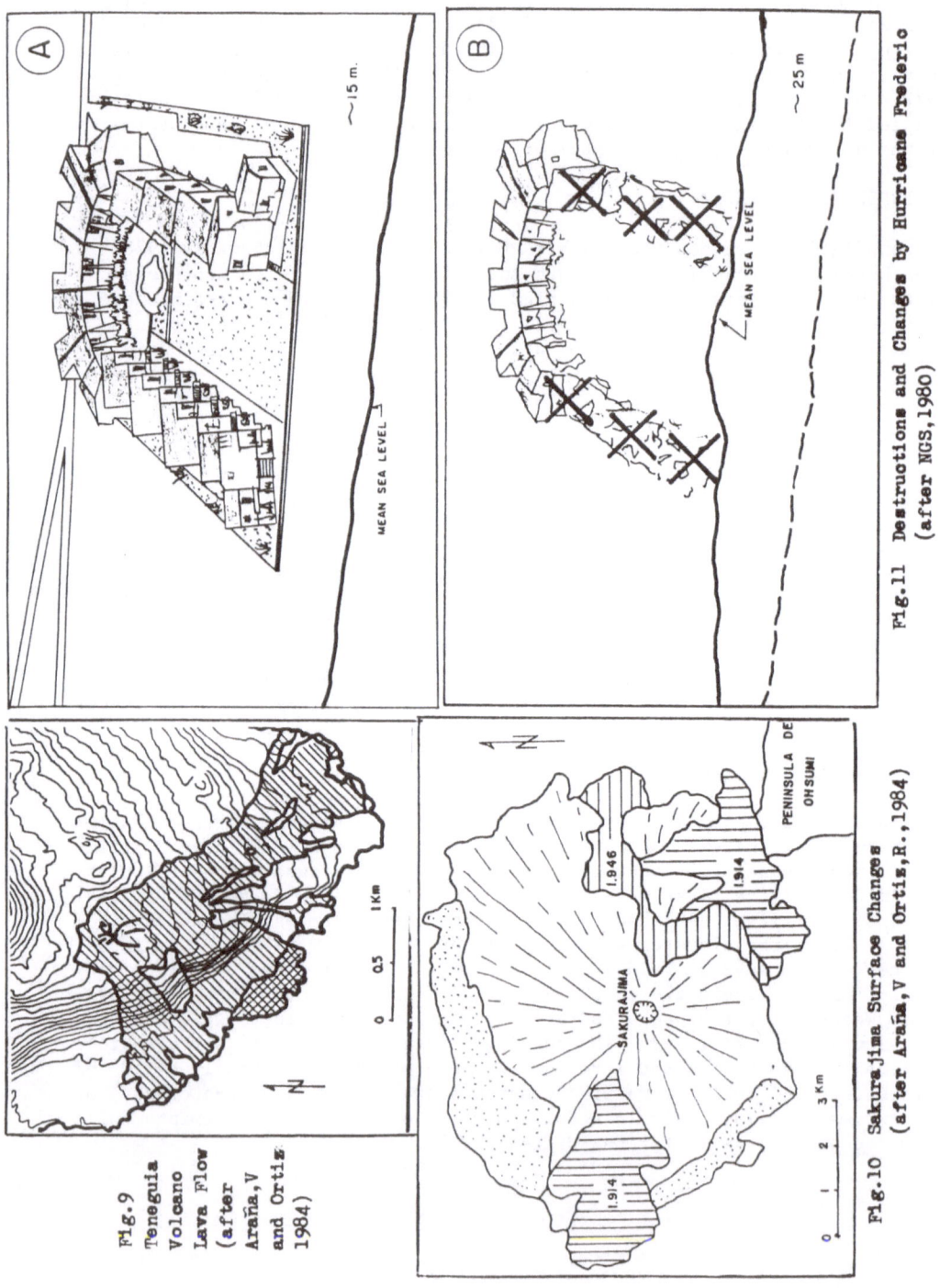

Fig.9 Teneguia Volcano Lava Flow (after Araña,V and Ortiz 1984)

Fig.10 Sakurajima Surface Changes (after Araña,V and Ortiz,R.,1984)

Fig.11 Destructions and Changes by Hurricane Frederic (after NGS,1980)

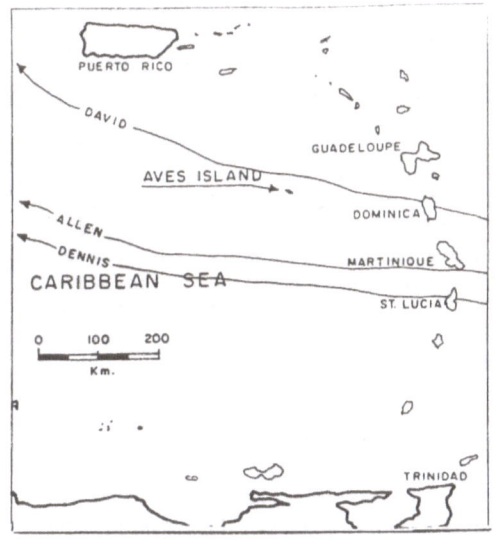

Fig.12 Location of Aves Island
        (after Schubert,C.and Laredo,M.,1984)

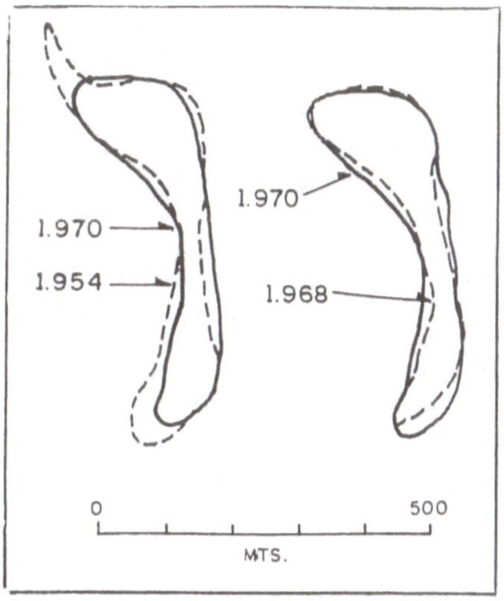

Fig.13 Size Changes of Aves
        Island(after Pantin H.,1971)

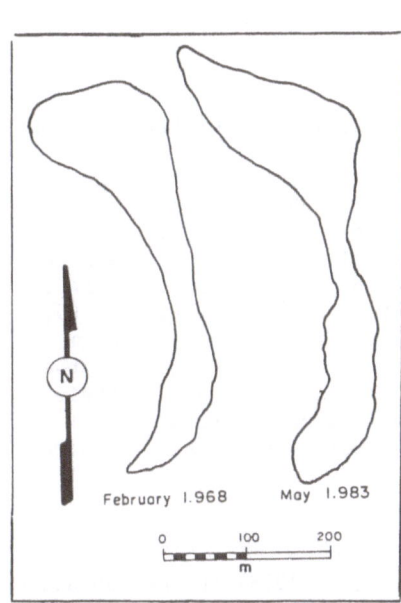

Fig.14 Size Changes of Aves Island(after
        Schubert and Laredo,1984)

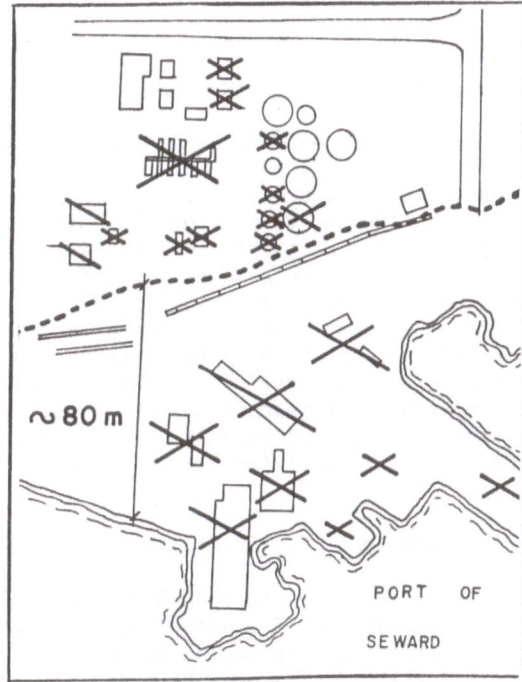

Fig.15 Earthquake and Tsunami Effects
        in Seward (after NGS,1964)

9. RECENT CRUSTAL MOVEMENTS, NEOTECTONICS, DEFORMATION

In the formerly described Alaskan region, where the big earthquake occured, but a little more to the south-east in the region of Glacier Bay between Yakutat and Sitka, a strong deformation of the earth's crust was detected. The measurements resulted in rising rates up to 4 cm per year near Juneau, 2 cm/yr near Skagway and 1 cm/yr near Yakutat. Fig 17 shows this deformation pattern (after Whitten, 1973). This changes, if interpreted as post-glacial uplift, are more than the rates in the Hudson Bay region or in Fennoscandia. Naturally there too can be influences of seismic correlated deformations. It is interesting to show the generally accepted deformation and movement relation "Displacement-Time" correlated to earthquakes, where we have quite always preseismic, co-seismic and post-seismic displacements (fig. 18). Other preliminary rates of elevation changes up to -4 mm/yr south of Chesapeake Bay (B.Meade,1975),and in the Charlestom area of 2.5 mm/yr (Holdahl and Morrison, 1974) , show the magnitude of crustal deformations in that part of the eastern coast of the U.S. In Venezuela great efforts are being made and under way to detect movements and deformations along the Caribbean-South American plate boundary (fig. 19,20). This boundary area has a rather high seismic activity (fig.21). The geodetic networks to detect horizontal displacements in the Pilar fault region are seen in fig. 22 (Gulf of Cariaco) and in fig.23 (extended to Margarita Island). Different networks along the plate boundary are planned to be linked with satellite methods. An example of horizontal displacements of the crust is given through the results of repeated triangulations in Japan (figures 24,25). Fig 24 show difference vectors between 1886-89 and 1948-55 and fig.25 between 1948-55 and 1968-72 (after T.Harada and M.Shimura, 1979).

10. MARINE ABRASION AND DUNES IN COASTAL AREAS

Marine abrasion is a destructive action of the sea especially seen on high and steep coasts. The force of waves accompanied by diminute   solid parts in suspension and thrown violently against the rocks causes a constant excavation of the coastal walls which produces the break down of upper layers (fig.26). The repetition of this action during years and centuries causes the gradual retreat of the coastal line. A typical example of this action is seen in fig. 27, where a station of the Maracaibo Lake Triangulation, located on the Aves island of the lake, installed in 1957 broke down after 20 years. The slab in that place had a width of 2.06 m. The strong wind action in coastal areas with dune formations produces the wandering of this sand masses, covering and cleaning areas along their path. A classical example of this phenomenon is the destruction during a 60 year period of the little church at Kunzen located on the Baltic shore (fig.28).

11. MUDSLIDES AND SEDIMENTATION

The research study of D.B.Prior and J.N.Suhayda (1979) on the submarine mudslide in the Mississippi Delta region gives a real impressive insight into this phenomenon. The velocity of a leading lobe was detected   to 1.9 Km/yr (fig.29) and the measured profiles in 1975 and 1976 show changes in vertical and horizontal component interpreted as subsidence, prograda-

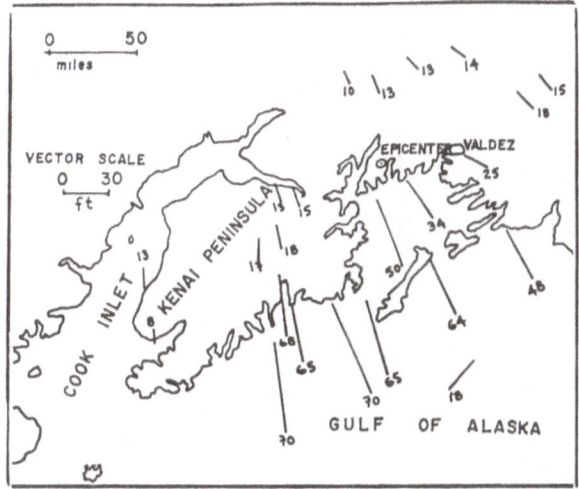

Fig.16 Apparent Horizontal Displacements in Alaska (after Whitten, Ch., 1973)

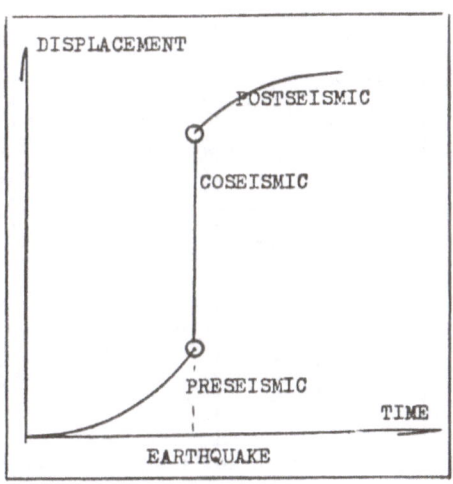

Fig.18 Displacement - Time Diagram

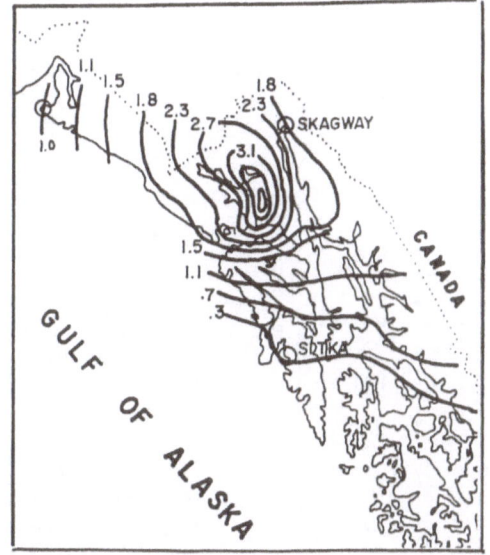

Fig.17 Deformation in Alaska (Whitten, 1973)

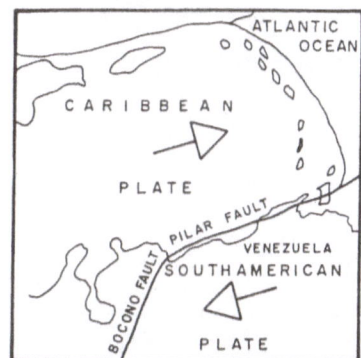

Fig.19 Caribbean-South American Plate Boundary

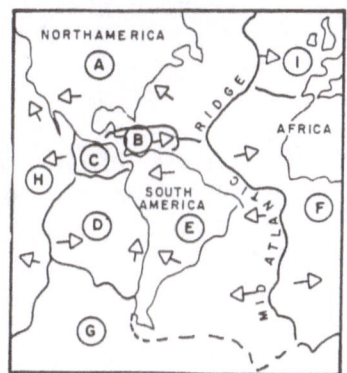

Fig.20 The Tectonic Plates around America
A-North American, B-Caribbean, C-Cocos,
D-Nazca, E-South American, F-African,
G-Antarctic, H-Pacific, J-Eurasian.
Arrows show the directions of the
main relative movements.

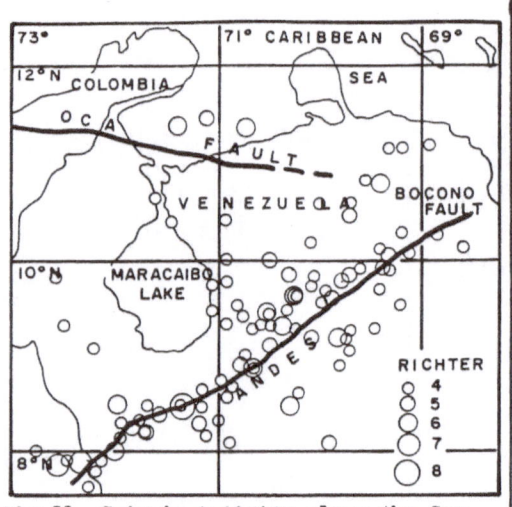

Fig.21 Seismic Activity along the Car.-
South Am.Plate Boundary,Bocono
Fault Zone.Location of Oca Fault
in the Venezuela Gulf Area

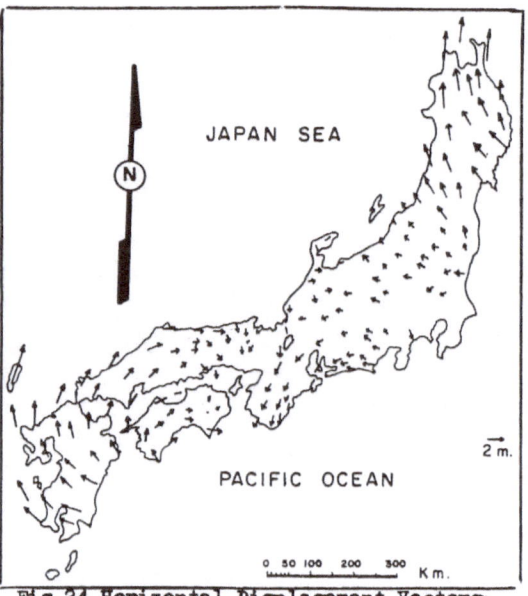

Fig.24 Horizontal Displacement Vectors
in Japan(after Harada and Shimura
1979)

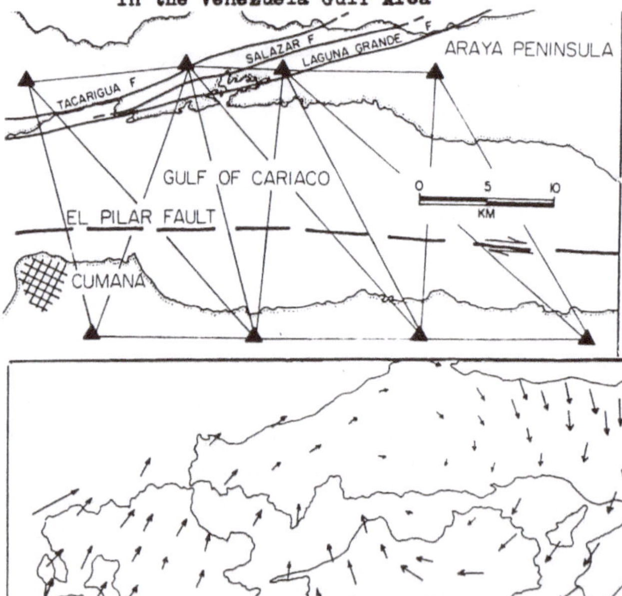

Fig.22 Pilar Fault Network,
Gulf of Cariaco,Venezuela

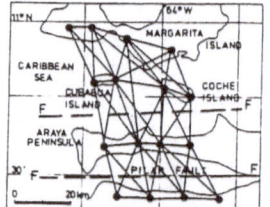

Fig.23 Network crossing
the Plate Boundary
between Cariaco Gulf
and Margarita Island

Fig.25 Horizontal Displacement Vectors in Ja-
pan(after Harada
and Shimura,1979)

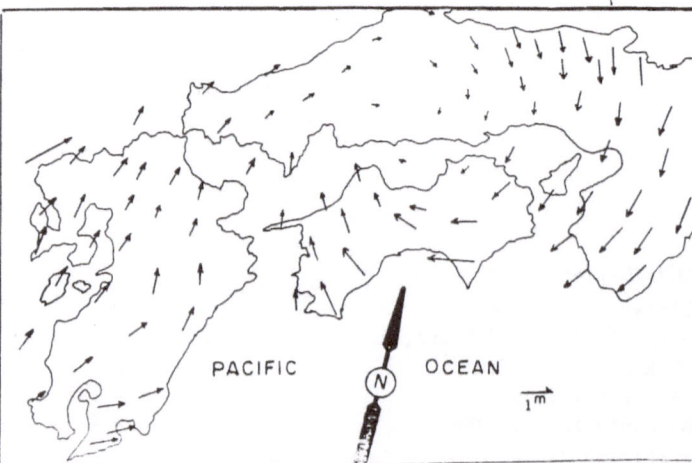

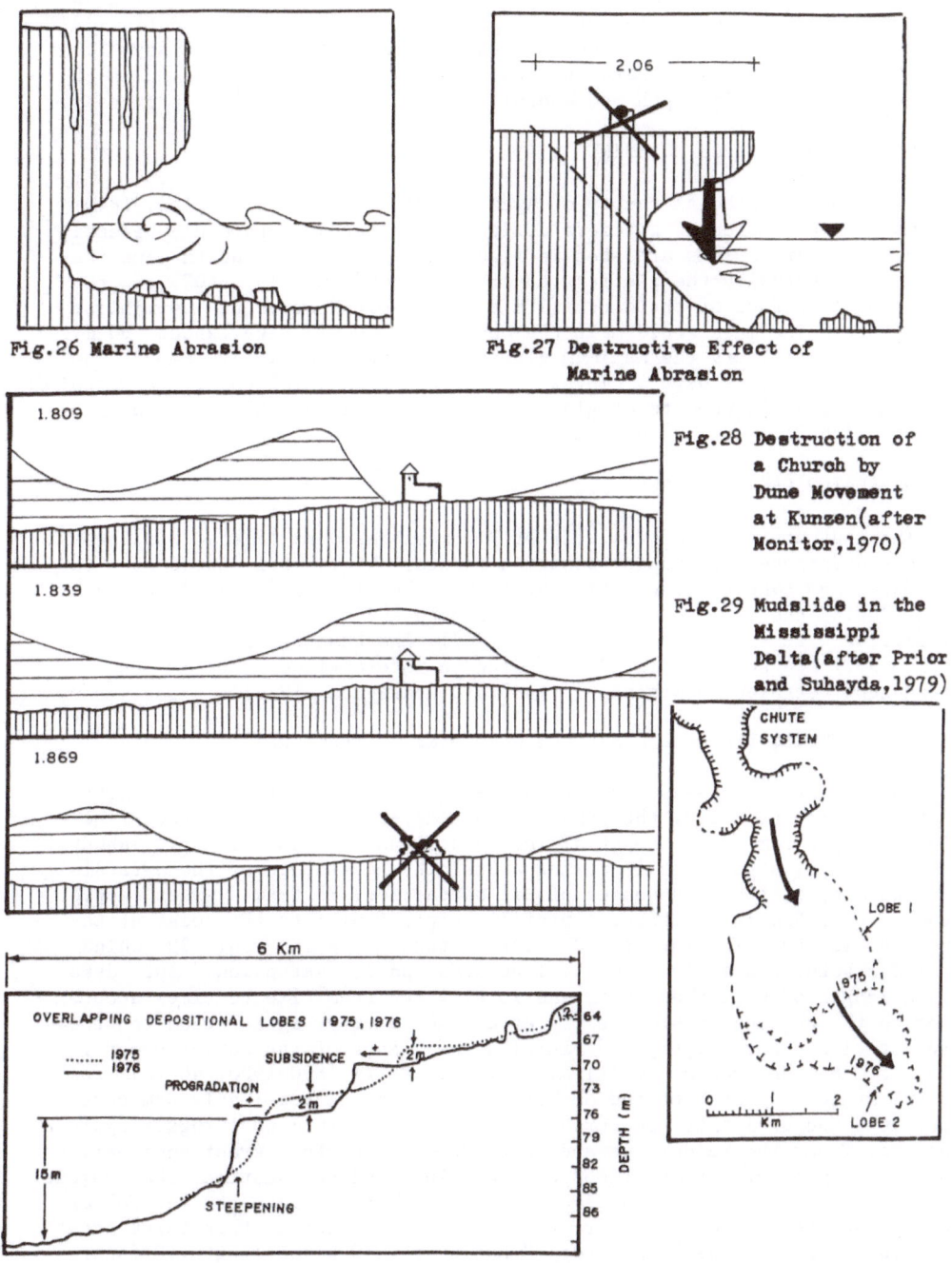

Fig.26 Marine Abrasion

Fig.27 Destructive Effect of
Marine Abrasion

Fig.28 Destruction of
a Church by
Dune Movement
at Kunzen(after
Monitor,1970)

Fig.29 Mudslide in the
Mississippi
Delta(after Prior
and Suhayda,1979)

Fig.30 Changing of the Seafloor caused by Mudslides(after Prior and Suhayda,1979)

tion and steepening of the sea floor (fig.30). The sedimentation in coas
tal embayments was studied extensively by B.F. Molnia (1979) in the north
eastern Gulf of Alaska. The sedimentation rates vary but are as high as
3.75 m/yr. Fig.31 shows rates of sedimentation in Yakutat Bay and fig. 32
the situation at Taylor Bay. The outwash plain in front of the Brady Glacier has grown about 4 km between 1926 and 1977. This are two examples of
tremendous changes in depth and shorelines due to sedimentation processes.

12. GLACIACION AND ICE MOVEMENT

Based on the glacial theory of Agazziz (Etudes sur les glacier, 1840)
the maximum extension of glaciers during the quaternary era could be esti
mated. These extensions are seen in figures 33 and 34 for North America
and Europe. Actually the glaciers cover 14.300.000 Km^2, Apr. 10% of the
continental surface of the earth with a total volume of about 24 million
km^3. If this ice mass would melt the ocean level would rise about 60 m.
During the Quaternary the maximum glaciation covered 42 mill. km^2. The
continental glaciers have a convex shield form with thicknesses up to
2.000 m. The displacements of glaciers cause the Marine Ablation of ice
masses which float as icebergs (fig. 35).

13. MARACAIBO LAKE ENTRANCE

The southern coast of the Gulf of Venezuela as part of the Caribbean
sea has undergone substantial position changes. Of special interest is
the Maracaibo Lake entrance. Fig.36, A shows the land and sea distributions during the interglaciation 2nd Yarmouth and fig. 36,B during the
glaciation 4 th Wisconsin (after Graf 1969). The changes in recent years
(1755-1937) are seen in fig.37 according to informations from the
Bataajsche Petroleum Maatschappij, 1952 (G.Rodriguez, 1973).

14. SUBSIDENCES, VERTICAL DISPLACEMENTS, DYKES, STRUCTURES

This chapter includes the subsidence area on the east coast of Maracaibo Lake in Venezuela, the dyke as man made element to avoid the floo-
ding, the problem of the coast of The Netherlands and some remarks about
the control of structures in marine areas. The subsidence of the Maracaibo Lake area is the consequence of oil extraction. Maximum values have
reached over 5 m in 50 years. A profile perpendicular to the coast at Lagunillas is shown in fig. 38 and a more extended one in fig. 39 which
includes both, the marine and the land area and its extension. The dyke
has a length of more than 60 km and is elevated from time to time according to the ongoing subsidence (rates around 15 cm/yr in relation to the
exploitation process). Fig. 40 exhibits the change of the Dutch coast in
the last thousand years: 40 (1) A.D. 1.000, 40 (2) 1800-1900, 40 (3) the
actual state. A fourth of the Netherlands lies below sea level. Dam cons-
tructions and drainages recovered land through a permanent struggle against
the sea. Today the Zuider Zee and Delta Projects perform great engineering
works of dam construction, flood control, and land reclamation. The project
inludes seven huge dams, four fronting the sea, three inshore. The Eastern
Schelde barrier alone is requiring more building matereal than three great
pyramids of Egypt. The "Afsluitdijk" has nearly 30 km of length. Offshore
structures as platforms for industrial purposes must be controlled for po
sitional changes, inclinations, vibrations and deformations in the highest
accuracy range due to their delicate response to wave and wind actions,
settlements of foundations and piles and internal tensions.

Fig.31.
Rates of Sedimentation(after Molnia,1979)

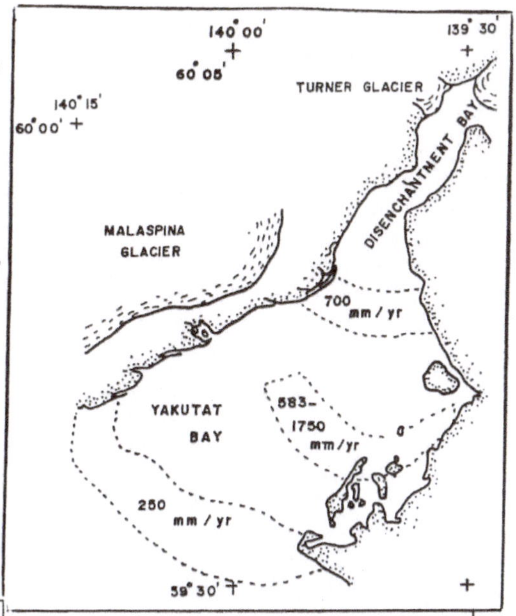

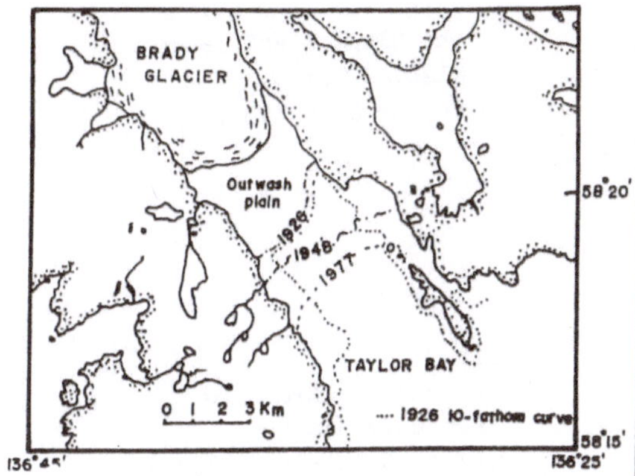

Fig.32   Growth of the Outwash Plain of
          Brady Glacier(after Molnia,1979)

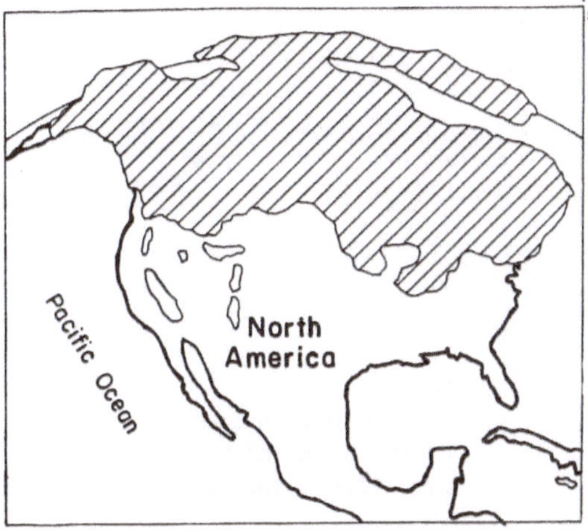

Fig.33 Glaciation during the Quaternary
        Era in America

Fig.34 Glaciation during the Quaternary
        Era in Europe

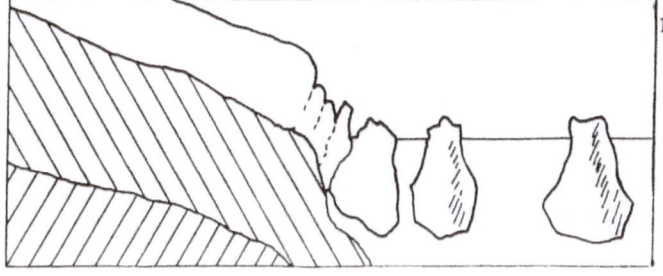

Fig.35 Marine Ablation
        of Glaciers

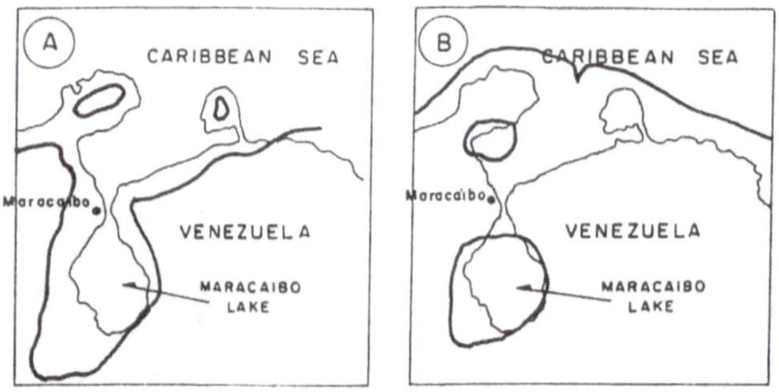

Fig. 36
**Situation of the Vene-
zuela Gulf and the
Maracaibo Lake during
the Glacial and Inter
Glacial Time.**
(after Graf, 1969)

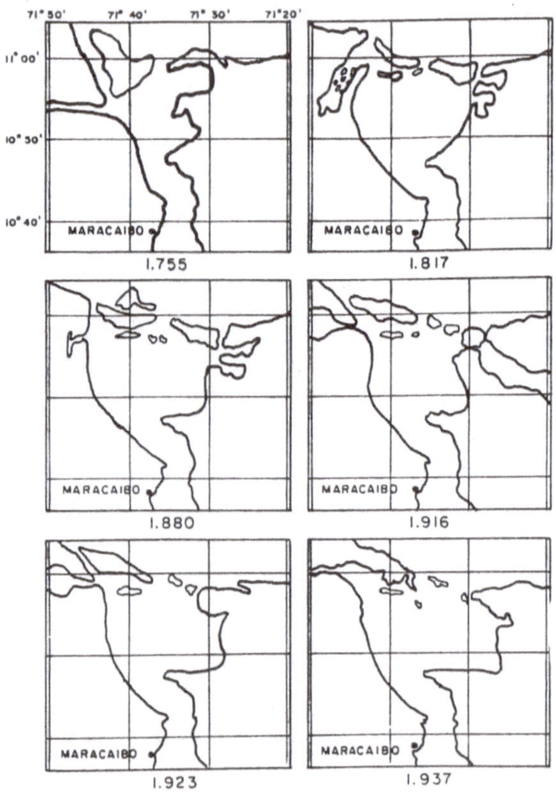

Fig. 37
**Changes of Maracaibo
Lake Entrance**
(after G. Rodriguez, 1973)

Fig. 38
**Profile perpendicular
to the Coast at
Lagunillas, Venezuela**

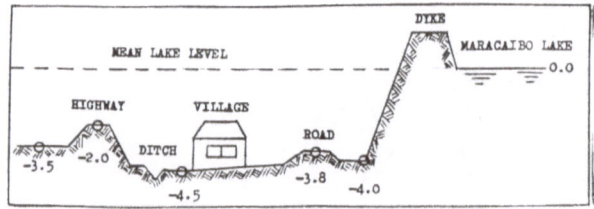

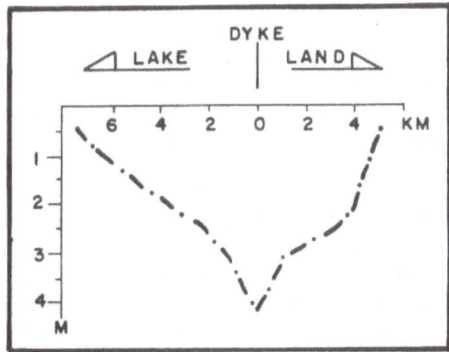

Fig.39
Subsidence Profile
at Lagunillas –
Venezuela

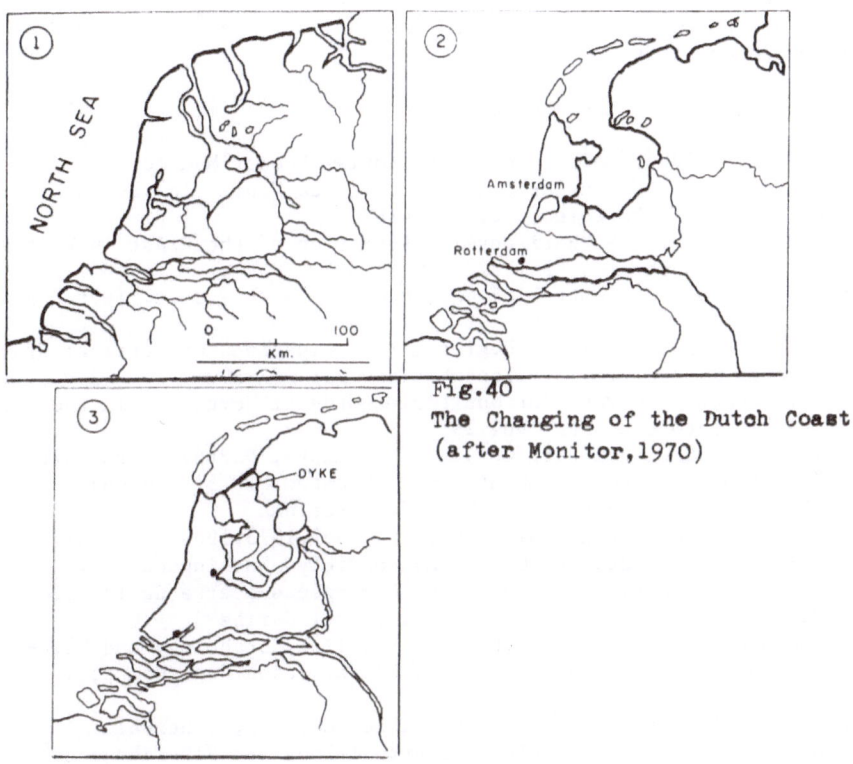

Fig.40
The Changing of the Dutch Coast
(after Monitor,1970)

15. OTHERS

We have seen that position changes in marine environments occur in all parts of coasts and seas. There are a lot of other changes to be considered and mentioned e.g.: dredgings, creation of artificial islands, underwater tunnels, orogenic and epirogenic details and so on.

16. CONCLUSION

The discussion and demostration of each example does not include the methods of measurement of displacements and deformations. Through other papers the actual symposium is giving a lot of information on instrumentation , measurements , analysis, methods, results, precision and accuracies. Each of the mentioned phenomena of the author's papers has to be qualified and quantified according to the necessities, possibilities, and desired accuracies. The field of action is enormous. The need of coordination and interchange of experience and multidisciplinary cooperation is evident. Karl Rinner in his key paper "The Past and Present of Marine Geodesy" (Maracaibo 1983) exposed the roots, the history, and the traditional activities of marine geodesy, its present state and its brilliant future.Marine geodesy always will be a challenge to science and profession, but it is a task where great solutions are waiting, and to bring about these solutions is our task.

REFERENCES

Araña,V. and Ortiz,R.,1984."Volcanology". Editorial Rueda, Madrid
Graf,1969. "Estratigrafia Cuaternaria del Noroeste de Venezuela".Bol.Inf. Asoc.Venez.Geol.Min.Petroleo, 12, 393-416
Harada, T and Shimura,M.,1979."Horizontal Deformation of the Crust in Western Japan...". Tectonophysics, 23: 373-390
Henneberg,H 1983. "Geodetic Control of Neotectonics in Venezuela."Tectonoph., 97: 1-15
Holdahl,S.R.and Morrison,N.L.,1974."Regional Investigations of Vertical Crustal Movements in the U.S." Tectonophysics, 23: 373-390
Meade, B.,1975."Geodetic Surveys for Monitoring Crustal Movements in the United States". Tectonophysics, Vol.29
Molnia,B.F.,1979. "Sedimentation in Coastal Embayments, Northeastern Gulf of Alaska". Offshore Technology Conference. Houston.OTC 3435,p.665
MONITOR.1970."Enciclopedia Universal". Madrid, Barcelona
National Geographic. NGS Cartographic Div.,1964,1965,1973,1980,1985.Vol.125, 126,127,144,158,167.National Geographic Society, Washington D.C.
Pantin Herrera J.,1971."Nuevas Observaciones Geologicas acerca de la Isla de Aves, Venezuela". VI Conferencia Geologica del Caribe
Prior, D.B. and Suhayda,J.N.,1979."Submarine Mudslide Morphology and Development Mechanism, Mississippi Delta". Offsh. Techn.Conf.Houston. OTC 3482 p.1055
Rodriguez,G.,1973."El Sistema de Maracaibo".IVIC, Caracas, Venezuela
Schubert,C and Laredo, M.,1984."Geology of Aves Island and Sibsidence of Aves Ridge,Caribbean Sea". Marine Geology, 59:305-318
Small,J,1973."Subsidence in the Texas Gulf Coast Area".Reports on Geodetic Measurements of Crustal Movements 1906-1971.U.S.Department of Commmerce, Rockville.
Thorarinsson, S.,1975"Volcanoes of Iceland".Solarfilma,Reykjavik,Iceland.
Whitten,Ch.,1973."Crustal Movement from Geodetic Measurements".Reports on Geodetic Meas. of Crustal Movements 1906-1971.US Dep.of Commerce, Rockville.

WORKSHOP I

Potential, Problems, and Projected Directions in the
Oceanographic Applications of Sea Level Measurements
from Satellite Altimetry

Prepared by Jim L. Mitchell
Naval Ocean Research and Development Activity
NSTL, MS   39529

INTRODUCTION AND AGENDA

The potential, problems, and future projected directions of satellite
altimetry was the three-fold focus of INSMAP'86 Working Group I. The discussion
of this working group was structured around nine "mini-sessions" with a
designated presenter/dicussion leader (three mini-sessions from each of the
focus areas). The agenda for the workshop was:

Part 1: Oceanographic Potential of Satellite Altimetry

  (a) GEOSAT Oceanographic Results and the GEOSAT-ERM
      Presenter/Discussion Leader- J. Mitchell, NORDA

  (b) Potential for Mesoscale Oceanography
      Presenter/Discussion Leader- T. Rossby, URI

  (c) Intermediate Scale Oceanography (El Nino)
      Presenter/Discussion Leader- G. Maul, AOML

Part 2: Problems with Satellite Altimetry

  (a) Geoid/Reference Surface Determination
      Presenter/Discussion Leader- J. Marsh, GSFC

  (b) Orbit Determination Error
      Presenter/Discussion Leader- B. Tapley, UT

  (c) Space/Time Sampling Problems
      Presenter/Discussion Leader- D. Grant, JAYCOR

Part 3: Projected Directions in Satellite Altimetry

  (a) The TOPEX Altimeter Mission
      Presenter/Discussion Leader- G. Born, UC

  (b) Data Assimilation into Ocean Models
      Presenter/Discussion Leader- H. Hurlburt, NORDA

  (c) Multi-Beam Altimetry
      Presenter/Discussion Leader- C. Kilgus, JHU/APL

M. Kumar and G. A. Maul (directors), Marine Positioning, 433–442.
© *1987 by the Marine Technology Society.*

WORKSHOP RECOMMENDATIONS

The following concensus recommendations were the result of the workshop:

(1) Gravity field improvement efforts are critical inorder to meet the need for better orbit determination and for better geoidal reference surfaces. The proposed NASA GRM mission is strongly endorsed.

(2) The GPS receiver should be flown on TOPEX as a formal backup to the nominal tracking system (TRANET II). Presently, NASA plans fly the TOPEX GPS receiver only as a Class C demonstration.

(3) The multibeam altimeter system appears to have great promise for studying the oceanic mesoscale. A thorough feasibility/conceptual design study should be carried out for this system.

(4) One of the greatest utilities of satellite altimetry is its promise as a highly desired data type for assimilation into ocean numerical models. All issues associated with the assimilation of altimetry into ocean models need concentrated study.

(5) Satellite altimetry can be greatly enhanced by carefully designed complementary in situ measurements. As a minimum, the following in situ data should be collected for study with concurrent satellite altimetry:

   (a) Extensive AXBT sections along satellite tracks,

   (b) Surface drifter and velocity sections,

   (c) Continuous data records from inverted echo sounders.

   (d) At least for GEOSAT (which has no onboard radiometer)
       water vapor monitoring for pathlength corrections is
       desirable for some applications.

(6) The inverted echo sounder plays a central role in providing complementary in situ data to the satellite altimeter. Several problems associated with the interpretation of IES records (e.g., surface waveheight biases) warrant further study. To facilitate the analysis of IES data it is recommended that actual waveform data be recorded onboard the instrument for subsequent analysis.

SUMMARY OF PRESENTATIONS/DISCUSSIONS

The presentation/discussion of each of the nine mini-sessions is summarized briefly in this section.

Part 1(a): GEOSAT Oceanographic Results and the GEOSAT-ERM
(J. Mitchell)

Though the primary mission for the U.S. Navy's GEOSAT satellite has been the generation of a global mean sea level surface for approximating the marine geoid for military applications, GEOSAT has also returned a wealth of oceanographic information in the form of sea surface topography (i.e., sea level minus tides and geoid), surface waveheights, and surface wind speeds. GEOSAT data from the initial 18-month "geodetic" mission has been used in both the ONR-sponsored basic research program called REX (the Regional Energetics Experiment) and in the applied developmental GEOSAT Ocean Applications Program (GOAP) sponsored by the Oceanographer of the Navy. Preliminary results from each of these programs were presented.

The REX focuses on the use of GEOSAT-derived surface topography and topographic fluctuations in the NW Atlantic/Gulf Stream region along with appropriate complementary in situ data from extensive AXBT surveys and bottom-moored Inverted Echo Sounders with Pressure Gauges (IES/PG's) to study the mesoscale dynamics and energetics through the assimilation of these data into regional eddy-resolving numerical models. Some of the REX modeling results are described by Grant and Hurlburt below. Favorable comparisons of GEOSAT-derived surface topography and in situ temperature sections derived from AXBT's dropped during aircraft underflights of the satellite were presented. Additionally, comparisons of surface drifter-derived crosstrack currents and GEOSAT topographic gradient-derived crosstrack geostrophic currents were presented and discussed (see T. Rossby's discussion below).

Plans for the initiation of the oceanographic phase of GEOSAT's mission, the GEOSAT-Exact Repeat Mission (GEOSAT-ERM) (which was just getting underway at the time of INSMAP'86), were discussed at length. At the time of this writing (Dec, 1986) the GEOSAT-ERM has been very successfully initiated and plans have been made for the wide distribution of GEOSAT-ERM data to the oceanographic research community.

Part 2(b): Altimetric Potential for Mesoscale Oceanography)
(T. Rossby)

Since I did not give a formal presentation, perhaps I should refrain from writing here, but there are some observations I can add:

(1) Mesoscale processes certainly have the strongest geostrophically

balanced altimetric signal. So the signal to noise is as large as it can get! Those of us working in the Gulf Stream have the added advantage of working with very energetic processes. If there is anywhere the altimeter should be quantitative, that is where! My impression is that SEASAT was still primarily qualitative. RMS maps look pretty good, but there's a lot of smoothing in them. The Cheney/Marsh analysis of SEASAT passes over the Gulf Stream continues to haunt me. I strongly support Marsh's suggestion to reanalyse the SEASAT Gulf Stream passes. I am really interested in the upcoming GEOSAT data for the Gulf Stream.

(2) I got the impression that there is some uncertainty in how well the water vapor corrections are handled. Along the edge of the Gulf Stream there can be very strong contrasts in the atmospheric water content; similarly there can be very large differences in sea state.

(3) The best way to test the accuracy of altimetric slopes is by means of the surface drift measurements like those that J. Mitchell (see above) has tried. I would really make an effort to do that systematically. Using AXBT's will give qualitative agreement, but, as pointed out by Mitchell, one won't see the barotropic signal, and I do know it can be quite substantial. One could wind up spending a lot of energy for data that won't really meet the measure or make the grade. Using surface drifters is the way to go.

(4) I have no quarrel with G. Maul's comments about the need for more detailed information on what the IES really measures (see general recommendation #6), but I think he was being unnecessarily pessimistic. A lot of people have been doing fine work with the IES's.

Part 2(c): Intermediate Scale Oceanography (El Nino)
(G. Maul)

Reports from the National Academy of Science and the National Science Foundation emphasize the need for measuring the horizontal flux of heat and mass across certain critical areas within the ocean. These critical areas include major current systems such as the Gulf Stream, the Kuroshio, the Brazil Current, the Somali Current, the Antarctic Circumpolar Current, and the East Australian Current; outflows from semi-enclosed seas such as the Mediterranean and Norwegian Seas; and cross-equatorial flows in the eastern Pacific and central Atlantic Oceans. Measuring absolute dynamic topography across the equatorial current systems in the eastern central Pacific Ocean was discussed to explore the role of satellite altimetry in determining the horizontal fluxes.

The signal of dynamic topography in the eastern equatorial Pacific increases with increasing latitude from a minimum of approximately + 5 dynamic centimeters at the equator to approximately + 20 dyn-cm at 10 degrees north latitude. Significant oscillations with periods ranging from 20-80 days and space scales of 100-500 km are thought to exist

based on in situ observations. Detailed measurements at space scales
of tens of kilometers and sub-weekly time scales are needed to define
the variability in the North Equatorial Counter Current/South
Equatorial Current system, and its role in ENSO events.

Error analysis of satellite altimetry suggests that the precision
requirements in determining absolute dynamic topography across
equatorial current systems requires certain in situ oceanic
observations and observations of atmospheric water vapor variability
to achieve 1-3 dyn-cm accuracy levels. Studies have shown that
inverted echo sounder/pressure gauges placed at suborbital crossover
sites can provide the in situ verification data to correct most of the
long wavelength altimeter errors. Shorter wavelength altimeter errors,
mostly associated with media variability such as atmospheric water
vapor, will require additional observations either by passive
microwave radiometry, additional in situ oceanic observations, or
both. A generalized scheme to achieve the requisite accuracy includes
using collinear orbits, precision tracking, in situ oceanic
observations, ancillary atmospheric observations, and space-time
objective analysis.

Part 2(a): Geoid/Reference Surface Determination
(J. Marsh as reported by J. Mitchell)

The effects of the earth's unknown gravity field on the utility of
satellite altimetry for determination of sea surface topography is
two-fold:

 (a) any unknown geoid undulation can easily mask the relatively
weaker surface gradients of the ocean topography and,

(b) uncertainties in the earth's gravity field coefficients lead to
inaccuracies in the dynamic modeling of the satellite's orbit and
hence to inaccuracies in the derived altimetric sea level.

J. Marsh described the ongoing gravity field improvement program at
GSFC (which is being partially sponsored in preparation for NASA's
TOPEX/POSEIDON altimetric mission).

For additional discussion of this topic see B. Tapley's contribution
below.

Part 2(b): Orbit Determination Error
(B. Tapley)

During the last decade, the development of satellite altimetry
techniques for the active sensing of ocean surface topography has been
one of the principal drivers for improvements in orbit determination
abilities. The improvement in satellite sensor technology, which
currently measures the height from the satellite to the ocean surface
with a precision of 2 cm and an accuracy of less than 5 cm, gives a

strong impetus to determine the satellite's orbit with comparable accuracy. To achieve the orbit accuracy required to utilize such measurements for general ocean circulation investigations, an intense effort is required to develop appropriate software systems and an improved model for the satellite's motion to support the orbit determination functions. The primary force models which influence the satellite's motion include the earth's gravity field, atmospheric drag, direct and reflected solar radiation pressure, the effects of solid earth and ocean tides, and the attractions of the sun, moon and planets. The primary limitation on the accuracy with which current satellite orbits can be computed lies in errors in the model for the earth's gravity field. The orbit accuracies for current oceanographic satellites are limited to the order of 50 cm, at best, by gravity model error. As the laser and radiometric tracking systems improve, the ability to compute orbits with an accuracy of 10 cm or greater will be possible provided the model for the earth's gravity field is improved. This limitation provides a justification for the current efforts to improve the gravity model for TOPEX and other oceanographic satellites and indicates an extremely strong need for the Geopotential Research Mission (GRM), whose objective is to improve the global gravity field model. The GRM will improve not only the long wavelength components of the gravity field which influence the satellite orbit, but also will improve the shorter wavelength gravity field errors which limit the accuracy of the ocean surface geoid.

One of the more significant recent developments for satellite orbit determination is the potential for using the satellites of the Department of Defense Global Positioning System (GPS) for orbit determination purposes. The GPS receiver will provide a global set of pseudo-range measurements, accurate at the 2 cm level. This data from each of 21 GPS satellites will yield the ability to compute the orbit of satellites such as TOPEX with a radial accuracy on the order of 5 cm. Furthermore, the global nature of the data collected from this system will allow a significant improvement in the long wavelength components of the earth's gravity field and marine geoid. Finally, the deployment of the GPS tracking system on oceanographic satellites such as TOPEX, NROSS, ERS-1, or EOS will give the ability to routinely determine orbits at the sub-10 cm level with only a few days delay.

Further modeling activities are required to understand the effects of atmospheric drag on near-earth satellites. Improvement in the models for the atmospheric density at satellite altitudes up to 1500 km is required. Further research in developing variable area models for the effects of atmospheric drag are warranted also.

In summary, the primary limitations in orbit computation methodology at present reside in the accuracy of the force models, primarily the model for the earth's gravity field, and in the accuracy and geographical distribution of the tracking data. The GRM is of critical importance in removing the gravity model error. Satellite-borne GPS receivers should be developed to provide accurate global tracking data for current oceanographic satellite missions.

Part 2(c): Space/Time Sampling Problems
(D. Grant)

Altimetry is the most promising source of global, synoptic ocean surface data for the oceanographic community. However, in using alitmeter data there is an unavoidable tradeoff between the crosstrack spatial resolution obtained and the temporal resolution obtained. For example, an altimeter that provides new data every 17 days has tracks 164 km apart at the equator, but reducing the track spacing to 80 km would more than double the repeat period to 36 days. Both the author and J. Kindle (see Feb. 86 JGR Oceans) have examined analytically and through modeling, respectively, the spatial and temporal sampling requirements of altimetry for it to resolve mesoscale ocean features.

Both approaches indicated that the NRMS error in mapping a stationary eddy decreases quadratically with the number of tracks per eddy, and that 2-3 tracks per eddy generally suffice. Furthermore, Kindle shows that when the height field is adequately resolved both the along and crosstrack geostrophic velocity components can be calculated with accuracy comparable to that of the height field.

The analytical results indicate that the minimum NRMS error in mapping an eddy increases quadratically with the relative frequency of the track, relative frequency being the fraction of the track spacing that an eddy moves between when a track is taken and when it is used. This is a linear function of the track's age, so the error in mapping an eddy of constant propagation speed also increases quadratically with the age of the track.

Modeling and statistics can significantly reduce the spatial and temporal resolution required of altimeter data, but even with modeling and statistics one single beam altimeter does not have the spatial and temporal resolution necessary to successfully map all mesoscale ocean features, although I believe a multibeam altimeter with three beams would (see discussion of C. Kilgus below).

Part 3(a): The TOPEX Altimeter Mission
(G. Born)

The design of the Ocean Topography Experiment (TOPEX), a precise altimeter system to map global ocean circulation was initiated in 1979 after failure of the SEASAT spacecraft. After 7 years this mission has finally been approved by Congress for a project start in 1987 and will launch late in 1991. During the past seven years TOPEX has evolved into a joint mission (TOPEX/POSEIDON) between the US and France with the US supplying the spacecraft, altimeter, laser retroflector, radiometer, Tranet beacon, and an experimental GPS receiver. The French will launch the spacecraft on the Ariane and will contribute another altimeter and a precision tracking system to the payload.

The TOPEX/POSEIDON system will provide ocean topography measurements

with unprecedented accuracy. Complementary wind field data will be
collected by the NASA scatterometer to be flown on the Navy's N-ROSS
spacecraft and in situ data will come from the international World
Ocean Circulation Experiment (WOCE). In addition, the European space
agency will orbit the Earth Remote Sensing Satellite (ERS-1) carrying
an altimeter, a scatterometer/synthetic aperture radar and a
radiometer during this period. This diverse system of satellites and
in-water data will allow us for the first time to attack the problem
of global ocean circulation measurement and modeling.

The TOPEX/POSEIDON altimeter is a dual frequency instrument to allow
correction for ionospheric refraction to the sub-centimeter level.
The three channel microwave radiometer will provide the atmospheric
water vapor measurement necessary for the wet tropospheric range
correction. Precise tracking of the satellite will come from the
Tranet system complemented by laser ranging, a GPS receiver and a
French precision tracking system (DORIS). The TOPEX/POSEIDON system
will provide altimeter range measurements of ocean topography precise
to 2 cm with absolute accuracy of 14 cm.

Thus, some 12 years after the pioneering mission of SEASAT, a
measurement system to monitor the global oceans with the necessary
space and in situ elements, will become reality.

Part 3(b): Data Assimilation into Ocean Models
(H. Hurlburt)

A global, eddy-resolving ocean prediction capability is
technologically feasible within the next decade (Hurlburt, 1984). For
this purpose the satellite altimeter is the most promising operational
source of oceanic data with global coverage. However, there are
several major issues that must be addressed in using altimeter data in
ocean models: (1) spatial and temporal sampling requirements, (2)
asynoptic data assimilation, (3) coping with uncertainty in the data,
especially inadequate knowledge of the geoid and longwave orbit
errors, and (4) extraction of subsurface information from surface
data.

Downward transfer of information turns out to be a critical issue
because there is no comparable source of subsurface data on the
horizon and because models for ocean monitoring and prediction need
subsurface information as demonstrated by Hurlburt (1986). Acoustic
tomography is an enticing source of subsurface data (Munk and Wunsch,
1982), but there is no indication that it or any other subsurface
observing system will provide sufficient eddy-resolving coverage to
meet the needs of ocean prediction during this century. There were
some early hints that surface to subsurface transfer of information
might be feasible in the ocean. Observations show a high correlation
between the sea surface height and the depth of the thermocline
(Cheney, 1982) and most oceanic energy contained in a few vertical
modes (Pochapsky, 1976; McWilliams, 1976; Richman et al., 1977). Ocean
models also provided some early hints because eddy-resolving numerical

models with low vertical resolution have shown a remarkable ability to
simulate many observed features of the ocean circulation and because
of the affinity between these models and altimeter data (Hurlburt,
1984).

Recently, modeling studies have offered new indications that downward
transfer of surface information is feasible (Hurlburt, 1986; Thompson,
1986; DeMey and Robinson, 1986; Grant and Hurlburt, 1985):

 (1) Both dynamical and statistical models were able to reconstruct
the subthermocline flow from simulated altimeter data when: (a) the
surface and subthermocline correlations were small, and (b) the
subthermocline flow had a major impact on the surface flow;

 (2) Knowing something about the subthermocline presssure field
greatly improved forecast skill, but subthermocline pressure fields
with RMS errors of 30-50% still yielded greatly enhanced forecast
skill for greater than one eddy cycle over those model runs using no
subthermocline knowledge at all;

 (3) Even models with severe imperfections demonstrated some forecast
skill (as did models with no subsurface information).

Part 3(c): Multi-Beam Altimetry
(C. Kilgus as reported by J. Mitchell)

C. Kilgus discussed the potential and problems associated with a
multibeam altimeter system being studied as a possible sensor on
follow-on missions to the N-ROSS. The baseline system described has
three beams: one directed along the nadir track and additional beams
directed 50 km off-nadir. The hardware configuration described
consists of two 1.5 m dish antennas mounted on a 10 meter boom. The
satellite is assumed to be flying at an orbital altitude of 800 km.
The center beam is two-frequency (K_u and C band) for ionospheric
correction, while the two side beams are K_u band only. Tracker noise
for this system is about 2.3 cm on the center beam and 5 cm on the two
side beams. The alongtrack effective resolution is 11.5 km.

The large (10 m) boom width is necessary to provide for narrow, easily
tracked return pulse profiles for the two side beams. However, this
boom width results in a bothersome sensitivity to both rigid body
rotation and flexible body bending. A modern roll rate gyro (e.g., the
NASA Standard Inertial Reference Unit-DRIRU II) coupled to a stiff
boom would appear to be able to solve these potential problems. For a
5 cm height (range) measurement precision the resulting attitude
determination requirement is 0.2 arcsec after removal of a 300 sec
(2000 km alongtrack) linear trend in the measured range. Note that it
is assumed that the prime objective of the multibeam system is
observation of the oceanic mesoscale.

As pointed out by D. Grant (see above), the enhanced space/time
sampling of a multibeam system like that described by Kilgus would

result in quasi-synoptic coverage of the oceanic mesoscale in most
regions of the world ocean. Thus, it is strongly recommended that a
development program to resolve the design issues associated with such
a system and to firmly establish its measurement capabilities be
undertaken.

WORKSHOP ATTENDEES

NAME	ORGANIZATION
T. Austin	Benthos Inc.
E. Berg	Univ. of Hawaii
G. Born	Univ. of Colorado
A. Davies	Hunting Surveys Ltd.
A. Gallegos	UNAM/Mexico
J. Giannini	JHU/APL
D. Grant	JAYCOR/NORDA
E. Groten	THD/IPG Germany
H. Hurlburt	NORDA
C. Kilgus	JHU/APL
J. Kindle	NORDA
M. Kumar	DMAHTC/GST
J. Marsh	NASA/GSFC
G. Maul	NOAA/AOML
J. Mitchell (Chairman)	NORDA
C. Mooers	INO
J. de Munck	Delft Univ. of Tech./Holland
T. Rossby	URI
M. Salzmann	Delft Univ. of Tech./Holland
N. Saxena	Univ. of Hawaii
L. Spielvogel	SEACO
B. Tapley	Univ. of Texas
S. Tucker	Naval Postgrad. School
P. Worcester	SIO

WORKSHOP II

PRECISE MARINE POSITIONING FOR GEODESY AND GEOPHYSICS

Compiled by

Günter Seeber (Chairman)
Institut für Erdmessung
Universität Hannover
Federal Republic of Germany

Eight panel members had agreed to cooperate within the workshop

   Prof. D. Egge, Seattle, USA
   Senior Eng. R. Ekseth, Stavanger, Norway
   Prof. H. Henneberg, Maracaibo, Venezuela
   Prof. J. de Munck, Delft, Netherlands
   Prof. N.K. Saxena, Honululu, USA
   Prof. F. Spiess, La Jolla, USA
   Prof. A. Stolz, Kensington, Australia
   Prof. D. Wells, Fredericton, Canada

Through this, a good international coverage was achieved. Four main topics
were discussed in such a way that after an introduction of one or more panel
members, a discussion between all participants from the panel and from the floor
followed up. Some fourty people attended the meeting.
   The following four main topics had been selected

1. Where are the Highest Demands

   After a review, given by N.K. Saxena some special aspects were treated, namely

- Geodesy and Geodynamics by A. Stolz
- Ocean Industry and Exploration by R. Ekseth
- Coastal Aspects by H. Henneberg

2. Competition of "Classical" Positioning Techniques in the Era of GPS

   The term "classical" was understood in the sense of already existing and not
in the meaning of already obsolete.
   A review was given to

- Acoustic Techniques by F. Spiess
- Radio Techniques by J. de Munck
- Integrated Techniques by D. Wells

M. Kumar and G. A. Maul (directors), Marine Positioning, 443–451.

3. Where are the Limiting Factors and what has to be Done to Overcome Them

Here an introduction was given by D. Egge.

4. Proposals and Recommendations for Future Activities

with an introduction by D. Wells.

   In the following I will shortly report on the main results of the presentations and the discussions. Some more detailed written comments from individual
panel members are added at the end of the report.
With respect to the highest demands, four regimes were identified.

   Seabed mapping: Considering systems like SEABEAM or HYDROSWEEP, the complete
mapping of the seabed in all areas of interest becomes a feasible prospect.
Precise positioning becomes a more and more important aspect for the evaluation
of such data.

   Mapping of subsurface structures, i.e. through 3D seismic, where hydrophone
arrays have to be positioned relatively in the 1 m range or better becomes another
important positoning demand.
   Most challenging demands are coming from requirements for the determination
of seabed tectonic motions. Most of the places of interest to this respect are
covered with water. For baseline changes, an accuracy in the 1 cm range is required. One or both endpoints of the baseline may be on the ocean floor. In that case,
a connection between ocean bottom positions and ocean surface positions in the
1 cm range has to be realized. This is still out of range with current techniques.
   A fourth aspect, related to coastal areas, are sea level datum variations and
connections. Regional height systems are referred to tidal stations, not to mean
sea level. A connection of tidal stations via space techniques will lead to absolute sea level differences, which is of importance to climatic studies and which
will improve the knowledge about the difference between the physical and the
mathematical height reference systems.
   With respect to the second topic, the competition of classical techniques in
the era of GPS,the main feeling of the group was, that GPS is a very powerful
and interesting new tool in the arsenal of the "Marine Positioner",but it won't
be the only one in the future. We still will need some of the existing techniques.
This is especially true for acoustic measurements which are the only ones to
relate subsurface markers or structures.
   Three regimes were identified: 0 - 1 km; 1 - 10 km; > 10 km. Only for short
ranges up to 10 km, a connection of bottom markers directly or via some intermediate vehicle is possible. For longer distances, i.e. measurements over trenches
or plate boundaries, a connection via surface platforms carrying satellite
sensors, is necessary. The 10 cm accuracy range seems to be feasible. The 1 cm,
which is desirable, will remain a challenging demand. Most critical is the
determination of sound velocity: 10^{-4} is possible; 10^{-5} is very difficult to
achieve.
   With respect to radio techniques, the meaning was that for the next 20 years
they will hold some importance, because of some advantages; for instance they may
be under national or individual control. Most development work has to be done in the
improvement of calibration. "High technology does not always win", was one statement; it will be the appropriate technique which wins.

The integrated navigation systems will not be obsolete through GPS, but use GPS as an additional very powerful sensor. The data logging aspect will become more and more important, and also the case of non-navigational sensors like depth, seismic and swath mapping. We see a challenging development with respect to integrating all ocean related information in such a hybrid integrated system where precise navigation is one result.

The limiting factors arise from different sources
- operation in marine environment (i.e. wave propagation, sensor motion)
- application of specific positioning sensors (i.e. calibration)
- application of specific data handling techniques (i.e. data transport, mathematical modeling)
- optimization aspects (i.e. cost and benefit)
- human aspects (i.e. gaps in understanding between geodesists, navigators, electronic engineers).

The last topic, recommendations cannot be a complete view, because 2,5 hours were too short for this type of discussion. The general feeling of the group was, that a 1 day workshop would have been more appropriate. With respect to the discussion during the workshop, four recommendations were formulated by a group of panel members directly after the meeting, concerning Plate Tectonics, Geophysical Exploration, Sea Bed Mapping and Sea Level Datum.

1. Most plate tectonic boundaries are under the ocean. Ocean plates are moving fastest. It is recommended to develop techniques to measure spreading rates to the highest achievable accuracy, to install benchmarks on the sea bottom in order to monitor such motions and to commence with first epoch measurements as soon as possible.

2. For development of oceanic resources and for geophysical exploration the 3D-seismic becomes to be a powerful tool. The performance task for relative positions of streamer elements is around 1 meter. This requirement is not fulfilled with classical techniques. New methods have to be developed. To this respect, differential GPS should be studied and developed.

3. Since precise bathymetric mapping plays an important role in the exploitation of the ocean resources, the scientific studies of Tsunamis and other related fields and since most of the ocean is still unsurveyed, areas of special interest have to be mapped efficiently. A key factor is the precise knowledge of the position of the survey element. We recommend that methods be developed which provide the appropriate positioning accuracy.

4. Since the determination of absolute sea level differences across the oceans is of importance for oceanographic and geophysical research such as the theory of climate, we recommend that appropriate methods be developed to establish a unified global vertical datum. Space methods, in particular the Global Positioning System (GPS Levelling) should be used to solve this problem.

With respect to further activities tools have to be developed in order to overcome the technical limitations and to achieve progress in concepts. To this respect also the agreement on standards could help.

In the sequel, some of the written statements of panel members are added.

Topic: <u>Positioning element in the geodynamic aspect of precise positioning</u>
       <u>(A. Stolz)</u>

Those that come to mind are shown in Table 1. I will address mainly the specific aspects of marine geoid determination and two particular issues of plate tectonics   which should be investigated.

<u>Marine Geoid</u>: Precise knowledge of marine geoid is essential for study of mantle
convection of oceanic litosphere. Requirement is high resolution gravity field
data of the order of 1 mgal with 25 x .25 km spatial resolution. Both space based
and surface measurements will contribute. Position element in surface measurement,
particularly height component must be precisely known. Second problem is that
measurements are made from moving vessels. East-west component of velocity must be
known to 10^{-3} m/sec. This is a most demanding exercise.

<u>Ocean Ridge Spreading Centers</u>: Oceanic plates such as Nazca and Pacific plates
move much faster than predominently continental plates. Maximum relative velocities of nearly 20 cm/yr. occur along Pacific-Nazca plate boundary. It would be
very valuable to measure this motion.

<u>Zones of Continental-Oceanic Convergence</u>: Great earthquakes occur in these re-
gions. During intervals between earthquakes, ocean lithospheric plate is fairly
stationary at trench boundary. Plate is under stress due to more-or-less steady
plate tectonic force, and state of compression may extend well into the interior
of plate.
    At the trench, the litosphere is held fixed by frictional forms along the
interface. But at the seaword side there is an upward bulging of the lithosphere.
With time, the stresses and deformations increase until a critical stage is
reached. One or both of two things occur.
    Firstly underthrusting of the oceanic plate may occur if the existing force
is exceeded. The boundary between the oceanic and restraining plate is broken
and a tempory decoupling occurs.
    Secondly, the bending moment of the oceanic plate may become excessive
resulting in a tensile fracture where the bulge is greatest. Simple models suggest
that this may occur at a distance of ~ 100 km from the trench.

<u>Deformation</u>: Later of underthrusting of downgoing slab and later of approach of
oceanic plate to trench will increase, while previously built up stress will be
released. On continental side sinking of litoshpere may occur. On oceanic side
there will be a reduction in elevation of litospheric bulge. Post-seismic deformation may be of the order of 1-5 cm/yr within 100 km of boundary.

Table 1: Observation Requirements: Geodynamics *

Observation	Science Problem	Measurement Characteristics
Gravity Field	Mantle convection Oceanic lithosphere	25x25 km resolution 1 mgal accuracy global coverage
Oceanic Geoid	Mantle convection Oceanic lithosphere	1 cm accuracy global oceanic coverage
Baseline changes	Plate motion and deformation	1 cm accuracy in each component

* specific to ocean

Topic: Positioning Demands in Ocean Industry and Exploration (R. Ekseth)

To find a short and safe way to an onshore point has traditionally determined demands to offshore positioning. The later decades' offshore oil activity has created a much higher accuracy demand to the navigation and positioning service. While some years ago we spoke about an accuracy of nautical miles, do we today speak about meters. Todays positioning technology is often a limiting factor in reaching a meter accuracy level. Many survey services would probably have been of much more help if the positioning had been with more confidence. The most demanding areas are 3-D seismic, sea bottom mapping for offshore installations and underwater pipelines, positioning of oilrigs, etc. Dynamic positioning of ships, that is to keep the ship at an exact position without moving, is another demanding area. Not only the oil industry have their powerful demands to navigation and positioning accuracy. A fisherman will for example have good coordinates on wrecks etc. This is to be able to make use of as much of the oceans as possible, without destroying equipment. For most of these cases the possibility to find back to a given point in the critical factor, and not the true absolute position in a global reference system. The high accuracy demands are therefore more based on repeat accuracies than on absolute accuracies.

In seismic and 3-D seismic particularly, we want to get an accurate position of each hydrophone on the cable for each shotpoint. Traditionally this has been solved by transferring the ship's position to the hydrophones with auxiliary compasses on the cable. This method does not give us an accuracy of 1 meter, which we often search in detailed surveys of oilreservoirs. The limiting factor in use of compasses is that they do not give us range in addition to direction. We have in this case to find the range with help of the cablelength; and this is not easy with invisible winding cables. In the later years one has started use Hydro Acoustic positioning systems in this work. These systems do measure both range and direction relative to the ship. The limiting factors are shipdynamics, and timing between dynamics and measurements. If these factors are under control, one will get adequate accuracies from this method. More direct, safe and easier methods are, however, of interest. A GPS receiver on the towing buoy would for instance of great help as an overall control.

Construction of offshore installations and underwater pipelines is depended of detailed sea bottom maps in large scale. This is to be able to place the equipment on the most suitable point. The accuracy of sea bottom mapping must therefore be within a few meters, and the least as good as the accuracy one can place a platform with. Final coordinates for permanent offshore installations should be down to the meter accuracy level absolute (in a global reference system). This to make them suitable as reference stations for future explorations.

Where to drill is dependent of the shape of the geological structures, and one should not miss with much more than 10 meters to get good results. 100 meters off may in worst case lead to wrong conclusion. This means that seismics, hydrography and platform positioning all should be performed with an accuracy of 10 meters or better relative to permanent reference stations.

Wells that are left because they are dry or not economic, must be positioned very accurate absolutely. A meter level is required to make them easy accessable, and to prevent other offshore industries from getting difficulties. This demand is also valid for pipelines and other underwater installations.

Topic: Acoustic Approaches   to Geodesy in Relation to Geophysics (F.N. Spiess)

    The material alluded to during our brief meeting is covered in greater detail
in:
    Spiess, F.N. et al. 1983. Seafloor referenced positioning:
      needs and opportunities. Panel on Ocean Bottom Positioning,
      Committee on Geodesy, National Research Council, National
      Academy Press, Washington, DC, 53 pp.

    Spiess,F. N. 1985. Suboceanic geodetic measurements.
      IEEE Transactions of Geoscience and Sensing, special issue
      on satellite geodynamics, L.S. Walter, ed., vol. GE-23,
      no. 4, pp. 502-510.

    Need for ability to measure strain buildup (more precisely: temporal change
in horizontal position) of seafloor points in geodynamic contexts exists on a
wide variety of scales, ranging from centimeters (monitoring individual fissures
at spreading centers or other tectonically active zones) to thousands of kilometers (relative motion of crustal plates). Optical techniques can be used at
the short-length end, out the somewhere between 100 m and 1000 m, although in
many environments turbidity can limit this to a few meters. Approaches using
acoustics are relevant, in one form or another, over this entire range.
    Acoustic methods have some inherent limits. First, sound is absorbed as it
travels through the water, and this absorption increases with frequency. If one
chooses something between one and ten cm as an upper limit on usable accuracy for
geodynamic purposes and asserts that one can make measurements of travel time
to between 0.1 and 0.01 cycle of the carrier (given good signal-to-noise ratios),
then one can build systems capable of adequate traveltime measurement out to
ranges of twenty to perhaps a hundred km. Unfortunately the ocean is not homogeneous and the knowledge of sound velocity over the transmission path will usually
be no better than a part in 10^5, thus this sets an upper limit of about 10 km.
    Given this environmental limit, acoustic systems will fall into two categories:
pure (for ranges to about 10 km) and hybrid (for longer ranges).
    Within the pure systems there are, again, two classes: short range (< ~ 1 km),
in which the acoustic reference beacons can communicate with one another, and
long range (> ~ 1 km) in which, because of refractive effects, one is better off
using some intermediate vehicle. It should be noted that, unless one makes very
accurate current measurements (1.5 cm/sec), one should use two-way travel paths
to cancel out these effects.
    The most promising prospects for situations requiring more than 10-km paths
lie in the use of acoustics to tie a near-surface point to the seafloor, GPS to
tie an above-surface point to some other reference, and some auxiliary system to
relate the GPS antenna to the nearby acoustic transducer. Again the most obvious
limits are the acoustic paths, which must traverse the variable near-surface
water. By proper selection of sites and system geometries, sub-decimeter capabilities show promise of attainability.
    Actual hardware to achieve these goals exists at best in laboratory tested,
breadboard form. If we are to begin actual field programs it is necessary that
key elements be built, their performance validated and then that they be deployed
in appropriate areas. The technological situation has, however, moved forward
only very slowly. Growing interest in the scientific and engineering communities,
however, gives hope that the level of activity may escalate.

Topic: Integrated Navigation (D. Wells)

The intention of this introduction is to spare discussion on basic questions regarded integrated navigation

    a) what do we mean by "integrated navigation"?

    b) why might we want to have an integrated system?

    c) what methods can be used to merge navigation data in an
       integrated system?

I would like to present my personal views on these questions.

    a) Definition: An integrated navigation system is a computer-based system
       with two or more input sensors, at least one of which is a navigation
       input.
    b) There are two classes of reasons for integrated system
       - to overcome deficiencies in navigation,
         e.g. to obtain more LOP's and better redundancy,
         to provide better reliability (checking cycle slips) because single
         sensors available do not provide a complete navigation capability,
         (e.g. log, gyro, Transit),
         to allow correction for biases (clock, phaselag),
         to provide filtering/smoothing of navigational/observational noise
       - for non-navigational reasons,
         e.g. to combine navigation with other sensors (depth, seismic, swath
         mapping...),
         to provide homogeneous data (impose standard format, time tagging,
         etc. on all data).
    c) There are a variety of possibilities. In general, the performance of the
       method becomes more and more dependent on the stochastic modelling, with
       more sophisticated demands. The ocean environment is variable enough
       to make such modelling very difficult.

Topic: Limiting factors and how to overcome them (Delf Egge)

The limiting factors arise from different sources:

    a) operation in marine environment;
    b) application of specific positioning sensors;
    c) application of specific data handling techniques;
    d) optimization aspects;
    e) human aspects.

According to these groups, most important factors well be addressed below.

    a) Operation in Marine Environment

The range limitation of specific positioning techniques is due to the wide area of the oceans where positioning may be required, and the curvature of the earth. This limitation can be overcome with long-range techniques or "bridging" with intermediate land-, sea-, or space-based stations.

The sea surface creates two groups of techniques for applications above and below this surface. For some applications there is a need to link these two groups by a sensor network attached to floating objects at the surface.

The position determination of the network points in real time will be more problematic with sub-decimeter accuracy requirements. Only inertial or other dead
reckoning techniques would not be affected here.

The limited knowledge of the gravity field can affect inertial positioning.
Problems might be overcome, if high resolution gravity data were available in real
time to the inertial navigator, or if additional sensors were used (e.g. gravity
gradiometer).

The inevitable sensor motion in the marine environment necessitates a precise
time-tag and data processing with dynamic/kinematic models. This limitation might
also be overcome by construction of suitable stabilization mechanisms.

If positioning techniques make use of electromagnetic or acoustic waves, the
wave propagation has to be assessed. Currently this appears to be the most critical
limitation. For electromagnetic waves, "refractometers" and "radiometers" may
diminish the refraction uncertainty to less than 0.1 ppm . The determination
of sound velocity in water is now at a level of 10 ppm. Of same importance is
the elimination of multipath errors.

b) Application of Specific Positioning Sensors

Of basic concern is the precise definition of the reference point at a sensor
used for positioning. This problem may be overcome by calibration.

Many techniques may employ sensors with a specific direction of sensitivity.
For more flexibility, these may have to be made mobile, or alternatively, a
number of similar sensors may be set up in an array configuration.

The resolution limitation of specific sensors may only be overcome by technological refinement or new developments in the field.

Restrictions on the data acquisition rate used to be due to limited computer
resources. With more powerful microprocessors and computers, this limitation
appears to be overcome.

c) Application of Specific Data Handling Techniques

Limitations of data transport speed may be overcome by high speed data links,
optionally involving satellite communication. The information of large data
bases will be available in real time at sensor location by local mass storage
or by access to remote computers using high speed data communication.

Improvements are expected in the field of mathematical modeling and estimation techniques. These improvements are based on implementation of theoretical
research and adaptation of previous experience.

As an example of a current challenge in modeling and data processing, one
might consider the use of the GPS carrier phase in a dynamic environment, which,
if properly implemented,will bring a completely new dimension to marine geodetic
positioning.

Incorporation of a priori knowledge appears to be possible not only in the
field of estimation theory but also in a wider scope of artificial intelligence
applications. Restrictions due to limited computer power will no longer be a
factor.

d) Optimization Aspects

The aspects of cost and benefit may by itemized more thoroughly. The benefits might be analyzed according to
  - accuracy and reliability of positioning techniques;
  - logistic constrains;
  - productivity;

- automatization;
- miniaturization.

Costs are encountered for hardware, software, logistics, human resources.
Up to now, these aspects have been dealt with on a more or less empirical basis.

In the future, more rigorous solutions to the optimization problem in marine positioning should be available in various scopes. The solution would be a computer program indicating the optimal design of data acquisition and data reduction for a certain positioning task.

e) Human Aspects

A limitation appears to exist in the area of education of decision makers and individuals working in marine positioning.

There are still typical gaps between navigators, electronic engineers, hydrographic surveyors, geodesists, and geophysicists. This gap can only be overcome by dissemination of more information in these fields to the appropriate individuals.

WORKSHOP III

GPS APPLICATIONS IN OCEANOGRAPHY

Christopher N.K. Mooers
Institute for Naval Oceanography
NSTL, MS 39529-5005

ABSTRACT

Approximately twenty participants met for two-and-a-half hours as a working group to scope the subject topic. Substantial opportunities were noted for innovative applications of GPS precision positioning in oceanographic instrumentation design and in the conduct of oceanographic operations.

1. INTRODUCTION

In general, oceanographers need increased precision in positioning to support new in situ and acoustic remote sensing measurement systems for ocean currents and thermal structure. These systems include expendable, free-fall profilers; drifting buoys; acoustic doppler velocity profilers; and acoustic tomographic arrays. There is great interest in having "stripped down" GPS receivers for use on expendable probes, in use of GPS "translators" on surface drifters, and in a mechanism for disseminating real-time information on the status of GPS.

2. DISCUSSION

More specifically, several oceanographic applications for GPS, and issues concerning GPS, were identified:

2.1. Some new ocean measurement systems for ocean currents and thermal structure which need the increased precision in positioning that could be provided by GPS:

a. From Tom Rossby, URI - PROGRO, an expendable profiler a la the Richardson partial volume transport profiler or dropsonde could be deployed by research vessels (or ships of opportunity) doing oceanographic station work during programs like WOCE (World Ocean Circulation Experiment); it requires knowing where the probe descends from the surface and where it returns to the surface; a GPS receiver on the ship could be used to determine the relative positions (i.e., the displacement vector) or, better yet, a GPS "translator"(ca. $200 antenna pre-amp) could be installed on the probes (recently, a similar approach has been taken with atmospheric dropsondes by Vincent Malley, NCAR).

453

b. From Tom Rossby, URI - high quality nearsurface flow measurements are
needed in strong shear, frontal convergence zones and in convection (Langmuir,
etc.) cells; surface drifters equipped with "translators" might by very
effective.

c. From George Maul, AOML - ocean current profiling, the inverse of the
PEGASUS (the SUSAGEP or MAULER), where the pair of bottom-mounted acoustic
transponders would be replaced by a pair of transponders either hull-mounted
or towed.

d. From George Maul, AOML - GPS probably can play a role in separating
coastal sea level variations from coastal subsidence.

e. From John Anderson, PRL - based on his use of a variety of navigation
systems, an increased precision of drifter positioning could usefully
facilitate working with current meters suspended from drifters; also, more
precise positioning (ca. 50m) would be useful for tracking movement of glacial
and shorefast ice and conducting bathymetric surveys from drifters; if GPS
receivers could be available for $3-to-6K, there would be a competitive demand
for them.

f. From Larry Clark, NSF - ROV operations need accurate positioning in
order to return to bottom sites; the academic research fleet has seven GPS
receivers and is adding them as rapidly as possible; the major applications
are for use with (1) SEABEAM and SEAMARC bathymetric mapping, (2) acoustic
doppler velocity profilers (ADVPs), and (3) coregistration of data from
synoptic surveys of physical, chemical, and biological variables.

g. From Peter Worcester, SIO - in acoustic tomographic studies, travel
time fluctuations of 100 ms need to be measured with a precision of 1 ms over
ca. 100 km source/receiver separations; these fluctuations are then analyzed
with inverse techniques to infer the depth - horizontal structure of the
temperature field. To make such measurements from moving ships vice moored
arrays would have cost and operational advantages; there are three key, linked
factors: (1) positioning of the navigation antenna on the ship, (2) ship
motion which determines location of short baseline acoustic sensors on the
ship's hull, and (3) positioning of the "fish" (sensor package) towed at the
sound channel axis (ca. 1300 m depth in mid-latitudes) relative to the ship,
preferably to 1 m precision (but one can include the position estimate in the
general data inversion operation); the towed "fish" is supplied power and
commands (time marks) from the ship and the "fish" is "dumb", with computer
power on the ship. Alternative system configurations were considered; e.g., a
RELAYS buoy with a surface package for positioning and a suspended "fish",
which would require a sophisticated signal processing capability. Key
technological requirements include positioning to 10 m and velocity

determination of the tow body to 1 cm/s. For the time frames of application,
GPS and inertial navigation are needed for velocity determination. Systems
analysis of the many tradeoffs, including cheap bandwidth compression, may be
beneficial.

h.   From Phil Kies, NDBC - NDBC is interested in buoy positions; their approach is the integrated use of two or more systems; e.g., LORAN-C and OMEGA or GPS and OMEGA; he highlighted the fact that DMA/DOD operations of GPS will perturb the civil sector's utilization; and that, in the foreseeable future, USCG will be asked to operate GPS for the civil sector.

i.   From Dale Chayes, LDGO - GPS could be used to make direct measurements of strain in seafloor spreading centers a la Spiess.

2.2.   Strong interest was expressed in having real-time access to information on the status of GPS so that one would know whether an outage is due to receiver or system failure.

NOTE: Since the availability of GPS receivers to the deep sea research community began about five years ago, many new types of experiments have been developed and implemented in the context of intermittently available satellite constellations (many more are being planned).

Prudent use of scarce and expensive ship time requires careful planning plus the best possible real-time decision support information for adapting to unpredictable circumstances.

The present non-real-time predictions and historical data available from SAMSO by phone and via the U.S. Naval Observatory meet part of this need, but there is no generally available, near-real-time source of GPS monitor data.

2.3.   An interest was indicated in establishing training programs for developing countries, especially onboard U.S. ships in cooperative experiments.

2.4.   For oceanographers it would be useful to have more information on GPS; sources include:

a.   Institute of Navigation
b.   IEEE Trans. Geoscience and Remote Sensing,
     Vol. GE-23, July 1985
c.   Primer published by Prof. Art Stolz of the University of New South
     Wales, Australia
d.   Another primer to be published by Prof. Wells of the University
     of New Brunswick
e.   Centers of GPS expertise include APL, U of Texas, JPL, MIT, and NSWC.

WORKSHOP IV

REPORT OF INSMAP INSTRUMENTATION WORKSHOP
Chairman : Javad Ashjaee

JAVAD ASHJAEE
GPS Consultant
5000 Parkfield Ave.
San Jose, Ca 95129
(408)257-1831

and

Paul D. Perreault
Trimble Navigation
585 N. Mary Ave.
Sunnyvale, Ca 94086

ABSTRACT

The digest of the Instrumentation Workshop discussions is presented in this report. The discussions were focused on recent satellite positioning systems (GPS, GLONAS, STARFIX, and GEOSTAR).

Panel members:

Javad Ashjaee, Workshop Chairman, GPS Consultant.
Ron Hatch, Magnavox
Chuck Counselman III, MIT, Consultant
Phil Stutes, J. E. Chance & Assoc.
Peter McDoran, ISTAC
Paul Perreault, Trimble Navigation, Ltd.

The workshop started with 5 minutes of opening comments from each panel member about their activities and the instruments that they are involved with. The summary of opening statements are as follows:

McDoran: The Federal records show that the US Government will degrade the accuracy of the C/A code and encrypt the P code during the operational phase of GPS. Most survey applications require second order accuracy and are not concerned with millimeters. ISTAC products have demonstrated this capability and further, we hope to demonstrate soon a dynamic positioning capability with speeds up to 7 knots.

Stutes: We developed STARFIX in the face of the GPS competition because we felt that we could bring a totally commercial

457

system to the market long before GPS becomes fully operational. Starfix is being used today by over a dozen of clients. Twenty systems have been built and twenty more are under production. With STARFIX there is no government control, no licenses, no degradation of accuracy. This equipment uses 4 spread spectrum channels to receive signals from always in view communication satellites at geostationary orbits.

Hatch: The WM-101 uses the C/A code. It also uses the code to provide accurate timing for measurements, taking advantage of the high accuracy of GPS clocks. The WM-101 will perform differential positioning to beat the aforementioned denial of accuracy. The unit will contain orbit relaxation algorithms to improve the broadcast ephemerides. One area of future work is integration of GPS with INS and associated tracking bandwidth questions and receiver filter overshoot issues. WM-101 is currently using single frequency and provisions for dual frequency tracking is being implemented.

Counselman: Invented Macrometer, a codeless GPS receiver. But since only GPS C/A code will be available in the operational phase, the Litton Minimac GPS dual band receiver will use L1 C/A code and L2 codeless channels to correct for the ionosphere and to gain single point and dynamic positioning capability. Dual frequency is required for computing and removing the effect of ionospheric delay in 1-part-per-million survey applications. There is no other technique that could sufficiently remove the ionospheric effect in such applications.

Perreault: The dual capabilities of Trimble GPS receivers have proven to be especially useful to the marine positioning community. The 4000 series receivers can perform static geodetic surveys on one day, and be used in hydrographic surveys on a moving ship the next. Integrated doppler smoothing, developed and implemented by Dr. Ashjaee, and has been the subject of several papers this week, has been a standard feature. The 10X series receivers integrate GPS with LORAN and Dead Reckoning to give the mariner a real-time continuous navigation capability.

In the following session, panel members and the audience discussed topics on the agenda prepared by the workshop chairman. Their annotated comments are presented below.

1.   Cost/Performance:

GPS, upon completion of its 21 NAVSTAR satellites, will be the most cost effective positioning and navigation system. Glonas (Russian equivalent of GPS) will have similar capabilities. The panel members did not have enough detailed information about GEOSTAR and its future to comment about its cost effectiveness.

But the general consensus was that it is highly unlikely that it can compete with GPS from cost/performance and accuracy points of view. Phil Stutes provided the following information on STARFIX.

STARFIX - Continuous positioning system with STARFIX is available now. Footprint area is CONUS up to 50-60 degrees latitude and about 100 nmiles of east and west coasts. Three geostationary satellites are required. Fixed satellite usage cost is about $30K/month. Costs should decrease as the number of users increases. Presently the system is using tracking directional antennas but an omnidirectional antenna could be used when the number of users is large (2000 - 3000) to share the cost of the required higher XPNDR power. The accuracy of the system presently is 5 meters 2DRMS. STARFIX does not need the differential link used by real time GPS systems. STARFIX can not be used and is not intended for survey applications.

(Comments from floor) But there is a positioning degeneracy problem at latitudes of 18 degrees and lower. You need an out of plane satellite to provide necessary geometry. Also could use LORAN. Work is proceeding to provide an integrated STARFIX/GPS receiver to resolve these problems.

2. Status of New Positioning Systems.

The status of GPS, GEOSTAR, STARFIX, and GLONAS was discussed next.

2.1. Status of GPS:

Currently 7 NAVSTAR satellites are on orbit. The constellation was scheduled to be complete by 1989. Due to the shuttle tragedy and launch problems, the completion of the constellation will be delayed to 1991. Other methods of launches are being investigated (expendible launch vehicles) and contracts for the developments have been awarded.

2.2. Status of GLONAS:

At least 12 GLONAS satellites are in orbit and operating intermittently. The orbital period is about 12 hours and the orbits are inclined about 55 degrees with 3 satellites per plane. The altitudes of the vehicles are not uniform and their pattern repeats about every 8 days. The code rates are about 1/2 that of GPS with a 7/9 carrier ratio. There are different frequencies per channel. Codeless receivers can use the system with slight modifications, but Counselman and McDoran stated that ISTAC and Macrometer have no plan for such modifications and use of GLONAS.

2.3. Status of GEOSTAR:

    (general comments) System will use it's own geostationary
satellites, large mainframe computers and a large digital terrain
model of CONUS. Apparently funding is still continuing as is the
FCC process.   A two way communication link will be required for
the operation.   In the initial phase of this system LORAN transmitters will be used for position determination. The use of geostationary satellites will take several years to get implemented.

2.4. Status of STARFIX:

    Three satellites are being used today. It might expand in
future.

3. Government Policies:

    Denial of accuracy, also called Selective Availability, and
Selective Availability, also called Anti Spoof, were discussed.

    The C/A code will be degraded to 100m 2DRMS by, among other
techniques, degrading the broadcast ephemerides to 20 ppm. The P
code will be encrypted and available only to "authorized users".
Time of implementation should be about 1989 when continuous 2D
operation commences.

    The debate over which receiver implementation (codeless or
code correlating) would be successful in the post Denial of Accuracy era continued. From recent government statements one can
conclude that the denial of accuracy implementation will not impact the differential and survey applications, provided a post-
mission accurate ephemerides data is used for post processing of
the data.

4. Single vs. Multiple Channels

ISTAC - can receive all satellites in view
STARFIX - has 4 channels but only needs three for a fix
WM-101 - has a fast sequencing receiver and thus can track all
         satellites in view above 15 degrees (7-8)
Macrometer - tracks all in view.
Trimble 4000SX - has 5 independent channels and tracks 5
                 satellites. Needs a minimum of 3 for a fix

5. GPS Receiver or Data Intercompatibility:

    Judged by various users represented in the audience that
receiver compatibility will not be a problem since most survey
houses can now decode most receiver data output.

6.   Certification

The   Federal Geodetic Control Commission tests receivers for
static geodetic applicability.   A strong need was voiced   for   a
similar   test or evaluation of dynamic GPS receivers.   NOAA has
made some progress in this area as Tom Bartholomew (recently   at
TASC) pointed out.

CONCLUSIONS:

GPS   has   now   been   established   without   a   doubt   as   the
positioning system in the survey world.   But until full time continuous operational status is achieved GPS offers only   incremental   navigation capability to the marine community.   However research and exploration survey companies are already pushing   back
the   frontiers   and   are   performing surveys that heretofore have
never been possible.   The need for and unbiased evaluation of GPS
marine receivers, possibly by the government,   was   expressed to
help the smaller user.

This   report   was   generated from the notes collected during
the workshop. We made all the   effort   to   reflect   the   correct
quotations and views   of   the panel members and audiences. We
apologize for any possible error.

WORKSHOP V

POSITIONING IN MARINE MAPPING AND CHARTING

Gary W. Hill
U.S. Geological Survey
12201 Sunrise Valley Drive
National Center 915
Reston, Virginia 22092

The workshop on "Positioning in Marine Mapping and Charting" (INSMAP 86 Workshop #5) was attended by 32 participants from several defense, governmental, and academic institutions. For mapping and charting at different scales, several specific topics were discussed. The different scales were defined as: (1) local - point sampling such as coring, dredging, photography, etc; (2) area - local studies restricted to areas smaller than several hundred to a few thousand square nautical miles; (3) regional - areas of substantial size such as the Exclusive Economic Zone off the west coast of the United States (approximately 250,000 square nautical miles) or the Gulf of Mexico; (4) global - areas equivalent to ocean basins or mapping programs associated with global tectonic problems. Topics for discussion at the various scales included: (1) what are the greatest demands (e.g., accuracy, range, lifespan) and limitations (e.g., cost, political boundaries) for the positioning system; (2) what are the unique demands asked of a positional system such as positioning of secondary systems (e.g., streamers, camera sleds, side-scan sonar fish, etc; (3) what are the real world factors limiting successful positioning; (4) what can be done to overcome particular factors; and (5) what recommendations can be made to INSMAP 86 regarding research and developmental activities which will meet future mapping and charting positional requirements.

Several problem areas were identified. Ship positioning was found not to be a problem due to the many systems available (each addressing different problems of accuracy, range, or lifespan) and that the coverage of existing systems was steadily increasing. It was the consensus of the working group, however, that a great deal of work needs to be done regarding positioning of scientific systems which are towed behind vessels, lowered to the seafloor, or employ acoustic sampling methods. Other problems identified which need to be addressed by technical and scientific groups include: (1) lack of marine positioning standards for the research community; (2) need for a national thrust into research specific to positioning problems; (3) lack of calibrations between different positioning systems; (4) lack of adequate positional data in national marine digital data bases; (5) inadequate educational opportunities relative to positional systems/theory/practise exist for the general "user" community; and (6) lack of sufficient forums such as INSMAP 86 to discuss problems and advances in positioning systems. Participants in the workshop agreed that INSMAP 86 was an interesting and worthwhile meeting and that future meetings should be held.

M. Kumar and G. A. Maul (directors), Marine Positioning, 463.
© 1987 by the Marine Technology Society.

LIST OF REGISTERED PARTICIPANTS

Albuquerque, Erico
Hydrographic Department
Rua Bareo de Jace Guay
Rio de Janeiro-Niteroi
24040-BRAZIL

Anderson, John O.
Polar Research Lab
6309 Carpinteria Avenue
Carpinteria, CA 93013
USA

Anderson, Kenneth
DMAHTC
2 Mastenbrook Ct.
Gaithersburg, MD   20879
USA

Anderson, William W.
USN/Code 8412, Bldg. 1002
NSTL, MS   39522
USA

Ashjaee, Javad
GPS Consultant
5000 Parkfield
San Jose, CA   05129
USA

Austin, Tom
Benthos
Edgerton Drive
Falmouth, MA   02556
USA

Ayres, James Edward
Int'l Hydrographic Org.
B.P. 445
MONTE CARLO, MC   98011
PRINCIPALITY OF MONTE CARLO

Baker, Sam
Satellite Surveying Systems
P.O. Box 34225
Bethesda, MD 20817
USA

Bartholomew, Tom
TASC
8301 Greensboro Drive
McLean, VA 22102
USA

Berg, Eduard
Hawaii Inst. of Geophysics
University of Hawaii at Manoa
Honolulu, HI 96822
USA

Berkovich, Efim
Sony Corp.
15 Essex Road
Paramus, NJ 07652
USA

Berman, Alan
RSMAS
6645 SW 118 Street
Miami, FL 33156
USA

Booker, Robert
DMAHTC
US Naval Observatory Bldg. 56
Washington, D.C. 20305
USA

Born, George
University of Colorado
CB 431
Boulder, CO 80309
USA

Bossler, J.D.
C&GS/NOAA
6001 Executive Blvd.
Rockville, MD 20852
USA

Boudreau, H.A.
Bedford Inst. of Oceanography
Dartmouth, Nova Scotia
CANADA B2Y 4A2

Bradley, Betty
NOAA/NOS
3401 Toledo Terr. E2
Hyattsville, MD 20782
USA

Buhl, Peter
Lamont Doherty Geolog. Labs
Columbia University
Palisades, NY 10964
USA

Burke, Kenneth
DMAHTC
7001 Highview Terrace #104
Hyattsville, MD 20702
USA

Casey, Michael J.
CDN Hydrographic Service
153 Glen Avenue
Ottawa, Ontario,
CANADA

Chayes, Dale
Lamont Doherty Geolog. Observ.
Columbia University
Palisades, NY 10964
USA

Clark, Del
Oceano Instruments USA
12737 28th St., NE
Seattle, WA 98125
USA

Clark, Larry
Nat'l Science Foundation
Ocean Technology Room 613
Washington, D.C. 20550
USA

Clynch, James R.
Applied Research Laboratories
University of Texas
Austin, TX 78713-8029
USA

Comoglico, Rick
Eastport Institute
501 Prince Georges Blvd.
Upper Marlboro, MD 20772
USA

Converse, Vince
DMAHTC/GST
6500 Brookes Lane
Washington, D.C. 20315
USA

Counselman, Charles C., III
Massachusetts Inst. of Tech.
123 Radcliffe Rd.
Belmont, MA   02178
USA

Dano, Paul K.
Del Norte Tech. Inc.
1100 Pamela Drive
Euless, TX 76040
USA

Davies
Hunting Surveys Ltd.
Ecstrpe Way Borshamwood
Herts,   WD6 15B
ENGLAND

De Bow, Samuel
NOAA
18520 Cape Jasmine Wy.
Gaithersburg, MD 20879
USA

DeLoach, Stephen
U.S. Army Corps of Engineers
ETL-TD-EA
Fort Belvoir, VA 22060
USA

De Matos, Euclides Janot
Hydrographic Department
Rua Bareo De Jace Guay
Rio De Janeiro-Niteroi
24040-BRAZIL

De Munck, J.C.
Delft Univ. of Technology
2628 Be Delft
THE NETHERLANDS

Dennis, Arthur
Analytical Tech. Lab., Inc.
301 Wells Fargo Drive Suite 5
Houston, TX 77090

USA

Denaro, R.
TAU Corp.
485 Alberto Way Blvd.
Los Gatos, CA 45030
USA

Drummond, Scott
SEACO, INC.
2560 Huntington Avenue, Suite 205
Alexandria, VA 22303-1482

USA

Dupin, Bill
Trimble Navigation
585 N. Mary Ave.
Sunnyvale, CA 74022
USA

Durboraw, I. Newton
Motorola, Inc.
2100 E. Elliot Road
Tempe, AZ 85282
USA

Egge, Delf
Dept. of Civil Engineering
University of Washington
Seattle, WA 98195

Ekseth, Roger
Norwegian Hydrographic Service
P.O. Box 60
4001 Stavanger
NORWAY

Escowitz, E.D.
U.S. Geological Survey
915 National Center
Reston, VA 22090
USA

Ethridge, Max M.
NOAA/NOS
17504 Ira Court
Derwood, MD 20855
USA

Evenden, G. I.
USGS
Woods Hole, MA
USA

Ferguson, J. Scott
URI/SEA BEAM Office
P.O. Box 527
Saunderstown, RI 02874
USA

Flor, Thomas H.
Naval Civil Engineering Lab.
Port Hueneme, CA 93043
USA

Flynn, Randall E.
DMAHTC
11581 North Shore Drive
Reston, VA 22090
USA

Fort, E. Roy
Del Norte Tech.
P.O. Box 696
Euless, TX 76039
USA

Fuechsel, John C.
COMSAT
22300 Comsat Drive
Clarksburg, MD 20871
USA

Fujimoto, Hiromi
Ocean Research Institute
1-15-1 Minamidai, Nakano
Tokyo
JAPAN

Furuta, Toshio
Ocean Research
Univ. of Tokyo
Nalcame
Tokyo 164
JAPAN

Gallegos, Artemio
UNAM
Coyozcan 04510
Mexico City,
MEXICO

Gibson, Richard T.
CAST, Inc.
5450 Katella Ave.
Los Alamitos, CA 90720
USA

Glenn, Morris
DMAHTC
1805 Sword Lane
Alexandria, VA 22308
USA

Gomez, Pedro
Peruvian Navy
Direecion De Hidrograния
Lima
PERU

Groten, E.
Institute of Physical Geodesy
Technical Universitv
Darmstadt
FEDERAL REPUBLIC OF GERMANY

Guenther, Gary C.
NOAA/NOS
6001 Executive Blvd.
Rockville, MD 20852
USA

Haladay, Martin
DMAHTC
6500 Brookes Lane
Washington, D.C. 20315
USA

Hampson, John C.
U.S. Geological Survey
Woods Hole, MA   02543
USA

Handschumacher, David
NORDA/CODE 360
NSTL, MS 39529
USA

Harrell, D. Martin
Eastport International, Inc.
501 Prince Georges Blvd.
Upper Marlboro, MD 20772
USA

Hatch, Ron
Magnavox
1142 Lakme
Wilmington, CA 90744
USA

Haykin, Simon
McMaster University
1280 Main St.
W. Hamilton, Ont.
CANADA

Henderson, Laurel
NRL
4555 Overlook Ave., S.W.
Washington, D.C. 20375
USA

Henneberg, Heinz
University of Zulia
Apartado 6
Maracaibo
VENEZUELA

Higgins, Michael
Eastport International
501 Prince Georges Blvd.
Upper Marlboro, MD 20772
USA

Hill, Gary W.
U.S. Geological Survey
915 National Center
Reston, VA 22092
USA

Homick, Michael H.
USCG
6125 Leesburg Pike #521
Falls Church, VA 22241
USA

Humphrey, John
National Ocean Service
NOAA-NOS N/CG241
Rockville, MD 20852
USA

Johler, Ralph J.
CO. Research & Prediction Lab
16796 W. 74th Pl
Golden, CO 80403
USA

Johnson, James E.
DMAHTC
11699 Bennington Woods
Reston, VA 22094
USA

Jones, Theodore P.
Navy
Bldg. 212 WNY
Washington, D.C. 20374
USA

Joshi, C.S.
Survey of India
17 E. C. Road
Dehradun 248001
INDIA

Kageyama, Koji
University of Wisconsin
7733 Carrington Dr. #C
Madison, WI 53719
USA

Kearse, Charles
N/M013, NOAA
WSC#1, Room 221
6001 Executive Blvd
Rockville, MD 20852
USA

Ketrick, Paul K.
Navy (CHESDIV)
Bldg. 212 NWY
Washington, D.C. 20374
USA

Khosla, K.L.
Surveyor General of India (Rtd.)
2425 Vineyard Lane
Crofton, MD 21114
USA

Kies, Phillip J.
MTS
National Data Buoy Center
NSTL, MS 39529
USA

Kreamer, John
U.S. Naval Oceanographic
Box 588
FPO Miami, FL 34059
USA

Kumar, Muneendra
DMAHTC
10625 Wayridge Drive
Gaithersburg, MD 20879
USA

Kutzleb, Mike
Steadfast Engineering
6269 Leesburg Pike
Falls Church, VA 22044
USA

Kutzleb, R.E.
Maritime Group
6311 Beachway Drive
Falls Church, VA 22044
USA

Lachapelle, G.
Nortech Surveys (Canada), Inc.
Calgary, Albert T2P OC5
CANADA

Ladd, Jonathan
Aero Service
8100 Westpark Drive
Houston, TX 77063
USA

Lally, Vincent E.
Nat'l Ctr. for Atmospheric Res
4330 Comanche Drive
Boulder, CO 80303
USA

Lauer, Bernard
DMAHTC
6500 Brookes Lane
Washington, D.C.   20315
USA

Leigh, Leon George
NOAA, Atlantic Marine Center
6316 Dartmouth Way
Virginia Beach, VA   23464
USA

Lohrenz, Maura Connor
Naval Ocean R&D Activity
511 Lake Sardis Drive
Slidell, LA   70461
USA

MacDoran, Pete
ISTAC, INC.
444 No. Altadena Suite 101
Pasadena, CA   91107
USA

Mahar, Dennis
national Data Buoy Center
808 Constitution Drive
Slidell, LA   70458
USA

Malison, Alex
Eastport Wt
721 Mass. Ave. N.E
Washington, D.C. 20002
USA

Mansfield, Karen C.
Hawaii Inst. of Geophy.
2525 Correa Road
Honolulu, HI   96822
USA

Maul, George A.
NOAA/AOML/PHOD
4301 Rickenbocker Cswy.
Miami, FL   33149
USA

McGregor, Bonnie A.
U.S. Geological Survey
915 National Center
Reston, VA   22090
USA

McIntyre, Marie
Scripps Inst. of Oceanography
Marine Physical Lab A-005
LaJolla, CA   92093
USA

Meng-Frecker, John
OMAHA Dist. Corp of Engineers
215 N. 17th St.
Omaha, NE   68101-1294
USA

Mertikas, Stelios
Dept. of Surveying Engineers
University of New Brunswick
Fredericton, (NB) E3B 5A3
CANADA

Meyerhoff, Stanley L.
Naval Surface Weapons Center
Box 642
Dahlgren, VA   22448
USA

Miller, Joyce
Sea Beam - Univ. of RI
P.O. Box 527
Saunderstown, RI   02874
USA

Minkel, David H.
N/CG 2x2, NOAA
6001 Executive Blvd.
Rockville, MD   20852
USA

Mink, Adam W.
DMAHTC
P.O. Box 5031
Annapolis, MD   21403
USA

Mitchell, Jim
NORDA/Dept. of Navy
Code 321
NSTL, MS 39529
 USA

Mooers, Christopher
Inst. for Naval Oceanography
Room 311 Bldg. 1100
NSTL, MS 39529
USA

Moore, Meg
NOS/NOAA
1006 Trebing Lane
Upper Marlboro, MD 20772
USA

Moran, James
DMAHTC
6500 Brookes Lane
Washington, D.C. 20315
 USA

Normark, William R.
U.S. Geological Survey
345 Middlefield Road
Menlo Park, CA 94025
USA

Muller, Richard
Krupp Atlas Elektronik
1453 Pinewood Street
Rahway, NJ 07065
USA

Nagengast, Steven J.
MGS/MTS
Pacific Missile Test Center
Point Mugu, CA 93042-5000
 USA

Nelius, Carl
DMAHTC
6500 Brookes Lane
Washington, D.C. 20315
USA

Nygren, Harley
NOAA (Retd)
1438 Carrington Lane
Vienna, VA 22180
USA

Oyvind, Stene
Norwegian Hydrographic
P.O. Box 60
4001 Stavanger
NORWAY

Penton, Ron
Ferranti O.R.E., Inc.
P.O. Box 709
Falmouth, MA 02540
USA

Perreault, Paul D.
Trimble Navigation
585 N. Mary Avenue
Sunnyvale, CA 94086
USA

Pitts, Ed
K.G. Pollock & Assoc., Inc.
85 Fair Haven Road
Fair Haven, NJ 07701
USA

Pollock, Ken
K.G. Pollock & Assoc., Inc.
952 Placid Ct.
Arnold, MD 21012
USA

Pope, Joan
Coastal Engineering Research
P.O. Box 631
Vicksburg, MS 39180
USA

Power, Andrew
MC Elhanne Service
17 Greenoch Drive
Dartmouth, NOVA SCOTIA
CANADA

Puccini, Donald
COMNAVOCEANCOM
105 Basswood
Pass Christian, MS   39571
 USA

Read, Chung Hye
DMAHTC
6500 Brookes Lane
Washington, D.C.   20315
 USA

Remondi, Benjamin
NOAA/NGS, N/CG14
6001 Executive Blvd.
Rockville, MD 20852
USA

Renzetti, Nicholas A.
Jet Propulsion Laboratory
4800 Oak Grove Drive
Pasadena, CA   91109
USA

Rogoff, Mortimer
The Digital Directions Co.
4201 Cathedral Ave. N.W.
Washington, D.C.   20016
USA

Rossby, Tom
GSO/URI
Kingston, RI   02881
USA

Rossi, Frank P.
US Naval Observatory
34th & Massachusetts Ave. N.W.
Washington, D.C.   20350-2000
USA

Rulon, Tim
NOAA
6001 Executive Blvd.
Rockville, MD 20852
USA

Salamonowicz, Paul
Defense Mapping Agency
8121 Maplewood Drive
Manassas, VA   22111
USA

Salzmann, Martin A.
Delft Univ. of Technology
Department of Geodesy
2629 Ja Delft
THE NETHERLANDS

Sasaki, Minoru
Hydrographic Dept. of Japan
Chome Shinjuku-Ku
Tokyo 160,
JAPAN

Saxena, Narendra
University of Hawaii
611 Hahaione Street
Honolulu, HI 96825

USA

Schoonmaker, W.
USGS
National Mapping Division
915 National Center
Reston, VA 22092

USA

Seeber, Gunter G.
University of Hanover
Roeddinger Street 23
3008 Garbsen 1
FEDERAL REPUBLIC OF GERMANY

Seesholtz, J.R.
Naval Observatory, Bldg. 1
34th and Massachusetts Ave., NW
Washington, D.C.   20390
USA

Sender, John
Hawaii Inst. of Geophysics
2525 Correa Road
Honolulu, HI 96822
USA

Sennott, James
Bradley University
11500 Pine-Cone Ct.
Reston, VA 22091
USA

Sheard, Steve
Morotola
2100 E. Elliot Road
Tempe, AZ 85282
USA

Shelton, Robert F.
AC&M
3217 Brookforest
Tallahasee, FL 32312
USA

Sibold, A. P.
Collins Avionics
400 Colins Road N.E.
Cedar Rapids, IA 52302
USA

Siry, Joseph W.
NASA/Goddard
4438 42nd St., N.W.
Washington, DC 20016
USA

Slater, James
DMAHTC
19018 Coltfield Court
Gaithersburg, MD 20879
USA

Spielvogel, Les
SEACO, Inc.
146 Hekili St.
Kailua, HI 96734

Spiess, Fred
Univ. of Calif./Scripps
LaJolla, CA 92093
USA

Sprent, Anthony
School of Surveying
GPO Box 252C
Hobart, Tasmania 7001
AUSTRALIA

Stembel, Oren
NOS/NOAA
127 Eastmoor Drive
Silver Spring, MD 20901
USA

Stolz, A.
School of Surveying
University of NWS
P.O. Box 1
Kensington, NSW
AUSTRALIA 2033

Stout, Gary J.
AT&T
412 Mt. Kemble Avenue
Morristown, NJ 07960

Stuart, Jeffrey
NOS/NOAA
128 Spring Street
Gaithersburg, MD 20877
USA

Stutes, Phil
John E. Chance & Assoc.
200 Dulles Drive
Lafayette, LA 70506
USA

Swale, Stephen L.
Tau Corporation
485 Albert Way Bldg. D
Los Gatos, CA 95030
USA

Tabe, Kazushi
Sony Corp.
1-7-4 Konan, Minato-Ku
Tokyo
JAPAN

Tapley, Byron D.
Center for Space Res.
University of Texas
Austin, TX 78712
USA

Terauchi, Toshiro
Info. System Res. Ctr., Sony Corp.
1-7-4 Konan, Minato-Ku
Tokyo
JAPAN

Tucker, Stevens P.
Dept. of Oceanography
Naval Postgraduate School
Monterey, CA 93943
USA

Van Opstal, L. H.
Chief of Hydrographic
Koninklijke Marine
P.O. Box 90794
2509 Ls's - Gravenhage
THE NETHERLANDS

Von Alt, Christopher
WHOI
P.O. Box 603
Waqvoit, MA 02543
USA

Watkins, Carlos A.
Venezuelan Coast Guard
Av. Volmer Sanbernardino
Caracus
VENEZUELA

Weems, Lynn
Sercel, Inc.
17155 Park Row
Houston, TX 77084
USA

Wells, D. E.
Dept. of Surveying Engineering
University of New Brunswick
Fredericton, NB E3B 5A3
CANADA

Westmeyer, William
U.S. Congress
Office of Tech. Assessment
Washington, D.C. 20373
USA

Worcester, Peter
Scripps Inst. of Oceanography
Univ. of California, San Diego
La Jolla, CA 92093
USA

Young, Larry
Jet Propulsion Lab.
630 Groveview Lane
La Canada, CA 91011
USA

Ziegler, R.E.
DMAHTC
1419 Wolftrap Run Road
Vienna, VA 22180
USA

REPORT DOCUMENTATION PAGE

1a. REPORT SECURITY CLASSIFICATION Unclassified	1b. RESTRICTIVE MARKINGS
2a. SECURITY CLASSIFICATION AUTHORITY	3. DISTRIBUTION/AVAILABILITY OF REPORT
2b. DECLASSIFICATION/DOWNGRADING SCHEDULE	

4. PERFORMING ORGANIZATION REPORT NUMBER(S) Marine Geodesy Committee, Marine Technology Society	5. MONITORING ORGANIZATION REPORT NUMBER(S)

6a. NAME OF PERFORMING ORGANIZATION Marine Geodesy Committee	6b. OFFICE SYMBOL (If applicable)	7a. NAME OF MONITORING ORGANIZATION Marine Technology Society
6c. ADDRESS (City, State, and ZIP Code) 2000 Florida Ave., N.W. Washington, D.C.20009		7b. ADDRESS (City, State, and ZIP Code) 2000 Florida Ave., N.W. Washington, D.C. 20009

8a. NAME OF FUNDING/SPONSORING ORGANIZATION USGS, NOAA, ONR	8b. OFFICE SYMBOL (If applicable)	9. PROCUREMENT INSTRUMENT IDENTIFICATION NUMBER

8c. ADDRESS (City, State, and ZIP Code) USGS, Reston, VA 22092 NOAA, Rockville, MD 20852 ONR,	10. SOURCE OF FUNDING NUMBERS			
	PROGRAM ELEMENT NO.	PROJECT NO.	TASK NO.	WORK UNIT ACCESSION NO.

11. TITLE (Include Security Classification)

Proceedings, International Symposium on Marine Positioning (INSMAP 86)

12. PERSONAL AUTHOR(S)

Symposium Directors: Muneendra Kumar and George A. Maul

13a. TYPE OF REPORT proceedings	13b. TIME COVERED FROM 10/14/86 TO 10/17/86	14. DATE OF REPORT (Year, Month, Day) January 1987	15. PAGE COUNT

16. SUPPLEMENTARY NOTATION

17.	COSATI CODES		18. SUBJECT TERMS (Continue on reverse if necessary and identify by block number)
FIELD	GROUP	SUB-GROUP	

19. ABSTRACT (Continue on reverse if necessary and identify by block number)

   This symposium was conceived by the Marine Geodesy Committee at OCEANS 84, Washington, DC. It became clear at that time, that time is appropriate to focus attention on individual specific problem areas under the broad umbrella of Marine Geodesy. These proceedings contain the written versions of the talks, presentations, and reports as submitted by the authors or workshop chairpersons. The sections are divided into: Instrumentation in the Marine Environment, Global Positioning System in Marine Positioning, Marine Mapping and Charting, Positioning in Oceanography, Calibration and Intercomparison, Applications/Requirements, and summaries of the five workshops.

20. DISTRIBUTION/AVAILABILITY OF ABSTRACT ☒ UNCLASSIFIED/UNLIMITED ☐ SAME AS RPT. ☐ DTIC USERS	21. ABSTRACT SECURITY CLASSIFICATION unclassified	
22a. NAME OF RESPONSIBLE INDIVIDUAL Dr. M. Kumar	22b. TELEPHONE (Include Area Code) 227-2152	22c. OFFICE SYMBOL DMAHTC/GST

DD FORM 1473, 84 MAR

83 APR edition may be used until exhausted.
All other editions are obsolete.